PUBLICATIONS OF THE NEWTON INSTITUTE

Security Market Imperfections in
Worldwide Equity Markets

Publications of the Newton Institute

Edited by H.K. Moffatt

Director, Isaac Newton Institute for Mathematical Sciences

The Isaac Newton Institute of Mathematical Sciences of the University of Cambridge exists to stimulate research in all branches of the mathematical sciences, including pure mathematics, statistics, applied mathematics, theoretical physics, theoretical computer science, mathematical biology and economics. The four six-month long research programmes it runs each year bring together leading mathematical scientists from all over the world to exchange ideas through seminars, teaching and informal interaction.

SECURITY MARKET IMPERFECTIONS IN WORLDWIDE EQUITY MARKETS

edited by

Donald B. Keim

University of Pennsylvania

and

William T. Ziemba

University of British Columbia

PUBLISHED BY THE PRESS SYNDICATE OF THE UNIVERSITY OF CAMBRIDGE
The Pitt Building, Trumpington Street, Cambridge CB2 1RP, United Kingdom

CAMBRIDGE UNIVERSITY PRESS
The Edinburgh Building, Cambridge CB2 2RU, United Kingdom
40 West 20th Street, New York, NY 10011-4211, USA
10 Stamford Road, Oakleigh, Melbourne 3166, Australia

First published 2000

Printed in the United Kingdom at the University Press, Cambridge

Typeset in 12pt Computer Modern

A catalogue record for this book is available from the British Library

ISBN 0 521 57138 3 hardback

DEDICATION

To Sandra with thanks for all the help and support on this project.
(WTZ)

To Susan – Thank you
(DBK)

CONTENTS

Biographical Notes of Contributors vii

Preface
William T. Ziemba xiii

Security market imperfections: an overview
Donald B. Keim and William T. Ziemba xv

PART I. AN OVERVIEW OF CROSS SECTIONAL PATTERNS
IN STOCK RETURNS

2. The cross-section of common stock returns: a review of the evidence and
some new findings
Gabriel Hawawini and Donald B. Keim 3

3. Beta and book to market: is the glass half full or half empty?
S.P. Kothari and Jay Shanken 44

4. The psychology of over-reaction and under-reaction in world equity
markets
Werner F.M. De Bondt 65

5. A view of the current status of the size anomaly
Jonathan B. Berk 90

6. The demise of size
Elroy Dimson and Paul Marsh 116

7. Direct evidence of non-trading of NYSE and AMEX stocks
Stephen R. Foerster and Donald B. Keim 144

PART II. SEASONAL PATTERNS IN STOCK RETURNS
AND OTHER PUZZLES

8. Is there still a January effect?
Donald G. Booth and Donald B. Keim 169

9. Anticipation in the January effect in the US futures markets
Chris R. Hensel and William T. Ziemba 179

10. How does Clinton stand up to history? US investment returns and
presidential party affiliations

Chris R. Hensel and William T. Ziemba 203

11. A long term examination of the turn-of-the-month effect in the S&P500
Chris R. Hensel, Gordon A. Sick and William T. Ziemba 218

12. The closed-end fund puzzle
Carolina Minio–Paluello 247

13. Stock splits and ex-date returns for Nasdaq stocks: the effects of investor trading and bid-ask spreads
Mark Grinblatt and Donald B. Keim 276

PART III. INTERNATIONAL EVIDENCE

14. Canadian security market anomalies
George Athanassakos and Stephen Foerster 297

15. Seasonal anomalies in the Italian stock market, 1973–1993
Elio Canestrelli and William T. Ziemba 337

16. Efficiency and anomalies in the Turkish stock market
Gülnur Muradoğlu 364

17. Efficiency and anomalies in the Finnish stock market
Teppo Martikainen 390

18. Characteristics-based premia in emerging markets: sector-neutrality, cycles, and cross-market correlations
Sandeep A. Patel 416

19. Anomalies in Asian emerging stock markets
Seng-Kee Koh and Kie Ann Wong 433

20. Japanese security market regularities, 1990–1994
Luis R. Comolli and William T. Ziemba 458

21. Predicting returns on the Tokyo Stock Exchange
Sandra L. Schwartz and William T. Ziemba 492

22. High stock returns before holidays: international evidence and additional tests
Alonso Cervera and Donald B. Keim 512

CONTRIBUTORS

George Athanassakos is Professor of Finance and Director of the Financial Planning program at the School of Business and Economics of Wilfrid Laurier University, Waterloo, Canada. He received his PhD in Finance from York University. His research has dealt with the institutional attributes of the Canadian capital markets, the effect institutional trading has on stock and bond market performance, the relation between stock returns and predetermined variables such as market capitalization and calendar turning points, and bond and equity valuation issues. He is a member of the Editorial Advisory Board of the Canadian Investment Review, a member of the Research Advisory Board of the Canadian Securities Institute and a VP-Membership of the Multinational Finance Society.

Address The Mutual Group Financial Services Research Centre, School of Business and Economics, Wilfrid Laurier University, Waterloo, Ontario, Canada, N2L 3C5.

Jonathan B. Berk is an Assistant Professor of Finance at the Haas School of Business, University of California, Berkeley and a Faculty Research Fellow at the National Bureau of Economic Research. He received his PhD in finance from Yale University. His research has covered a broad range of topics including the theoretical relation between stock returns and fundamental variables (market value, book-to-market, interest rates, etc.), firm investment, valuing firm growth potential, experimental evidence of rationality and job discrimination. His research has won numerous awards including the best paper in the *Review of Financial Studies*, the Graham and Dodd Award of Excellence, *Financial Analyst's Journal* and the Roger F. Murray Prize, The Institute for Quantitative Research in Finance

Address Haas School of Business, University of California, Berkeley CA 94720, USA.

David G. Booth is Chairman and Chief Executive Officer of Dimensional Fund Advisors Inc., a registered investment advisor with over $25 billion under management. The company is headquartered in Santa Monica, California, with offices in Chicago, Connecticut, London and Sydney. Dimensional Fund Advisors has developed a reputation for successfully applying leading-edge academic research to money management. Mr. Booth has an MS, BA from the University of Kansas and in 1971 received his MBA from the University of Chicago. Before founding Dimensional Fund Advisors Inc. in 1981, he was Vice President, Pension Investment Consulting at A.G. Becker, Inc. Mr. Booth has written numerous articles. Best known among these is the article, 'Diversification Returns and Asset Contributions' which he co-authored with Professor Eugene F. Fama. This paper earned the authors the Graham and Dodd Award for best article in the *Financial Analysts' Journal* in 1992. Mr. Booth is on the Board of Governors of the UCLA Foundation and is a member of the Foundation's Investment Committee. He is a Trustee of St. Matthew's Parish School. Most recently, he was elected to the Board of Trustees of the American Academy in Rome.

Address Dimensional Fund Advisors, Santa Monica, CA 90401, USA.

Elio Canestrelli is full Professor of Financial Mathematics and Head of the Department of Applied Mathematics at the University of Venice (Italy). His major area of research interest is dynamic stochastic programming applied to financial modelling. He was the organizer of the EURO Working Group on Financial Modeling in Venice in the fall of 1997.

Address Department of Applied Mathematics & Computer Science, University of Venice, Dorsoduro 3825/E, 30123, Venice, Italy.

Alonso Cervera is an Associate Economist for Latin America at Donaldson, Lufkin & Jenrette in New York. He holds an MBA degree from the Wharton School of the University of Pennsylvania. He won Mexico's National Award in Economic Research (Banamex) in 1994 and the National Award in Financial Research (IMEF) in 1993. His academic research interests include calendar anomalies in global stock markets, portfolio diversification possibilities and the design of the regulatory framework for Mexican mutual and pension funds.

Address DLJ, 277 Park Avenue, New York, NY 10172, USA.

Luis R. Comolli is a PhD candidate in Geophysics at the University of California, Berkeley.

Address Graduate Group in Biophysics University of California Berkeley, CA 94720, USA.

Werner F.M. De Bondt is Professor of Business and Frank Graner Professor of Investment Management at the University of Wisconsin–Madison. He studies the psychology of investors and financial markets, and is one of the founders of behavioral finance. Werner De Bondt holds a PhD in business administration from Cornell University (1985) and also holds degrees in engineering and public management. In recent years, he has held professorships at Cornell University and at universities in Belgium, the Netherlands and Switzerland.

Address School of Business, University of Wisconsin-Madison, 975 University Avenue, Madison, WI 53706, USA.

Elroy Dimson is Professor of Finance at London Business School where he is Director of the school's Investment Management Program, and was formerly Dean of MBA Programs and Chair of the Finance Department. Dr Dimson has held visiting positions at the Universities of Chicago, California (Berkeley) and Hawaii, and at the Bank of England. He has served as editor of the *Advances in Finance, Investment & Banking* book series; and as associate editor of *Journal of Finance, Journal of Banking & Finance*, and several other journals. Dr Dimson has been elected to membership of the Financial Economists' Roundtable, and is President-designate of the European Finance Association. With Paul Marsh, he created and produces the Hoare Govett Smaller Companies index series for the UK. His publications cover a variety of topics in investment management, including a number of papers, coauthored with Professor Marsh, on the small firm effect.

Address London Business School, Sussex Place, Regent's Park, London NW1 4SA, UK.

Stephen R. Foerster is an Associate Professor of Finance at the Richard Ivey School of Business, the University of Western Ontario. He received his PhD in finance from the University of Pennsylvania. His research has dealt with the tests of asset pricing models and related test statistic properties, global listings and capital raising, the behavior of international common stocks related to the US presidential cycle, and price momentum equity trading strategies. He is a member of the advisory boards of the Canadian Investment Review and FEN-Course.

Address Richard Ivey School of Business, The University of Western Ontario, London, Ontario, Canada, N6A 3K7.

Mark Grinblatt is a Professor at UCLA's Anderson School, where he has held an appointment since 1981. He held a visiting position at the Wharton School from 1987-1989 and worked as a Vice President for Salomon Brothers, Inc. from 1989–1990. Dr. Grinblatt received a PhD in Economics from Yale university in 1982 and a BA in Economics and Mathematics from the University of Michigan in 1977. He currently serves as an Associate Editor of the *Review of Financial Studies* and the *Journal of Financial and Quantitative Analysis* and as a Director on the board of Salomon Swapco, Inc. As an author of over 25 papers published in finance and economics journals, he has written extensively on the topic of performance evaluation. His past research, both theoretical and empirical, has also explored asset pricing, rational expectations equilibria, corporate finance,derivatives valuation, and agency theory. In addition to pursuing his current research interests, Dr. Grinblatt has a textbook for MBA's, *Financial Markets and Corporate Strategy*, published by Irwin–McGraw Hill.

Address The Anderson School at UCLA, 110 Westwood Plaza, Los Angeles, CA 90095, USA.

Gabriel Hawawini is the Henry Grunfeld Chaired Professor of Investment Banking at INSEAD, the European Institute of Business Administration. He is currently Associate Dean in charge of the school's Development Campaign and its Doctoral Program. He has also taught finance at New York University, Columbia University, and the Wharton School of the University of Pennsylvania, where he won the Helen Kardon Moss Anvil Aware for Excellence in Teaching. Professor Hawawini is author of ten books and over 60 research papers. Besides serving as Vice President of the French Finance Association, he has organized, directed and taught in management development programs at INSEAD and around the world.

Address INSEAD, 77305 Fontainebleau Cedex, France.

Chris R. Hensel has an MBA in finance from the University of Chicago and has published research papers in *Management Science*, the *Journal of Portfolio Management* and the *Financial Analysts' Journal*. This research was conducted while he was Senior Practice Consultant with Frank Russell Company in Tacoma, Washington.

Address 6406 60th Place South, Seattle, WA 98118-3025, USA.

Donald B. Keim is the John B. Neff Professor of Finance at the Wharton School of the University of Pennsylvania. He received his PhD in finance and economics from the University of Chicago. His research has dealt with the relation between stock returns and predetermined variables (market capitalization, earnings/price ratios, and calendar turning points), tests of asset pricing models, the junk bond market, the behavior of real estate-related common stocks, and market microstructure issues relating to institutional investor equity trading. He is an Associate Editor for the *Journal of Financial and Quantitative Analysis*, and Co-Editor of the *European Finance Review*.

Address Wharton School, University of Pennsylvania, Philadelphia PA 19104, USA.

Seng-Kee Koh is the Deputy Dean of the Faculty of Business Administration and Associate Professor in the Department of Finance and Accounting at the National University of Singapore. He is an associate editor of the *Asia Pacific Journal of Finance* and a board member of NUS ASPF Pension fund. His research interests are focused primarily on emerging financial markets and mutual funds. He has a PhD in Finance from the Wharton School. He is currently the Director of the Financial Management Program in NUS where he oversees the training of management executives in various aspects of finance. Prior to joining academia, he was a corporate banker at Chase Manhattan Bank N.A.

Address Department of Finance and Accounting, National University of Singapore, 10 Kent Ridge Crescent, Singapore 119260.

S.P. Kothari is a Professor at the Massachusetts Institute of Technology's Sloan School of Management. He received his PhD from the University of Iowa. His research has focused on tests of market efficiency and asset-pricing models, the relation between security prices and financial information, corporate uses of derivatives, and the use of earnings and prices in compensating managers. Professor Kothari's research appears in leading academic journals such as the *Journal of Financial Economics*, the *Journal of Finance*, the *Journal of Accounting & Economics*, and the *Journal of Accounting Research*. He is the Editor of the *Journal of Accounting & Economics*.

Address Sloan School of Management, E52–325, Massachusetts Institute of Technology, 50 Memorial Drive, Cambridge, MA 02142, USA.

Paul Marsh is Esmée Fairbairn Professor of Finance at London Business School where he is Chair of the Finance Department and Director of the School's Institute of Finance & Accounting. He is currently Academic Director of the School's Masters in Finance Program, and was formerly Faculty Dean, Deputy Principal, and an elected Governor of London Business School. He has a wide range of publications in the *Journal of Finance*, the *Journal of Financial Economics*, the *Journal of Banking & Finance*, *Harvard Business Review*, etc., spanning issues in corporate finance, risk measurement and investment management. He is a Director of M&G Group plc, Majedie Investments plc and Hoare Govett Indices Ltd, and is a Governor of the Examining Board of The Securities Institute.

Address London Business School, Sussex Place, Regent's Park, London NW1 4SA, UK.

Teppo Martikainen is Professor of Finance at the Helsinki School of Economics and Business Administration, Finland. His publications cover a variety of topics in investment management, including asset valuation, financial statement

analysis and risk measurement in thinly traded and international markets. He
has authored or co-authored more than 100 research articles in internationally
refereed journals such as *Management Science, Journal of Futures Markets* and
Journal of Banking and Finance. He is a board member of several Finnish
investment companies, a member of the editorial board of many journals and
President elect of the Multinational Finance Society.

Address Department of Accounting and Finance, University of Vaasa, PO Box
700, FIN-65101 Vaasa, Finland.

Carolina Minio-Paluello is Edward Jones Scholar at London Business School.
She has taught at the Universities of Geneva and London and has presented
her research at academic and professional conferences in a number of countries.
Educated in Italy, Belgium and London, Dr. Minio-Paluello received her PhD
in Finance from the London Business School. Her research interests include the
performance of closed-end funds, style analysis and factor models of returns. She
received the 1997 StyleADVISOR prize from Zephyr Associates for her research
using returns-based style analysis, and has been sponsored by the Research
Foundation of The Institute of Chartered Financial Analysts to co-author a
monograph on closed-end funds.

Address London Business School, Sussex Place, Regent's Park, London NW1
4SA, UK.

Gülnur Muradoğlu is a lecturer in Finance at Manchester School of Accounting
and Finance. She has been a Visiting Fellow at the Warwick Business School
and Associate Professor of Finance at Bilkent University. She has published articles on market efficiency, financial forecasting and behavioral finance in journals
including *Applied Financial Economics, European Journal of Finance, Journal
of Forecasting, International Journal of Forecasting, European Journal of Operations Research, Developing Economies, Middle East Business and Economic
Review,* and the *Istanbul Stock Exchange Review.*

Address Manchester School of Accounting and Finance, The University of
Manchester, M13 9PL, UK.

Sandeep A. Patel is at Tudor Investment Corporation and an Adjunct Associate Professor at the New York University. Previously, he was at J.P. Morgan
where he designed and implemented security, country, and currency selection
models in emerging markets, modeled transactions costs and equity selection in
the US, built mortgage prepayment model, and developed a risk model for US
investment-grade corporate bonds. His recent work includes an examination of
country and industry influences on European equities. Dr. Patel received his
PhD in Finance from the Wharton School of the University of Pennsylvania and
has published research in the *Journal of Finance, Financial Analyst's Journal*
and the *Emerging Markets Quarterly.*

Address 60 West 66th Street, 30G, New York, NY 10023, USA.

Sandra L. Schwartz has taught policy analysis, decision analysis, strtegic management and applied economics courses at the University of California, Berkeley,
UCLA, the University of Tsukuba, Simon Fraser University, the University of

British Columbia, the Open University and at Royal Roads College. Her PhD is
from the University of British Columbia. She has co-authored two books on the
Japanese economy and financial markets, co-edited two books on energy policy,
and written a number of research papers on economic policy and management
issues.

Address 6 Tamath Crescent, Vancouver, BC, V6N 2C9, Canada.

Jay Shanken is a Professor of Finance at the Simon School of the University of
Rochester. He has a masters degree in mathematics and received his PhD in
economics from Carnegie–Mellon University. His research interests include the
theory and testing of asset-pricing models, the predictability of stock returns,
tests of performance evaluation and market efficiency, and applied econometrics.
Professor Shanken was a Batterymarch Fellow and is an associate editor for the
Journal of Finance, the *Journal of Financial Economics* and the *Review of
Quantitative Finance and Accounting*.

Address William E. Simon Graduate School of Business Administration University of Rochester, Rochester, NY 14627, USA.

Gordon A. Sick is Professor of Finance and Co-Director of the Mathematical
Finance Laboratory at the University of Calgary. He has previously been a
faculty member at Yale University and at the University of British Columbia
where he also received his PhD. In 1999 he is a fellow at the Netherlands Institute
for Advanced Study. His research interests include real options, mathematical
finance, investments, corporate finance and bank cost functions. He has been
on the editorial board of *Management Science* and a director of the Western
Finance Association and the Financial Management Association.

Address Faculty of Management, University of Calgary, Calgary, Alberta, Canada, T2N 1N4.

Kie Ann Wong, is Professor of Finance and Head of Department of Finance and
Accounting of the National University of Singapore. He has published widely
in the behaviour of securities markets. He is also Chief Editor of the *Asia
Pacific Journal of Finance* and Founding President of the Asia Pacific Finance
Association.

Address Department of Finance and Accounting, National University of Singapore, 10 Kent Ridge Crescent, Singapore 119260.

William T. Ziemba is the Alumni Professor of Management Science at the University of British Columbia. He has published a number of articles and books
including the North Holland *Handbook of Finance* and the companion to the
present volume, *Worldwide Asset and Liability Modeling*. He has been a visiting professor at Chicago, Stanford, University of California, Berkeley, UCLA,
London School of Economics, Cambridge, Warwick and Tsukuba. His research
interests include financial modeling, portfolio theory and management, Asian financial markets, stochastic programming, and, the subject of this book, security
market imperfections.

Address Faculty of Commerce, University of British Columbia, Vancouver, BC,
Canada V6T 1Z2.

Preface

Are asset prices predictable or do current prices fully reflect underlying value? Does taking more risk lead to higher expected returns? To study this area, I organized a week long set of research seminars under the general theme 'Worldwide anomalies and behavioral finance' on May 8–13, 1995 at the Isaac Newton Institute for Mathematical Science on the campus of the University of Cambridge. This research program was followed by an institutional investor workshop on Saturday May 20th. This week's activities formed part of the six-month Financial Mathematics Symposium held at the Newton Institute from January to June 1995. I organized this part of the program under the general direction of the financial mathematics seminar organizers Mark Davis, Stewart Hodges, Ioannis Karatzas and Chris Rogers. This volume consists of twenty-one original papers arising from this program. Most of the papers appearing here were presented in Cambridge with a few added to round out the volume.

The papers included are broadly concerned with the predictability of equity stock returns. This predictability is commonly referred to as security market imperfections, or fundamental and seasonal regularities, or anomalies. The research presented addresses the difficult question of the existence of true market imperfections versus changing risk. Such imperfections may be broadly categorised as cross sectional, where the aim is to predict the differences in returns of individual stocks or groups of stocks, and time series where the aim is to predict time periods in which particular indices such as small or large capitalized equities have high or low returns. The studies discuss many markets worldwide including the US, Japan, Asia, and Europe. They discuss the measurement of risk and prediction models that have been used by institutional investors. Several papers discuss the high returns of equities in January and the relationship of small and large capitalized stocks around the turn-of-the-year. The coverage also includes papers discussing the psychology of financial markets, closed end country funds, liquidity aspects, stock splits, the holiday, the turn-of-the-month, and US presidential election effects.

The seminar in Cambridge took place in the efficient and most pleasant facilities of the Isaac Newton Institute on the campus of the University of Cambridge. The staff of the Institute, particularly Anne Cartwright, Florence Leroy, the Associate Director John Wright and Director Michael Atiyah were most helpful before and during our pleasant stay in Cambridge. Financial mathematics seminar chairman Chris Rogers was most supportive and helpful throughout this activity. This programme was sponsored by the Issac Newton Institute for Mathematical Science and the Program in Financial Modeling at the University of British Columbia.

I was pleased to have Donald Keim join me as a co-editor of this volume. Besides co-authoring the introduction with me, Don added his special insights gained from years of outstanding research to improve the papers in this volume as well as contributing five outstanding co-authored papers based on his own pioneering work. Our editor David Tranah has been most helpful and patient in the preparation of this volume. My research work in security market imperfections has been supported by the Social Science and Humanities Research Council of Canada, the Centre for International Business Studies at the University of British Columbia and the Frank Russell Company. Thanks go to Chris Hensel for his encouragement and our joint work, which is represented by three papers in this volume. Finally special thanks go to my wife Sandra Schwartz for much encouragement and help on the seminar in Cambridge and in the preparation of this volume.

William T. Ziemba

Vancouver

Security Market Imperfections: An Overview

Donald B. Keim and William T. Ziemba

The area of academic and practitioner research in financial economics that has generated the most excitement and attracted the most attention over the past twenty years concerns the predictability of stock returns. Hundreds of papers have documented persistent cross-sectional and time series patterns in returns that are not predicted by existing theory. Academics have intense interest in these findings because the results question long-standing theories of how the markets price securities. Practitioners find the results intriguing because of the prospect of using them to design investment strategies to earn abnormal profits. Even individual investors have tuned into this literature, perhaps unwittingly, as they tilt their personal portfolios toward small cap or value stocks, or by attempting to exploit the January effect.

These findings arouse curiosity in such a diverse audience because, quite simply, they are not well understood. Although the best minds in financial economics have exerted considerable energy to explain them, most of these phenomena remain puzzles – imperfections (warts) in the perfect world that is so often depicted in the models used to describe equity markets. As such, these findings are often classified as anomalies, a term that can be traced to Thomas Kuhn (1970). Kuhn maintains that research activity in any normal science revolves around a central paradigm and that experiments are conducted to test the predictions of the underlying paradigm and to extend the range of phenomena it explains. Eventually, results are found that don't conform to the paradigm. Kuhn [1970, pp. 52–53] terms this stage 'discovery': "Discovery commences with the awareness of *anomaly*, i.e., with the recognition that nature has somehow violated the paradigm-induced expectations that govern normal science" (emphasis added).

This book contains 21 new and original papers that collectively represent our current understanding of the nature of, and explanations for, these anomalies. One might view this volume as a sequel to the excellent collections in the 1983 *JFE* special issue and Dimson (1988) that first catalogued these phenomena in the 1980s and increased our awareness of them.

The plethora of research that has been generated over the succeeding years has important benefits. First, we now are able to provide substantial out-of-sample evidence to confirm the results presented earlier because of the additional years that have been added to sample period. Perhaps more importantly, the increased availability of reliable, machine-readable data sources for equity returns in markets around the globe permits important 'independent' confirmation across equity markets. These markets have widely differing structures along the dimension of trading, taxes, and institutional investor participation and, thereby, provide valuable new insights into the nature of these phenomena.

The papers presented herein are arranged into three sections. The first section addresses cross-sectional patterns in stock returns, such as market capitalization (size) and value vs. growth. The second section deals with the time series predictability, with a particular focus on predictability relating to calendar turning points like the beginning of the year or the beginning of the week. The papers in the final section discuss the existence of both cross-sectional and time series predictability in a broad cross section of international equity markets—both developed and emerging. We now discuss each section and the papers in more detail.

1   The Cross Section of Stocks Returns

The capital asset pricing model (CAPM) has occupied a central position in financial economics for the 35 years since its original appearance. Given certain simplifying assumptions, the CAPM states that the rate of return on any security is linearly related to that security's systematic risk (or beta) measured relative to the market portfolio of all securities. Hence, according to the CAPM, the cross-sectional relation between return and risk can be expressed as

$$E(R_i) = a_0 + a_1\beta_i$$

If the model is correct *and* security markets are informationally efficient, security returns will *on average* conform to this linear relation. Persistent departures represent violations of the joint hypothesis that both the CAPM and the efficient market hypothesis (EMH) are correct.

The early empirical tests of the model in the 1960s and early 1970s did not unambiguously support the CAPM. As a result, researchers formulated alternative models. Many developed equilibrium models by relaxing the CAPM assumptions. For example, Mayers (1972) allows for non-marketable assets such as human capital, and Brennan (1970) and Litzenberger and Ramaswamy (1979) relaxed the no-tax assumption. However, others examined *ad hoc* alternatives to the CAPM. For example, Basu (1977) and Banz (1981) found that the ratio of price to earnings and the market capitalization of common equity, respectively, provided considerably more explanatory power than beta. These two seminal studies served as a springboard for much subsequent research that has confirmed the ability of *ad hoc* variables to explain cross-sectional differences in returns. Absent in this literature, though, is any supporting theory to justify the choice of variables. Nevertheless, these findings collectively represent a set of stylized facts that stand as a challenge for alternative asset pricing models. The papers in this section address the more important contributions to these stylized facts.

The first paper in this section by **Hawawini and Keim** serves as an introduction to the cross-sectional evidence. The paper surveys the large quantity of research across a wide range of international equity markets on cross-sectional patterns in returns that has been produced over the past twenty years. The focus is on those variables whose relation with returns has proven to be most robust: market capitalization; ratio of book to market value; earnings yield; and price momentum. Hawawini and Keim report that, in the end, the debate on the interpretation of the significance of these variables boils down to two possibilities—the premia related to these variables represent either compensation for risk or evidence of an inefficient market. Although there is not much support for an inefficient market story (however, see the discussion in DeBondt below), Hawawini and Keim argue that the evidence in favor of a risk-factor story is not strong either. Indeed, some authors find that it is the variables themselves, and not the sensitivities to 'risk factors' associated with the variables, that drive the cross-sectional relation with returns. Further, Hawawini and Keim present some additional findings of their own: (1) the premia associated with the ad hoc variables derive primarily from the month of January; and (2) although the return premia associated with these variables are significant in most international markets, the premia are uncorrelated across markets. This prompts Hawawini and Keim to ask: If these return premia occur primarily in January and are uncorrelated across major international markets that are well-integrated, is it reasonable to characterize the premia as compensation for risk?

The prevalent reaction to the collected cross-sectional findings, both from the practitioner audience and from a surprisingly large segment of the academic audience, is that the CAPM (and by extension beta) is 'dead' and variables like size and book-to-market represent far more accurate estimates of the risks for which investors should be compensated. Like Hawawini and Keim, **Kothari and Shanken** take exception to this interpretation of the results. They maintain that many observers tend to ignore positive evidence on the importance of beta and over-emphasize the importance of book-to-market. They argue that even the most widely cited results on the 'insignificance' of beta can just as likely be interpreted as being consistent with the market factor being important and relevant. They go on to show that alternative estimates of beta can significantly explain cross-sectional variation in returns. Although they find evidence of a significant book-to-market effect for their sample period, they also find that the book-to-market effect is not uniformly significant for all stocks (it does not appear to be present for large-cap stocks). In the end, they caution that the popular acceptance of size and book-to-market as risk factors is premature.

While most economic models rely on rational human behavior, a growing number of financial economists are attempting to explain the observed patterns in returns with 'behavioral' models. In these models market participants often exhibit irrational behavior and make investment decisions that violate central axioms of rationality such as transitivity and dominance. In these models, investors are often overconfident about their own abilities; or they overreact to private information but under-react to public information. **DeBondt** provides a comprehensive survey of this rapidly growing literature, and makes a strong case that such investor behavior can translate to momentum and mean reversion in stock returns. He illustrates these concepts empirically in the context of the profitability of contrarian and momentum strategies in international equity markets, and concludes that the data support the behavioral models.

As the above papers attest, the debate concerning the cross-sectional pattern in returns has focused on the question of whether the size premium and the value premium represent compensation for risk or evidence of an inefficient market (investor irrationality). In the context of the size effect, **Berk** questions this focus, arguing instead that the size effect is not an anomaly at all. Berk maintains that the only scenario in which the size effect is an anomaly is if it can be shown to be inconsistent with the theoretical model of asset pricing. Using simple valuation principles, however, he argues that "any theory (e.g., the CAPM) that successfully explains any cross-sectional variation in expected returns must also predict an inverse relation between expected return and market value (i.e., the size effect)." Thus, the size effect is not an anomaly. Berk provides empirical results consistent with this view, and in the end concludes that while the development of factor models based on size (or book-to-market) 'might be useful in helping to explain the cross-section of stock returns, such models provide no information on the underlying economic cause of the variation.'

A characteristic of these cross-sectional patterns in returns is that their magnitude varies considerably over time. For example, the value premium, although positive on average when measured over long time periods, is often negative (growth outperforming value) over consecutive years. Indeed, we are currently in the midst of a four-year period during which value has consistently under-performed growth in US and European markets. There are many possible explanations for such cyclicality, ranging from the notion that the premia are indeed risky to the existence of cycles in investor behavior. **Dimson and Marsh** provide a discussion of these issues in the context of the size effect in the UK. After providing an extensive historical perspective on the UK size premium, they document its reversal precisely at the time when it was originally publicized and when investment companies were formed to exploit the premium. A similar reversal is also noted for the US size premium by Booth and Keim in this volume. Although Dimson and Marsh speculate on the possible causes for such a reversal, they are unable to conclude that it is attributable to the increasing participation of institutional investors in the markets for small-cap stocks. They correctly point out that even though the size *premium* has been negative recently, one must conclude that the size *effect* is still alive and well—small stocks continue to behave differently than large stocks.

As is often noted in the papers referenced in this volume, the stocks that display the most interesting return patterns – usually small-cap, low-price stocks – are the most thinly traded or illiquid stocks. Much has been written that theoretically links infrequent trading with average stock returns, return autocorrelations, and estimated betas. **Foerster and Keim** empirically document the amount of non-trading that is exhibited by stocks with different market capitalization. They find that non-trading is a pervasive phenomenon among small-cap stocks during the time periods studied in the size effect literature. Their paper also serves as a useful bridge to the next section of the book on time series and seasonal patterns in returns as Foerster and Keim also document a distinct relation between non-trading and the calendar turning points that are associated with unexplained returns.

2   Seasonal Patterns in Stock Returns and Other Puzzles

At about the same time that Banz and Basu were documenting the cross-sectional patterns in returns related to size and E/P, other researchers were finding time series predictability in stock returns that violated the random walk model which maintains that past stock return history cannot be used to predict future stock returns. The most serious violations seemed to be associated with calendar timing points—the beginning of the year, the beginning of the month, the beginning of the week. For example, Rozeff end Kinney (1976) found that NYSE stock market indexes displayed significantly higher returns in January than in the rest of the months, and Keim (1983) identified that this January seasonal was entirely a small stock phenomenon—large-cap stocks do not display a January seasonal. Cross (1973) and French (1980) documented a Monday effect whereby broad stock market indexes (e.g., the S&P 500) exhibit negative average returns on Mondays that are significantly lower than the other days of the week. Ariel (1987) showed that average returns for NYSE and AMEX stocks are positive only during the first half of every month, and that average returns in the first half of the month are significantly larger than the average returns in the second half of the month. And Ariel (1990) found that over one-third of the return accruing in the US stock market over the 1963–82 period was earned on the trading days preceding the eight holidays that result in market closings each year.

The papers in this section deal with these calendar-related patterns as well as several other puzzles. The first two papers address the durability and exploitability of the January effect, the phenomenon that Fama refers to as the 'premier anomaly' in his 1991 sequel to his classic 1970 paper on efficient markets. In the first paper in this section, **Booth and Keim** ask the question: Is there still a January effect? Awareness of the January effect, as well as institutional presence in the small cap markets, increased in the 1980s and 1990s *after* the publication of the papers describing the January effect. This led many authors to speculate that the forces of competitive markets would effectively eliminate the pattern. Using a fourteen-year out-of-sample period, Booth and Keim show that the January effect as described in the original research still exists. Specifically, the stocks in the smallest decile of market cap significantly outperform the stocks in the largest decile of market cap in January (although not in the remaining eleven months, on average.) Interestingly, the returns for a small-cap mutual fund, a benchmark series used by many academics and investment professionals, do not exhibit a January effect. Booth and Keim show that this is not due to the effect of large transactions costs associated with management of that portfolio of very liquid stocks. Rather, the lack of a January seasonal in the mutual fund returns is due to the exclusion of the lowest-price and lowest market-cap stocks from the portfolio, these being the stocks that exhibit the most pronounced January effect. They conclude that the January effect is alive and well, but difficult to capture in a portfolio of equities.

The search for alternative, lower-cost investment vehicles to capture the January effect motivates the paper by **Hensel and Ziemba**. Ever since the introduction of stock index futures trading in 1982, both individual and institutional investors have been using futures contracts based on underlying diversified portfolios of small-cap and large-cap stocks to try to capitalize on the January small stock premium. Not only do investments in futures positions provide exposure to indexes of equities, but they do so at very low cost. These features have attracted growing numbers of investors interested in exploiting the January size premium and, as Hensel and Ziemba report, these turn-of-the-year investments in stock index futures have been profitable. Hensel and Ziemba show that investors' desire to get into their positions ahead of the next investor has led to increased pressure on index futures prices in the second half of December. This has resulted in an uncoupling of the theoretical relation between the futures prices and the underlying stock index price, such that the 'January effect' appears in the stock index futures pries in the second half of December before it appears in the underlying stock index values. Hensel and Ziemba argue that this December anticipation of the January effect can be profitably exploited, but not on a large scale because of the limited liquidity in the small-stock index futures market.

Turning to the more general turn-of-the-month effect, of which the turn-of-the-year effect is a special case, **Hensel, Sick and Ziemba** augment existing results in the literature with an exhaustive examination of daily returns for the S&P Composite Index for the period 1928–1996. They confirm previous results, showing that average S&P returns are positive in the first half of the month and significantly larger than the S&P returns in the second half of the month (which are on average negative.) They find that these results are stable across ten-year subperiods in their sample but find substantial differences in the pattern across months within the year. For example, the turn-of-the-month seasonal is much stronger in January, March, May and July in comparison with the other months; and May, September November had statistically significant losses in the last half of the month. They simulate investment strategies that exploit these cross-month differences and report profitable results.

In an interesting twist on the return differential between small-cap and large-cap stocks, **Hensel and Ziemba** analyze returns in the periods following US presidential elections won by Democratic presidents and by Republican presidents for the period 1928–1997. They find that the small-cap premium is significantly larger during Democratic administrations than in Republican administrations. This finding is attributable to significantly larger small stock returns in Democratic years, as average large-cap returns are statistically indistinguishable across Democratic and Republican administrations. They also document that equity returns were substantially higher in the last two years, in comparison to the first two years, of both Democratic and Republican administrations. This effect seems to be related to the desire of the party in power to pursue favorable economic policies to gain reelection. Hensel and Ziemba investigate asset allocation strategies that are conditioned on the political affiliation of the occupant of the White House and find that those strategies outperform other more traditional asset mixes (e.g., a fixed 60–40 stock bond mix).

The last two papers in this section address long-standing puzzles in financial economics—the closed-end fund discount and the abnormally large ex-date returns associated with stock splits and stock dividends. In the first of these two papers, **Minio-Paluello** provides a comprehensive survey of the research documenting the closed-end fund puzzle. The puzzle here is that closed-end funds, which trade like shares on an exchange, typically trade at prices significantly lower than their NAV, a composite value reflecting the prices of the publicly-traded shares held in the fund. Because the NAV is an observable and easily quantifiable value, it is puzzling that closed-end funds would trade systematically at a price other than NAV. Minio-Paluello provides a discussion of both the empirical literature that has documented this anomaly and the array of possible explanations that include agency costs (poor managerial performance, incentive-incompatible management fee structures) and costs associated with market imperfections (liquidity costs, 'hidden' tax liabilities associated with unrealized capital gains in the fund's portfolio).

In the last paper in this section, **Grinblatt and Keim** investigate the puzzling *ex-date* behavior of stock splits and stock dividends. Stock splits and stock dividends are purely cosmetic events that do not affect the after-tax cash flows of firms, nor do they affect a shareholder's personal taxes or proportionate ownership of the firm. Nevertheless, a large body of research has shown that stock splits and stock dividends affect shareholder returns in a significantly positive fashion on ex-dates. Grinblatt and Keim investigate a market microstructure explanation for this anomaly. Because of significant changes in investor buying and selling behavior around the ex-date for stock splits and stock dividends, Grinblatt and Keim find a disproportionate number of the trade prices surrounding the ex-date occur at the ask price. This tendency for stock prices to close at the ask on the ex-date can contribute to the abnormally large returns observed on that date, but that component of the return is illusory because speculators wanting to exploit the ex-date price movement cannot sell at the ask. Grinblatt and Keim decompose ex-date returns into an 'intrinsic' component and a microstructure component and find that even after elimination of the microstructure component (which is statistically significant), there remains an upward revision in the value of splitting firms around the ex-date. Thus, the puzzle is only partially solved.

3   International Evidence

The final section of the book surveys the evidence on the cross-sectional and time series patterns in returns in equity markets around the world. We can learn much about these phenomena by investigating their existence worldwide. As mentioned previously, the increased availability of reliable machine-readable data sources for equity returns in markets around the globe permits independent confirmation of the phenomena across boundaries that are defined not only by geographical, economic, and political distinctions, but also by differences in stock market trading mechanisms, financial institutions, and cultural idiosyncrasies. Such differences can provide valuable new insights into the nature of these phenomena and, hopefully, expedite the search for explanations.

Development of reliable stock price data bases in many 'developed' markets (e.g., France, Germany, Japan, UK) in the 1980s led to a flurry of 'anomaly' research activity in those markets into the early 1990s. Much of that evidence has been synthesized in Hawawini and Keim in this volume for the cross-sectional patterns and in Hawawini and Keim (1995) and Ziemba (1994) for the time series patterns. Although this section of the book contains surveys of results in several developed markets (Canada, Italy and Japan), we focus our coverage on emerging markets. Thus, this section includes papers devoted to two of the smallest European markets (Finland and Turkey), a paper that investigates the size and value premium in each of the markets represented in the International Finance Corporation's Global Index of emerging markets, a paper devoted to both cross-sectional and time series anomalies in Asian emerging stock markets, and a paper that extends Ariel's (1990) evidence of a holiday effect to seventeen non-US markets worldwide.

In the first paper in this section, **Athanassakos and Foerster** provide a comprehensive survey of the empirical evidence on stock returns on the Toronto Stock Exchange. Athanassokos and Foerster first provide a useful overview of the Canadian economic environment, tax treatment of income, and the returns on stocks, long-term bonds and short-term bills that covers much of this century. They then discuss seasonal patterns in stock returns (weekend and turn-of-the-year effects), cross-sectional patterns in returns (size and value effects), and the relation between stock market performance and election cycles (cf. Hensel and Ziemba in this volume for the US evidence). Despite differences in taxation and institutional/market structures between Canada and the US, Athanassakos and Foerster find that the anomalous patterns are similar in the two markets. They also posit a time series model of expected returns as a function of inflation, slope of the yield curve, and default risk and report that a simple tactical asset allocation model based on the results outperforms a buy-and-hold equity strategy.

Canestrelli and Ziemba focus on seasonal anomalies in the Italian stock market. To illustrate these patterns for Italy, they use the returns (without dividends) for the COMIT index of all securities traded on the Milan Stock Exchange for the 1974–93 period. Canestrelli and Ziemba examine the weekend, turn-of-the-month, turn-of-the-year, monthly and holiday patterns in Italian stock returns and find statistically significant evidence for each of these phenomena during their sample period.

There are two papers that investigate stock return predictability in Japan. In the first of them, **Comolli and Ziemba** examine calendar-related anomalies for two equity indexes composed of stocks on the Tokyo Stock Exchange. The interesting angle here is that Comolli and Ziemba examine the 1990–94 period, an out-of-sample period not included in the previous literature on Japanese equity markets, particularly Ziemba (1991). Thus, the findings represent independent observations useful to test the robustness of previous results. On the flip side, however, the short sample period may make it difficult to draw precise conclusions because the statistical tests will have low power. Comolli and Ziemba find that most of the anomalies that existed in the Japanese equity market in earlier periods (e.g., the turn-of-the month effect, the January effect, the Golden week effect, the holiday effect) were still evident in equity returns in the 1990–94 period, but were not statistically significant (largely due to the small number of observations.) Because of the elimination of Saturday trading during the 1990–94 period the Japanese day-of-the-week effect bears a closer resemblance to the pattern observed in the US – average returns are lowest on Mondays and highest on Fridays for the TOPIX index.

Schwartz and Ziemba examine the Japanese equity markets from a cross-sectional perspective, in the spirit of the papers described in Hawawini and Keim that use variables like market capitalization and price-earnings ratios to separate stocks into high expected return vs. low expected return categories. Schwartz and Ziemba develop a model using thirty firm characteristics (ranging from market cap to book-to-market ratio to dividend yield to earnings growth to price momentum) to rank stocks on expectations of prospective returns. Using these rankings, they construct long-short portfolios (long the stocks with highest expected returns, short the lowest expected returns) and test the model's results in a hold out sample period. Consistent with the results for simpler models using only size and/or value growth characteristics, Schwartz and Ziemba report abnormal returns for their long-short portfolio simulations.

Founded in 1986, the Istanbul Stock Exchange is one of the world's youngest stock markets. **Muradoglu** provides an introduction to the exchange with a discussion of the history of its development, the market structure and operation, the investor base, trading and microstructure issues, and the integration of the Turkish stock market with other developed markets. The primary objective in the paper, though, is describing the risk and return characteristics of Turkish stocks and investigating the existence of both cross-sectional and time series patterns in Turkish stock markets. Muradoglu reports that many of the cross-sectional and time series patterns found in developed markets are also evident during the brief history of the Istanbul Stock Exchange. However, the reader is appropriately cautioned that in the volatile and ever-changing environment of emerging markets, past performance is not always an accurate forecast of future performance.

Like the Istanbul Stock Exchange, the Helsinki Stock Exchange in Finland is one of the smallest in Europe. As is the case for emerging markets generally, the infrequent trading that is characteristic for most stocks on the Helsinki Exchange make estimation of risk measures, return autocorrelations and portfolio mean returns difficult; and drawing statistical inferences based on such estimates can be challenging. Nevertheless, there is a large body of empirical research that has been conducted on the Finnish markets and **Martikainen** provides a comprehensive survey of this work. While the primary focus of the paper is on equities, Martikainen also gives extensive coverage to the markets for stock index derivatives. After a discussion of the institutional structure, market trading mechanisms and other microstructure issues for both the equity and derivative markets, Martikainen surveys the evidence for both cross-sectional and calendar-related patterns. He finds that most of the anomalies identified in major markets also are evident in Finnish markets – there are significant day-of-the-week, turn-of-the-month, and January effects, as well as a significant size premium and value premium.

The paper by **Patel** extends the evidence of cross-sectional patterns in stock returns to the 22 markets represented in the International Finance Corporation's Global Index of emerging markets. Patel examines the relation between stock returns and pre-determined characteristics like price-to-book ratio, price-to-earnings ratio, and market capitalization. He finds evidence of significant size and value premia in most of the emerging markets, results that provide added confirmation of the robustness of these unexplained premia. Patel finds that although the premia are significant in the separate markets, they are uncorrelated across markets, a result similar to that reported in Hawawini and Keim in this volume for developed markets. Finally, and from a more practical perspective, Patel finds that *sector-neutral* small-cap and value portfolios strategies—that is, portfolios that are constrained to diversify across economic sectors to mimic the sector weightings of the local composite market index—are less risky and outperform unconstrained (naive) small-cap and value strategies.

In the paper by **Koh and Wong**, the research focus is narrowed to just the emerging stock markets in Asia. They include Hong Kong, Malaysia, Philippines, Singapore, South Korea, Taiwan and Thailand in this category. Koh and Wong provide a comprehensive discussion of the evidence on the January effect, the day-of-the-week effect, the holiday effect, and the size and value premia in these markets. Similar to most of the other papers described here, they find that these phenomena are surprisingly robust across the markets they examine that reflect widely varying trading mechanisms and economic environments. One exception is the January effect which is evident in only two of the markets surveyed – Malaysia and Singapore. Interestingly, neither of these countries imposes taxes on capital gains, the reason most often used to explain the January effect.

The last paper in the volume, by **Cervera and Keim**, also narrows the focus, but this time by focusing on only one anomaly. Cervera and Keim examine the holiday effect for a sample of seventeen non-US markets for the period 1980–94. Consistent with research referenced above, they find that stock market indexes worldwide tend to exhibit significantly higher (5.5 times higher) than average returns on the trading days preceding holidays that result in market closure. The Netherlands is the only market in which pre-holiday returns are lower than the average return over all days; the effect is strongest, on average, for non-European markets. Cervera and Keim are unable to provide an explanation for the holiday effect, but they do provide evidence consistent with a partial explanation. They find that a large component of the international holiday effect is due to large returns preceding holidays that are common to most countries, a finding consistent with a common global market factor in increasingly integrated worldwide markets. However, after accounting for this common effect, a significant holiday effect remains.

4   What's next?

The objective of this volume is to report what we currently know about 'anomalies' in equity markets around the world. We feel it has accomplished that objective. Unfortunately, the explosion of research that has focused on these stock price irregularities also serves to remind us just how incomplete our understanding of these phenomena is. Yes, we have substantially more documentation about the behavior of the phenomena than we did ten years ago, but it is not obvious that we are much closer to explanations for them. Invoking Kuhn's *Structure of Scientific Revolutions* once again, it is at this stage in the research cycle that researchers' focus should be on explanation rather than further documentation. How long until the shift to a new paradigm is complete? It is no easier to predict when we will have alternative theories that explain the empirical regularities than it is to predict tomorrow's closing value of the S&P 500 index. What we can safely predict is that the coming years will witness a continuing stream of research in this area and, eventually, a better understanding of the causes of the phenomena.

References

Ariel, R. (1987) 'A Monthly Effect in Stock Returns', *Journal of Financial Economics* **18** 161–74.

Ariel, R. (1990) 'High Stock Returns Before Holidays: Existence and Evidence on Possible Causes', *Journal of Finance* **45** 1611–26.

Banz, R. (1981) 'The Relationship between Return and Market Value of Common Stock', *Journal of Financial Economics* **9** 3–18.

Basu, S. (1977) 'Investment Performance of Common Stocks in Relation to their Price-Earnings Ratio: a Test of the Efficient Market Hypothesis', *Journal of Finance* **32** June, 663–682.

Brennan, M.J. (1970) 'Taxes, Market Valuation, and Corporate Financial Policy', *National Tax Journal* **23** 417–27.

Cross, F. (1973) 'The Behavior of Stock Prices on Fridays and Mondays', *Financial Analysts Journal* **29** 67–69.

Dimson, E. (ed.) (1988) *Stock Market Anomalies*, Cambridge University Press.

French, K. (1980) 'Stock Returns and the Weekend Effect', *Journal of Financial Economics* **8** 55–69.

Hawawini, G., and D. Keim (1995): 'On the Predictability of Common Stock Returns: World-Wide Evidence'. In *Finance: Handbooks in OR & MS*, R. Jarrow V. Maksimovic and W. Ziemba (eds.), Volume 9, Chapter 17, North Holland.

Keim, D. (1983) 'Size-Related Anomalies and Stock Return Seasonality: Further Empirical Evidence', *Journal of Financial Economics* **12** 13–32.

Kuhn, T. (1970) *The Structure of Scientific Revolutions*, University of Chicago Press.

Litzenberger, R. and K. Ramaswamy (1979) 'The Effects of Personal Taxes and Dividends on Capital Asset Prices : Theory and Empirical Evidence', *Journal of Financial Economics* **7** (2), 163–195.

Mayers, D. (1972): 'Nonmarketable Assets and Capital Market Equilibrium under Uncertainty'. In *Studies in the Theory of Capital Markets*, M.C. Jensen (ed.), Praeger, New York.

Reinganum, M. (1981) 'A Misspecification of Capital Asset Pricing: Empirical Anomalies Based on Earnings Yields and Market Values', *Journal of Financial Economics* **9** 19–46.

Rozeff, M. and W. Kinney (1976) 'Capital Market Seasonality: The Case of Stock Returns', *Journal of Financial Economics* **3** 379–402.

Ziemba, W.T. (1991) 'Japanese Security Market Regularities: Monthly, Turn-of-the-month and Year, Holiday and Golden Week Effects', *Japan and the World Economy* **3** 119–146.

Ziemba, W.T. (1994) 'Worldwide Security Market Regularities', *European Journal of Operational Research* **74** 198–229.

Part I
An Overview of Cross Sectional Patterns in Stock Returns

The Cross Section of Common Stock Returns: A Review of the Evidence and Some New Findings*

Gabriel Hawawini and Donald B. Keim

Abstract

A growing number of empirical studies suggest that betas of common stocks do not adequately explain cross-sectional differences in stock returns. Instead, a number of other variables (e.g., size, ratio of book to market, earnings/price) that have no basis in extant theoretical models seem to have significant predictive ability. Some interpret the findings as evidence of market inefficiency. Others argue that the Capital Asset Pricing Model is an incomplete description of equilibrium price formation and these variables are proxies for additional risk factors. In this paper we review the evidence on the cross-sectional behavior of common stock returns on the US and other equity markets around the world. We also report some new evidence on these cross-sectional relations using data from both US and international stock markets. We find, among other results, that although the return premia associated with these *ad hoc* variables are significant in most international stock markets, the premia are uncorrelated across markets. The accumulating evidence prompts the following question: If these return premia occur primarily in January and are uncorrelated across major international equity markets, is it reasonable to characterize them as compensation for risk?

1 Introduction

In this paper we review the evidence on the cross-sectional behavior of common stock returns in the US and other equity markets around the world.[1]

---

*We thank Marshall Blume, Jay Shanken, Bill Sharpe, Bill Ziemba and participants at the Fall 1996 Q-Group Seminar for helpful comments. We are responsible for any errors or ommissions.

[1]This paper draws on our earlier paper ([60]), updating its content and extending the section that reviews the recent evidence regarding the cross-sectional behavior of common stock returns. This paper, as our earlier one, is not meant to be an exhaustive compilation of the findings on the cross-sectional predictability of stock returns. The focus is on the subset of findings whose existence has proved most robust with respect to time and the stock markets in which they have been observed.

3

Since the early 1980s, a growing number of empirical studies have documented the presence of persistent cross-sectional patterns in stock returns that do not support one of the fundamental tenets of modern finance: expected stock returns are determined by their level of beta risk through a positive and linear relation known as the capital asset pricing model, or CAPM ([99], [80], [86], [106]).

The evidence suggests that betas of common stocks do not adequately explain cross-sectional differences in stock returns. Instead, a number of other variables that have no basis in extant theoretical models seem to have significant predictive ability. These other variables include firm size (measured by market capitalization of the firm's common stock), the ratio of book to market values (the accounting value of a firm's equity divided by its market capitalization), earnings yield (the firm's reported accounting net profits divided by price per share), and the firm's *prior* return performance.

Interpretation of the cross-sectional explanatory power of such *ad hoc* variables presents a challenge to the profession. Recall that tests of asset pricing models involve the joint null hypothesis that security markets are informationally efficient and expected returns are described by a prespecified equilibrium model (e.g., the CAPM). If the joint hypothesis is rejected, we cannot specifically attribute that rejection to one or the other branch of the hypothesis. Indeed, a lively debate continues in the literature regarding the interpretation of these results. Some interpret the findings as convincing evidence of market inefficiency: if stock returns can be predicted on the basis of historical factors such as market capitalization, book-to-market value and prior return performance, then it is difficult to characterize stock markets as informationally efficient. On the other hand, the rejection may be due to a test design that is based on an incorrect equilibrium model. The fact that so many of these regularities have persisted for more than thirty years suggests that perhaps our benchmark models are incomplete descriptions of equilibrium price formation.[2]

There are additional findings that challenge the notion that these *ad hoc* variables proxy for additional risk factors. First, the evidence indicates that the relation between returns and variables like firm size and book-to-market ratio is typically significant *only* during the month of January. Why would a risk-based factor manifest itself during the month of January and not the

---

[2]Another explanation that has not been adequately addressed in the literature is that these *ad hoc* variables proxy for betas that are not properly estimated. Put differently, the failure of the empirical evidence to provide unambiguous support for the CAPM is not necessarily proof of the model's invalidity, but may be the manifestation of our inability to accurately measure beta risk. For example, one can argue that stocks with lower ratios of P/B have higher average returns than stocks with higher ratios of P/B because they are indeed riskier in a beta sense. If we could measure beta risk with less error, then the reported negative relation between P/B and *beta-adjusted* returns may disappear.

rest of the year? If investors expect higher returns for holding stocks with a particular characteristic, then it is reasonable to expect the market to deliver that premium uniformly throughout the year. If that premium is compensation for risk, is there reason to believe that the market is systematically more risky in January than the rest of the year?

Second, some new international evidence on the premia associated with size, E/P, CF/P and P/B that we report in Section 6 is difficult to reconcile with an international version of the risk story. If these premia are compensation for additional risks that are priced in the context of an international asset pricing model under conditions of integrated international capital markets, then the premia should be correlated across markets in much the same way that the market risk premium is significantly correlated across markets. Inconsistent with this hypothesis, we find that the premia correlations are insignificant across the markets in our sample.

The rest of the paper is organized as follows. In the next section we briefly review the early tests of the CAPM which were generally supportive of that model. In Section 3 we survey the recent empirical evidence which is at odds with the predictions of the CAPM. In Section 4 we examine evidence that sheds light on whether the explanatory power of the *ad hoc* variables reflects the underlying influence of one or several underlying phenomenon. In Section 5 we discuss the research that attempts to sort through the effects to determine which ones have the greatest explanatory power. Section 6 reviews potential interpretations of the evidence on the cross-sectional behavior of returns, and reports some new international premia evidence that appears to be inconsistent with a risk-based story. Section 7 concludes the paper.

2 Tests of the (Single-Beta) Capital Asset Pricing Model

The capital asset pricing model has occupied a prominent position in financial economics for the thirty years since its origins in the papers by Treynor [106], Sharpe [99], Lintner [80] and Mossin [86]. Given certain simplifying assumptions, the model states that the expected rate of return on any security is positively and linearly related to that security's systematic risk (or beta) measured relative to the market portfolio of all marketable securities. Hence, according to the CAPM, the relation between the expected return $E(R_i)$ and the systematic risk β_i of security i can be expressed as:

$$E(R_i) = a_o + a_1\beta_i \qquad (2.1)$$

If the model is correct and security markets are efficient, stock returns should on average conform to this linear relation. Persistent departures from positive linearity would represent violations of the joint hypothesis that both the CAPM and the efficient market hypothesis are valid.

The early tests (e.g., [19], [20], and [50]) found evidence of a significant relation between average returns and estimated betas, but the estimated intercept was higher (a_o in equation (2.1)), and the estimated slope (a_1 in equation (2.1), representing an estimate of the market risk premium) was lower than predicted by the Sharpe-Lintner CAPM and only marginally important in explaining cross-sectional differences in average stock returns. The results of these studies were interpreted as being consistent with the Black, [18], version of the CAPM according to which the relation between expected returns and systematic risk should be flatter than that predicted by the standard CAPM if a risk-free security is not available to investors.

Although the early tests lend some support for the model, subsequent research was not always as accommodating. For example, in his critique of existing tests of the CAPM, Roll, [92], argued that tests performed with any 'market' portfolio other than the true market portfolio are not tests of the CAPM and, therefore, cannot be interpreted as evidence for or against the model. In response to Roll's criticism of the earlier tests, Stambaugh, [101], constructed broader market indexes that included bonds, real estate and consumer durables and found that tests of the model with these broader indexes were not very sensitive to the breadth of the definition of the market proxy.

3   Evidence Inconsistent with the CAPM

The central prominence of beta in the asset pricing paradigm came into question with the first tests of ad hoc alternatives to the CAPM in the late 1970s. The earliest of these tests are those of Basu, [13], and Banz, [9]. They found that the price-to-earnings ratio (P/E) and the market capitalization of common equity (firm size), respectively, provided considerably more explanatory power than beta. Other studies have extended the list of predictive factors to include, among others, the ratio of book-to-market value, price per share, and prior return performance. Combined, these studies have produced convincing evidence of cross-sectional return predictability that greatly transcends the marginal explanatory power of beta found in the earlier studies. Notably absent in this literature, though, is any supporting theory to justify the choice of factors. Nevertheless, the findings collectively represent a set of stylized facts that stand as a challenge for alternative asset pricing models.

In this section we present a sample of the more important contributions to these stylized facts. The basic set of findings have been reported in a variety of manner. For example, some researchers employ cross-sectional regression techniques similar to those originally used by Fama and MacBeth, [50]:

$$R_i = a_0 + a_1\beta_i + a_2\Sigma c_{ij} + e_i \tag{3.1}$$

where c_{ij} represents characteristic j (size, earnings yield, price-book ratio, etc.) for stock i. Researchers have also documented these phenomenon by examining the returns of portfolios constructed on the basis of these characteristics c_{ij}.

3.1 Data and empirical methods

To maintain a unifying thread throughout the following discussion of cross-sectional return predictability, we augment much of our reporting of the original results in the literature with some basic summary statistics that document the findings with a common data set for the same time period using the same empirical methods. We believe this approach avoids some of the apples-and-oranges comparisons that bog down literature surveys of disparate studies employing widely varying samples, time periods and empirical methods. We portray our new evidence in portfolio form because we feel the returns to feasible portfolio strategies provide a useful perspective on the economic significance of the results. We report our findings using monthly value-weighted portfolio returns to avoid the potential statistical biases associated with measuring these effects with daily portfolio returns (e.g., [93], [90], [22]). The use of monthly data also avoids biases in estmated betas due to the infrequent or nonsynchronous trading of securities (e.g., [44], [97]). The data are drawn from the monthly return file of the Center for Research in Security Prices (CRSP) and the Compustat annual industrial and research files. In each of the 'simulation' experiments reported below, portfolios are created on March 31 of each year using prices and shares outstanding on March 31 and accounting data for the year ending on the previous December 31 (with portfolios containing only December 31 fiscal closers). Aside from new listings and delistings, which are added to or dropped from the portfolios as they occur during the year, the portfolio composition remains constant during the following twelve months over which the portfolio returns are calculated. As such, the simulated portfolios display little trading, and represent feasible strategies.

3.2 The size effect

Much of the research on cross-sectional predictability of stock returns has focused on the relation between returns and the market value of common equity, commonly referred to as the size effect. Banz, [9], was the first to

document this phenomenon. For the period 1931 to 1975, he estimated a
model of the form:

$$R_i = a_0 + a_1 b_i + a_2 S_i + e_i, \qquad (3.2)$$

where S_i is a measure of the relative market capitalization ('size') for firm i.
He found that the statistical association between returns and size is negative
and of a greater order of magnitude than that between returns and beta
documented in the earlier studies of the CAPM.

The first set of columns in Table 1 reports the average monthly returns for
ten value-weighted size-portfolios of NYSE and AMEX stocks for the period
April 1962 to December 1994, along with corresponding values for portfolio
beta and average market capitalization of the stocks in the portfolio. The
negative relation between size and average returns is clearly evident. The
annualized difference in returns between the smallest and largest size deciles
is 8.8%. Note that the portfolio betas decline with increasing size, but the
differences are small. Thus, consistent with research that finds significant negative coefficients on size in equation (3.2) after adjusting for the explanatory
power of beta, the difference in estimated OLS betas between the smallest
and the largest size portfolios is insufficient to explain the difference in returns
between the two extreme portfolios in Table 1.

Additional evidence in Reinganum, [91], suggests that the relative price behavior of small and large firms may differ for over-the-counter (OTC) stocks.
Using data for the 1973–1988 period, he finds that small OTC shares have
significantly lower returns than NYSE and AMEX firms with the same size,
and that the small-firm premium for OTC stocks is much lower than for
NYSE and AMEX stocks. Reinganum, motivated by earlier work of Amihud and Mendelson, [4], argues that the differences are related to differences
in liquidity between the two markets, suggesting differential costs of trading
small stocks in these two types of markets.[3] [4] The implication is that market
structure may be an important influence on the measured size effect. If so,
the analysis of the international evidence on the size effect, where we observe
very different market organizations and structures, should reveal significant
differences in the magnitude of the size premium across markets.

Following the discovery of a size premium in the US equity markets, numerous studies have documented its existence in most stock markets around

---

[3]Loughran, [83], finds, however, that of the 5.7% difference in returns between NYSE
and NASDAQ stocks in the bottom five size deciles (based on NYSE ranking), 60% is
due to the poor (long-run) performance of initial public offerings (IPO's) on NASDAQ. A
difference of only 2.3% remains after purging NASDAQ returns of an IPO effect (IPO's are
much more heavily concentrated on NASDAQ than on the NYSE).

[4]Fama, French, Booth and Sinquefield, [51] show that small NYSE stocks have substantially lower ratios of price to book value than comparably-small NASDAQ stocks. They
argue that the higher returns for small NYSE stocks are related to the lower price/book
ratios, a relation that appears to persist independently of the size-return relation.

Table 1. Monthly percentage returns (Standard errors) and betas for value-weighted portfolios of NYSE and AMEX stocks formed on the basis of size (market capitalization), earnings-to-price ratios, cash flow-to-price ratios, price-to-book ratios, and prior return[1] (Apr. 1962 to Dec. 1994).

Portfolio[2]	Size (Market cap.)			Earnings-to-price ratio			Cash flow-to-price ratio[3]			Price-to-book ratio			Prior return (%)		
	Size ($ million)	Return (%)	Beta	E/P	Return (%)	Beta	CF/P	Return (%)	Beta	P/B	Return (%)	Beta	Prior Return (%)	Return (%)	Beta
1	$10	1.56 (0.37)	1.11	19.39	1.21 (0.27)	1.01	52.08	1.47 (0.33)	1.00	0.57	1.43 (0.28)	1.04	53.1	1.18 (0.29)	1.13
2	$26	1.41 (0.34)	1.14	12.88	1.25 (0.23)	0.93	27.75	1.32 (0.29)	0.90	0.84	1.42 (0.25)	0.97	24.9	1.24 (0.26)	1.05
3	$48	1.25 (0.31)	1.10	11.26	1.08 (0.23)	0.88	23.02	1.17 (0.29)	0.91	1.02	1.06 (0.23)	0.92	16.7	1.09 (0.24)	1.02
4	$83	1.23 (0.31)	1.15	10.09	1.02 (0.23)	0.95	19.91	0.94 (0.31)	0.99	1.18	1.05 (0.21)	0.84	11.2	1.03 (0.23)	1.02
5	$104	1.22 (0.28)	1.10	9.08	0.96 (0.23)	0.94	17.37	1.14 (0.31)	1.01	1.35	1.00 (0.22)	0.90	6.5	0.88 (0.22)	0.96
6	$239	1.12 (0.26)	1.04	8.14	0.77 (0.23)	0.99	15.05	0.87 (0.30)	0.99	1.56	0.79 (0.22)	0.91	2.3	0.91 (0.22)	0.93
7	$402	1.09 (0.25)	1.06	7.19	0.83 (0.22)	0.96	12.96	1.12 (0.30)	1.03	1.86	0.84 (0.23)	0.98	-1.9	0.85 (0.23)	0.95
8	$715	1.09 (0.24)	1.05	6.13	0.89 (0.24)	1.04	10.85	1.05 (0.32)	1.08	2.30	0.91 (0.24)	1.03	-6.6	0.92 (0.24)	0.96
9	$1,341	1.03 (0.23)	1.03	4.78	0.88 (0.25)	1.06	8.40	0.89 (0.31)	1.06	3.10	0.82 (0.25)	1.11	-13.1	0.62 (0.27)	1.05
10	$5,820	0.83 (0.21)	0.95	2.49	0.82 (0.26)	1.08	4.77	0.80 (0.33)	1.07	10.00	0.90 (0.25)	1.07	-29.6	0.83 (0.28)	1.15

1. Portfolios are created on March 31 of each year using prices and shares outstanding on March 31 and accounting data for the year ending the previous December 31 (with portfolios containing only December 31 fiscal closers). Prior return is calculated from the beginning of October to the end of February immediately preceding the March formation month. Aside from new listings and delistings, which are added to or dropped from the portfolios as they occur during the year, the portfolio composition remains constant during the following twelve months over which portfolio returns are calculated. The data reported in the table are average values over the sample period. Betas are computed relative to the S&P 500 Index.
2. Portfolio 1 (portfolio 10) is the portfolio with the smallest (largest) market capitalization, highest (lowest) E/P and CF/P, lowest (highest) P/B, and largest (smallest) 6-month prior returns.
3. All results for the Cash Flow-to-Price portfolios are for the period April 1972–December 1994.

the world. Models similar to (3.2) have been estimated for Belgium ([62]), Canada ([24]), France ([61]), Ireland ([34]) Japan ([58], [28]), Mexico ([63]), Spain ([96]), Switzerland ([37]) and the United Kingdom ([36]). In all these countries, except Mexico, there is no relation, on average, between return and beta risk when all months of the year are considered (i.e., a_1 is statistically indistinguishable from zero). There is, however, a significant negative relationship between returns and portfolio size in all countries except Canada and France (i.e., a_2 is significantly less than zero).

The portfolio evidence from international equity markets is summarized in Table 2 for the stock markets of Australia, New Zealand, Canada, Mexico, Japan, Korea, Singapore, Taiwan, and eight European countries. The monthly size premium is defined as the difference between the average monthly return on the portfolio of smallest stocks and the average monthly return on the portfolio of largest stocks. In all countries, except Korea, the size premium is positive during the reported sample periods (which, in most cases, are significantly shorter than the 32 years of data we use in Table 1 to estimate the size effect in the US market). As expected, the size premium varies significantly across markets[5]: It is most pronounced in Australia (5.73%) and Mexico (4.16%), and least significant in Canada (0.44%) and the United Kingdom (0.40%).[6] As is the case for US data, differences in beta across size portfolios cannot explain differences in returns.

There are, however, significant differences across the fifteen markets in the spread between the size of the largest and smallest portfolios as indicated by the ratios of the average market capitalization of the largest portfolio to that of the smallest one, reported in Table 2. For example, in Spain, the largest size-portfolio is 228 times larger than the smallest one, whereas in the case of Taiwan it is only 17 times larger. There does not seem to be

---

[5]Although we hypothesized that the magnitudes of the size effect across markets should reflect differences in the organizations and structures of these markets, the results reported in Table 2 are also likely to be sensitive to differences in sample dates and lengths.

[6]Although Levis, [78], finds that the size effect on the London Stock Exchange (LSE) is not statistically significant, others report a significant size premium. Banz, [10], provides evidence of a significant size effect on the LSE. His analysis is based on 29 years of monthly returns (1955–1983) taken from the London Share Price Data base (LSPD). With ten value-based portfolios, he reports a compounded annual return of 39.9% for the smallest portfolio versus 13.0% for the largest. Dimson and Marsh, [45], also report evidence of a size effect on the portfolios constructed from a sample of stocks taken from the LSPD. Over the period 1977–1983, the portfolio of smallest stocks earned a compound annual return of 41% and the portfolio of largest stocks realized a compound annual return of 18%. In [10], the compound annual return on the smallest portfolio exceeded that of the largest by 27%. Dimson and Marsh, [45], report that the difference is 23%, both before adjustment for risk. More recently, Strong and Xu, [105], report an average monthly size-premium of 0.61% (7.3% annually) for the extreme portfolios drawn from decile portfolios formed on size during the period July 1973 to July 1992 using the London Share Price Data base, a result that is not explained by differences in betas.

Table 2. The Size Effect: International Evidence. (Size Premium $= \overline{R}_{\text{Small}} - \overline{R}_{\text{Large}}$)

	Monthly Size Premium (%)	Test Period	Number of Portfolios	Largest Size/[2] Smallest Size
Australia	1.21	1958–81	10	NA
Belgium	0.52	1969–83	5	188
Canada	0.44	1973–80	5	67
Finland	0.76	1970–81	10	133
France	0.90	1977–88	5	NA
Germany	0.49	1954–90	9	NA
Ireland	0.47	1977–86	5	NA
Japan	1.20	1965–87	10	NA
Korea	−0.40	1984–88	10	62
Mexico	4.16	1982–87	6	37
New Zealand[3]	0.51	1977–84	5	60
Singapore[4]	0.41	1975–85	3	23
Spain[3]	0.56	1963–82	10	228
Switzerland	0.52	1973–88	6	99
Taiwan[5]	0.57	1979–86	5	17
UK	0.61	1973–92	10	182
US	0.61	1951–94	10	490

1. **Sources**: *Australia*: Brown et al. (1983); *Belgium*: Hawawini et al. (1989); *Canada*: Berges et al. (1984); *Finland*: Wahlroos et al. (1986); *France*: Louvet et al. (1991); *Germany*: Stehle (1992); *Ireland*: Coghlan (1988); *Japan*: Ziemba (1991); *Korea*: Kim et al. (1992); *Mexico*: Herrera et al. (1994); *New Zealand*: Gillan (1990); *Singapore*: Wong et al. (1990); *Switzerland*: Cornioley et al. (1991); *Taiwan*: Ma et al. (1990); *United Kingdom*: Strong and Xu (1995).

2. Ratio based on average market values (median in Singapore) over sample period, except for Great Britain where it is calculated in 1975 and Finland in 1970. NA = Not Available.

3. Returns, for this country, are not raw returns but risk-adjusted, *excess* returns estimated with the Capital Asset Pricing Model.

4. Note that Wong et al. (1990) report that the size effect in Singapore appears to be of secondary importance when compared with the E/P effect.

5. *Excess* returns estimated by subtracting from each individual security the return of the portfolio to which they belong. Note that another study of the size effect in Taiwan (Chou and Johnson (1990)) finds no evidence of a significant size effect after controlling for an E/P effect.

a relation between the magnitude of the size premium and the size ratio. However, because the size and number of portfolios as well as the sample periods differ across markets, it is difficult to gauge whether the magnitude of the size premium is indeed significantly different across these countries, although, as pointed out earlier, we suspect that the differences in the size premium are unlikely to be explained by these factors alone. Differences in market structures and organizations may account for some of the reported variation in the size premium across markets.

3.3   The earnings-yield effect

Earnings-related strategies have a long tradition in the investment community. The most popular of these strategies, which calls for buying stocks that sell at low multiples of earnings, can be traced back at least to Graham and Dodd, [55], who proposed that "a necessary but not a sufficient condition [for investing in a common stock is] a reasonable ratio of market price to average earnings" (p. 533). They advocated that a prudent investor should never pay as much as 20 times earnings and a suitable multiplier should be 12 or less.

Ball, [6], argues that earnings-related variables like the earnings-to-price ratio (E/P) are proxies for expected returns. In that case, if the CAPM is an incomplete specification of priced risk, then we would expect E/P to explain the portion of expected return that is in fact compensation for risk variables omitted from the tests. A valid question, then, is whether a documented relation between average returns and E/P is due to the influence of E/P, or whether E/P is merely proxying for other explanators of expected returns.

Nicholson, [87], published the first extensive study of the relation between P/E multiples (the reciprocal of the earnings yield) and subsequent total returns, showing that low P/E stocks consistently provided returns greater than the average stock. Basu, [13], introduced the notion that P/E ratios may explain violations of the CAPM and found that, for his sample of NYSE firms, there was a significant negative relation between P/E ratios and average returns in excess of those predicted by the CAPM. If one had followed his strategy of buying the quintile of lowest P/E stocks and selling short the quintile of highest P/E quintile stocks, based on annual rankings, the average annual abnormal return would have been 6.75% (before commissions and other transaction costs) over the 1957 to 1975 period. Reinganum, [89], analyzing both NYSE and AMEX stocks, confirmed and extended Basu's findings to 1979.

In the second set of columns in Table 1 we report the relation between average monthly returns and E/P for the 1962–94 period using the same data file of NYSE and AMEX stocks used to examine the size effect. The portfolio returns in Table 1 confirm the E/P effect documented in previous studies.[7] The difference in returns between the highest and lowest E/P portfolios is, on average, 0.39% per month ($T = 1.68$).[8]

There is less evidence of an E/P effect in markets outside the United States. This is partly due to a lack of computerized accounting databases available for academic research. The evidence is also more varied than that for the size effect. Countries in which an E/P effect has been examined include the United Kingdom, Japan, Singapore, Taiwan, Korea, and New Zealand. In the UK, Levis, [79], documents a significant E/P effect for the period April 1961 to March 1985. He reports an average monthly premium of 0.58% (7.0% annually). The magnitude of the E/P effect in the United Kingdom is confirmed by the more recent work of Strong and Xu, [105]. They report an average monthly premium of 0.60% for the extreme portfolios drawn from decile portfolios formed on E/P ratios over the period July 1973 to July 1992.[9] This premium is of the same order of magnitude as that observed in the US and reported in Table 1. Adjusting portfolio returns for differences in systematic risk does not modify this conclusion.

Aggarwal, Hiraki and Rao, [1], provide evidence of a significant E/P effect for a sample of 574 firms listed on the first section of the Tokyo Stock Exchange during the period from 1974 to 1983. Only firms with positive earnings were included in the sample. Portfolios of high E/P stocks outperformed those with low E/P stocks even after controlling for differences in systematic risk and size across portfolios. In the case of Singapore, Wong and Lye, [108], show that there is a significant E/P effect on that country's stock market for the same sample of firms that revealed the presence of a significant size effect in Table 2. They conclude that the E/P effect is stronger than the size effect, although not independent of firm size.

---

[7]Some have argued that because firms in the same industry tend to have similar E/P ratios, a portfolio strategy that concentrates on low E/P stocks may indeed benefit from higher than average returns, but at a cost of reduced diversification. These arguments also suggest that the E/P effect may in fact be an industry effect. For example, during the 1980s financial firms and utilities comprised anywhere from 45 to 86% of the highest E/P quintile constructed from our sample of firms. Peavy and Goodman, [88], address this potential bias and examine the P/E ratio of a stock relative to its industry P/E (PER). They find a significant negative relation between PER's and abnormal returns over the 1970–1980 period. A portfolio strategy that bought the quintile of lowest PER stocks and sold short the highest PER quintiles would have yielded an annualized abnormal return of 20.80% over the period, although this number does not account for transactions costs.

[8]The table reports total returns that are not adjusted for risk. Since the betas are not substantially different across the portfolios, inferences drawn from total returns should not diverge in a meaningful way from inferences drawn from returns adjusted for beta risk.

[9]Levis, [79], also reports a size effect (see the evidence in Table 2 for the case of a slightly different sample characteristics), but it is weaker than the P/E effect. He also finds a large degree of interdependency between the two effects with the P/E effect tending to subsume the size effect.

For the Taiwanese stock market, Chou and Johnson, [32], report a significant P/E effect during the period 1979–1988 for a comprehensive sample of shares with positive earnings. They show that the average monthly return of the lowest quintile P/E portfolio exceeds that of the highest quintile P/E portfolio by 2.27% (27.2% annually), and find that after adjusting for differences in systematic risk, the P/E premium is still significant with an average monthly return of 1.88% (22.6% annually). Ma and Shaw, [85], report a weaker but still significant Taiwanese P/E effect for a smaller sample of stocks over the period 1979 to 1986. Dividing their sample into 5 portfolios, they found a significant average risk-adjusted monthly P/E premium of 0.85% (10.2% annually).

Finally, in New Zealand, Gillan, [54], finds no evidence of a P/E effect during the period 1977 to 1984 for the same sample as the one described in Table 2 for which he reports a significant size effect. A similar conclusion is reached by Kim, Chung and Pyun, [73], for Korea, based on the same sample of firms used to examine the size effect reported in Table 2. They find no evidence of a P/E effect on that market during the period 1980–1988 for a sample of up to 224 stocks.

In summary, the evidence from six markets outside the United States indicates that in the United Kingdom, Japan, Singapore and Taiwan there is a significant P/E effect similar to that found in the US market. There is no evidence, however, of a significant P/E effect in New Zealand and Korea. Given the small size and relatively short sample period for the cases of Taiwan, New Zealand and Korea, it is difficult to draw definitive conclusions from the evidence regarding these markets.

3.4   Variations on the E/P effect: cash flow-to-price and sales-to-price ratios

One alternative to the E/P ratio is the ratio of cash flow to price, where cash flow is defined as reported accounting earnings plus depreciation. Its appeal lies in the fact that accounting earnings may be a misleading and biased estimate of the economic earnings with which shareholders are concerned. Cash flow per share is less manipulable and, therefore, possibly a less biased estimate of economically important flows accruing to the firm's shareholders. The distinction between reported earnings and cash flow is important when examining these effects across countries with different accounting practices regarding the reporting of earnings. In some countries, such as Japan, firms are required to use the same depreciation schedule to calculate earnings reported to shareholders and earnings subject to corporate taxes. As a result, virtually all Japanese firms use accelerated depreciation for financial reporting (to reduce their tax liability) which creates large distortions in reported earnings for firms with large capital investments. In other countries, such as the United States, firms can use accelerated depreciation for tax purposes (which reduces taxable profits) and straight-line depreciation for reporting purposes (which produces relatively higher reported earnings to shareholders). Such accounting differences explain why there is a narrower difference between Japanese and American P/CF ratios compared to the much larger difference in the P/E ratios prevailing in these countries. For example, in August 1990, the market P/CF was 7.6 in the United States and 10.6 in Japan, whereas the market P/E was 15.8 in the United States and 35.3 in Japan (Goldman Sachs Research, August 1990).[10]

There is evidence of a CF/P effect in the United States and Japan. Chan, Hamao and Lakonishok, [28], find evidence of a significant relation between average returns and CF/P for Japanese stocks. The US evidence is summarized in the third set of columns in Table 1 which reports average returns and other portfolio characteristics for ten decile portfolios based on annual rankings (at March 31) of NYSE and AMEX securities on the ratio of cash flow per share to price per share (CF/P) for the period 1972 to 1994. Cash flow (CF) is defined as 'reported earnings plus depreciation'. The average difference in returns between the two extreme CF/P decile portfolios is 0.67% per month ($T = 2.32$). It is larger than the 0.56% per month ($T = 1.93$) obtained for the E/P effect for the same time period. This translates to an average annual difference between the two effects of more than 1.25%.

An alternative to both the E/P and CF/P ratios is the price-to-sales (P/S) ratio. Compared to earnings and cash flow, sales revenues are probably least influenced by accounting rules and conventions. There is evidence of a P/S effect in both the United States ([98], [64]) and Japan ([3]). For example, during the period 1968 to 1983, a portfolio of Japanese stocks with the lowest P/S ratio had an average monthly return of 1.86% compared with 1.13% for the portfolio of stocks with the highest quintile of P/S.

3.5 The price-to-book effect

Although less research, both in the United States and other countries, has examined the ability of other variables to predict cross-sectional differences in security returns, the ratio of price-per-share to book-value-per-share (P/B) deserves mention because of its significant predictive power. As is the case for the other variables discussed above, there is no theoretical model which predicts that P/B should be able to explain the cross-sectional behavior of

---

[10]French and Poterba, [53], adjust the E/P ratio for the Japanese and US markets for differences in accounting techniques and report adjusted E/P ratios of 22.8 for Japan and 14.5 for the United States. Thus, holding accounting techniques constant does not eliminate the difference between the estimates. The remaining differences may be explained by a lower level of interest rate and a faster economic growth rate in Japan compared to the United States during that period.

stock returns. However, investment analysts (e.g., [55]) have long argued that the magnitude of the deviation of current (market) price from book price per share is an important indicator of expected returns.

A succession of papers ([102], [95], [43], [71], and [47]) have documented a significant negative relation between P/B and stock returns. To provide some perspective on the magnitude of the P/B effect, the fourth set of columns in Table 1 reports average monthly returns and other portfolio characteristics for ten decile portfolios drawn from the same data we used to examine the size, E/P, and CF/P effects in the US market. The average monthly returns in Table 1 indicate a significant negative relation between P/B and returns. The monthly difference in returns between the extreme P/B portfolios (0.53%, $T = 2.31$) is higher than the corresponding differential return for the E/P effect (0.38%, $T = 1.68$), but lower than that for the size effect (0.72%, $T = 2.35$) for the 1962–94 period.

There is some evidence of a P/B effect outside the United States. A P/B effect has been documented for stocks trading on the Tokyo Stock Exchange ([2], [28] and [25]), the London Stock Exchange ([25] and [105]), and also on stock exchanges in France, Germany, and Switzerland ([25]). In [25], Capaul, Rowley and Sharpe report the following average monthly values for the difference in returns between lowest and highest P/B portfolios: 0.53% in France; 0.13% in Germany; 0.50% in Japan; 0.31% in Switzerland; and 0.23 in the United Kingdom.

3.6   Prior return (reversal and momentum) effect

There is evidence that prior returns can explain the cross-sectional behavior of subsequent stock returns. The literature documents two (seemingly) unrelated phenomena. The first is the existence of return reversals (past 'losers' become 'winners' and vice versa) over both *long-term* horizons (3 to 5 years) as well as *very short-term* periods (a month and shorter). The second is the presence of an opposite effect over horizons of *intermediate* lengths: When prior returns are measured over periods of 6 to 12 months, 'losers' and 'winners' retain their characteristic over subsequent periods. There is, in this case, return *momentum* rather than reversal.

DeBondt and Thaler, [42] and [43], find that NYSE stocks identified as the biggest losers (winners) over a period of 3 to 5 years earn, on average, the highest (lowest) market-adjusted returns over a subsequent holding period of the same length of time. This reversal effect does not seem to disappear when returns are adjusted for size and risk ([31]). Ball, Kothari and Shanken, [8], show that the abnormal returns associated with these strategies are sensistive to the portfolio formation date, which they attribute to microstructure-related biases that are most pronounced at the calendar year end. Specifically, they find negative abnormal returns when the strategy is initiated in June, in contrast to the positive abnormal returns when the strategy is initiated in December, the typical initiation date in this literature. Evidence of long-term return reversals has also been reported in a number of markets outside the United States, including Belgium ([107]), Japan ([40]), Brazil ([38]) and the United Kingdom ([33] and [46]). The reversal effect is not evident on the Toronto Stock Exchange ([75]).

There is also evidence of short-term return reversals. Jegadeesh, [67], Keim, [69], and Lehmann, [77], show that a 'contrarian strategy' that selects stocks on the basis of return performance over the previous week or month earn significant subsequent returns. Chang, Mcleavey and Rhee, [30] report the presence of the same phenomenon in Japan, after adjusting returns for both beta risk and size. But contrary to long-term reversals, that are often interpreted as 'market overreaction' due to irrational investors, short-term reversals may possibly reflect a lack of market liquidity ([68]) and the delayed reaction of stock prices to common factors ([82]).

Return momentum has been documented by Jegadeesh and Titman, [68], Chan, Jegadeesh and Lakonishok, [29], and Asness, [5]. In [68], Jegadeesh and Titman show that buying past winners and selling past losers generates significant positive returns over holding periods of 3 to 12 months, an investment strategy that does not seem to be due to differences in risk or delayed stock price reactions to common factors.

The last set of columns in Table 1 provides evidence of return momentum in our sample (we do not examine reversal strategies). Consistent with the procedures used in previous studies, we measure prior returns over the six months prior to the portfolio formation month, but exclude from this calculation the return from the last month of the six-month period to avoid possible contamination of the results from (microstructure-induced) return reversals for the adjacent monthly returns that bracket our portfolio formation date. Specifically, our prior return is measured over the 5-month period from the beginning of October to the end of March. Consistent with previous research, portfolios with the highest prior returns (the winners) earn, on average, higher subsequent returns. Also, portfolios with the lowest prior returns (the losers) earn, on average, the lowest subsequent returns. The difference in monthly returns between the extreme portfolios is 0.34% ($T = 1.66$) for the 1962–1994 period.

4 One Effect or Many?

4.1 Price as a common denominator

Size, E/P, CF/P and P/B are computed using a common variable: price per share. Blume and Stambaugh, [22], and Stoll and Whaley, [104], explored the relation between size and price and reported evidence suggesting a high rank

correlation between size and price.[11] Keim, [71], Jaffe, Keim and Westerfield, [65], and Fama and French, [47], have all recently raised this possibility
regarding the other effects. We examine the association among these variables using pairwise Spearman rank correlations. The rank correlations and
their associated T-values are computed as follows. Each year at the end of
March, all NYSE and AMEX stocks are ranked independently on size, E/P,
P/B, CF/P, prior return, and price (i.e., six separate rankings are produced).
Each variable was computed using price per share at March 31 and, when
applicable, accounting numbers for the previous year, with only December
fiscal closers included in the rankings. Pairwise Spearman rank correlations
are then computed. This procedure is repeated in each year for the period
1962 to 1994, and mean rank correlations and standard errors are computed
for the entire time series of values. Table 3 reports the average rank correlations and associated T-values. The estimated rank correlations are generally
large and significant. The pairwise correlations among the size, E/P, CF/P,
P/B, prior return, and price rankings are significantly different from zero,
indicating some commonalities among the effects. Consistent with previous
work ([22] and [104]), we find that the rank correlation between market capitalization and price was by far the strongest (0.78).

4.2   Correlation of the premia

The difference in return between the extreme portfolios in Table 1 (e.g., P/B
portfolio 1 minus P/B portfolio 10) can be loosely interpreted as risk premia,
if these variables are sorting out securities based on risks that are not defined
by extant asset pricing models. Under the scenario that the five variables
discussed above are proxies for five separate risk 'factors', then the premia
should be uncorrelated across variables. In Table 4 we report the pairwise
correlations between the monthly premia. All of the correlations are large
(in absolute value) and are significantly different from zero. Interestingly, the
prior return premia are *negatively* correlated with the premia associated with
the other four variables, suggesting that prior return is capturing a characteristic of stock returns that is quite different from the other variables. Note
that the correlations are also significantly different from 1.0, indicating that
the variables are not all proxying for the same underlying characteristic. Nevertheless, the significant correlations indicate a high degree of commonality
among the effects.

4.3   The premia are concentrated in January

The significant correlation of the premia reported in Section 4.2 is in part
a reflection of the fact that these effects are most pronounced in January.

---

[11]Results in [21], [104] and [22] also reveal a significant cross-sectional relation between
price per share and average returns.

Table 3. Average rank correlations (*t*-statistics) between several predetermined characteristics for NYSE and AMEX stocks (1962–1994)

	Market Capitalization	Earnings/ Price	Cash Flow/ Price	Price/ Book	Prior Return
Earnings/Price	−0.10				
	(−4.14)				
Cash Flow/Price	−0.11	0.68			
	(−4.72)	(45.00)			
Price/Book	0.32	−0.43	−0.48		
	(17.31)	(−25.23)	(−24.32)		
Prior Return	0.06	−0.13	−0.14	0.16	
	(1.72)	(−6.42)	(−7.27)	(7.54)	
Price per Share	0.78	−0.07	−0.15	0.34	0.16
	(104.21)	(−3.80)	(−6.44)	(19.49)	(5.01)

Correlations are computed annually using ranks of individual stocks. All rankings are conducted at the end of March, using prices at the time and accounting numbers for the previous fiscal year. Stocks with negative earnings, cash flows, or book values are excluded. Correlations computed with CF/P are for 1972-1994.

Table 4. Premia Correlations (April 1962–December 1994)

	E/P	CF/P	P/B	Prior Return
Size	0.265	0.444	0.472	−0.017
E/P		0.727	0.590	-0.230
CF/P			0.760	-0.212
P/B				-0.172

Correlations involving CF/P are for 1972-94.

Specifically, the average premia during January tend to be positive and are usually significantly larger than the average premia measured during the rest of the year. This seasonality was first documented for the size premia ([70])[12], and subsequently for the E/P premia ([35], [65]), the P/B premia ([71], [47], and [84]), and the momentum premium ([5]).

Internationally, most of the evidence on January seasonality has been related to the size premium. A significant January seasonal has been reported

---

[12]In [22], Blume and Stambaugh show that, after correcting for an upward bias in average returns for small stocks, the size-premium is evident only in January.

in Belgium ([62]), Finland ([15]), Taiwan ([85]) and Japan ([110] and [59]).[13] Countries in which the January size premium is insignificant include France ([56]), Germany ([103]) and the United Kingdom ([78]). In [28] Chan, Hamao and Lakonishok report a significant January seasonal for E/P, P/B, and CF/P in Japan but, surprisingly, no January seasonal for size.

A January seasonal in the size premium that is computed with our value-weighted size portfolios is summarized in the first set of columns in Table 5 where the 'all months' returns generated by the ten size portfolios in Table 1 have been separated into 'January' returns and 'rest-of-the-year' returns. The results in Table 5 show that average returns are significantly larger in January than the rest of the year for all size-portfolios except the largest one (portfolio ten). Further, the results show a clear negative relation between January returns and market capitalization: The smallest size portfolio has a January return of 12.11% whereas the largest size portfolio has an average return of only 1.94%. The average January size premium is thus 10.17%. This is significantly larger than observed in markets outside the United States and reported in the studies mentioned above. For example, the January size premium is 2.4% in Belgium (data from 1969 to 1983 for five size portfolios), 3.4% in Finland (data from 1970 to 1983 for five size portfolios) and 7.2% in Japan (data from 1965 to 1987 for ten size portfolios), one of the largest January size premium reported outside the US.

Table 5 also reports on January seasonality in the premium associated with earnings yield (E/P), cash flow yield (CF/P), price-to-book ratio (P/B) and 6-month prior return. The results are similar to those reported for the size premium, but are not as strong (see also Figure 1). Average returns are higher in January than in the rest of the year, but the difference is only significant for portfolios with the highest E/P and CF/P, and lowest P/B and 6-month prior returns (the deciles for which the January effect is significant are identified by asterisks). Also, unlike the other four variables, the relation between returns and prior returns is negative in January and positive in February to December (see also [52]). That the relation between prior returns and subsequent returns switches signs in January precisely when the other effects are more exaggerated probably explains why the prior return premium has a negative correlation with the other premia (see Section 4.2). Second, the magnitude of the January premium varies depending on the sorting variable used to construct the portfolios. The January premium is largest for portfolios based on size (10.17%) but is insignificant for portfolios based on 6-month prior return.

---

[13]Note, however, that in [28] Chan, Hamao and Lakonishok do not find a significant January size premium in their Japanese data.

Table 5. Monthly percentage returns during January and the rest of the year for value-weighted portfolios of NYSE and AMEX stocks formed on the basis of size (market capitalization), earnings-to-price ratios, cash flow-to-price ratios, price-to-book ratios, and prior return[1]

(April 1962 to December 1994)

Portfolio[2]	Size (Market cap.)		Earnings-to-price ratio		Cash flow-to-price ratio[3]		Price-to-book ratio		Prior Return	
	January	Rest of year	January	Rest of year	January	Rest of year	January	Rest of year	January	Rest of year
1	12.11*	0.63	5.37*	0.84	6.28*	1.04	7.23*	0.92	2.82	1.03
2	8.87*	0.75	4.30*	0.98	4.94*	1.00	5.38*	1.07	1.72	1.20
3	7.58*	0.69	3.59*	0.86	4.68*	0.87	4.36*	0.77	1.73	1.04
4	6.80*	0.74	2.73*	0.86	2.91	0.76	3.69*	0.82	2.19	0.92
5	5.74*	0.82	2.60*	0.81	2.65	1.01	3.13*	0.81	2.91*	0.70
6	4.67*	0.81	1.91	0.67	2.74	0.71	2.03	0.68	2.86*	0.73
7	3.77*	0.86	2.21	0.70	2.84	0.97	1.90	0.75	3.31*	0.64
8	3.52*	0.87	1.78	0.81	2.64	0.91	2.58*	0.76	3.29*	0.71
9	3.01*	0.85	1.61	0.82	2.23	0.78	1.91	0.72	3.65*	0.36
10	1.94	0.75	2.28	0.70	1.01	0.78	1.37	0.86	3.37*	0.61

1. Portfolios are created on March 31 of each year using prices and shares outstanding on March 31 and accounting data for the year ending the previous December 31 (with portfolios containing only December 31 fiscal closers). Prior return is calculated from the beginning of October to the end of February immediately preceding the March formation month. Aside from new listings and delistings, which are added to or dropped from the portfolios as they occur during the year, the portfolio composition remains constant during the following twelve months over which portfolio returns are calculated. The data reported in the table are average values over the sample period. An asterisk indicates that January average returns are significantly larger than the average returns during the rest of the year at the 0.05 level.
2. Portfolio 1 (portfolio 10) is the portfolio with the smallest (largest) market capitalization, highest (lowest) E/P and CF/P, lowest (highest) P/B, and largest (smallest) 6-month prior returns.
3. Results for the Cash Flow-to-Price portfolios are for the period April 1972–December 1994.

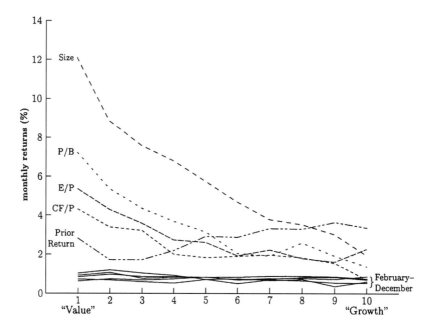

The portfolio labeled 'Value' ('Growth') is that with the smallest (largest) market capitalization, highest (lowest) E/P and CF/P, lowest (highest) P/B, and largest (smallest) prior
return. Stocks with negative values for P/B, E/P or CF/P are excluded from the sample.

Figure 1. The January Effect. Monthly percentage returns, computed separately
in January and in February–December, for portfolios constructed on the basis of
market cap, E/P, CF/P, P/B, and prior return. (April 1962 – December 1994)

5   Can we disentangle the effects?

The evidence in Section 4 suggests a great deal of commonality among the
various effects. In this section we discuss the research that has attempted to
disentangle the interrelation between the effects in an effort to assess whether
the results are due to one or several different factors.

5.1   Interactions between the effects: a somewhat chronological perspective

The earlier studies tended to examine interactions between two effects at a
time. Much of the initial research in the area focused on the interrelation
between the E/P and size effects, and a variety of techniques have been used,
ranging from simple analysis of average portfolio returns to sophisticated regression techniques. The disparate methods used often make comparisons

difficult. In the end, the results are less than conclusive. For example, Rein-ganum, [89], argues that the size effect subsumes the E/P effect: Once we control for size, there is no marginal E/P effect. Recall that the same phe-nomenon was reported in New Zealand ([54]). In [14] Basu argues just the opposite. So do Wong and Lye, [108], in the case of Singapore, and Chou and Johnson, [32], in the case of Taiwan. Peavy and Goodman, [88], and Cook and Rozeff, [35], perform meticulous replications of, and extensions to, the methods of Basu and Reinganum, and reach surprisingly different conclu-sions. Peavy and Goodman's results are consistent with Basu's while Cook and Rozeff conclude that no one effect dominates the other. In [11], Banz and Breen find a size effect but no independent E/P effect, a result similar to Reinganum.

In [65], Jaffe, Keim and Westerfield argue that the inability to disentangle the two effects may be attributable to the relatively short, and sometimes nonoverlapping, periods used in the above studies (they range from 8 to 18 years) as well as the failure of the studies (with the exception of Cook and Rozeff, [35]) to account for potential differences between January and the other months. Using data covering a 36 year period, Jaffe, Keim and Westerfield, [65], find that after controlling for size there is a significant E/P effect in both January and the other months; controlling for the E/P effect, there is a significant size effect only in January.[14] They also conclude that the results of the earlier studies conflict because the magnitude of the two effects is period specific. In [47], Fama and French reach similar conclusions regarding the joint significance of size and E/P effects (see the regression results in their Table 3).

Stattman, [102], and Rosenberg, Reid and Lanstein, [95], were the first to examine the interaction between size and P/B for NYSE and AMEX stocks. Stattman examines average risk-adjusted portfolio returns for April 1964 to April 1979 and concludes that "even after taking account for the size effect, there remains a positive relationship between B/P [the inverse of P/B] and subsequent returns". Rosenberg, Reid and Lanstein examine market model residuals of P/B portfolios that are constructed to be orthogonal to size and other influences. They also find a significant relation between abnormal re-turns and P/B for the 1973–1984 period.

A number of studies have examined the interaction between (contrarian) return reversals and both size and beta risk. Keim, [69], and Jegadeesh, [67], find that short-term (monthly) reversals provide explanatory power for aver-age returns beyond the influence of size. Chan, [26], and Ball and Kothari [7] argue that the abnormal risk-adjusted returns reported for long-term (3 to 5 year) contrarian investment strategies are due to inadequate adjustment for

---

[14]This result is consistent with other studies which have found that when examined alone, the size effect is significant only in January (e.g., [22]). See Section 3.2.

risk. That is, a loser firm whose stock price (and therefore market capitalization) has declined, in the absence of a concomitant decline in the value of the debt, becomes more leveraged and, other things equal, more risky. Traditional methods for computing betas may underestimate the CAPM-implied risk for these firms and, therefore, overstate the abnormal return. In this vein, Zarowin, [109], shows that loser firms tend to be small firms and winner firms tend to be large firms. After controlling for the size effect, he finds insignificant evidence of contrarian profits. More recently, Ball, Kothari and Shanken, [8], show that contrarian results are sensitive to the timing of portfolio formation in the empirical simulations of contrarian portfolio strategies. They show that when portfolios are formed by grouping based on past performance periods ending in mid-year rather than in December, the reversal effect is significantly reduced. In contrast to the above studies, Chopra, Lakonishok and Ritter, [31], find that after adjusting for size and beta there is an economically important overreaction effect, especially for small stocks. They find, though, that these contrarian profits are "heavily concentrated in January," suggesting they are related to the January effect. They conclude that their findings are not due to tax-loss-selling, but the prominent role played by small stocks in their findings suggests that the buying and selling behavior of individual investors may be important. Clearly more work is needed to sort out these issues.

The consensus from the research detailed above is that the relation between market capitalization and average returns is quite robust.[15] In addition, variables such as E/P and P/B seem to provide explanatory power for cross-sectional differences in average returns beyond the influence of size. Fama and French, [47], attempt to sort out the relative explanatory power of the three variables. They perform their test on a sample of individual stocks[16] trading on the NYSE and AMEX over the 1963–1990 period and the NASDAQ market for the 1973–1990 subperiod. They estimate equation (3.1) with beta, market cap, book-to-market ratio (B/P) and earnings yield (E/P). Their regression results indicate that beta and E/P are insignificant while both size and B/P have predictive power. Based on their findings, they

---

[15]An exception to the above findings is the recent study by Chan, Hamao and Lakonishok, [28], who find that in the Japanese stock market P/B and CF/P are sufficient to characterize cross-sectional differences in expected returns. That size is unimportant in explaining expected returns appears to be unique to the Japanese and Korean markets.

[16]Most previous research has demonstrated that estimates of beta do not enter significantly into models like the one represented in equation (3.1) in the presence of other explanatory variables such as size and E/P (an exception is [27]). Thus, Fama and French argue that methodologies that use portfolios to avoid estimation error in individual beta estimates unduly forfeit the valuable information in the cross section of individual security characteristics such as market value or E/P. This latter point was emphasized earlier in Litzenberger and Ramaswamy, [81], in their analysis of the relation between stock returns and dividend yields.

conclude that size and B/P are sufficient to characterize cross-sectional differences in expected returns. They conclude that three 'risk' factors – market, size, and P/B – are sufficient to explain the cross section of expected stock returns (see also [48]). However, in a subsequent paper ([49]), they report that size and B/P cannot explain the momentum effect. In the next subsection, we reexamine the interrelation between size, B/P and momentum with the data from Section 3.

5.2 Interactions between the size, P/B and prior-return effects: another look

To facilitate the comparison of the interaction between size, the P/B effect and the 6-month prior-return performance, we use our sample of NYSE and AMEX stocks to compute size-adjusted returns for portfolios created on the basis of both P/B and prior returns. Briefly, at the end of March in each year from 1962 to 1994 we sort all NYSE and AMEX stocks by P/B and form five groups of equal numbers of securities based on their P/B ranking.[17] The book value in our ratio is a December 31 value, the market cap is from March 31. Within each of these groups we again rank the stocks according to the magnitude of their prior returns (calculated from the beginning of October to the end of February) and create five subgroups within each of the five P/B groups. Individual stock returns are adjusted for the influence of the size effect by simply subtracting the return corresponding to the size decile (see Table 1) of which that security is a member. The composition of the portfolios remains constant over the next twelve months and value-weighted size-adjusted portfolio returns are computed each month.

The results are reported in Table 6. First, there is a positive relation between prior returns and subsequent returns within each of the five P/B quintiles, except for the highest P/B quintile in which subsequent returns drops from 0.28% to 0.02% when we move from the fourth to the fifth prior-return quintile. Second, there is evidence of a negative relation between P/B and subsequent returns within each of the five prior-return quintiles, although this relation is strongest for the quintile of stocks with negative prior returns. As prior returns increase, the negative relation between P/B and subsequent returns is confined to the first three P/B quintiles.

Note that the P/B effect occurs in the prior-return categories in which the average price of the securities in the portfolios increases with P/B[18] (except for the highest prior-return category in which average price first rises to $40.7

---

[17]For purposes of this experiment, we eliminate from the sample all stocks with negative P/B values.

[18]In this regard, recall that we found a significant rank correlation between price and P/B (0.33).

Table 6. Size adjusted monthly returns (%) and other characteristics for 25
portfolios of NYSE and AMEX stocks ranked first by price-to-book ratio (P/B)
and then by prior return; *April 1962 to December 1994 (393 observations).*

	Lowest	2	3	4	Highest
		Prior Return (beginning of October to end of February of preceding year)			
A. Monthly percentage returns (Standard error)					
Low P/B	0.15 (0.22)	0.35 (0.15)	0.27 (0.15)	0.48 (0.13)	0.44 (0.16)
2	−0.12 (0.15)	0.04 (0.15)	0.01 (0.13)	0.13 (0.12)	0.39 (0.15)
3	−0.12 (0.14)	−0.09 (0.13)	−0.08 (0.10)	−0.10 (0.11)	0.16 (0.14)
4	−0.26 (0.13)	−0.28 (0.12)	−0.08 (0.10)	−0.05 (0.11)	0.32 (0.15)
High P/B	−0.40 (0.14)	−0.09 (0.12)	0.11 (0.11)	0.28 (0.13)	0.02 (0.17)
B. Price-to-book ratio					
Low P/B	0.66	0.69	0.71	0.71	0.70
2	1.08	1.09	1.09	1.09	1.09
3	1.44	1.44	1.44	1.45	1.45
4	2.04	2.05	2.04	2.06	2.07
High P/B	7.36	5.78	5.83	7.17	6.79
C. Six-month prior returns (%)					
Low P/B	−25.66	−8.74	−0.08	8.92	29.47
2	−18.32	−4.16	3.07	10.87	32.16
3	−18.14	−3.19	4.37	12.93	34.24
4	−19.41	−2.68	6.33	16.11	40.58
High P/B	−20.98	−1.01	9.00	19.50	50.40
D. Price ($)					
Low P/B	13.5	16.6	17.9	18.9	17.9
2	21.0	24.9	26.3	26.0	23.4
3	22.5	44.9	30.0	29.4	40.7
4	27.1	32.2	34.5	34.3	32.8
High P/B	30.4	40.8	46.9	48.3	39.3
E. Market capitalization ($ million)					
Low P/B	321	299	360	418	290
2	638	758	888	830	535
3	660	1144	1079	1082	715
4	811	1403	1195	1173	693
High P/B	1075	1847	2210	1947	872

The size-adjusted monthly return for a security is defined as the return for that
security minus the monthly portfolio return for the size decile in which the security
is a member. P/B and prior return portfolios in the table are value-weighted
combinations of these monthly size-adjusted returns. All portfolios are formed on
March 31 of each year using year-end accounting values and March 31 market
prices. Stocks with negative P/B values are excluded from the sample.

and then drops to \$32.8 to rise again to \$39.3). Although the experiment controls for the influence of size (market capitalization) on average returns, it does not explicitly control for price, which has also been shown to influence returns. Thus, the high average returns for low P/B stocks may reflect some underlying relation between returns and low price.

An alternative hypothesis involves the possibility that low P/B stocks are simply stocks whose prices have dropped relative to book values that vary little through time. Firms whose stocks have recently declined in price, in the absence of a concomitant decline in the value of the debt, have become more leveraged and, other things equal, more risky. Traditional estimation methods may underestimate 'true' beta risk for such firms and, therefore, overstate 'risk-adjusted' returns.[19] As a result, stocks that have recently suffered a substantial decline in price will tend to have underestimated betas and low ratios of P/B. Hence, P/B may be a more accurate proxy for 'true' beta risk than traditional estimates of beta due to the measurement error in the traditional estimates. Given their unobserved higher levels of risk, the subsequent higher average returns that compensate for this risk appear anomalous.[20]

Previous research described above shows that the size and P/B effects are most concentrated in the month of January. In Table 7, therefore, we report the average returns for the 25 portfolios in Table 6 separately for January and for the remaining eleven months. Consider the P/B effect first. The P/B effect in January (Panel A) is more pronounced than when computed over all months as in Table 6. In February–December, though, the relation between P/B and returns is flat. Thus, after controlling for size and momentum effects, the P/B effect is evident in the data only during the month of January. The story is quite different for the momentum effect. After controlling for both size and P/B, the relation between prior and subsequent returns has the same significant positive relation as noted in Table 6 for all months. In January, though, the 'momentum' effect has the appearance of a reversal effect in that subsequent returns increase as prior returns decrease. That is, stocks that have recently declined (and therefore more likely candidates for trading at the

---

[19]Traditional methods (e.g., OLS) that have been used for estimating betas in most cross-sectional analyses use four or more years of monthly returns data and (implicitly) apply equal weights to all observations in the time series. Clearly, the most relevant observations, the ones that should be given the most weight in the estimation, are those occurring closest to the period of analysis (e.g., portfolio formation date). Thus, the betas used in such studies are 'stale' in that they are estimated using information that, in large measure, is not relevant. This estimation shortfall also applies to studies that use 'future' betas estimated from data occurring after the analysis interval since structural changes that impact firm risk can also affect the post-analysis observations, thereby rendering them less relevant for assessing the firm's risk in the analysis interval.

[20]DeBondt and Thaler, [43], show that such price reversals are most extreme for low P/B stocks.

Table 7. Size-adjusted monthly returns (%) for 25 portfolios of NYSE and AMEX
stocks ranked first by price-to-book ratio (P/B) and then by prior return;
April 1962 to December 1994 (393 observations)

	Prior Return				
	Lowest	2	3	4	Highest
A. January					
Low P/B	3.71	3.13	2.13	2.15	1.34
	(1.32)	(0.69)	0.76	0.53	0.55
2	0.96	1.28	1.41	1.10	0.38
	(0.57)	(0.50)	(0.53)	(0.54)	(0.52)
3	0.35	0.66	0.05	−0.16	−1.07
	(0.63)	(0.71)	(0.36)	(0.44)	(0.52)
4	0.92	−0.36	−0.03	−0.92	−0.83
	(0.53)	(0.56)	(0.40)	(0.47)	(0.74)
High P/B	−0.08	−0.06	−1.03	−1.02	−0.84
	(0.57)	(0.44)	(0.45)	(0.55)	(0.55)
B. February–December					
Low P/B	−0.16	0.10	0.10	0.33	0.36
	(0.20)	(0.15)	(0.15)	(0.13)	(0.16)
2	−0.21	−0.06	−0.11	0.04	0.39
	(0.15)	(0.15)	(0.13)	(0.12)	(0.15)
3	−0.16	−0.16	−0.09	−0.10	0.27
	(0.15)	(0.13)	(0.11)	(0.11)	(0.14)
4	−0.36	−0.28	−0.08	0.03	0.42
	(0.14)	(0.12)	(0.10)	(0.12)	(0.15)
High P/B	−0.43	−0.09	0.21	0.40	0.10
	(0.15)	(0.12)	(0.11)	(0.13)	(0.17)

The size-adjusted monthly return for a security is defined as
the return for that security minus the monthly portfolio return
for the size decile in which the security is a member. P/B and
prior return portfolios in the table are value-weighted
combinations of these monthly size-adjusted returns. All
portfolios are formed on March 31 of each year using year-end
accounting values and March 31 market prices. Stocks with
negative P/B values are excluded from the sample.

end of the year based on taxes or window dressing) have the largest returns
in January, particularly if they are low-price stocks. Thus, in January, it
appears that end-of-year trading patterns tend to offset the momentum effect
that, unlike the other effects, persists throughout the rest of the year.

6   In Search of an Explanation

In [47], Fama and French argue that the significant relation between returns and variables like size, P/B and prior return is evidence of compensation for additional sources of risk that are not included in extant asset pricing models. A plethora of papers have appeared recently that question this interpretation. They fall into four categories.

6.1   Beta-adjusted excess returns are evidence of market inefficiency

Several studies interpret the reported excess returns as evidence of market inefficiency. For example, in [76] Lakonishok, Schleifer and Vishny argue that investors are irrational because they avoid 'value' stocks that they mistakenly consider too risky, even though the evidence indicates that this does not appear to be the case (at least for conventional measures of risk). According to Haugen, [57], institutional investors avoid buying 'value' stocks because these investors' performance is measured against indexes of mostly large, 'glamour' stocks. Again, individuals who buy these neglected 'value' stocks outperform the indexes of large, 'glamour' stocks in the long run.

6.2   The results are due to statistical biases or peculiarities in the data sample

Recent papers by Kothari, Shanken and Sloan, [74], and Breen and Korajczyk, [23], suggest that the P/B results may be due to survivor biases in samples of stocks drawn from the Compustat files. Compustat tends to include (and add) stocks in their files only after the stock has demonstrated a successful track record. Thus, small firms with low P/B ratios that subsequently perform poorly (or fail) are unlikely to appear on the files. However, Davis, [41], and Kim, [72], provide evidence that appears contrary to this hypothesis. In [41], Davis finds evidence of P/B, cash flow and E/P effects for a survivor-bias-free sample in the 1940–1963 period prior to the Compustat-specific period used in many studies, although his estimates of the P/B effect are half those estimated by Fama and French, [47]. In [72], Kim shows that survivor bias does not significantly reduce the P/B effect.

Also in this second category are papers that show that the magnitude of the P/B effect is sensitive to the types of securities analyzed. For example, in [84] Loughran argues that the relation between P/B and returns is a manifestation of the low returns for Amex and Nasdaq growth stocks used in those studies. He shows that the P/B effect is not significant for large-cap stocks or small-cap NYSE stocks during the 1963–1991 period. Based on his results, Loughran concludes that "for the majority of money managers, the empirical findings of Fama and French, [47], are of little economic importance for predicting future portfolio returns".

Finally, Berk, [16] and [17], claims that the reported relation between firms' returns and their market value is not at all a 'size' effect. He makes the claim in light of his findings that there is no significant relation between average returns and four measures of firm size (book value of assets, book value of gross fixed assets, annual sales, and number of employees). Berk concludes that his "results are evidence in favor of the hypothesis that the size effect is due to the endogenous identity relating the market value of a firm to its discount rate".[21]

6.3   Three factors are not enough

Some argue that more than three variables are required to characterize the multidimensional nature of risk. This notion has some precedence in the literature. A number of earlier papers (e.g., [94], [100], [64]) estimate cross-sectional regression models similar to Fama and French but using a larger and richer set of independent variables. The renewed willingness of the academic community to entertain such analyses since the publication of Fama and French has resulted in much activity in this area. For example, Barbee, Mukherji and Raines, [12], examine the explanatory power of the ratio of sales to price when added to size, book-to-market ratio and leverage (debt-to-equity ratio) in a multiple regression, and find that it has significant explanatory power. Jagannathan and Wang, [66], suggest that the lack of empirical support for the CAPM may be due to: (1) the exclusion of human capital in the market portfolio; and (2) the systematic variability of beta-risk over the business cycle. They account for human capital by estimating the sensitivity of portfolio returns to the growth rate in per capita labor income. The variation in beta is taken into account by estimating the sensitivity of portfolio returns to the yield spread on low- and high-grade corporate bonds When added to beta and size, the two new variables emerge as the only variables with significant predictive power. In [5], Asness examines the marginal contribution of (intermediate-term) price momentum and (long-term) price reversals to the size and B/P effects in a multiple regression. He finds that both prior return variables are significant in explaining the cross-sectional behavior of subsequent returns, beyond the contribution of size and B/P.

---

[21]Berk argues that since "a firm's market value is endogenously determined in equilibrium as the discounted value of expected future cash flows, it depends on the discount rate. For example, if two firms have the same expected cash flow, the one with the larger discount rate will have the lower market value. Consequently, according to this view expected returns will always be negatively correlated with market value, *ceteris paribus*".

6.4 The findings are real, but cannot be attributed to risk

In [84], Loughran shows that for NYSE stocks, the size and P/B ratio explain none of the cross-sectional variation in returns if January is excluded from the sample (see also [71] and Table 5). It is difficult to characterize the relation between *ad hoc* factors and returns as risk-based when it manifests itself primarily in January. In a similar vein, Daniel and Titman, [39], show that the significant coefficients on size and P/B in regressions like (3.1) are simply correlation between returns and those characteristics, and not implied risk premia associated with risk factors. Similarly, in the following section we report some new evidence on size, P/B, E/P and CF/P premia across several developed equity markets that appear to be inconsistent with a risk story.

6.4.1 Are the premia compensation for risk: Some more international evidence

Previous sections surveyed evidence of size, E/P, CF/P and P/B effects on many international stock markets. In this section we present some new evidence to confirm these findings. In the same spirit as the evidence we presented for the US market, we present findings for a common time period using the same empirical methods across all markets. We again report our evidence in portfolio form because we feel the returns to feasible portfolios provide a useful perspective on the economic significance of the results.

The data for the international equity markets are from Morgan Stanley Capital International (MSCI) for the period January 1975 to December 1994.[22] The data include all stocks in the MSCI indexes for France, Germany, Japan and the UK. The stocks in the sample are the most actively traded and popular stocks, and represent the bulk of the market capitalization, in their respective markets. In addition, stocks with negative or missing values for the ranking variables and stocks that have extremely small market capitalization relative to the local market are excluded.[23] Returns include dividends and are adjusted for the local investor tax credit by country.[24] Finally, returns are computed in home currencies. Note that for the premia results, in which we are most interested, any currency influences will be effectively canceled, allowing for clean cross-country comparisons.

In each market we construct four sets of portfolios based on independent

---

[22]We are grateful to Brandywine Asset Management, especially Steve Tonkovich, for providing access to these data.

[23]For example, a stock was excluded from the portfolios formed at December 1991 if its market cap, stated in millions of US dollars, was $56 in France, $64 in Germany, $91 in the UK, and $449 in Japan. These lower bounds changed from year to year as market conditions varied.

[24]The adjustment is conservative, implying the maximum that foreign investors might pay: 25% in UK; 36% in Germany; 33% in France; and 0% in Japan.

rankings on size (market capitalization), E/P (ratio of earnings to price), CF/P (ratio of cash flow to price), and P/B (ratio of price per share to book value per share). Rankings are conducted at December 31 of each year using end-of-year market cap and price and the most recently-reported accounting values (not restated). Note that all of these values are recorded in the published versions of the MSCI data as of the portfolio formation date so that the values used to construct the portfolios are known to market participants. Thus, there is no look-ahead, fill-in or survivor bias in the portfolios or the associated tests. The portfolios are buy-and-hold portfolios that are equal-weighted at the time of portfolio formation.[25] The portfolios are held for twelve months, at which time the sorting and portfolio rebalancing procedure is repeated. The process is repeated each year from December 1973 to December 1994 resulting in a 20-year time series of monthly returns. The US portfolios are value-weighted (quintile) portfolios of NYSE and AMEX stocks based on the US data described in previous sections.

Monthly premia associated with each of the sorting variables are constructed as the difference between the extreme quartile portfolio returns in each month. For example, the E/P premium is computed as the difference in return between the highest E/P quartile stocks and the lowest E/P quartile stocks. Table 8 contains the average and standard deviation of the premium for each ranking variable within each country. The premia are economically and statistically significant for all the variables across all the countries (one exception is the size effect in Germany). Further, the magnitude of the effects in the four international markets is comparable to that computed for the US and reported in previous sections.

In [47], Fama and French argue that premia like those reported in Table 8 are compensation for risk. If that is the case, under conditions of integrated international capital markets, one expects that the premia will be correlated across markets (e.g., the size premia should be correlated across markets). To this end, Table 8 reports, within each panel, the correlations across countries for a particular variable. Our objective is to assess the degree of correlation among premia *across* countries.[26] The correlations are mostly insignificantly different from zero.[27] This lack of correlation of the premia across markets

---

[25]Since there is not substantial diversity in market capitalization values within quartiles in the MSCI data, the equal weighting of the MSCI portfolios at the portfolio formation date will not produce results that differ in a meaningful way from value-weighted portfolios.

[26]We have also examined whether *different* ranking variables produce similar results *within* a country. The within-country correlations are low, suggesting that the different variables seem to be picking up different sources of variability in returns within each country. The correlations tend to be highest among the P/E, P/B, and P/CE premia. The size premia displays a lower correlation with the other premia within a particular market. These results are consistent with existing research for US equities.

[27]Evidence from two ten-year subperiods show that this lack of correlation is not period specific.

Table 8. Mean Monthly Premia (%) and Correlations Between Premia Across
Five International Markets, 1975–1994.

	Mean Premia	Standard Deviation	Correlations			
			UK	Germany	Japan	US
A. Size Effect						
France	0.47	3.66	0.07	0.15	−0.00	0.07
UK	0.60	3.26		0.02	0.12	0.24
Germany	−0.01	3.03			0.09	0.07
Japan	0.35	4.11				−0.02
US	0.76	4.62				
B. E/P Effect						
France	0.75	3.17	0.12	0.11	0.07	0.15
UK	0.47	2.35		0.02	0.07	0.10
Germany	0.34	2.90			0.02	0.10
Japan	0.49	2.88				0.01
US	0.49	3.33				
C. CF/P Effect						
France	0.77	3.96	0.22	0.24	0.08	0.03
UK	0.57	2.89		0.11	0.04	0.14
Germany	0.45	2.71			0.02	0.11
Japan	0.40	4.00				0.10
US	0.50	3.45				
D. P/B Effect						
France	0.80	4.04	0.14	0.13	−0.00	0.16
UK	0.57	3.10		0.20	0.17	0.29
Germany	0.19	2.96			0.09	0.25
Japan	0.84	3.30				0.10
US	0.57	3.60				

The premia are computed independently in each market. The premia are defined as
the difference in returns between equal-weighted portfolios containing stocks from,
respectively, the smallest and largest quartiles of market cap in panel A, the highest
and lowest quartiles of E/P in panel B, the highest and lowest quartiles of CF/P in
panel C, and the lowest and highest quartiles of P/B in panel D. Quartile cutoffs are
determined separately for each market.

contrasts with the significant correlation of the *market* risk premium across
international markets that has been widely documented. The lack of correlation in the size, E/P, P/B, and CF/P premia across markets calls into
question risk-based explanations for these premia, at least in the context of
extant international asset pricing models.[28]

[28]The low correlations also suggest the possibilty of substantial benefits from international diversification of value and size portfolio strategies.

7 Concluding Remarks

A large body of evidence has accumulated that collectively suggets that betas do not adequately explain cross-sectional differences in average stock returns. Instead, several variables that have no basis in theory sem to have significant explanatory power. In addition, many papers have attempted to explain these findings. In this paper we provide a synthesis of the evidence and explanations. One interpretation is that the beta-adjusted excess returns represent evidence that a multidimensional model of risk and return is necessary to explain the cross section of stock returns. That is, beta by itself is insufficient to characterize the risks of common stocks. Prominent in this category are Fama and French, [47], who suggest a three-factor equity-pricing model to replace the CAPM. Their three-factor model adds two empirically-determined explanatory factors: size (market capitalization) and financial distress (B/M). Others proposes an additional factor, prior return.

Our findings suggest that such conclusions may be premature. Aside from the absence of a theory that predicts such variables have a place in the risk-return paradigm, the evidence strongly indicates that the statistical relation between returns and variables like size and B/M derives primarily from the month of January. It is difficult to tell an asset pricing story where risk manifests itself only during one month of the year. In addition, we find that the premia are uncorrelated across international equity markets. It is difficult to reconcile this finding with a risk-based story in which, under conditions of integrated international capital markets, the risk premia from a multidimensional international asset pricing model are correlated across markets in the same way the market risk premium is correlated across markets.

These points notwithstanding, a significant contribution of this line of research is that is has sharpened our focus on potential alternative sources of risk, and future theoretical work should certainly benefit. However, the evidence, in and of itself, does not constitute proof that the CAPM is 'wrong'. For example, no one has yet conclusively shown that variables like size and P/B are not simply proxies for measurement error in betas. Are we certain that cross-sectional variation in P/B is not picking up variation in leverage that is not reflected in betas that are typically estimated with sixty months of prior – and arguably stale – prices? The book is not closed; we think more research is necessary to resolve these issues.

There is also the question of believability. That is, is the evidence as robust as the sheer quantity of results would lead us to believe? First, there is the issue of data snooping – many of the papers we have cited in this article were predicated on previous research that documented the same findings with the same data. Degrees of freedom are lost at each turn and several authors have warned about adjusting tests of significance for these lost degrees of freedom.

Also, the existence of these patterns in our experiments does not necessarily imply that they exist in the returns of implementable portfolios – that is, returns net of transactions costs. (e.g., market illiquidity and transactions costs may render a small stock strategy infeasible). Finally, that many of these effects have persisted for nearly 100 years in no way guarantees their persistence in the future. How many years of data are necessary to construct powerful tests? Research over the next 100 years will, we hope, settle many of these issues.

References

[1] Aggarwal, R., T. Hiraki and R. Rao (1988) 'Earning/Price Ratios, Size, and Seasonal Anomalies in the Japanese Securities Market,' working paper, John Carroll University.

[2] Aggarwal, R., R. Rao and T. Hiraki (1989) 'Price/Book Value Ratios and Equity Returns on the Tokyo Stock Exchange: An Empirical Study,' working paper, John Carroll University.

[3] Aggarwal, R., R. Rao and T. Hiraki (1990) 'Equity Return Regularities Based on the Price/Sales Ratio: An Empirical Study of the Tokyo Stock Exchange.' In *Pacific-Basin Capital Markets Research*, S.G. Rhee and R.P. Chang (eds.), North Holland.

[4] Amihud, Y. and H. Mendelson (1986) 'Asset Pricing and the Bid-Ask Spread,' *Journal of Financial Economics* **17** 223–250.

[5] Asness, C. (1995) 'The Power of Past Stock Returns to Explain Future Stock Returns,' unpublished manuscript, Goldman Sachs Asset Management.

[6] Ball, R. (1978) 'Anomalies in Relationships Between Securities' Yields and Yield-Surrogates,' *Journal of Financial Economics* **6** 103–126.

[7] Ball, R. and S.P. Kothari (1989) 'Nonstationary Expected Returns: Implications for Tests of Market Efficiency and Serial Correlation in Returns,' *Journal of Financial Economics* **25** 51–74.

[8] Ball, R., S.P. Kothari and J. Shanken (1995) 'Problems in Measuring Portfolio Performance: An Application to Contrarian Investment Strategies,' *Journal of Financial Economics* **38** 79–107.

[9] Banz, R. (1981) 'The Relationship between Return and Market Value of Common Stock,' *Journal of Financial Economics* **9** 3–18.

[10] Banz, R. (1985) 'Evidence of a Size-Effect on the London Stock Exchange,' unpublished manuscript, INSEAD.

[11] Banz, R. and W. Breen (1986) 'Sample Dependent Results using Accounting and Market Data: Some Evidence,' *Journal of Finance* **41** 779–794.

[12] Barbee, W., S. Mukherji and G. Raines (1996) 'Do Sales-Price and Debt-Equity Explain Stock Returns Better than Book-Market and Firm Size?,' *Financial Analysts Journal* **52** 56–60.

[13] Basu, S. (1977) 'Investment Performance of Common Stocks in Relation to their Price-Earnings Ratio: A Test of the Efficient Market Hypothesis,' *Journal of Finance* **32** 663–682.

[14] Basu, S. (1983) 'The Relationship Between Earning's Yield, Market Value and the Returns for NYSE Common Stocks: Further Evidence,' *Journal of Financial Economics* **12** 129–156.

[15] Berglund, T. (1985) 'Anomalies in Stock Returns in a Thin Security Market: The Case of the Helsinki Stock Exchange,' Doctoral Thesis, The Swedish School of Economics and Business Administration, Helsinki.

[16] Berk, J. (1995) 'A Critique of Size Related Anomalies,' *Review of Financial Studies* **8** 275–286.

[17] Berk, J. (1997) 'A View of the Current Status of the Size Anomaly,' working paper, University of Washington.

[18] Black, F. (1972) 'Capital Market Equilibrium with Restricted Borrowing,' *Journal of Business* **45** 444–455.

[19] Black, F., M. Jensen and M. Scholes (1972) 'The Capital Asset Pricing Model: Some Empirical Tests.' In *Studies in the Theory of Capital Markets*, M. Jensen (ed.), Praeger, New York, 79–121.

[20] Blume, M. and I. Friend (1973) 'A New Look at the Capital Asset Pricing Model,' *Journal of Finance* **28** 19–33.

[21] Blume, M. and F. Husic (1973) 'Price, Beta and Exchange Listing,' *Journal of Finance* **28** 283–299.

[22] Blume, M. and R. Stambaugh (1983) 'Biases in Computed Returns: An Application to the Size Effect,' *Journal of Financial Economics* **12** 387–404.

[23] Breen, W. and R. Korajczyk (1994) 'On Selection Biases in Book-to-Market Based Tests of Asset Pricing Models,' working paper, Northwestern University.

[24] Calvet, A. and J. Lefoll (1989) 'Risk and Return on Canadian Capital Markets: Seasonality and Size Effect,' *Journal of the French Finance Association* **10** 21–39.

[25] Capaul, C., I. Rowley and W. Sharpe (1993) 'International Value and Growth Stock Returns,' *Financial Analysts Journal* **49** 27–36.

[26] Chan, K.C. (1988) 'On the Contrarian Investment Strategy,' *Journal of Business* **61** 147–163.

[27] Chan, K.C. and N.F. Chen (1988) 'An Unconditional Asset-Pricing Test and the Role of Firm Size as an Instrumental Variable for Risk,' *Journal of Finance* **43** 309–325.

[28] Chan, L., Y. Hamao and J. Lakonishok (1991) 'Fundamentals and Stock Returns in Japan,' *Journal of Finance* **46** 1739–1764.

[29] Chan, L., N. Jegadeesh and J. Lakonishok (1995) 'Momentum Strategies,' working paper, University of Illinois at Urbana-Champaign.

[30] Chang, R., D. McLeavey and G. Rhee (1995) 'Short-Term Abnormal Returns of the Contrarian Strategy in the Japanese Stock Market,' *Journal of Business Finance and Accounting* **22** 1035–1048.

[31] Chopra, N., J. Lakonishok and J. Ritter (1992) 'Measuring Abnormal Returns: Do Stocks Overreact?,' *Journal of Financial Economics* **31** 235–268.

[32] Chou, S.R. and K. Johnson (1990) 'An Empirical Analysis of Stock Market Anomalies: Evidence from the Republic of China in Taiwan.' In *Pacific-Basin Capital Markets Research*, S.G. Rhee and R.P. Chang (eds.), North-Holland, 283–312.

[33] Clare, A. and S. Thomas (1995) 'The Overreaction Hypothesis and the UK Stock Market,' *Journal of Business Finance and Accounting* **22** 961–973.

[34] Coghlan, H. (1988) 'Small Firms versus Large on the Irish Stock Exchange: An Analysis of the Performances,' *Irish Business and Administrative Research* **9** 10–20.

[35] Cook, T. and M. Rozeff (1984) 'Size and Earnings/Price Anomalies: One Effect or Two?,' *Journal of Financial and Quantitative Analysis* **13** 449–466.

[36] Corhay, A., G. Hawawini and P. Michel (1988) 'The Pricing of Equity on the London Stock Exchange: Seasonality and Size Premium.' In *Stock Market Anomalies*, E. Dimson (ed.), Cambridge University Press, 197–212.

[37] Corniolay, C. and J. Pasquier (1991) 'CAPM, Risk Premium Seasonality and the Size Anomaly: The Swiss Case,' *Journal of the French Finance Association* **12** 23–44.

[38] daCosta, N.C.A. (1994) 'Overreaction in the Brazilian Stock Market,' *Journal of Banking and Finance* **18** 633–642.

[39] Daniel, K. and S. Titman (1995) 'Evidence on the Characteristics of Cross Sectional Variation in Stock Returns,' working paper, University of Chicago.

[40] Dark, F. and K. Kato (1986) 'Stock Market Overreaction in the Japanese Stock Market,' working paper, Iowa State University.

[41] Davis, J. (1994) 'The Cross-Section of Realized Stock Returns: The Pre-Compustat Evidence,' *Journal of Finance* **49** 1579–1593.

[42] DeBondt, W. and R. Thaler (1985) 'Does the Stock Market Overreact?,' *Journal of Finance* **40** 793–805.

[43] DeBondt, W. and R. Thaler (1987) 'Further Evidence on Investor Overreactions and Stock Market Seasonality,' *Journal of Finance* **42** 557–581.

[44] Dimson, E. (1979) 'Risk Measurement when Shares are Subject to Infrequent Trading,' *Journal of Financial Economics* **7** 197–226.

[45] Dimson, E. and P. Marsh (1984) 'The Impact of the Small Firm Effect on Event Studies and the Performance of Published UK Stock Recommendations,' *Journal of Financial Economics* **17** 113–142.

[46] Dissanaike, G. (1997) 'Do Stock Market Investors Overreact?,' *Journal of Business Finance and Accounting* **24** 27–49.

[47] Fama, E. and K. French (1992) 'The Cross Section of Expected Stock Returns,' *Journal of Finance* **47** 427–466.

[48] Fama, E. and K. French (1993) 'Common Risk Factors in the Returns of Stocks and Bonds,' *Journal of Financial Economics* **33** 3–56.

[49] Fama, E. and K. French (1996) 'Multifactor Explanations of Asset Pricing Anomalies,' *Journal of Finance* **51** 55–84.

[50] Fama, E. and J. MacBeth (1973) 'Risk, Return and Equilibrium: Empirical Tests,' *Journal of Political Economy* **71** 607–636.

[51] Fama, E., K. French, D. Booth and R. Sinquefield (1993) 'Differences in the Risks and Returns of NYSE and NASD Stocks,' *Financial Analysts Journal* **49** 37–41.

[52] Fant, F. and D. Peterson (1995) 'The Effect of Size, Book-to-Market Equity, Prior Returns and Beta on Stock Returns: January versus the Remainder of the Year,' *Journal of Financial Research* **18** 129–142.

[53] French, K. and J. Poterba (1991) 'Were Japanese Stock Prices too High?,' *Journal of Financial Economics* **29** 337–364.

[54] Gillan, S. (1990) 'An Investigation into CAPM Anomalies in New Zealand. The Small Firm and Price Earnings Ratio Effects,' *Asia Pacific Journal of Management* **7** 63–78.

[55] Graham, B. and D. Dodd (1940) *Security Analysis: Principles and Technique*, McGraw-Hill Book Company, Inc., New York.

[56] Hamon, J. (1986) 'The Seasonal Character of Monthly Returns on the Paris Bourse,' *Journal of the French Finance Association* **7** 57–74.

[57] Haugen, R. (1995) *The New Finance*, Prentice-Hall, New Jersey.

[58] Hawawini, G. (1991) 'Stock Market Anomalies and the Pricing on the Tokyo Stock Exchange.' In *Japanese Financial Market Research*, W. Ziemba, W. Bailey and Y. Hamao (eds.), North-Holland, 231–250.

[59] Hawawini, G. (1993) 'Market Efficiency and Equity Pricing: International Evidence.' In *International Finance: Contemporary Issues*, D.K. Das (ed.), Routledge, London and New York.

[60] Hawawini, G. and D.B. Keim (1995) 'On the Predictability of Common Stock Returns: World-Wide Evidence.' In *Handbooks in OR and MS*, R. Jarrow, V. Maksimovic and W. Ziemba (eds.), North-Holland, 497–544.

[61] Hawawini, G. and C. Viallet (1987) 'Seasonality, Size Premium and the Relationship Between the Risk and Return of French Common Stocks,' working paper, INSEAD, Fontainebleau, France.

[62] Hawawini, G., P. Michel and A. Corhay (1989) 'A Look at the Validity of the Capital Asset Pricing Model in Light of Equity Market Anomalies: The Case of Belgian Common Stocks.' In *A Reappraisal of the Efficiency of Financial Markets*, R.C. Guimaraes, B.G. Kingsman and S. Taylor (eds.), Springer-Verlag, NATO ASI Series.

[63] Herrera, M. and L. Lockwood (1994) 'The Size Effect in the Mexican Stock Market,' *Journal of Banking and Finance* **18** 621–632.

[64] Jacobs, B. and K. Levy (1988) 'Disentangling Equity Return Regularities: New Insights and Investment Opportunities,' *Financial Analysts Journal* **44** 18–43.

[65] Jaffe, J., D. Keim and R. Westerfield (1989) 'Earnings Yields, Market Values and Stock Returns,' *Journal of Finance* **45** 135–148.

[66] Jagannathan, R. and Z. Wang (1996) 'The Conditional CAPM and the Cross-Section of Expected Returns,' *Journal of Finance* **51** 3–53.

[67] Jegadeesh, N. (1990) 'Evidence of Predictable Behavior of Security Returns,' *Journal of Finance* **45** 881–898.

[68] Jegadeesh, N. and S. Titman (1993) 'Returns to Buying Winners and Selling Losers: Implications for Stock Market Efficiency,' *Journal of Finance* **48** 65–92.

[69] Keim, D.B. (1983a) *The Interrelation Between Dividend Yields, Equity Values and Stock Returns: Implications of Abnormal January Returns*, unpublished dissertation, University of Chicago.

[70] Keim, D.B. (1983b) 'Size-Related Anomalies and Stock Return Seasonality: Further Empirical Evidence,' *Journal of Financial Economics* **12** 13–32.

[71] Keim, D.B. (1988) 'Stock Market Regularities: A Synthesis of the Evidence and Explanations.' In *Stock Market Anomalies*, E. Dimson (ed.), Cambridge University Press, 16–39.

[72] Kim, D. (1997) 'A Reexamination of Firm Size, Book-to-Market and Earning Price in the Cross-Section of Expected Stock Returns,' *Journal of Financial and Quantitative Analysis* **32** 463–489.

[73] Kim, Y.G., K.H. Chung and C.S. Pyun (1992) 'Size, Price-Earnings Ratio and Seasonal Anomalies in the Korean Stock Market.' In *Pacific-Basin Capital Markets Research*, S.G. Rhee and R.P. Chang (eds.), North-Holland.

[74] Kothari, S.P., J. Shanken and R. Sloan (1995) 'Another Look at the Cross-Section of Expected Stock Returns,' *Journal of Finance* **50** 185–224.

[75] Kryzanowski, L. and H. Zhang (1992) 'The Contrarian Investment Strategy Does Not Work in the Canadian Markets,' *Journal of Financial and Quantitative Analysis* **27** 383–395.

[76] Lakonishok, J., A. Schleifer and R.W. Vishny (1994) 'Contrarian Investment, Extrapolation and Risk,' *Journal of Finance* **49** 1541–1578.

[77] Lehmann, B. (1990) 'Fads, Martingales and Market Efficiency,' *Quarterly Journal of Economics* **105** 1–28.

[78] Levis, M. (1985) 'Are Small Firms Big Performers?,' *The Investment Analyst* **76** 21–27.

[79] Levis, M. (1989) 'Market Size, PE Ratios, Dividend Yield and Share Prices: The UK Evidence.' In *A Reappraisal of the Efficiency of Financial Markets*, R.C. Gumaraes, B.G. Kingsman and S.J. Taylor (eds.), Springer-Verlag, NATO ASI Series.

[80] Lintner, J. (1965) 'The Valuation of Risk Assets and the Selection of Risky Investment in Stock Portfolios and Capital Budgets,' *Review of Economic Statistics* **47** 13–37.

[81] Litzenberger, R. and K. Ramaswamy (1979) 'The Effects of Personal Taxes and Dividends on Capital Asset Prices: Theory and Empirical Evidence,' *Journal of Financial Economics* **7** 163–195.

[82] Lo, A. and C. MacKinlay (1990) 'When are Contrarian Profits due to Stock Market Overreaction,' *Review of Financial Studies* **3** 175–205.

[83] Loughran, T. (1993) 'NYSE vs NASDAQ Returns: Market Microstructure or the Poor Performance of IPOs?,' *Journal of Financial Economics* **33** 241–260.

[84] Loughran, T. (1996) 'Is There a Book-to-Market Effect?,' working paper, University of Iowa.

[85] Ma, T. and T.Y. Shaw (1990) 'The Relationships between Market Value, P/E Ratio, Trading Volume and the Stock Return of Taiwan Stock Exchange.' In *Pacific Basin Capital Markets Research*, S.G. Rhee and R.P. Chang (eds.), North-Holland, 313–335.

[86] Mossin, J. (1966) 'Equilibrium in a Capital Asset Market,' *Econometrica* **34** 768–783.

[87] Nicholson, F. (1960) 'Price-Earnings Ratios,' *Financial Analysts Journal*, 43–50.

[88] Peavy, J. and D. Goodman (1983) 'Industry-Relative Price-Earnings Ratios as Indicators of Investment Returns,' *Financial Analysts Journal* **39** 60–66.

[89] Reinganum, M. (1981) 'A Misspecification of Capital Asset Pricing: Empirical Anomalies Based on Earnings Yields and Market Values,' *Journal of Financial Economics* **9** 19–46.

[90] Reinganum, M. (1982) 'A Direct Test of Roll's Conjecture on the Firm Size Effect,' *Journal of Finance* **37** 27–35.

[91] Reinganum, M. (1990) 'Market Microstructure and Asset Pricing: An Empirical Investigation of NYSE and NASDAQ Securities,' *Journal of Financial Economics* **28** 127–148.

[92] Roll, R. (1977) 'A Critique of the Asset Pricing Theory's Test, Part 1: On Past and Potential Testability of the Theory,' *Journal of Financial Economics* **4** 129–176.

[93] Roll, R. (1981) 'A Possible Explanation of the Small Firm Effect,' *Journal of Finance* **36** 879–888.

[94] Rosenberg, B. and V. Marathe (1979) 'Tests of the Capital Asset Pricing Hypotheses.' In *Research in Finance*, H. Levy (ed.), JAI Press, 115–223.

[95] Rosenberg, B., K. Reid and R. Lanstein (1985) 'Persuasive Evidence of Market Inefficiency,' *Journal of Portfolio Management* **11** 9–17.

[96] Rubio, G. (1988) 'Further International Evidence on Asset Pricing: The Case of the Spanish Capital Market,' *Journal of Banking and Finance* **12** 221–242.

[97] Scholes, M. and J. Williams (1977) 'Estimating Betas from Non-Synchronous Data,' *Journal of Financial Economics* **5** 309–328.

[98] Senchack, A. and J. Martin (1987) 'The Relative Performance of the PSR and the PER Investment Strategies,' *Financial Analysts Journal* **43** 45–56.

[99] Sharpe, W. (1964) 'Capital Asset Prices: A Theory of Market Equilibrium under Conditions of Risk,' *Journal of Finance* **19** 425–442.

[100] Sharpe, W. (1982) 'Factors in NYSE Stock Returns,' *Journal of Portfolio Management*, 5–19.

[101] Stambaugh, R. (1982) 'On the Exclusion of Assets from the Two-Parameter Model: A Sensitivity Analysis,' *Journal of Financial Economics* **17** 237–268.

[102] Stattman, D. (1980) 'Book Values and Expected Stock Returns,' Unpublished MBA Honors Paper, University of Chicago.

[103] Stehle, R. (1992) 'The Size Effect in the German Stock Market,' Unpublished manuscript, University of Augsburg.

[104] Stoll, H. and R. Whaley (1983) 'Transactions costs and the Small Firm Effect,' *Journal of Financial Economics* **12** 57–80.

[105] Strong, N. and X.G. Xu (1995) 'Explaining the Cross-Section of UK Expected Returns,' working paper, University of Manchester.

[106] Treynor, J. (1961) 'Toward a Theory of Market Value of Risky Assets,' unpublished manuscript.

[107] Vermaelen, T. and M. Versringe (1986) 'Do Belgians Overreact?,' working paper, Catholic University of Louvain, Belgium.

[108] Wong, K.A. and M.S. Lye (1990) 'Market Values, Earnings Yields and Stock Returns,' *Journal of Banking and Finance* **14** 311–326.

[109] Zarowin, P. (1989) 'Does the Stock Market Overreact to Corporate Earnings Information?,' *Journal of Finance* **44** 1385–1399.

[110] Ziemba, W. (1991) 'Japanese Security Market Regularities: Monthly, Turn-of-the-Month and Year, Holiday and Golden Week Effects,' *Japan and the World Economy* **3** 119–146.

Beta and Book-to-Market: Is the Glass Half Full or Half Empty?*

S.P. Kothari and Jay Shanken

Abstract

We review recent empirical work on the determinants of the cross-section of expected returns. This literature, which includes the influential work by Fama and French (1992, 1993), tends to ignore the positive evidence on beta and to overemphasize the importance of book-to-market. Kothari, Shanken and Sloan ((1995) show that beta significantly explains cross-sectional variation in average returns, but that size also has incremental explanatory power. We find that, while statistically significant, the incremental benefit of size given beta is surprisingly small economically. Book-to-market is a weak determinant of the cross-sectional variation in average returns among large firms and, as others have documented, it fails to account for return differences related to momentum and trading volume.

1 Introduction

Since the publication of the influential work of Fama and French (henceforth FF) (1992, 1993) on the determinants of the cross-section of expected returns, there has been a tendency in the literature to ignore the positive evidence on beta and negative evidence on book-to-market (B/M). In this paper, we review some of the recent empirical work and provide a balanced view. With regard to beta, the focus here is on the practical issue of whether betas defined with respect to commonly-employed market proxies provide useful information about expected returns. The possibility that a more comprehensive market index might yield different results is recognized, but not pursued. We expand on the evidence in Kothari, Shanken and Sloan (henceforth KSS) (1995) concerning the ability of beta and size to explain cross-sectional variation in average returns for 100 portfolios ranked on size and then beta. In particular, the extent to which size is incrementally helpful in explaining average portfolio returns is assessed. While statistically significant, the incremental economic benefit of size, given beta, is surprisingly small.

*We gratefully acknowledge financial support from the John M. Olin Foundation and the Bradley Policy Research Center at the Simon School, University of Rochester.

KSS also present empirical evidence suggesting that the book-to-market (B/M) effect documented in previous research is exaggerated due to selection biases. While recent results in Kim (1997) indicate that selection biases do not significantly influence the measured B/M effect, the evidence inconsistent with B/M pricing remains. We review this evidence and some of the more recent work of others on the relation between returns and B/M.

While much of the evidence undoubtedly supports the existence of an important B/M effect, several empirical observations raise significant doubts about the reliability of B/M-based predictions:

(i) the weakness of the effect in large firms;

(ii) the period-specific nature of the risk-adjusted (Jensen alpha) returns for the related contrarian strategy;

(iii) the failure of B/M (as well as size and beta) to account for return differences related to trading volume (Brennan, Chordia and Subrahmanyan, 1999) and momentum (Fama and French, 1996a).

2 Return-Beta Relation: The Glass is Not Empty

2.1 Beta and CAPM deviations: peaceful coexistence?

Evidence that betas estimated from time-series regressions on stock indices like the CRSP value-weighted portfolio fail to fully account for observed cross-sectional differences in average returns is not new. In perhaps the most striking example, Banz (1981) reports a statistically significant risk-adjusted difference in expected returns of about 1.5% per month (18% annualized) between very small and very large firms. This difference controls for beta risk with respect to the CRSP value-weighted index, over the period 1931–75. Given this background, how is one to account for the furor that the FF (1992) evidence has created? Like Roll and Ross (1994), we attribute much of the reaction to the FF conclusion that, not only does beta fail to fully account for all differences in expected returns, but "the relation between market β and average return is flat, even when β is the only explanatory variable."

Prior to FF, instructors in modern portfolio theory could point to the positive empirical relation between expected return and beta, while acknowledging that other variables like size or dividend yield, that are related to liquidity (Amihud and Mendelson, 1986) or taxes (Brennan, 1970, and Litzenberger and Ramaswamy, 1979), might also be needed to fully explain expected stock returns. Since size is highly correlated with estimated beta, one could also speculate that size might even be a better proxy for the true beta than the estimate itself (see, for example, Chan and Chen (1988)).[1] It is hard, though, to find a silver lining in the observation that beta tells us *nothing* about differences in expected stock returns. Hence, the relevance of the very basic observation in KSS that the *ex ante* market risk premium is about as likely to be 6% per year as zero (point estimate 0.24%, standard error 0.23% per month), *given the FF (1992) evidence.*

Despite this simple observation, much current discussion in the finance literature proceeds as if beta's complete lack of predictive power for expected returns has been firmly established and, by implication, can be expected to continue in the future. This is even more difficult to understand in light of the KSS finding that an alternative beta estimate, derived from annual return observations, is significantly positively related to average return. In some ways, this finding is consistent with a 'world view' similar to that of a decade ago in which beta and size peacefully coexisted.[2] Although the relation between B/M and estimated beta is weaker than the size-beta correlation, it may account for some of the explanatory power of B/M. Moreover, insofar as much of the B/M effect is driven by 'mispricing', as several papers suggest (e.g., Lakonishok, Shleifer and Vishny (1994)), a reasonable conjecture is that the effect will be smaller in the future.

2.2   Reconciling the differences between Fama & French (1996b) and KSS

FF (1996b) find that the simple regression t-statistics for the cross-sectional coefficients on monthly and annual betas are small and nearly identical. This is due to the fact that their post-formation betas (for decile portfolios based on past beta rankings) estimated from either monthly or annual returns are nearly perfectly cross-sectionally correlated. These results contrast sharply with those in KSS, reproduced here in Table 1, which show annual betas to be significantly priced for a variety of portfolios. One wonders then how KSS could find such different results for monthly and annual betas when FF do not. The following two considerations appear to explain the differences between the results.

---

[1]The advantage size enjoys over beta is that size reflects the market's latest assessment based on available information. By contrast, beta is estimated using a time series of return observations. Thus, a single estimate of beta might not accurately measure the systematic risk characteristic of a security at the end of each period (year).

[2]A recent paper by Kim (1995) finds that betas estimated from past monthly data are also significantly related to average returns and the significance of size is reduced when a maximum likelihood procedure is applied to individual securities. We experimented with an alternative errors-in-variables approach (Litzenberger and Ramaswamy 1979 and Shanken 1992), but found that results based on the 100 size-then-beta portfolios were altered only slightly.

Table 1. Cross-sectional regressions of monthly returns on beta and firm size: Equal-weighted market index. Taken from Kothari, Shanken and Sloan (1995).

Portfolios	γ_0	γ_1	γ_2	Adj R^2	γ_0	γ_1	γ_2	Adj R^2
	t-stat	t-stat	t-stat		t-stat	t-stat	t-stat	
		Panel A: 1927–90				Panel B: 1941–90		
20, beta ranked	0.76	0.54		0.32	0.95	0.36		0.33
	3.25	1.94			4.69	1.63		
	1.76		−0.16	0.27	1.61		−0.10	0.28
	2.48		−2.03		2.31		−1.49	
	1.68	0.09	−0.14	0.35	1.70	−0.03	−0.10	0.36
	3.82	0.41	−2.57		3.49	−0.18	−2.00	
20, size ranked	0.30	1.02		0.32	0.54	0.76		0.32
	−0.18	3.91			0.82	3.69		
	1.73		−0.18	0.33	1.73		−0.14	0.34
	3.70		−3.50		4.03		−3.28	
	−0.05	1.15	0.03	0.40	0.32	0.85	0.02	0.44
	−0.85	4.61	0.76		−0.15	4.35	0.56	
100, beta and size ranked independently	0.63	0.66		0.07	0.87	0.42		0.07
	1.67	3.65			2.95	3.33		
	1.72		−0.17	0.09	1.70		−0.13	0.10
	3.92		−3.71		4.29		−3.40	
	1.21	0.40	−0.11	0.12	1.43	0.20	−0.10	0.13
	3.74	2.63	−2.83		4.63	2.12	−2.89	
100, first beta, then size ranked	0.57	0.73		0.12	0.82	0.49		0.12
	1.43	3.49			2.76	3.07		
	1.73		−0.18	0.12	1.73		−0.14	0.13
	3.70		−3.48		3.99		−3.22	
	1.12	0.45	−0.10	0.16	1.35	0.26	−0.09	0.17
	3.43	2.83	−2.65		4.35	2.20	−2.78	
100, first size, then beta ranked	0.58	0.71		0.12	0.81	0.49		0.12
	1.54	3.39			2.75	3.12		
	1.72		−0.18	0.12	1.71		−0.13	0.13
	3.66		−3.43		3.96		−3.17	
	1.14	0.43	−0.10	0.16	1.32	0.27	−0.09	0.17
	3.78	2.58	−2.87		4.39	2.38	−2.77	

Time-series averages of estimated coefficients from the following monthly cross-sectional regressions from 1927–90 (Panel A) and from 1941–90 (panel B), associated t-statistics in parentheses, and adjusted R^2s are reported (with and without Size included in the regressions):

$$R_{pt} = \gamma_{0t} + \gamma_{1t}\beta_p + \gamma_{2t}\mathrm{Size}_{pt-1} + \varepsilon_{pt},$$

where R_{pt} is the buy-and-hold return on portfolio p for one month during the year from July 1 of year t to June 30 of year $t + 1$; β_p is the full-period post-ranking beta of portfolio p, i.e., the slope coefficient from a time-series regression of annual buy-and-hold post-ranking portfolio returns on the returns for an equal-weighted portfolio of all the beta-size portfolios; Size_{pt-1} is natural log of the average market capitalization in millions of dollars on June 30 of year t for the stocks in portfolio p; γ_{0t}, γ_{1t}, and γ_{2t} are regression parameters; and ε_{pt} is the regression error. Portfolios are formed in five different ways: (i) 20 portfolios by grouping on beta alone; (ii) 20 portfolios by grouping on size alone; (iii) taking intersections of 10 independent beta or size groupings to obtain 100 portfolios; (iv) first, grouping stocks into 10 portfolios on beta and then into 10 portfolios on size within each beta group; and (v) first, grouping stocks into 10 portfolios on size and then into 10 portfolios on beta within each size group. When ranking on beta, the beta for an individual stock is estimated by regressing 24 to 60 monthly portfolio returns through June of each year on the CRSP equal-weighted portfolio. The t-statistic below γ_0 is for the difference between the average γ_{0t} and the average risk-free rate of return over the 1927–90 or 1941–90 period. The t-statistics below γ_1 and γ_2 are for their deviations from zero.

First, unlike KSS, who also include AMEX firms from July 1964, FF restrict their sample to NYSE stocks. This has the effect of substantially reducing the range of estimated betas. For example, using equal-weighted annual index returns, the post-formation betas of NYSE decile portfolios formed by ranking stocks on past betas range from 0.78 to 1.27 (FF, 1995, table I). In comparison, the corresponding range is 0.52 to 1.35 in KSS (1995, table I, average of vitile portfolios 1 and 2, and 19 and 20, respectively), an increase of nearly 70%. KSS obtain a much lower estimate of the large firm betas because the inclusion of AMEX stocks makes small stocks far more important in the CRSP equal-weighted index; thus, large stocks are relatively less risky compared to the index.

The second difference between FF (1996b) and KSS or this study is that FF compound equal-weighted monthly portfolio returns to obtain annual returns. That is, the portfolios are rebalanced each month, yielding annual returns that differ from buy-and-hold annual returns obtained by averaging the compounded annual returns on individual securities in a portfolio.

Since the difference between monthly and annual betas increases as the betas deviate from one (Levhari and Levy 1977, and Handa, Kothari and Wasley 1989), it is not surprising that the cross-sectional results in KSS are much more sensitive to the horizon over which beta is estimated. The reduction in range in FF, exacerbated by the fact that only ten portfolios are used in the cross-sectional regressions, appears to explain the low t-statistics for the coefficients on beta in Part B of Table I in FF (1996b). The t-statistics cluster around 1.3, despite the fact that the estimated market risk premium is, in the case of the equal-weighted index, 65 basis points per month (7.8% per year). This is higher than the corresponding estimate in KSS Table II (54 basis points), which has a t-statistic of 1.94. Thus, the low FF t-statistics are driven by an inherent lack of precision.

We are inclined to take a more positive view of cross-sectional beta-return regressions than FF (1992, 1996b). Both Roll and Ross (1994) and Kandel and Stambaugh (1995) emphasize that the equivalence between the mean-variance efficiency of an index and exact linearity of expected returns in the betas breaks down when the index is only 'approximately' efficient. Roll and Ross argue, for example, that given a set of assets, it is conceivable that there are indices 'close' to the efficient frontier that produce no relation between beta and mean return. Kandel and Stambaugh, on the other hand, focus on the assets. They show that, when an index is approximately efficient with respect to a set of asset returns, the assets may be 'repackaged' into (potentially unusual-looking) portfolios such that the return/beta relation is arbitrarily weak.[3] While these insights are relevant to the interpretation

---

[3]The argument is, in some ways, reminiscent of that in Shanken (1982), where assets are repackaged to form portfolios with an arbitrary factor space, thereby revealing the

of tests of asset pricing theories, they do not change the fact that a cross-sectional regression of returns on betas does provide information about the relation between the expected returns on the given assets and their betas on the given index. In many applications, this is the central issue.

2.3   Average residuals from return-regressions

We agree with FF on the importance of examining average residuals from cross-sectional regressions. Average residuals are the basis for evaluating whether betas or other firm characteristics completely capture cross-sectional variation in expected returns. Although some of the deviations in Part B of Table III in FF appear to be quite large, the extent to which they are driven by random noise is not obvious. The multivariate literature on testing asset pricing relations addresses the statistical issue of whether observed residuals merely represent random ex post fluctuations or, instead, are manifestations of a violation of the ex ante expected return relation.[4] Without minimizing the importance of this question, we prefer here to address an easier question – one which is more relevant to the issues raised by FF.

FF point to the large "CAPM pricing errors" for the extreme size and beta deciles and view their bivariate regression results (on size and beta), as well as those in KSS, as evidence that "size always adds substantially to β's description of average returns." Yet no information is provided on the extent to which size actually reduces the average residuals seen in their Table III! As KSS note, the incremental contribution that size makes to the prediction of returns depends not only on the size coefficient, but also on the magnitude of the residuals from an auxiliary regression of size on beta. Since beta and size are strongly (negatively) correlated, these residuals need not be very large.

Before we discuss average residuals from annual return models, we briefly consider summary statistics for 100 equal-weighted size-then-beta ranked portfolios from the CRSP universe of stocks. The portfolios are formed every July from 1927 to 1992 (66 years). Each year, all available securities are ranked first on their market capitalization of equity on June 30 and then on their betas estimated by regressing 24 to 60 months of past returns on the CRSP equal-weighted index returns. Betas as of July 1927 are estimated using 18 monthly returns. The post-formation betas are estimated using each portfolio's time series of annual buy-and-hold returns. The index used is the equal-weighted buy-and-hold annual return on all the CRSP securities available in that year.

Table 2, panels A–C report average annual returns, average firm size in millions of dollars, and post-formation betas of the 100 size-beta portfolios.

---

limitations of the traditional exact view of the arbitrage pricing theory.

[4]See the recent review article by Shanken (1996).

Average values for each size and beta decile portfolio are also reported in the last row and column of the table. There is a substantial spread in betas, size and average returns among the 100 portfolios. Average returns generally increase with beta and decrease with size. Average return is 14% per annum on the lowest beta decile and 20.4% for the highest beta decile. The spread in average returns across the size deciles is even more impressive, ranging from 13.3% for the largest decile to 25.4% for the smallest. The average returns on the beta portfolios within each size portfolio also exhibit considerable variation that is consistent with a positively-sloped risk-return relation. For example, the lowest-beta portfolio in the largest size decile earns 11.5% average annual return compared to 15.3% earned by the highest-beta portfolio in the largest size decile. While returns for the higher beta portfolios in the smallest size decile are erratic, they line up well with the post-formation annual betas in the bottom panel.

Table 3, panels A–C provide average residuals from annual Fama–MacBeth return models that include only size or (annual) beta, as well as the bivariate model with size and beta. The Fama–MacBeth regression models are:

$$R_{pt} = \gamma_{0t} + \gamma_{1t}\beta_p + \gamma_{2t}\text{Size}_{pt-1} + \varepsilon_{pt} \qquad (2.1)$$

where R_{pt} is the buy-and-hold return on portfolio p for one year beginning from July 1 of year t to June 30 of year $t+1$; β_p is the full-period post-ranking beta of portfolio p, i.e., the slope coefficient from a time-series regression of annual buy-and-hold post-ranking portfolio returns on the returns for an equal-weighted portfolio of all the beta-size portfolios; Size_{pt-1} is the natural log of the average market capitalization in millions of dollars on June 30 of year t of the stocks in portfolio p; γ_{0t}, γ_{1t} and γ_{2t} are regression parameters; and ε_{pt} is the regression error. For each portfolio, average residual is calculated by averaging the estimated ε_{pt}s for portfolio p across time. We refer to a portfolio's average residual as its 'deviation' from the model. The basic properties of (cross-sectional) regression residuals guarantee that the average deviation across the 100 portfolios is zero.

Table 3 shows that for both the beta-only (size-beta) model, four (one) of the portfolios have deviations that exceed 4% and two (four) others exceed 3%. In contrast, for the size-only model, one deviation exceeds 10% and seven others exceed 5%. Most of these occur either in the lowest size portfolio or the smallest beta portfolio. The standard deviation of the 100 portfolio average residuals from the size-only model is 2.9%, compared to 1.5% for the size-beta model, and 1.8% for the beta-only model. These numbers suggest that the incremental effect of size on expected return (fitted value), given that beta is already considered, is small.

Table 2. Descriptive statistics for 100 equal-weighted NYSE-AMEX portfolios formed on size and then beta: 1927–92

	Low β	2	3	4	5	6	7	8	9	High β	Avg
	Panel A: Average annual return in %										
Small	17.1	23.7	24.1	26.4	22.4	29.8	29.8	25.0	34.6	21.5	25.4
2	15.9	19.8	16.6	21.8	21.3	26.4	23.5	26.3	21.5	25.4	21.9
3	17.1	19.5	18.2	22.6	19.7	19.6	22.7	22.0	21.7	19.3	20.2
4	12.9	19.3	16.6	19.5	19.1	19.0	22.3	23.5	19.2	22.5	19.4
5	14.4	15.2	19.1	16.3	18.0	16.2	19.4	22.7	18.7	19.7	18.0
6	13.9	15.1	16.7	19.0	17.2	16.5	21.8	18.3	18.8	20.9	17.8
7	12.6	14.1	18.9	18.0	18.7	14.5	20.4	22.8	20.7	22.3	18.3
8	13.9	16.1	16.5	14.6	16.0	17.1	16.8	17.4	19.2	17.1	16.5
9	10.9	14.4	12.7	15.0	14.9	18.0	17.7	15.6	18.6	20.2	15.8
Big	11.5	11.1	13.2	12.0	13.0	12.4	13.1	16.1	15.0	15.3	13.3
Avg	14.0	16.8	17.3	18.5	18.0	19.0	20.8	21.0	20.8	20.4	
	Panel B: Size, average market value of equity in millions of dollars										
Small	3.9	4.1	4.0	4.1	3.9	4.0	3.8	3.9	3.8	3.7	3.9
2	9.9	9.7	9.7	9.9	9.9	9.7	9.6	9.8	9.7	9.6	9.7
3	18.1	17.8	18.0	17.8	17.8	17.8	17.9	17.9	17.8	17.6	17.9
4	29.1	29.4	29.1	29.3	29.1	29.4	29.4	29.3	29.4	29.5	29.3
5	47.0	47.1	46.1	46.9	46.7	46.9	46.6	47.0	46.5	46.1	46.7
6	76.1	76.2	76.7	77.6	76.9	76.0	76.0	75.3	75.9	76.0	76.3
7	131.6	130.1	129.5	133.4	133.4	132.8	130.3	129.7	129.7	127.5	130.8
8	236.2	239.6	236.8	237.5	242.4	241.6	238.6	237.3	237.3	234.4	238.2
9	490.4	494.3	500.4	494.7	501.6	497.0	488.0	485.5	488.6	465.1	490.5
Big	2459.3	2930.0	3232.7	2746.4	2478.2	2412.5	2056.2	1725.2	1509.1	1407.9	2295.8
Avg	350.2	397.8	428.3	379.8	354.0	346.8	309.7	276.1	254.8	241.7	
	Panel C: Post-formation equal-weighted, annual return beta										
Small	0.78	1.35	1.30	1.39	1.35	2.01	1.95	1.52	2.47	1.25	1.54
2	0.85	1.05	0.98	1.15	1.03	1.74	1.55	1.81	1.37	1.80	1.33
3	0.77	1.02	0.81	0.97	1.67	1.14	1.54	1.46	1.54	1.58	1.20
4	0.56	0.69	0.93	0.77	0.89	1.03	1.13	1.75	1.01	1.66	1.04
5	0.45	0.65	0.88	0.83	0.85	0.64	1.07	1.37	1.14	1.34	0.95
6	0.31	0.51	0.58	0.77	0.83	0.79	1.08	1.15	1.17	1.47	0.86
7	0.35	0.48	0.69	0.93	0.90	0.81	1.19	1.30	1.39	1.55	0.96
8	0.24	0.53	0.60	0.59	0.74	1.02	0.90	1.08	1.14	1.16	0.80
9	0.26	0.53	0.44	0.62	0.61	0.78	0.93	0.87	1.04	1.29	0.74
Big	0.24	0.37	0.48	0.43	0.57	0.58	0.67	0.84	0.85	0.97	0.60
Avg	0.48	0.72	0.77	0.84	0.89	1.05	1.20	1.31	1.31	1.44	

All securities with CRSP monthly return data available [at the beginning of July of each year from 1927 to 1992] for at least 24 prior months (18 months in case of the first year, 1927) are grouped on firm size (market value of equity) into ten portfolios. Each year, stocks in each size portfolio are grouped into ten portfolios based on their CRSP equal-weighted index betas estimated using up to 60 monthly returns. Equal-weighted annual buy-and-hold returns are calculated for each of the 100 size, then beta ranked portfolios using returns for the July-to-June period. This yields a time series of 66 annual returns for each portfolio, which is used to estimate post-formation betas for the 100 portfolios. The market index used to estimate the post-formation betas is the equal-weighted index of annual buy-and-hold returns on all the stocks available at the beginning of July of each year.

Size is the average market value of equity, in millions of dollars, for the stocks in a portfolio. The figures in the table are the time-series averages of size for each portfolio.

The Avg column at the right of each panel is the equal-weighted average of the parameter values for the beta decile portfolios within each size decile portfolio.

The Avg row at the bottom of each panel is the equal-weighted average of the parameter values for the size decile portfolios within each beta decile portfolio.

Table 3. Average residuals from Fama–MacBeth cross-sectional regressions of
annual returns on beta, size, or both beta and size: 100 equal-weighted
NYSE-AMEX portfolios formed on size and then beta, 1927–92

	Low β	2	3	4	5	6	7	8	9	High β	Avg
					Panel A: Beta-only model						
Small	0.4	1.9	2.7	4.3	0.6	2.1	2.6	1.7	2.7	0.6	2.0
2	0.9	0.7	−1.9	1.9	2.4	1.2	−0.1	0.3	−0.4	−0.4	0.2
3	0.5	0.7	1.3	4.2	−0.4	−0.3	−0.8	−0.8	−1.9	−4.6	−0.2
4	−1.8	3.5	−1.4	2.9	1.4	0.2	2.6	−1.9	0.4	−2.0	0.4
5	0.7	−0.2	1.6	−0.8	0.7	0.8	0.0	0.8	−1.2	−4.7	−0.2
6	1.4	0.8	2.0	2.7	0.1	−0.3	2.5	−1.7	−1.4	−1.9	0.4
7	−0.2	0.1	3.0	0	1.0	−2.4	0.1	1.5	−1.5	−1.3	0.0
8	2.1	1.7	1.5	−0.3	−0.3	−1.7	−0.9	−2.0	−0.7	−3.0	−0.4
9	−1.0	0	−0.9	−0.1	−0.2	1.3	−0.3	−1.9	−0.4	−1.0	−0.5
Big	−0.3	−1.9	−0.7	−1.6	−1.7	−2.4	−2.5	−1.1	−2.3	−3.1	−1.8
Avg	0.1	0.7	0.7	1.3	0.4	−0.2	0.3	−0.5	−0.7	−2.1	
					Panel B: Size-only model						
Small	−6.9	−0.2	0.2	2.7	−1.5	6.1	6.0	1.6	10.6	−2.2	1.6
2	−6.1	−2.1	−5.4	−0.2	−0.7	4.4	1.6	4.3	−0.5	3.4	−0.1
3	−3.7	−1.3	−2.6	1.7	−1.2	−1.3	1.8	1.1	0.8	−1.7	−0.6
4	−7.0	−0.8	−3.3	−0.6	−0.9	−1.0	2.4	3.5	−0.8	2.5	−0.6
5	−5.0	−4.0	−0.1	−3.0	−1.2	−3.0	0.1	3.1	−0.6	0.4	−1.3
6	−4.6	−3.4	−1.9	0.5	−1.3	−2.1	3.3	−0.2	0.2	2.4	−0.7
7	−5.0	−3.6	1.2	0.3	1.2	−3.2	2.8	5.1	2.9	4.6	0.6
8	−2.7	−0.5	−0.1	−2.2	−0.7	0.5	0	0.7	2.5	0.5	−0.2
9	−4.4	−0.8	−2.6	−0.3	−0.2	2.7	2.4	0.3	3.3	4.9	0.5
Big	−0.5	−0.7	0.4	−0.3	0.1	0.2	0.9	3.6	1.9	2.1	0.8
Avg	−4.6	−1.7	−1.4	−0.1	−0.6	0.3	2.1	2.3	2.0	1.7	
					Panel C: Size-beta model						
Small	−2.1	0.6	1.1	3.0	−0.8	1.9	2.1	0.6	3.2	−0.9	0.9
2	−2.9	−0.5	−3.3	0.8	1.1	1.1	−0.5	0.4	−1.1	−0.4	−0.5
3	−0.7	−0.2	0.1	3.3	−1.0	−1.0	−0.8	−0.9	−1.8	−4.5	−0.7
4	−3.0	2.4	−2.1	2.0	0.7	−0.3	2.3	−1.1	−0.1	−1.4	−0.1
5	−0.5	−1.1	1.2	−1.3	0.2	0.0	0.0	1.2	−1.2	−3.8	−0.5
6	0.3	0.0	1.3	2.3	−0.1	−0.6	2.6	−1.4	−1.0	−1.1	0.2
7	−0.9	−0.4	2.9	0.3	1.2	−2.4	0.8	2.4	−0.5	0.0	0.3
8	1.6	1.7	−1.6	−0.3	−0.0	−0.9	−0.3	−1.1	0.3	−1.9	0.1
9	−1.0	0.5	−0.6	0.4	0.5	2.3	0.9	−0.8	1.0	0.8	0.4
Big	1.2	−0.5	0.8	−0.2	−0.1	−0.5	−0.6	1.0	−0.3	−1.0	0.0
Avg	−0.8	0.3	0.3	1.0	0.2	−0.0	0.7	0.0	−0.2	−1.4	

All securities with CRSP monthly return data available [at the beginning of July of each year 1927 to 1992] for at least 24 prior months (18 months in case of the first year, 1927) are grouped on firm size (market value of equity) into ten portfolios. Each year, stocks in each size portfolio are grouped into ten portfolios based on their CRSP equal-weighted index betas estimated using up to 60 monthly returns. Equal-weighted annual buy-and-hold returns are calculated for each of the 100 size, then beta ranked portfolios using returns for the July-to-June period. This yields a time series of 66 annual returns for each portfolio, which is used to estimate post-formation betas for the 100 portfolios. The market index used to estimate the post-formation betas is the equal-weighted index of annual buy-and-hold returns on all the stocks available at the beginning of July of each year.

Size is the average market value of equity, in millions of dollars, for the stocks in a portfolio.

Panel A contains the average residuals from 66 annual Fama–MacBeth cross-sectional regressions of annual returns on the post-formation betas of the 100 equal-weighted size and then beta-ranked portfolios. The average residuals in panels B and C are from Fama–MacBeth regressions using only size, and size and beta, respectively.

The Avg column at the right of each panel is the equal-weighted average of the residuals for the beta decile portfolios within each size decile portfolio.

The Avg row at the bottom of each panel is the equal-weighted average of the residuals for the size decile portfolios for each beta decile portfolio.

A more direct measure of the incremental effect of size is given in Table 4, panels A–B, for each portfolio. Panel A contains the difference between the absolute value of the portfolio deviation from the beta-only model and that for the size and beta model. Only two of these differences are greater than 2% in magnitude (maximum 2.1%). In contrast, seven portfolios have differences that exceed 4% (maximum 7.4%) when beta is added to the size-only model (see panel B). The standard deviation of the differences in panel B is 1.9% compared to only 0.8% in panel A. These observations substantiate our claim that the estimated incremental impact of size in explaining expected returns is fairly small, despite its statistical significance. We do not advocate ignoring this information, however, particularly in the applied context of trying to estimate a predictive model.

We repeated the above analysis using monthly, rather than annual, portfolio returns as the dependent variable in the Fama–MacBeth cross-sectional regressions in eq. (2.1). The tenor of the results using monthly returns is not different. The size-only model slightly outperforms the beta-only model in that the average residuals are less extreme. The cross-sectional standard deviation of the average monthly residuals from the size-only model is 0.20% compared to 0.23% from the beta-only model, and 0.17% from the size-beta model. Similarly, the variability of the difference between absolute deviations from the beta-only model and size-beta model (i.e., numbers like those in Table 4, panel A, except using monthly return observations) is 0.12%, compared to 0.10% when beta is added to the size-only model (i.e., numbers corresponding to those in Table 4, panel B).

To gain additional insight into the economic significance of the above findings, we repeated the analysis excluding the smallest 20% market capitalization stocks each year. In the literature, these smallest 20% stocks are generally referred to as 'small stocks.' The variability of the average residuals and the differences between absolute average residuals declines dramatically with the exclusion of (the economically less important) small stocks. Using annual returns, the variability of the average residuals from the beta-only model continues to be considerably smaller than that of the average residuals from the size-only model. This difference is small, but in the opposite direction using monthly portfolio returns.

2.4   Why annual betas?

We have presented evidence that estimates of beta from annual returns are significantly related to expected stock returns, both economically and statistically. A number of important questions arise in thinking about annual betas, however. For one thing, annual betas necessarily have a large estimation error component. Why, then, are these betas superior in 'explaining' the

Table 4. Absolute average residuals from the beta-only minus those from
size-&-beta model, and absolute average residuals from the size-only minus those
from size-&-beta model: Fama–MacBeth cross-sectional regressions of annual
returns on beta, size, and size and beta using 100 equal-weighted NYSE-AMEX
portfolios formed on size and then beta, 1927–92

	Low β	2	3	4	5	6	7	8	9	High β	Avg
				Panel A: Beta-only model minus Size-&-Beta model							
Small	−1.7	1.3	1.6	1.3	−0.2	0.2	0.5	1.1	−0.5	−0.2	0.3
2	−1.5	0.2	−1.4	1.1	1.3	0.1	−0.0	−0.1	−0.7	0.0	−0.2
3	−0.2	0.5	1.2	0.9	−0.6	−0.7	0.0	−0.1	0.1	0.1	0.1
4	−1.3	1.1	−0.7	0.9	0.8	−0.2	0.3	0.8	0.3	0.6	0.3
5	0.2	−0.8	0.4	−0.5	0.5	0.8	0.1	−0.4	0.0	0.9	0.1
6	1.1	0.8	0.7	0.4	−0.0	−0.3	−0.1	0.3	0.3	0.8	0.4
7	−0.7	−0.2	0.2	−0.3	−0.2	0.0	−0.7	−0.9	1.0	1.3	−0.1
8	0.5	−0.0	−0.1	0.0	0.3	0.8	0.6	0.9	0.4	1.1	0.5
9	0	−0.5	0.3	0.3	−0.3	1.0	−0.5	1.1	−0.6	0.2	−0.2
Big	−0.9	1.3	−0.0	1.4	1.6	1.9	2.0	0.2	2.1	2.1	1.2
Avg	−0.4	0.2	0.5	0.3	0.2	0.3	0.2	0.2	0.3	0.2	0.7
				Panel B: Size-only model minus Size-&-Beta model							
Small	4.8	−0.4	−0.9	−0.3	0.7	4.2	3.9	0.9	7.4	1.4	2.2
2	3.1	1.6	2.2	−0.6	−0.5	3.3	1.1	3.9	−0.6	3.0	1.6
3	3.0	1.1	2.5	−1.6	0.2	0.3	1.1	0.3	−1.0	−2.9	0.1
4	4.0	−1.6	1.2	−1.3	0.3	1.7	0.1	2.4	0.7	1.1	0.8
5	4.4	2.9	−1.1	1.7	1.0	3.0	0.1	2.3	−0.5	−3.4	1.0
6	4.4	3.4	0.6	−1.8	1.1	1.5	0.6	−1.2	−0.8	1.3	0.9
7	4.1	3.2	−1.7	0.0	−0.0	0.7	2.0	2.7	2.5	−1.5	1.8
8	1.2	−1.3	−1.5	1.9	0.6	−0.5	−0.3	−0.4	2.3	−1.5	0.1
9	3.4	0.3	2.0	−0.2	−0.3	0.4	1.6	−0.5	2.4	4.1	1.3
Big	−0.7	0.2	−0.4	0.1	0.0	−0.3	0.4	2.6	1.6	1.1	0.5
Avg	3.2	0.9	0.3	−0.2	0.3	1.3	1.0	1.3	1.4	0.9	1.0

All securities with CRSP monthly return data available (at the beginning of July of each year from 1927 to 1992) for at least 24 prior months (18 months in case of the first year, 1927) are ranked on firm size (market value of equity) into ten portfolios. Each year, stocks in each size portfolio are grouped into ten portfolios based on their CRSP equal-weighted index betas estimated using up to 60 monthly returns. Equal-weighted annual buy-and-hold returns are calculated for each of the 100 size, then beta ranked portfolios using returns for the July-to-June period. This yields a time series of 66 annual returns for each portfolio, which is used to estimate post-formation betas for the 100 portfolios. The market index used to estimate the post-formation betas is the equal-weighted index of annual buy-and-hold returns on all the stocks available at the beginning of July of each year.

Size is the average market value of equity, in millions of dollars, for the stocks in a portfolio.

Panel A contains the difference between the absolute values of the average residuals from the size-&-beta model and the beta-only model. Size-&-beta model: 66 annual Fama–MacBeth cross-sectional regressions of annual returns on size and post-formation betas of the 100 equal-weighted size and then beta-ranked portfolios. Beta-only model has only post-formation beta in the regressions. In panel B, the difference is between the absolute average residuals from the size-&-beta model and the size-only model.

The Avg column at the right of each panel is the equal-weighted average of the residuals for the beta decile portfolios within each size decile portfolio.

The Avg row at the bottom of each panel is the equal-weighted average of the residuals for the size decile portfolios for each beta decile portfolio.

cross-section of expected returns? It has long been recognized that if there are delays in the complete incorporation of information in security prices then betas estimated using high-frequency returns (e.g., daily returns) will be biased (Scholes and Williams 1977, Cohen *et al.* 1983). Lo and MacKinlay (1990) and Mech (1993) suggest that the time it takes for security prices to accurately incorporate available information is surprisingly long. For example, in the case of small capitalization stocks, the non-synchronous trading problem continues to be important using a monthly return measurement interval.

The bias in estimated betas due to non-synchronous trading is not reduced by simply expanding the time series of returns. It can be reduced, however, by increasing the length of the return measurement interval used in forming returns. Thus, it may be that the relative performance of annual and monthly betas is a situation in which a noisy but less biased annual beta estimate is better than a less variable but biased monthly beta estimate. FF (1992) use an alternative method (specifically, the Dimson (1979) beta obtained by summing the slopes on the contemporaneous and prior month's market return) to deal with nonsynchroneity employing monthly return data. Why the betas estimated using their method exhibit a flatter relation between average returns and betas than using the annual betas is not clear and deserves further attention.

An alternative hypothesis is that the relation to annual betas is spurious, somehow reflecting a correlation between the measurement error in beta and average returns. This is conceivable given that we use full-period post ranking betas in order to increase the precision of our annual beta estimates. We offer two observations that raise doubts about this hypothesis, however. First, using size portfolios and annual betas estimated using 15 years of past return data, Handa, Kothari and Wasley (1989) find that annual beta dominates both monthly beta and size in explaining expected returns on size portfolios. Second, we experimented with an alternative estimation procedure for annual betas that should be largely free of any spurious measurement error problem. For each year in which cross-sectional regressions are estimated, the estimates of beta employed are computed using all annual data except the returns for that year. Thus, the betas are re-estimated each year, as opposed to using the full-period post-ranking estimates. We found that the significance of beta estimated in this manner was reduced only slightly.

3   Book-to-Market Effect: The Glass is Not Full

3.1   Evidence for various portfolios

As noted earlier, a recent paper by Kim (1997) demonstrates that survivor biases cannot account for much of the B/M effect documented in Fama and French (1992).[5] The Kim (1995) study notwithstanding, empirical evidence

---

[5]FF cite the paper by Chan Jegadeesh and Lakonishok (1995) as evidence against a substantial survivor bias. That paper's careful documentation of the fact that many of the

presented by KSS continues to pose a challenge to the consistency of the B/M effect. We now review that evidence. First, consider those firms that are on the CRSP tapes but are not on COMPUSTAT. If one looks at the lowest decile of such firms, in terms of market capitalization, one finds that their deviation (alpha) from the Fama and French (1993) three-factor (market, size and B/M factors) model is quite large, about −7%. These are extremely small and volatile firms, however, and the *t*-statistics for the significance of this deviation are only −1.65 and −2.18 when the lowest decile is split into two subportfolios. This finding may be related to Loughran's (1997) observations concerning the very low returns on small growth firms.

The statistical significance of these deviations from the three-factor model is not overwhelming, and it is possible that microstructure-related measurement problems color the results for such small firms. Ball, Kothari and Shanken (1995), for example, find that eliminating very low-priced stocks substantially reduces the raw returns on the DeBondt and Thaler (1987) five-year contrarian strategy. At the other end of the market capitalization spectrum, KSS examine cross-sectional regressions for the 500 largest stocks. They find that the coefficient relating expected return to B/M is about 40% smaller than that obtained using all COMPUSTAT stocks and the *t*-statistic is 1.96, as compared to 5.71 in FF (1992) using all stocks. Breen and Korajczyk (1994) and Davis (1994), La Porta (1993) and Loughran (1997) also report reduced B/M effects.

Cross-sectional results in KSS, based on Standard and Poor's industry portfolios, are even less encouraging for B/M proponents. Despite the considerable variation in B/M across industries, the *t*-statistic for the coefficient on B/M is 1.02 for the period 1947–87 and just 0.15 for the post-62 period. The latter result is particularly surprising since KSS do find a strong B/M effect using value-weighted industry portfolios of all COMPUSTAT firms. A recent paper by Loughran (1997) echoes our findings for large firms. For the largest size quintile, which accounts for about three-fourths of all market value, he concludes that B/M has no reliable predictive power for returns in the 1963–94 period.[6]

---

firms on CRSP, but not on COMPUSTAT, are omitted for reasons unrelated to survival inclines us to believe that these biases might be less than we previously suspected. On the other hand, the direct evidence on bias in Chan et al. is for the largest 20% of NYSE-AMEX firms on CRSP, a sample where the survival stories we tell for distressed firms are not likely to be very relevant.

[6]A word of caution is in order, since the composition of the S&P industry portfolios changes somewhat over time. It is possible that the disappearance of the relation is attributable to this changing composition. However, we have used these same industry portfolios in a cross-sectional analysis of the relation between returns in a given year and earnings growth over the next year (Collins, Kothari, Shanken and Sloan 1994). In this context, a strong and highly significant positive relation is observed, as expected. Thus, the S&P results remain a puzzle which we hope will be sorted out with further research.

In Table 5, we provide additional perspective on the weakness of the large-firm B/M effect by comparing results for value-weighted B/M deciles to those for equal-weighted B/M decile portfolios. The latter are taken from FF (1992), with returns multiplied by 12 for ease of comparison. Like FF, we split the lowest and highest deciles into two sub-portfolios. Average returns for the first six deciles are essentially flat at 11–12% with value-weighting. B/M ratios for these deciles exhibit a substantial range from 0.17 for portfolio 1A to 0.90 for portfolio 6 and firm size declines as well. Over a similar range of B/M values, equal-weighting yields average returns that increase from about 8% to 17%. Consistent with the weaker cross-sectional results for large firms, the first six B/M deciles have the largest average market capitalizations, ranging from 1.3 billion to 695 million dollars. The flat returns are quite surprising since the reinforcing effects on expected returns of an increase in B/M and the decline in size lead one to expect a steep increase in average returns from decile 1 to 6. Daniel and Titman (1997) provide some evidence that average return is more directly related to the level of the B/M ratio, than to loadings on the B/M factor (i.e., return on the high minus low portfolio) in the FF three-factor model. The smaller cross-sectional coefficient on B/M for large firms in KSS, along with the results in Table 5 on equal and value-weighting, suggest that a ratio-driven approach is not adequate either. This is an important topic for further analysis.

3.2   Stability over time

A second area of concern is the stability of the B/M effect over time. FF report very similar effects in both halves of their 1963–90 sample using individual-security cross-sectional regressions. Using portfolios based on B/M rankings, we find a very large and highly significant effect over the 1963–77 subperiod, but a much smaller and statistically insignificant effect for 1978–92. Huson (1995) actually reports a significantly negative B/M effect in the OTC market over the period 1974–90. We do not think it wise to make too much of results for such short periods, but cite these results to provide a balanced view of the B/M evidence.

KSS note the obvious parallel between the book-to-market strategy and a contrarian strategy based on past-return rankings. In general, when forming beliefs about the future, it makes sense to condition those beliefs on all relevant information. This idea can be seen, for example, in a paper by Stambaugh (1996) who considers investments whose histories differ in length – say, the US market and emerging markets. He finds that when the asset returns are correlated, the longer histories provide important information about the moments of the shorter-history assets. In our context, the performance of the DeBondt and Thaler (1985) contrarian strategy, based on five-year ranking returns, has been evaluated going back to 1931 (ranking period 1926–30),

Table 5. Returns on value- and equal-weighted portfolios formed on
Book-to-Market Equity (B/M): July 1963 to 1990

Value-weighted Book-to-Market Portfolios				Equal-weighted Book-to-Market Portfolios, adapted from FF (1992), Table IV		
Portfolio	Book-to-market	Average market capital-ization	Value-weighted portfolio return	Book-to-Market	Average market capital-ization	Equal-weighted portfolio return
1A	0.17	1328	11.4	0.11	93	3.7%
1B	0.29	1159	9.1	0.22	107	8.3
2	0.41	852	11.6	0.33	109	11.0
3	0.54	987	11.6	0.47	96	12.3
4	0.65	664	11.2	0.60	87	13.2
5	0.77	719	12.0	0.73	80	15.0
6	0.90	695	12.6	0.87	69	16.8

Portfolio formation procedure for the value-weighted and equal-weighted portfolios
is similar to that employed by Fama and French (1992, Table IV). At the end of
each year, Fama and French form 12 portfolios on the basis of ranked values of
B/M. Portfolios 2–9 cover deciles, whereas the bottom and top two portfolios (1A,
1B, 10A, and 10B) split the bottom and top deciles. Like Fama and French, we
exclude negative B/M stocks. All CRSP-COMPUSTAT stocks with B/M and
return data are included. We include NYSE-AMEX stocks, whereas Fama and
French include NYSE, AMEX, and NASDAQ stocks.

B/M is the ratio of book value of equity to the market capitalization of equity. We
report time series average of the portfolio B/M ratios. For the Fama–French
portfolios, we report $e^{\log(B/M)}$, where $\log(B/M)$ is the time-series average of the
natural log of the portfolio B/M. We report the average market capitalization of the
stocks in each B/M ranked portfolio. For the Fama–French portfolios, we estimate
the average portfolio market capitalization from the time-series average of the
natural logarithm of portfolio market values reported in their table IV.

Return is the time series average of the equal- or value-weighted annual portfolio
returns in percent. Annual returns for the Fama–French portfolios are calculated by
compounding the monthly average portfolio returns reported in their Table IV.

while data on the B/M factor only go back to 1963. Fama and French (1996)
find that the three-factor model adequately captures the returns on the long-
term contrarian strategy. Not surprisingly, the loading on the B/M factor
declines monotonically as one moves from the 'loser' decile up to the 'winner'
decile. Thus, returns on the contrarian strategy over the longer time period
since 1931 should have a bearing on our beliefs about future performance of
the B/M strategy.

KSS report excess-return (one-factor) market model alphas for losers and
winners based on five-year ranking periods. Interestingly, their numbers imply
that the Jensen alpha for one-year post-ranking returns on this contrarian
strategy is negative and more than 13 percentage points lower in the pre-

1963 period than the post-1962 period (–6.9% vs. 6.5%)![7] This striking observation, in conjunction with the contrarian-B/M parallel, raises doubts that the future performance of the B/M strategy will meet the standard of the past 30 years. On the other hand, additional relevant information is provided by Davis (1994), who finds a positive and statistically significant B/M effect over the period 1941–1962. While this is good news for B/M, it is also important to recognize that the Davis study is limited to a sample of large firms for which the coefficient on B/M, like ours for the 500 largest firms, is about half that of FF (1992). A more formal analysis of this broader conditional perspective on strategies and factor premia is clearly called for, perhaps incorporating the analytical approach of Stambaugh. At a minimum, we hope our discussion engenders a healthy skepticism toward strategies that have performed impressively over a given 30–year period.

3.3   Relation of B/M to other determinants of expected returns

Having played 'devil's advocate' on the B/M issues, we still come away with the impression that there is an effect, although one that is smaller and less consistent than suggested by the early evidence. We now offer some comments on its relation to the older size effect, and the likelihood of the B/M effect persisting in the future.

First, we note that size and B/M can be very highly negatively correlated for a set of assets. The size-B/M correlation for the 100 size-then beta portfolios used earlier is –0.95 over the period 1963–92. Moreover, the correlation between size and estimated beta for these portfolios is not that much higher than the correlation between B/M and beta, –0.61 as compared to 0.51. Thus, as has been argued previously for size (Chan and Chen 1988), part of the B/M effect may be due to its proxying for the true, but imprecisely estimated, market beta. On the other hand, this conclusion is sensitive to the particular method of portfolio formation. For example, looking at the B/M-sorted deciles in FF (1992), there is a strong positive relation between average return and B/M, but no apparent relation to beta.

Huson (1995) considers a possible relation between liquidity and B/M. He finds that for portfolios of NYSE stocks formed on bid-ask spread and prior five-year return, inclusion of the five-year prior return in cross-sectional regressions eliminates the statistical significance of size, B/M, and spread, while prior return has a t-statistic of –3.23 over the period 1961–90. This, in combination with our earlier observations concerning the five-year contrarian strategy raises further doubts about the B/M effect. A recent paper by

---

[7]Similar results for five-year post-ranking performance are presented in Ball, Kothari and Shanken (1995).

Brennan, Chordia and Subrahmanyan (1999) documents a strong negative relation between trading volume and average returns, even after adjusting for risk using the FF three-factor model. This is consistent with a liquidity premium explanation.

3.4 Risk or mispricing?

FF have explored the possibility that a B/M factor might be related to shifts in the investment opportunity set, a form of risk that would be priced in the intertemporal CAPMs of Merton (1973) and others. Some support for a risk-based view is provided in Lewellen (1998). Others, notably Lakonishok, Shleifer and Vishny (1994), Haugen (1995), MacKinlay (1995) and Daniel and Titman (1997), argue against a risk-based explanation for the cross-sectional significance of B/M, however.[8] Further doubts are raised by Hawawini and Keim (1999), who point to the concentration of B/M premia in January. Consistent with mispricing stories, Kothari and Shanken (1997) find that the time-series variation in expected market index returns implicit in regressions on B/M is so large as to cast doubt on explanations based solely on investor rationality. Their post-1940 evidence appears more consistent with market efficiency, however.

The documentation of B/M effects in several countries (Chan, Hamao and Lakonishok 1991, and Capaul, Rowley and Sharpe 1993) is an impressive finding that would seem, at first glance, to support the mispricing story. This is further supported by the fact that the effects are not very highly correlated across countries (Hawawini and Keim 1999). However, insofar as markets are not well integrated, these observations are also consistent with the possibility that B/M is proxying for other rational, but mismeasured or omitted, determinants of expected returns. Although difficult to sort out, a combination of risk-based and mispricing effects may well be the reality, making extreme views counterproductive, and wrong.

4   Conclusions

Annual betas 'work' over the periods 1941–1990 and 1927–1990, in the sense that they are significantly related to average stock returns. Although other variables like size also have explanatory power, the differences in expected returns based solely on annual beta, and those based on size and beta, are surprisingly small. We're not sure why annual beta works, however, and understanding this is an important topic for future research. In the meantime, though, we think this observation provides a ray of hope for those impressed

---

[8]See FF (1996a) for a defense of the risk perspective against some of the challenges.

with the compelling logic of modern portfolio theory. Another important related issue is how to best estimate betas from past data in a way that, while retaining the mysterious benefits of annual beta, can be used in a predictive manner for financial decision making.

B/M is also significantly related to expected returns and, over the 1963–92 period, thoroughly dominates any estimate of beta that we've seen. However, the B/M effect is much weaker and far less reliable, statistically, in large firms. In addition, the inconsistent performance of the related contrarian strategy over a longer history raises doubts that the very strong B/M effect observed for the broader universe of stocks in recent history will persist undiminished in the future.

Some evidence (Daniel and Titman 1997) points to the B/M ratio as the more fundamental determinant of expected return, in comparison with the B/M risk factor loading. However, our evidence and that of Loughran (1997) concerning the weaker cross-sectional relation for large firms suggests that sole reliance on the ratio itself is not advisable. Evidence on the related issue of the rationality of the B/M effect is mixed. If the B/M effect is indeed related to risk, liquidity, etc., it can reasonably be expected to persist to some degree in the future. If it has been driven largely by mispricing, though, it obviously is liable to self-destruct, particularly given the enormous attention it has received. If this happens, we will likely be right back where we 'started' – with beta.

It is important to remember that beta captures the marginal contribution of an investment to portfolio risk whether it is related to expected return or not. Therefore, investors who believe that systematic mispricing has driven expected returns and will continue to do so in the future should load up on low-beta, high B/M securities, and increase their expected returns while reducing their portfolio risks. Just be sure to be among the first to 'get through the door.'

References

Amihud, Yakov and Haim Mendelson (1986) 'Asset pricing and the bid-ask spread', *Journal of Financial Economics* **17** 223–249.

Ball, Ray, S.P. Kothari and Jay Shanken (1995) 'Problems in measuring portfolio performance: An application to contrarian investment strategies', *Journal of Financial Economics* **38** 79–107.

Banz, Rolf W. (1981) 'The relationship between return and market value of common stocks', *Journal of Financial Economics* **9** 3–18.

Breen, William J. and Robert A. Korajczyk (1994) 'On selection biases in book-to-market based tests of asset pricing models'. Working paper no. 167, Northwestern University.

Brennan, Michael J. (1970) 'Taxes, market valuation and corporate financial policy', *National Tax Journal* **23** 417–427.

Brennan, Michael J., Tarun Chordia and Avanidhar Subrahmanyan (1999) 'Alternative Factor Specification, Security Characteristics, and the Cross-Section of the Expected Stock Returns', *Journal of Financial Economics* **49** 345–363.

Campbell, John Y. (1987) 'Stock returns and the term structure', *Journal of Financial Economics* **18** 373–400.

Capaul, C., I. Rowley and W. Sharpe (1993) 'International value and growth stock returns', *Financial Analysts Journal* January-February, 27–36.

Chan, K.C. and Nai-fu Chen (1988) 'An unconditional asset-pricing test and the role of firm size as an instrumental variable for risk', *Journal of Finance* **43** 309–325.

Chan, Louis K.C., Yasushi Hamao and Josef Lakonishok (1991) 'Fundamentals and stock returns in Japan', *Journal of Finance* **46** 1739–1764.

Chan, Louis K.C., Narasimhan Jegadeesh and Josef Lakonishok (1995) 'Issues in evaluating the performance of the value versus glamour stocks', *Journal of Financial Economics* **38** 269–296.

Chan, Louis K. C. and Josef Lakonishok (1993) 'Are the reports of beta's death premature?', *Journal of Portfolio Management* **19** 51–62.

Chopra, Navin, Josef Lakonishok and Jay Ritter (1992) 'Measuring abnormal performance: Do stock prices overreact?, *Journal of Financial Economics* **31** 235–268.

Collins, Daniel W., S.P. Kothari, Jay Shanken and Richard G. Sloan (1994) 'Lack of timeliness and noise as explanations for the low contemporaneous return-earnings association', *Journal of Accounting and Economics* **18** 289–324.

Cohen, K., G. Hawawini, S. Maier, R. Schwartz and D. Whitcomb (1983) 'Frictions in the trading process and the estimation of systematic risk', *Journal of Financial Economics* **12** 263–278.

Daniel, Kent and Sheridan Titman (1997) 'Evidence on the characteristics of cross-sectional variation in stock returns', *Journal of Finance* **52** 1–33.

Davis, James L. (1994) 'The cross-section of realized stock returns: The pre-COMPUSTAT evidence', *Journal of Finance* **49** 1579–1593.

DeBondt, Werner F. M. and Richard H. Thaler (1985) 'Does the stock market overreact?', *Journal of Finance* **40** 793–805.

DeBondt, Werner F. M. and Richard H. Thaler (1987) 'Further evidence on investor overreaction', *Journal of Finance* **42** 557–581.

Fama Eugene F. and Kenneth R. French (1992) 'The cross-section of expected returns', *Journal of Finance* **47** 427–465.

Fama Eugene F. and Kenneth R. French (1993) 'Common risk factors in the returns on stocks and bonds', *Journal of Financial Economics* **33** 3–56.

Fama Eugene F. and Kenneth R. French (1995) 'Size and book-to-market factors in earnings and returns', *Journal of Finance* **50** 131–155.

Fama Eugene F. and Kenneth R. French (1996a) 'Multifactor explanations of asset pricing anomalies', *Journal of Finance* **51** 55–84.

Fama Eugene F. and Kenneth R. French (1996b) 'The CAPM is wanted, dead or alive', *Journal of Finance* **51** 1947–1958.

Handa, Puneet, S.P. Kothari and Charles E. Wasley (1989) 'The relation between the return interval and betas: Implications for the size effect', *Journal of Financial Economics* **23** 79–100.

Harvey, Campbell (1989) 'Time-varying conditional covariance in tests of asset pricing models', *Journal of Financial Economics* **24** 289–318.

Haugen, Robert (1995) *The New Finance: the Case against Efficient Markets* Prentice Hall.

Hawawini, Gabriel and Donald B. Keim (1999) 'The cross-section of common stock returns: A review of the evidence and some new findings'. This volume, 3–43.

Huson, Mark (1995) *Bid-ask Spreads and Expected Returns: Testing the Transaction Cost Asset-Pricing Model*, PhD thesis, University of Rochester.

Kandel, Shmuel and Robert F. Stambaugh (1995) 'Portfolio inefficiency and the cross-section of expected returns', *Journal of Finance* **50** 157–184.

Keim, Donald B. (1989) 'Trading patterns, bid-ask spreads and estimated security returns: The case of common stocks at calendar turning points', *Journal of Financial Economics* **25** 75–98.

Kim, Dongcheol (1995) 'The errors-in-variables problem in the cross-section of expected returns', *Journal of Finance* **50** 1605–1634.

Kim, Dongcheol (1997) 'A reexamination of firm size, book-to-market and earnings price in the cross-section of expected stock returns', *Journal of Financial and Quantitative Analysis* **32** 463–489.

Kothari, S.P. and Jay Shanken (1997) 'Book-to-market, dividend yield and expected market returns: a time-series analysis', *Journal of Financial Economics* **44** 169–203.

Kothari, S.P., Jay Shanken and Richard G. Sloan (1995) 'Another look at the cross-section of expected returns', *Journal of Finance* **50** 185–224.

La Porta, Rafael (1993) 'Survivorship bias and the predictability of stock returns in the COMPUSTAT sample'. Working paper, Harvard University.

Lakonishok, Josef, Andrei Shleifer and Robert W. Vishny (1994) 'Contrarian investment, extrapolation and risk', *Journal of Finance* **49** 1541–1578.

Levhari, David and Haim Levy (1977) 'The capital asset pricing model and the investment horizon', *Review of Economics and Statistics* **59** 92–104.

Lewellen, Jonathan (1998) 'The time-series relations among expected return, risk, and book-to-market', *Journal of Financial Economics*, forthcoming.

Litzenberger, Robert H. and Krishna Ramaswamy (1979) 'The effect of personal taxes and dividends on capital asset prices: theory and empirical evidence', *Journal of Financial Economics* **7** 163–195.

Lo, Andrew W. and A. Craig MacKinlay (1990) 'When are contrarian profits due to stock market overreaction?', *Review of Financial Studies* **3** 175–206.

Loughran, Tim (1997) 'Book-to-market across firm size, exchange and seasonality: is there an effect?', *Journal of Financial and Quantitative Analysis* **32** 249–268.

MacKinlay, A. Craig (1995) 'Multifactor models do not explain deviations from the CAPM', *Journal of Financial Economics* **38** 3–28.

Mech, Timothy S. (1993) 'Portfolio return autocorrelation', *Journal of Financial Economics* **34** 307–344.

Merton, Robert C. (1973) 'An intertemporal capital asset pricing model', *Econometrica* **41** 867–887.

Roll, Richard and Stephen A. Ross (1994) 'On the cross-sectional relation between expected returns and betas', *Journal of Finance* **49** 101–121.

Scholes, Myron and Joseph T. Williams (1977) 'Estimating betas from nonsynchronous data', *Journal of Financial Economics* **5** 309–327.

Schwert, G. William and Paul Seguin (1990) 'Heteroskedasticity in stock returns', *Journal of Finance* **45** 1129–1155.

Shanken, Jay (1982) 'The arbitrage pricing theory: is it testable?' *Journal of Finance* **37** 1129–1140.

Shanken, Jay (1990) 'Intertemporal asset pricing: an empirical investigation', *Journal of Econometrics* **45** 99–120.

Shanken, Jay (1992) 'On the estimation of beta-pricing models', *Review of Financial Studies* **5** 1–33.

Shanken, Jay (1996) 'Statistical methods in tests of portfolio efficiency: a synthesis'. In *Statistical Methods in Finance: Handbook of Statistics* **14**, North Holland, 693–711.

The Psychology of Underreaction and Overreaction in World Equity Markets

Werner F.M. De Bondt

Abstract

Two key elements of a new psychological theory of stock prices are the notions of 'mental frames' and 'heuristics'. How people seek out, interpret and act upon news depends on their beliefs. Investors' perceptions of value are socially shared. Popular models may be false, however, and yet resist change. We review the international evidence relating to the profitability of contrarian and momentum strategies in equity markets. Securities are sorted into portfolios on the basis of past share price performance or related criteria that proxy for investor sentiment. The data support the behavioral theory of over-and underreaction to news. Investor psychology is an important element in the dynamics of asset prices.

Introduction

Wherever we go these days, people want to discuss the latest twists and turns in world equity markets. Is America a 'bubble economy'? Is the current level of the Dow Jones justified by expected corporate earnings and low inflation, or is it driven by the public's unthinking forecast of perpetual economic bliss? Why are investors so bullish in Germany or Switzerland, even though these nations suffer from low growth and high unemployment? Why did a financial crisis abruptly break off the economic miracle in Asia? Yet, why did Asian stock markets quickly, but not fully, rebound after the initial fall?

Nearly everyone agrees that changes in stock prices, even very impressive changes, elude easy interpretation.[1] However, economists often approach the question differently than other people do. They stress the rationality of markets, whereas journalists, money managers, and even central bankers stress the psychology and (sometimes predictable) foolishness of traders. A centerpiece of modern finance is the efficient markets hypothesis in which prices do not deviate from intrinsic values in any systematic way. The marginal investor, it is believed, updates the probabilities of uncertain future events in rational Bayesian fashion and trades accordingly. Thus, it is not possible to

---

[1]For example, Robert Solow, a Nobel-prize winning economist, recently commented in the *International Herald Tribune* that "it is hard to explain why the Dow went from 6,000 to 9,000" (May 7, 1998).

contrive a trading rule that promises superior profits based on facts to which the public has access at no cost.

Finance theory almost completely ignores the complex cognitive and motivational factors that guide trading decisions. In contrast, this article explores 'why psychology matters' and why a behavioral approach is a productive way of thinking about issues of asset pricing. The central problem that I discuss is how the behavior of traders shapes the dynamics of stock prices—in particular, how investor sentiment sometimes causes price reversals and sometimes causes price momentum. Thus, the broad themes that motivate past research on the psychology of financial markets are reviewed. In addition, I present an overview of the evidence, in world equity markets, that documents the performance of trading rules that are based on contrarian and momentum strategies. Securities may be sorted into portfolios on the basis of past share price performance or related criteria, e.g., past earnings growth or book-to-market value ratios. As I argue, the international evidence largely supports a psychological interpretation of the dynamics of stock prices.

Three perspectives

What are the links between stock prices and new information? Over the decades, people have thought about this question in various ways. Three responses have emerged. The first response defines the efficient markets hypothesis: 'The price is right'. That is, on average, rational market prices correctly reflect all information about the firm (Fama, 1970).[2] The second response is that the relationship between prices and economic values, if it exists, does not mean very much. The market has a life of its own. In the words of John Maynard Keynes, prices are driven by "animal spirits."[3] The third response is the one that receives most support from empirical research in behavioral finance. It resembles Isaac Newton's law of universal gravitation: What goes up must come down. Applied to the stock market, this law means that over time prices tend to revert to value. In the short run, however, systematic disparities predicted by investor psychology may arise between the two.

The three perspectives on asset valuation have different implications for money management. The price-is-right answer suggests that an indexing strategy is best since "you can't beat the market." The animal-spirits view is fascinated by technical analysis and the study of investor sentiment (Pring, 1991, 1993). Newton's law suggests that one should pursue value-based investment strategies and fundamental analysis in the style of Benjamin Gra-

---

[2]Thus, predictable valuation errors are excluded but not random errors.

[3]In his 1998 book, George Soros dwells on this point and on what he calls the "reflexivity of financial markets." According to Soros, financial markets do not tend toward equilibrium but they are "given to excesses and if a boom/bust sequence progresses beyond a certain point [the pendulum] will never revert to where it came from" (p. xvi).

ham and David Dodd (1934). It is interesting that two of the three approaches recommend that investors pay careful attention to human behavior.

The empirical and theoretical challenge

Since the early 1960s, however, modern finance has counseled the opposite, i.e., that the details of investor behavior—i.e., what traders actually do—are not important for market behavior. Standard asset pricing models are based on the twin assumptions of 'perfect markets' in equilibrium and a representative agent who behaves like 'homo economicus', i.e., he adheres to the axioms of rationality that underlie expected utility theory, Bayesian learning, and rational expectations.[4]

Yet, there is a long list of empirical market anomalies that are inconsistent with the standard models. The theory fails in its predictions. There are various types of anomalies. Some have to do with the failure of dividend discount models. It seems that, for a century, the volatility of market indexes, such as the Standard & Poor's index, has not been validated by subsequent movements in dividends (Shiller, 1989).[5] Other anomalies have to do with the risk-return tradeoff. Over the long run, stocks outperform bonds by a surprisingly large margin (Mehra and Prescott, 1985).[6] The data also con-

---

[4]For details, see Fama and Miller (1972). Modern finance does not assume that every investor is fully rational. However, with arbitrage, individual irrationality does not have to lead to irrationality at the market level. Shleifer and Vishny (1997) discuss the limits to professional arbitrage. They state that "the efficient markets approach to arbitrage [is] based on a highly implausible assumption of many diversified arbitrageurs. In reality, arbitrage resources are ... concentrated in the hands of a few investors that are ... specialized in trading a few assets, and are far from diversified" (p. 52). Shleifer and Vishny believe that arbitrage may "become ineffective ... when prices diverge far from fundamentals" (p. 35) and that the ineffectiveness can account for the glamour/value anomaly in equity prices (p. 53). In earlier work, Grossman (1989) argues that "the assumptions that all markets, including that for information, are always in equilibrium and always perfectly arbitraged are inconsistent when arbitrage is costly" (p. 91). For discussion of how irrational noise traders may influence asset prices, see Shleifer and Summers (1990), Shefrin and Statman (1994), Palomino (1996), and Odean (1998).

[5]It could be, however, that the volatility in equity markets reflects how the returns that investors require to hold stocks vary through time, e.g., with movements in the business cycle. Also, the Shiller volatility tests may be misspecified. For a review of the debate, see LeRoy (1989).

[6]To explain the difference in returns, the representative investor has to be extraordinarily risk-averse—with a coefficient of relative risk aversion over 30. As implausible as it sounds, this level of risk-aversion means that one feels indifferent between (1) a coin flip that either pays $50,000 (heads) or $100,000 (tails) or (2) a certain payment of $51,209. There are other explanations for the equity premium puzzle, e.g., investor may have been rationally concerned with a catastrophe that did not happen, or there is 'ex post selection bias'. The estimates of the equity premium are for the United States. Foreign data may yield less extreme findings. Even for the US, the equity premium is not nearly as large when data for the 19th century are considered (Siegel, 1992, 1994). Real bond returns were much higher at that time than during the period after 1926. Unanticipated inflation and

tradict the notion of beta-risk defined by the capital asset pricing model. In the cross-section of firms, returns on equity move with market capitalization and with the ratio of market value to book value (Fama and French, 1992; Hawawini and Keim, this volume). No one has a good story why this happens, however. Finally, there is a long list of anomalies that relate to the time-series dynamics of asset returns, seasonalities, the reaction of prices to corporate financing decisions, the pricing of initial public offerings of equity (IPOs), the pricing of closed-end mutual funds, and so on.[7]

The standard theory also fails in its assumptions. For many years, psychologists have amassed experimental evidence that "economic man ... is very unlike a real man" (Edwards, 1954, p. 382). The literature abounds with laboratory settings where central axioms of rationality such as frame invariance, dominance, or transitivity are violated. These violations are systematic, robust, and fundamental, i.e., they require new theory. "Reason," psychologists conclude, is not an adequate basis for a descriptive theory of decision-making (Tversky and Kahneman, 1986). Studies of financial decision-making by individuals and households confirm this negative conclusion. De Bondt (1998) lists four classes of anomalies that have to do with investors' perceptions of the time-series process of asset prices, perceptions of asset value, portfolio choices, and trading practices. What is surprising is the failure of many people to infer basic investment principles from years of experience, e.g., the benefits of diversification.

How should we react to this state of affairs? The answer of many financial economists is to question the relevance of experiments. Laboratory research, they emphasize, may lack ecological validity and may not predict actual decision-making when much wealth is at stake. Also, in empirical studies, the quality of the data is sometimes suspect and the findings may be artifacts of research design errors (Ball, 1995; Ball et al., 1995). Fama (1998) calls attention to the fragility of some long-term return anomalies. The apparent findings of price momentum and reversals are sensitive to methodology and, in Fama's view, they may be chance results. In addition, it is often said that, if the data are many, one cannot exclude data mining. But, if the data are few, the rational theory has not been rejected and may still serve as a starting point. Finally, champions of modern finance formulate more complex rational theories that may yet account for the observed anomalies. While

---

other historical factors may explain the low interest rates after 1926. A final explanation relies on non-standard investor preferences, e.g., habit formation or myopic loss aversion (see, e.g., Benartzi and Thaler, 1995).

[7]Fama (1991, 1998) reviews the evidence. Merton Miller agrees, it appears, with my characterization of the evidence. In a 1994 interview with *The Economist* (April 23), he says that "the blending of psychology and economics ... is becoming popular simply because conventional economics has failed to explain how asset prices are set." Miller adds, however, that he believes that the new mix of psychology and finance "will lead nowhere."

these efforts are provocative, we should avoid false conclusions. It is not so that "if behavior can still be rationalized, it must be rational." Neither, of course, should we accept 'overly flexible' behavioral theories. If we explain everything, chances are that in fact we explain nothing.

Given the shortcomings of modern finance, I believe that the challenge is to develop new and better theories of asset pricing. Studying the psychology of investors is one alternative. Clearly, the behavioral approach that I advocate is quite different from a perspective that emphasizes perfect markets and perfect people. It also stands in contrast to the growing literature on market micro-structure. The institutionalist perspective tries to model market frictions but regards the marginal trader to be fully rational. (The notion of 'noise traders'—if present—is used to close the model, as a *deux ex machina.*) Thus, it is assumed that, if information is asymmetric and some people know more than others, those who know less are aware of this fact and they act strategically in full recognition that they know less. Behavioral finance, on the other hand, assumes that the less-informed or ignorant noise traders are indeed ignorant but do not appreciate that they are ignorant.

The psychology of beliefs

Psychologists have developed a series of new concepts under the general heading of "bounded rationality" (Simon, 1983). Under full rationality, a utility is assigned to each possible state-of-the-world and the economic agent chooses what is best. Under bounded rationality, the individual does not contemplate, in every instant, the whole range of possible actions that lie before him. Task complexity, attention, the cost of thinking, memory, habit, social influences, emotion and visceral responses all contribute to the decisions that are made. As a result, there are systematic differences between 'what people do' and 'what people should do'—from a normative perspective.

Various new theories of decision-making have been formulated. For example, in the psychology of choice, an alternative to expected utility theory is prospect theory (Kahneman and Tversky, 1979). The theory emphasizes the effects on choice of problem editing, reference points, loss aversion, and small probabilities. In the psychology of judgment, the dual notions of 'mental frames' and 'heuristics' are critical building blocks (Tversky and Kahneman, 1974). The central insight of all behavioral theories is that decision process influences decision outcome.[8]

The effect of judgment on asset prices is a product of the mental frames or beliefs about company value that investors hold. The interpretation of past events and the prediction of future scenarios always happens in a broader

---

[8]Slovic (1972), Earl (1990), De Bondt and Thaler (1995), Kahneman and Riepe (1998), and Rabin (1998) discuss a selection of psychological findings relevant to finance. For an introduction to cognitive and social psychology, I refer the reader to Nisbett and Ross (1980), Kahneman *et al.* (1982), and Fiske and Taylor (1991).

context. It is also of great importance how traders incorporate new information into the current frame. Because perceptions influence decisions, several attributes of beliefs are important to keep in mind. First, people do not create or use many frames that are uniquely their own. Concepts and frameworks are socially shared, e.g., through stories in the news media, conversation, and tips from friends and advisors. That is why we can speak over dinner about the troubles in Kosovo without ever having been there or even knowing anyone who has.[9] Second, beliefs differ greatly in sophistication. History teaches the power of flawed ideologies, false beliefs, and superstition (Mackay, 1841). For instance, when California housing prices were skyrocketing during the late 1980s, the man in the street explained what happened by a 'shortage illusion'. The illusion was that price increases and shortages would continue without limit, simply because "California is a good place to live" (Shiller, 1990). As a further illustration, I should mention how once I watched an afternoon television talk show about the savings and loan crisis. One of the people in the audience said, "The taxpayers shouldn't pay for this mess. The government should." Many financial economists surely overestimate the economic sophistication of the public. Finally, beliefs do not change easily—even in the face of conflicting evidence. People have an immense capacity to rationalize facts and fit them into a pre-existing belief system (Edwards, 1968). Confirmatory evidence is taken at face value while disconfirmatory evidence is subjected to skeptical scrutiny.[10] Forecasts of inflation demonstrate how powerful belief perseverance is. One way to interpret the low real returns on bonds in the 1970s is that most people, including experts, never imagined that inflation would rise as much as it did. Similarly, the high real returns during the 1980s may have resulted from the conviction that inflation was here to stay (De Bondt and Bange, 1992).

The psychological literature leaves little doubt that the quality of human inference can be improved (see, e.g., Nisbett and Ross, 1980, or Kahneman *et al.*, 1982). It may be thought that, in view of these shortcomings, people would exhibit appropriate caution concerning their judgmental abilities. However, many studies show that this is false. People are prone to overconfidence. This observation raises questions about the link between learning and experience.

Numerous factors restrict the ability to learn. It is well-known that prior

---

[9]This is also why people's (average) perceptions of the world differ from the (average) perception of their personal life space. For example, based on weekly data from ABC/*Money Magazine* polls (1986–1996), Mutz (1998) shows that Americans' perceptions of the state of the US economy have been "consistently and without exception" more negative than their views of personal financial conditions (p. 125).

[10]Indeed, the experiments of Lord *et al.* (1979) show how it may happen that two opposing parties draw support for their divergent views from the same corpus of findings. For further discussion, see Nisbett and Ross (1980).

expectations of relationships can lead to faulty inference, or 'illusory correlation'. For instance, our initial beliefs influence how we seek out and interpret new information. In addition, because of hindsight bias, outcomes often fail to surprise people as much as they should. People also tend to attribute success to skill and failure to bad luck. Finally, there are 'outcome irrelevant learning structures' which reinforce poor inferences with positive outcome-feedback. As a consequence, their validity is not questioned.

That shared beliefs affect market prices, often the wrong way, is evident from a careful study of business history. A good example is U.S. corporate restructuring. In retrospect, it is surprising that the merger and acquisition wave of the 1960s, when many companies diversified into new activities, was followed by the break-up wave of the 1980s and the 1990s. One wonders whether the initial M&A wave was largely in error. Profit data certainly suggest that it was. The management gurus of the 1960s loved diversification and saw it as a big plus for firm value whereas today, their buzzword is 'focus'. What is striking, however, is that the stock market apparently took the gurus seriously, not once but twice. Studies show that stock prices of bidder firms reacted favorably to acquisition news in the 1960s but unfavorably in the 1980s.[11] It is a perilous practice to judge the value of a long-term investment decision by the whim of a short-term price reaction.[12]

Overreaction

How do stock prices react to news? The answer depends in part on how investors' perceptions of company value and future earnings are influenced by new information. There are always two effects. The first effect has to do with the short-term impact of the news in light of the information already impounded in prices. The second effect depends on how the news changes the cognitive frame itself. At times, seemingly minor pieces of news trigger a change in mental frame and cause a big price reaction.

An important question in this context is the quality of financial forecasts. How valid are expert and amateur predictions of share prices, earnings, etc.? How do people go about making these forecasts? What type of information attracts the most attention? A recurring theme in the literature is the disposition to predict the future based on the recent past. People find it difficult

---

[11]Matsusaka (1990) reports the results for the 1960s and early 1970s. On average, bidder companies earned positive excess returns upon the public announcement of unrelated acquisitions.

[12]In April 1991, *The Economist* described the conglomerate merger wave as the "biggest collective error ever made by American business." Baker (1992) questions this assessment. He suggests that "acquirors were buying smaller companies that valued the ... resources that these acquiring firms could offer." The fact that many transactions were reversed later "is not evidence ... of foolishness ... [since] changes in financing technology and managerial sophistication ... can explain these reversals" (p. 1118).

to project anything that is greatly different from the apparent trend—even if over-optimistic forecasts and groundless confidence are the net result. Kahneman and Lavallo (1993) call this practice the "inside" view of forecasting. The inside view directs attention to the unique complexities of the case at hand and formulates specific future scenarios (e.g., obstacles and solutions). The forecasts overreact to facts that appear prominent in a narrow frame. In contrast, the 'outside' view directs attention to base-rate information, i.e., statistics for an ensemble of comparable cases. Even if decision-makers have access to base rates, they often ignore them. Non-Bayesian forecasting probably results from the use of the representativeness heuristic. This heuristic rule judges probability by similarity and causes predictable judgment errors (Tversky and Kahneman, 1974).[13]

Security analysts' earnings forecasts are a good example of non-Bayesian forecasting (De Bondt and Thaler, 1990). The forecasts are persistently wide off the mark. Nothwithstanding their large errors, analysts keep offering predictions that are too extreme. In addition, the data show optimism bias as well as serial correlation in forecasts errors.[14] A similar phenomenon applies to stock price forecasts made by individual investors. For several years, the American Association of Individual Investors has asked a random sample of its members for a stock forecast every week. The data show that, just like subjects in controlled experiments, most individuals are upbeat in bull markets and gloomy in bear markets. The forecasts, however, have little or no predictive power (De Bondt, 1993). A further example of how representativeness affects judgment in financial matters has to do with the inferior long-run performance of IPOs and the so-called 'next' syndrome. Investment bankers find it attractive to sell IPOs as the 'next' Microsoft or the 'next' Intel—especially if the start-up company is small and does not have a long historical earnings record. The data agree with the notion that "many firms go public near the peak of industry-specific fads" (Ritter, 1991).[15]

---

[13]One such error is the conjunction fallacy. A well-known experiment that demonstrates this error is the one about 'Linda'. Linda is described as "31 years old, single, outspoken, and very bright. She majored in philosophy. As a student, she was deeply concerned with issues of discrimination and social justice and also participated in antinuclear demonstrations." Subjects are asked what is more likely: (i) Linda is a bank teller; (ii) Linda is a bank teller and is active in the feminist movement. Most people choose the second alternative (since it "looks more like Linda") even though this answer violates a basic law of probability. The conjunction of two events can never be more likely than the probability of either event alone.

[14]De Bondt and Thaler (1990) fail to identify the sources of analyst overreaction. It is not evident, for instance, that analysts extrapolate past earnings growth into the future.

The same issues have also been investigated with data for the United Kingdom. See, Capstaff et al. (1995), Forbes and Skerratt (1996), O'Hanlon and Whiddett (1991), and others. Without exception, the results cast doubt on the rationality of earnings forecasts.

[15]A different interpretation of the poor performance of IPOs, endorsed by Alan Greenspan, is that investors buy them more or less like they buy lottery tickets. When

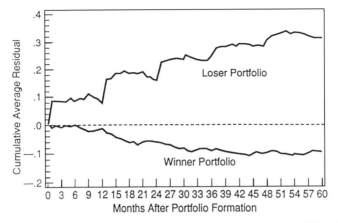

Figure 1. The Winner-Loser Effect in the United States, 1926–1982.

As far as I know, David Dreman (1982) is the originator of the phrase "overreaction". Dreman and others believed that there is exaggerated optimism in the stock market for firms with high price-earnings (PE) ratios and exaggerated pessimism for firms with low PE ratios. De Bondt and Thaler (1985) formulated a series of empirical tests that support the idea that overreaction bias affects stock prices. Figure 1 summarizes the initial, controversial study of the winner-loser effect. All companies listed on the New York Stock Exchange since December 1925 were examined. Thaler and I used past stock returns over a two- to five-year (portfolio rank) period as a proxy for investor sentiment and we predicted systematic price reversals. As Figure 1 shows, the portfolio of 50 NYSE stocks that did the worst over an initial five year period subsequently outperformed the 50 NYSE stocks that did the best. It seems that stock prices 'have a memory' since the difference in performance during the test period is, on average, about 8% per year, controlling for risk and other factors. The winner-loser effect was the first asset pricing anomaly predicted by a behavioral theory. Despite numerous past attempts to refute the winner-loser effect, hardly anyone doubts the statistical results. What continues is the struggle over the interpretation of the data.[16]

There is a related literature on 'value' and 'growth' (or glamour) investing

<hr>

they gamble, investors pay a premium for the small chance of a big gain—yet, they know that the expected payoff is negative.

[16]I will not review this debate here. See, however, two surveys by De Bondt and Thaler (1989, 1995) for a behavioral perspective. Over the years, it has been suggested that the winner-loser effect is due to tax effects, return seasonality, firm size, time-varying beta-risk, biases in computed returns, and other factors (see, e.g., Ball et al., 1995). Recent studies by Chopra et al. (1992), De Bondt (1992), Dreman and Berry (1995), Lakonishok et al. (1994), Loughran and Ritter (1996), and Shefrin and Statman (1997) have, on balance, strengthened the case for overreaction.

that can be traced back to Graham and Dodd (1934). Here, the universe of securities is sorted into portfolios on the basis of measures that compare intrinsic value with market price. Companies are ranked on their price-earnings ratios, book-to-market value of equity ratios, cash flow to price ratios, and so on. It is well-established that value stocks earn higher returns than growth stocks but it is not obvious that value stocks are more risky. See, e.g., Basu (1977), De Bondt and Thaler (1987), Fama and French (1992), Lakonishok *et al.* (1994), and many other studies.

Over time, the seeming profitability of contrarian strategies has been established for numerous countries and time periods. Table 1 summarizes the methods and results of thirteen different overreaction studies for ten world equity markets outside the US. An arbitrage portfolio that finances its purchases of past losers by selling past winners short earns positive returns in almost every case. The long-run predictability of returns even applies to country indexes. Vriezen (1996), Asness *et al.* (1997), De Bondt *et al.* (1998), and Richards (1998) sort the countries included in the *Morgan Stanley Capital International* indexes on the basis of past returns, capitalization measures, and so on. Countries that lag the world index over periods of three to five years tend to outperform the index during later years. Finally, Table 2 summarizes the results of three studies that look for international evidence on the relative performance of value and growth strategies (Brouwer *et al.*, 1996; Capaul *et al.*, 1993; Fama and French, 1997). The studies cover thirteen countries. The findings are unambiguous. On average, the value portfolio strongly outperforms the growth portfolio.[17] Hawawini and Keim, in this volume, also provide extensive international evidence on the relative merits of value vs. growth investing.

What causes the winner-loser effect? My favorite explanation is a theory of 'generalized overreaction'. At the center of this theory are mental frames that confuse attractive companies with attractive investments. The popular models are validated and reinforced by expert opinion, peer group consensus, and possibly the price action of securities—since traders detect imaginary trends. As it becomes public, news about the firm is blended into these mental frames and it causes further overreaction. Lastly, many investors simple-mindedly extrapolate past earnings trends into the future—even if, eventually, unusual runs in earnings growth, up or down, must end.[18]

---

[17]In a study of the Japanese market (1971–1988), Chan *et al.* (1991) also find that the book-to-market ratio has a reliably positive impact on expected returns.

[18]There are at least two more behavioral theories of the winner-loser effect. Both theories do not emphasize expectations of future cash flows. The first theory stresses how traders want to be paid for changing perceptions of risk. It may be, for instance, that investors require a 'regret' premium to purchase past loser companies because these firms 'look' more risky even if they are not.

The second theory is based on notions of herding and conformity. Perhaps investors

Table 1

CONTRARIAN STRATEGIES: EVIDENCE FOR 10 COUNTRIES			
Country	Period Sample	Length of Rank and Test Period Size of Extreme Portfolios	Arbitrage Portfolio Losers Minus Winners Annualized Returns
Australia (Brailsford, 1992)	1958–1987 330 Stocks	3 Years, Quintiles 3 Years, Deciles	3.6%* 5.6%*
Brazil (da Costa, 1994)	1970–1988 121 Stocks	2 Years, Quintiles	18.9%*
Canada (Kryzanowski/ Zhang, 1992)	1950–1988 From 137 (1950) to 1581 (1988) Stocks	3 Years, Deciles 5 Years, Deciles	4.9%* 6.4%*
Germany (Schiereck/De Bondt/ Weber, 1999) (Meyer, 1994)	1961–1991 ~210 Stocks 1961–1990 ~220 Stocks	5 Years, 20 Stocks 2 Years, Deciles 3 Years, Deciles 5 Years, Deciles	4.4%** 1.3%** 2.9%** 6.0%**
Malaysia (Ahmed/ Hussain, 1997)	1986–1996 166 Stocks	3 Years, Deciles	13.2%*
The Netherlands (Bos, 1991)	1985–1990 34 Major Stocks	2 Years, 5 Stocks	4.4%*
Spain (Alonso/Rubio, 1990)	1967–1984 ~80 Stocks	3 Years, 5 Stocks	12.3%***
Sweden (Karlsson/ Thoren, 1997)	1983–1996 ~90 Stocks	3 Years, 10 Stocks	−1.8% (1983–89)* 3.2% (1990–96)*
Switzerland (Dressendörfer, 1997)	1973–1996 197 Stocks	3 Years, Quintiles 5 Years, Quintiles	0.2%** 3.4%**
United Kingdom (Clare/Thomas, 1995) (Dissanaike, 1996)	1955–1990 1000 Stocks 1975–1991 500 Major Stocks	3 Years, Quintiles 4 Years, Deciles	1.6%* 24.7%** 10.8%***
(Forbes/ Kycriades, 1996)	1975–1993 1000 Stocks	3 Years, Deciles 5 Years, Deciles	2.3%* 17.5%** 6.2%* 7.6%**
*Rebalanced Raw Returns: **Buy-and-Hold Raw Returns: ***Beta-Adjusted Returns			

Table 2

VALUE VERSUS GROWTH STRATEGIES: EVIDENCE FOR 13 COUNTRIES			
Study	Period Countries	Criterion Variable Fraction of Sample in each Portfolio	Annualized Return of Value Minus Growth Stock Portfolio
Capaul/Rowley/ Sharpe (1993)	1981–1992 Europe[1] World[2]	B/M 50%	2.8%[1] 3.4%[2]
Brouwer/van der Put/Veld (1996)	1982–1993 F, G, NL, UK	E/P C/P B/M D/P 20%	5.0% 20.8% 10.0% 5.2%
Fama/French (1997)	1975–1995 A, B, CH, F, G, HK, I, J, NL, S, Si, UK, US	E/P C/P B/M D/P 30%	4.1% 6.6% 5.6% 2.8%

The countries are: A: Australia; B: Belgium; CH: Switzerland; F: France, G: Germany, HK: Hong Kong; I: Italy; J: Japan; NL: The Netherlands; S: Sweden; Si: Singapore; UK: United Kingdom, US: United States. Criterion Variable: B/M: Book/Market Ratio. E/P: Earnings-Price Ratio. C/P: Cash-flow-Price Ratio. D/P: Dividend-Price Ratio.

The value (growth) portfolio is the portfolio with companies that have high (low) E/P ratios, high (low) C/P ratios, high (low) B/M ratios, high (low) D/P ratios.

[1] Value-weighted portfolio of CH, F, G, and UK . [2] Value-weighted portfolio of CH, F, G, J, UK and US.

In a 1992 study, I first tested the overreaction-to-earnings growth hypothesis. I asked whether analysts' earnings forecasts could be used to earn abnormal profits. The period was 1976–1984 and I employed over 100,000 forecasts. Firms were ranked on the basis of analyst predicted earnings growth, over one-, two-, and five-year horizons. Apparently, an arbitrage strategy that buys the 20% of companies for which analysts are most pessimistic and finances the purchases by short-selling the 20% of companies for which analysts

---

spend most of their time observing and imitating the behavior of other traders (rather than gathering fundamental news). It seems that, ceteris paribus, investors are willing to pay more for what is familiar and comfortable. Many people keep large equity holdings in local firms (Huberman, 1997). Also, their portfolios lack international diversification (French and Poterba, 1991). Teh and De Bondt (1997) find that US equity returns depend on shareholder trading practices and identity. In the cross-section of companies, conventional stocks earn lower returns.

are most optimistic earns substantial profits. As predicted by the representativeness heuristic, the excess returns grow with the forecast horizon.[19]

Other aspects of the data are consistent with naive earnings extrapolation. For example, De Bondt and Thaler (1987) find that the earnings of winner (loser) firms, on average, show the same up and down (down and up) time-series pattern that is observed in stock returns.[20] Additionally, much of the differential price movement between winner and loser (or value and growth) stocks takes place within a window surrounding quarterly earnings announcements during the test period (Chopra *et al.*, 1992; La Porta, *et al.*, 1997). More evidence is gathered by Dreman and Berry (1995). These authors identify two types of earnings surprises: (i) 'trigger events' that go against prior expectations, and (ii) 'reinforcing events' that affirm prior expectations. For twenty years (1973–1993), Dreman and Berry rank companies by PE ratios—a proxy for market enthusiasm. They find that trigger events (good news for low PE stocks and bad news for high PE stocks) have a much larger absolute impact on prices than reinforcing events. Thus, bad earnings news damages stock prices but tends to be 'taken in stride' when expectations are low. Conversely, good news pushes prices up but less so for high PE companies.[21] Finally, Lakonishok *et al.* (1994) use the Gordon growth model to compare the growth forecasts for earnings and cash flow implied by market prices to the growth rates subsequently realized. The predictions of superior growth for glamour firms are borne out but only in the very short run. Beyond the first couple years, the growth rates of value and growth stocks are essentially identical.[22]

To repeat, many people equate a well-run firm with a good investment. The behavior of investors suggests stock market overreaction—more or less in the same way that voters approve or disapprove of politicians depending

---

[19]La Porta (1995) and Dechow and Sloan (1997) report similar findings based on similar data (i.e., analyst forecasts of five-year earnings growth provided by IBES Inc.) for the period 1981–1993. Contrarian investing is profitable because asset prices reflect analysts' long-term forecasts even though these predictions are systematically too extreme and too optimistic. Related findings appear in Abarbanell and Bernard (1992) and Bulkley and Harris (1997). In *The Intelligent Investor*, Benjamin Graham (1959) states that "no one really knows anything about what will happen in the distant future, but analysts and investors have strong views on the subject just the same" (p. 133). This comment agrees with the overreaction bias in earnings forecasts and stock returns.

[20]Tests for German firms listed on the Frankfurt Stock Exchange (1960–1991) yield similar results (Schiereck *et al.*, 1999).

[21]See also Basu (1978) and Dreman (1998). Bauman and Miller (1997) create portfolios of value and growth stocks for 1980–1993. They rank firms by PE-ratios, 5-year past growth rates in earnings, and other measures. On average, value stocks outperform growth stocks. Most earnings surprises are negative because of analyst optimism bias. Interestingly, however, the (standardized) earnings surprises are, on average, significantly less negative for value than for growth stocks.

[22]More than three decades ago, Rayner and Little (1966) made the same observation.

on the current state of the economy, or that other social fashions change. Thus, firms that enjoy rapid earnings growth or that somehow are glamorous enough to appear on the cover of major business magazines are seen as excellent investments. In contrast, companies that report losses or lose market share are ill-favored. Shefrin and Statman (1997) analyze the annual surveys of firm reputation published by *Fortune Magazine*. They find that, in the cross-section, reputation is inversely correlated with the ratio of book value to market value of equity, a statistic that is known to predict returns. In other words, on average, highly-reputed companies seem overpriced since they become poor stock market performers afterwards. Conversely, companies that look bad in the court of public opinion are bargains from an investment standpoint.[23]

Overreaction and underreaction

Once in place, popular models resist change. Studies of earnings announcements suggest market underreaction rather than overreaction. Bernard (1993) presents a comprehensive survey of the anomalous post-earnings-announcement drift in stock returns. If companies are ranked on the basis of standardized earnings surprises or of the returns that surround the earnings announcement window, companies with good earnings news are much better subsequent investments than are companies that report bad news (Bernard and Thomas, 1989; Foster *et al.*, 1984). The effect lasts for several months. Surprisingly, the strategy has consistently paid off for over a quarter of a century. The market behaves as if it discounts the earnings news—particularly at turning points. Around later announcements, prices react as if the market believes that earnings should mirror what they were for the corresponding quarter from the previous year (Bernard and Thomas, 1990).

The slow reaction to earnings announcements ('earnings momentum') is related to other evidence of stock price momentum reported by De Bondt and Thaler (1985), Jegadeesh and Titman (1993), and others. De Bondt and Thaler find that, for the 1926–1982 period, one-year past winners outperform one-year past losers by 7.6% per year. For 1965–1989, Jegadeesh and Titman find that a strategy which selects stocks based on their past six-month returns

---

[23]Two related papers are Clayman (1987) and Antunovich and Laster (1998). Clayman (1987) tracks the performance of the companies that were featured by Thomas Peters and Robert Waterman in their 1982 bestseller *In Search of Excellence*. Prior to 1980, so-called excellent companies scored high in return on sales, return on assets, asset growth, and other accounting measures of financial strength. However, these measures quickly reverted toward the mean during 1981–1985. The stock market performance of these companies was similarly disappointing. Antunovic and Laster (1998) analyze the same data as Shefrin and Statman (1997) but they reach opposite conclusions. They believe that "admired companies" outperform other companies and that the evidence agrees with market underreaction.

and holds them for six months, earns an average annualized return of 12.0%.[24] Again, as with the contrarian studies, the findings appear robust. Table 3 summarizes the methods and results of six momentum studies for thirteen world markets outside the US. Rouwenhorst (1998) examines twelve well-established equity markets. He finds that the returns to buying winners and selling losers are positive in each case. Nevertheless, the results of the momentum studies should be interpreted cautiously since the strategies are trading intensive.[25]

How do we square the overreaction results with the underreaction results? Are they contradictory? Logically, can both be true? The answer is yes. To repeat, large disparities between price and value may result from the wrong mental frame. Investors freely talk about 'growth firms' and 'declining industries' even though there is no evidence of any reliable time-series patterns in annual earnings changes (except in the tails of the distribution). All too often, the life-cycle metaphor proves persuasive. No wonder then that, when an earnings surprise hits, many investors refuse to believe it. A substantial part of the momentum effects in prices is concentrated around earnings announcements (Chan *et al.*, 1996). Thus, mental frames take time to adjust and the market responds only gradually to new information. Analysts seem to be particularly slow in adjusting their earnings forecasts.[26] Consistent with this interpretation of the data, it is found that, past stock market losers often experience positive earnings surprises. Similarly, past market winners report an unusual frequence of negative surprises (Chopra *et al.*, 1992). Another piece of evidence is that value strategies generally work well, except among very strong recent performers (Asness, 1997). The process is one of initial market mispricing and slow error correction. The market is slow to overreact.[27]

In recent years, alternative parsimonious models of the underreaction/

---

[24]It should be noted, however, that for short horizons (say, one day to one month) the evidence favors price reversals over momentum (De Bondt and Thaler, 1989). Internationally, there are similar results for Japan (Chang *et al.* 1995) and New Zealand (Bowman and Iverson, 1996). Market micro-structure effects (e.g., relating to the bid-ask spread or lead-lag effects between stocks) are the likely explanation.

[25]Chan *et al.* (1997) report momentum in stock price indexes for 23 countries. The study includes several emerging markets such as Thailand, Taiwan, or South Africa.

[26]Hong *et al.* (1998) find that the profitability of momentum strategies declines with firm size and that, holding size fixed, the strategies work well among stocks with low analyst coverage, particularly, past losers. These results agree with the slow diffusion of information. In addition, Lee and Swaninathan (1998) find that past trading volume predicts both the magnitude and persistence of price momentum. Moskowitz and Grinblatt (1998) observe that momentum has a strong industry component.

[27]I owe this insightful phrase to Robert Haugen (1999). Poteshman (1999) presents evidence from the option markets that agrees with the results for equity markets. It looks as if 'investors underreact to news that is preceded by a short period of similar news and overreact to news that is preceded by a long period of similar news'.

Table 3

MOMENTUM STRATEGIES: EVIDENCE FOR 13 COUNTRIES			
Country	Period Sample	Length of Rank and Test Period Size of Extreme Portfolios	Arbitrage Portfolios Winners Minus Losers Annualized Returns
Austria (Rouwenhorst, 1998)	1980–1995 60 Stocks	6 Months, Quintiles	11.2%**
Belgium (Rouwenhorst, 1998)	1980–1995 127 stocks	6 Months, Deciles	13.2%**
Canada (Kryzanowski/Zhang, 1992)	1950–1988 From 137 (1950) to 1581 (1988) Stocks	1 Year, Deciles	17.5%* 1.4%***
Denmark (Rouwenhorst, 1998)	1980–1995 60 Stocks	6 Months, Quintiles	13.1%**
France (Rouwenhorst, 1998)	1980–1995 427 Stocks	6 Months, Deciles	11.6%**
Germany (Bromann/Schiereck/ Weber, 1997)	1961–1991 ~210 Stocks	1 Year, 10 Stocks 1 Year, 40 stocks	7.9%** 3.2%**
(Meyer, 1994)	1961–1990 ~220 Stocks	1 Year, Deciles	9.2%**
(Rouwenhorst, 1998)	1980–1995 228 Stocks	6 Months, Deciles	8.6%**
Italy (Rouwenhorst, 1998)	1980–1995 223 Stocks	6 Months, Deciles	11.2%**
The Netherlands (Rouwenhorst, 1998)	1980–1995 101 Stocks	6 Months, Deciles	15.1%**
Norway (Rouwenhorst, 1998)	1980–1995 71 Stocks	6 Months, Quintiles	11.9%**
Spain (Rouwenhorst, 1998)	1980–1995 111 Stocks	6 Months, Deciles	15.8%**
Sweden (Rouwenhorst, 1998)	1980–1995 134 Stocks	6 Months, Deciles	1.9%**
Switzerland (Dressendörfer, 1997) (Rouwenhorst, 1998)	1973–1997 197 Stocks 1980–1995 154 Stocks	1 Year, Deciles 1 Year Quintiles 6 Months, Deciles	8.2%** 6.0%** 7.7%**
United Kingdom (Forbes/Kycriades, 1996)	1975–1993 1000 Stocks	1 Year, Deciles	8.4%* 10.4%**
(Rouwenhorst, 1998)	1980–1995 154 Stocks	6 Months, Deciles	10.7%**
*Rebalanced Raw Returns; ** Buy-and-Hold Returns			

overreaction findings and the dynamics of security prices have appeared. In bold theoretical papers, Odean (1998) and Daniel *et al.* (1998) start from the observation that investors are overconfident about their own ability. Daniel *et al.* believe that investors overreact to private information but underreact to public information. If it is assumed that public signals on average confirm private signals, a continuing overreaction may cause momentum in stock prices that is eventually corrected. Barberis *et al.* (1997) model how investors learn the stochastic process of earnings. They assume that earnings follow a random walk but that investors either believe that earnings are mean reverting or that earnings show trends. They update their beliefs in Bayesian fashion—even though their model of earnings is incorrect. This theory rationalizes some features of the data, e.g., that negative earnings surprises hit growth stocks more than other stocks but still insufficiently. Finally, Bloomfield *et al.* (1998) use experimental markets to test a model based on Griffin and Tversky (1992). Consistent with prior behavioral research, Bloomfield *et al.* find that people with evidence that is favorable but unreliable tend to overreact to information, whereas people with evidence that is somewhat favorable but reliable underreact. In particular, as it is assumed in Barberis *et al.* (1997) or De Bondt and Thaler (1985), investors overestimate what can be learned from a short sequence of earnings changes about the underlying earnings process.

Conclusions

In his classic 1970 paper, Fama stated that "research [on efficient markets] . . . did not begin with the development of a theory of price formation . . . rather the impetus came from the accumulation of evidence . . . that the behavior . . . of speculative prices could be well approximated by a random walk. Faced with the evidence, economists felt compelled to offer some rationalization. . . . In short, there existed a large body of empirical results in search of a rigorous theory." In many ways, thirty years later, we face the same search for rigor.

I have argued that "people are human" and that psychology plays a major role in the behavior of world financial markets. Modern finance, built on the logic of rational choice, helps our understanding of market behavior the most when the forces of arbitrage are strong. Consider, e.g., the explanatory power of the Black–Scholes option pricing formula. Yet, while we take pride in the progress of financial economics as a *science*, we should not forget that it is a *social* science. With costly arbitrage, behavioral factors are relevant. It would be unsound to model market behavior based on the assumption of common knowledge of rationality. " . . . The [stock] market is not a weighing machine, on which the value of each issue is recorded by an exact and impersonal mechanism," say Graham and Dodd. "Rather [it] is a voting machine, whereon countless individuals register choices which are the product partly of reason and partly of emotion" (1934, p. 23).

Hopefully, future research will throw more light on the inner workings of the 'voting machine' and on the links between market and decision-making anomalies. I agree with Fama (1998) that "any alternative model [to market efficiency] has a daunting task" and that "it must specify biases in information processing that cause the same investors to underreact to some types of events and overreact to others" (p. 284).[28]

Yet, I also believe that "a full understanding of human limitations will ultimately benefit the decision-maker more than will naive faith in the infallibility of his intellect" (Slovic, 1972). Much is learned by studying how people process data and solve problems. The behavioral approach reaffirms that good judgment is critical, in money management as well as in every other aspect of life. In the financial arena, there are usually no short-cuts, no simple ways to get rich quick, except with privileged inside information. Whether the techniques of fundamental (intrinsic value) analysis can yield abnormal profits is still somewhat unclear. Chances are that, for those who do not want to index their portfolios, the best advice is to live by Newton's Law. Nearly all the evidence that I know warns against buying glamour, i.e., against companies with high price-to-earnings ratios, highfliers in the stock market, and so forth. Similarly, most research finds wealth – if not virtue—in contrarian investing, and in going against the crowd.

Acknowledgements

I thank Donald Keim and William Ziemba for their comments and encouragement. I am also grateful to Jack Brush, William Forbes, Alex Fung, Andreas Kappler, Malou Lindholm, Esther Schweizer, and Marnix Vriezen. Earlier versions of this paper were presented at the 1995 Isaac Newton program in Financial Mathematics at Cambridge University, Hong Kong Baptist University, Credit Suisse in Zurich, the 1998 AIMR Annual Conference in Phoenix, and the 1998 IIR Behavioral Finance Conference in London. I thank the Institute for Empirical Research in Economics at the University of Zurich and the Frank Graner Professorship at the University of Wisconsin-Madison for financial support. Parts of this article are based on my previous manuscript, 'Investor Psychology and the Dynamics of Security Prices' (1995).

References

Abarbanell, J. and V. Bernard (1992) 'Tests of Analysts' Overreaction/ Underreaction to Earnings Information as an Explanation for Anomalous Stock Price Behavior', *Journal of Finance* **47** (3) 1181–1207.

---

[28]Every model has its own purpose, however. How much rationality we may reasonably assume will vary by context. Conlisk (1996) lists conditions such as deliberation costs, complexity, incentives, and market discipline.

Ahmed, Z. and S. Hussain (1998) 'Overreaction in KLSE Stock Returns', Working Paper, University of Newcastle.

Alonso, A. and G. Rubio (1990) 'Overreaction in the Spanish Equity Market', *Journal of Banking and Finance* **14** (3) 469–481.

Antunovich, P. and D.S. Laster (1998) 'Do Investors Mistake a Good Company for a Good Investment?', Working Paper, Federal Reserve Bank of New York.

Asness, C.S. (1997) 'The Interaction of Value and Momentum Strategies', *Financial Analysts Journal* **53** (2) 29–36.

Asness, C.S., J.M. Liew, and R.L. Stevens (1997) 'Parallels Between the Cross-Sectional Predictability of Stock and Country Returns', *Journal of Portfolio Management* **23** (2) 79–87.

Baker, G.P. (1992) 'Beatrice: A Study in the Creation and Destruction of Value', *Journal of Finance* **47** (3) 1081–1119.

Ball, R. (1995) 'The Theory of Stock Market Efficiency: Accomplishments and Limitations', *Journal of Applied Corporate Finance* **8** (1) 4–17.

Ball, R., S. Kothari, and J. Shanken (1995) 'Problems in Measuring Portfolio Performance: An Application to Contrarian Investment Strategies', *Journal of Financial Economics* **38** (1) 79–107.

Barberis, N., A. Shleifer, and R. Vishny (1998) 'A Model of Investor Sentiment', *Journal of Financial Economics* **49** (3) 307–343.

Basu, S. (1977) 'Investment Performance of Common Stocks in Relation to Their Price/Earnings Ratios', *Journal of Finance* **32** (3) 663–682.

Basu, S. (1978) 'The Effect of Earnings Yield on Assessments of the Association between Accounting Income Numbers and Security Prices', *Accounting Review* **53** (3).

Bauman W.S. and R.E. Miller (1997) 'Investor Expectations and the Performance of Value Stocks Versus Growth Stocks', *Journal of Portfolio Management* **23** (3) 57–68.

Benartzi, S. and R.H. Thaler (1995) 'Myopic Loss Aversion and the Equity Premium Puzzle', *Quarterly Journal of Economics* **110** (1) 73–92.

Bernard, V. (1993) 'Stock Price Reactions to Earnings Announcements: A Summary of Recent Anomalous Evidence and Possible Explanations', in *Advances in Behavioral Finance*, R.H. Thaler (ed.), Russell Sage Foundation, New York.

Bernard, V. and J. Thomas (1989) 'Post-Earnings-Announcement Drift: Delayed Price Response or Risk Premium?', *Journal of Accounting Research* **27**, Supplement, 1–36.

Bernard, V. and J. Thomas (1990) 'Evidence that Stock Prices Do Not Fully Reflect the Implications of Current Earnings for Future Earnings', *Journal of Accounting and Economics* **13** (4) 305–341.

Bloomfield, R., R. Libby, and M. Nelson (1998) 'Underreactions and Overreactions: The Influence of Information Reliability and Portfolio Formation Rules', Working Paper, Cornell University, April.

Bos, J. (1991) 'Overreactie in de Nederlandse Aandelenmarkt', *Risico & Rendement*, 1–20.

Bowman, R., and D. Iverson (1996) 'Do Share Prices Overreact?', *New Zealand Investment Analyst*, 7–13.

Brailsford, T. (1992) 'A Test for the Winner-Loser Anomaly in the Australian Equity Market: 1958–87', *Journal of Business Finance & Accounting*, 225–241.

Brouwer, I., J. Van der Put, and C. Veld (1996) 'Contrarian Investment Strategies in a European Context', Working Paper, Tilburg University.

Bulkley, G. and R. Harris (1997) 'Irrational Analysts' Expectations as a Cause of Excess Volatility in Stock Prices', *Economic Journal*, 359–371.

Capaul, C., I. Rowley, and W. Sharpe (1993) 'International Value and Growth Stock Returns', *Financial Analysts Journal* **49** (1) 27–36.

Capstaff, J., K. Paudyal, and W. Rees (1995) 'The Accuracy and Rationality of Earnings Forecasts by UK Analysts', *Journal of Business Finance and Accounting* **22** 67–85.

Chan, K., A. Hameed, and W. Tong (1997) 'Profitability of Momentum Strategies in the International Equity Markets', Working Paper, Arizona State University.

Chan, L., Y. Hamao, and J. Lakonishok (1991) 'Fundamentals and Stock Returns in Japan', *Journal of Finance* **46** (5) 1739–1764.

Chan, L., N. Jegadeesh, and J. Lakonishok (1996) 'Momentum Strategies', *Journal of Finance* **51** (5) 1681–1713.

Chang, R., D. McLeavey, and S. Rhee (1995) 'Short-Term Abnormal Returns of the Contrarian Strategy in the Japanese Market', *Journal of Business Finance & Accounting*, 1035–1048.

Chopra, N., J. Lakonishok, and J. Ritter (1992) 'Measuring Abnormal Performance: Do Stocks Overreact?', *Journal of Financial Economics* **31** (2) 235–268.

Clare, A., and S. Thomas (1995) 'The Overreaction Hypothesis and the UK Stock Market', *Journal of Business Finance & Accounting*, 961–973.

Clayman, M. (1987) 'In Search of Excellence: The Investor's Viewpoint', *Financial Analysts Journal* **43** (3).

Conlisk, J. (1996) 'Why Bounded Rationality?', *Journal of Economic Literature* **34** (2) 669–700.

Da Costa, N. (1994) 'Overreaction in the Brazilian Stock Market', *Journal of Banking and Finance* **18** (4) 633–642.

Daniel, K., D. Hirshleifer, and A. Subrahmanyam (1998) 'Investor Psychology and Security Market Under- and Overreactions', *Journal of Finance*, forthcoming.

De Bondt, W.F.M. (1992) *Earnings Forecasts and Share Price Reversals*, AIMR, Charlottesville, Virginia.

De Bondt, W.F.M. (1993) 'Betting on Trends: Intuitive Forecasts of Financial Risk and Return', *International Journal of Forecasting* **9** 355–371.

De Bondt, W.F.M. (1995) 'Investor Psychology and the Dynamics of Security Prices', in *Behavioral Finance and Decision Theory in Investment Management*, A.S. Wood (ed.), AIMR, Charlottesville, Virginia.

De Bondt, W.F.M. (1998) 'A Portrait of the Individual Investor', *European Economic Review*, 831–844.

De Bondt, W.F.M. and M.M. Bange (1992) 'Inflation Forecast Errors and Time Variation in Term Premia', *Journal of Financial and Quantitative Analysis* **27** (4) 479–496.

De Bondt, W.F.M., A.K. Fung, and K. Lam (1998) 'Can Investors Profit from Swings in World Equity Markets?', Working Paper, University of Wisconsin–Madison.

De Bondt, W.F.M. and R.H. Thaler (1985) 'Does the Stock Market Overreact?', *Journal of Finance* **40** (3) 793–808.

De Bondt, W.F.M. and R.H. Thaler (1987) 'Further Evidence on Investor Overreaction and Stock Market Seasonality', *Journal of Finance* **42** (3) 557–581.

De Bondt, W.F.M. and R.H. Thaler (1989) 'A Mean-Reverting Walk Down Wall Street', *Journal of Economic Perspectives* **3** (1) 189–202.

De Bondt, W.F.M. and R.H. Thaler (1990) 'Do Security Analysts Overreact?, *American Economic Review* **80** (2) 52–57.

De Bondt, W.F.M. and R.H. Thaler (1995) 'Financial Decision-Making in Markets and Firms: A Behavioral Perspective', in *Handbook of Finance*, R.A. Jarrow *et al.* (eds.), Elsevier-North Holland.

Dechow, P. and R. Sloan (1997) 'Returns to Contrarian Investment Strategies: Test of Naive Expectations Hypotheses', *Journal of Financial Economics* **43** (1) 3–27.

Dissanaike, G. (1994) 'Do Stock Market Investors Overreact?', Working Paper, University of Cambridge.

Dreman, D.N. (1982) *The New Contrarian Investment Strategy*, Random House, New York.

Dreman, D.N. (1998) *Contrarian Investment Strategies: The Next Generation*, Simon & Schuster, New York.

Dreman, D.N. and M. Berry (1995) 'Overreaction, Underreaction, and the Low-P/E Effect', *Financial Analysts Journal* **51** (4) 21–30.

Dressendörfer, M. (1997) 'Zyklische und Antizyklische Handelsstrategien am Schweizer Aktienmarkt', Working Paper, University of St. Gallen, December.

Earl, P. (1990) 'Economics and Psychology: A Survey', *Economic Journal* **100** (402) 718–755.

Edwards, W. (1954) 'The Theory of Decision-Making', *Psychological Bulletin* **41** 380–417.

Edwards, W. (1968) 'Conservatism in Human Information Processing,' in *Formal Representation of Human Judgment*, B. Kleinmuntz (ed.), Wiley, New York.

Fama, E.F. (1970) 'Efficient Capital Markets: A Review of Theory and Empirical Work', *Journal of Finance* **25** (2) 383–417.

Fama, E.F. (1991) 'Efficient Capital Markets II', *Journal of Finance* **46** (5) 1575–1617.

Fama, E.F. (1998) 'Market Efficiency, Long-Term Returns, and Behavioral Finance', *Journal of Financial Economics* **49** (3) 283–306.

Fama, E.F. and K.R. French (1992) 'The Cross-Section of Expected Stock Returns', *Journal of Finance* **47** (2) 427–465.

Fama, E.F. and K.R. French (1997) 'Value Versus Growth: The International Evidence', Working Paper, University of Chicago.

Fama, E.F. and M.H. Miller (1972) *The Theory of Finance*, Dryden Press, Hinsdale, Illinois.

Fiske, S.T. and S.E. Taylor (1991) *Social Cognition* (2nd ed.), McGraw-Hill, New York.

Forbes, W. and L. Skerratt, 1994) 'Overreaction in Analysts' Forecasts of Earnings: Evidence from UK Individual Analysts', Working Paper, University of Manchester.

Forbes, W. and M. Kycriades (1996) 'Stock Market Overreaction in the United Kingdom: Some Further Results', Working Paper, University of Manchester.

Foster, G., C. Olsen, and T. Shevlin (1984) 'Earnings Releases, Anomalies, and the Behavior of Security Returns', *Accounting Review* **59** (4) 574–603.

French, K.R. and J.M. Poterba (1991) 'Investor Diversification and International Equity Markets', *American Economic Review* **81** (2) 222–226.

Fung, A.K. (1999) 'Overreaction in the Hong Kong Stock Market', *Global Finance Journal*, forthcoming.

Graham, B. (1959) *The Intelligent Investor* (3rd ed.), Harper & Brothers.

Graham, B. and D. Dodd (1934) *Security Analysis*, McGraw-Hill, New York.

Griffin, D. and A. Tversky (1992) 'The Weighting of Evidence and the Determinants of Confidence', *Cognitive Psychology*, 411–435.

Grossman, S.J. (1989) *The Informational Role of Prices*, MIT Press, Cambridge, MA.

Haugen, R.E. (1999) *The New Finance: The Case Against Efficient Markets* (2nd ed.), Prentice-Hall, New Jersey.

Hawawini, G. and D.B. Keim (1999) 'The Cross-Section of Common Stock Returns: A Review of the Evidence and Some New Findings', this volume 3–43.

Hong, H., T. Lim, and J.C. Stein (1998) 'Bad News Travels Slowly: Size, Analyst Coverage and the Profitability of Momentum Strategies', *Journal of Finance*, forthcoming.

Huberman, G. (1997) 'Familiarity Breeds Investment', Working Paper, Columbia University.

Jegadeesh, N. and S. Titman (1993) 'Returns to Buying Winners and Selling Losers: Implications for Stock Market Efficiency', *Journal of Finance* **48** (1) 65–91.

Kahneman, D. and D. Lavallo (1994) 'Timid Choices and Bold Forecasts: A Cognitive Perspective on Risk Taking', in *Fundamental Issues in Strategy*, R.P. Rumelt *et al.* (eds.), Harvard Business School Press.

Kahneman, D., P. Slovic, and A. Tversky (eds.) (1982) *Judgment Under Uncertainty: Heuristics and Biases*, Cambridge University Press, Cambridge, UK.

Kahneman, D. and A. Tversky (1979) 'Prospect Theory: An Analysis of Decision Under Risk', *Econometrica* **47** (2) 263–291.

Kahneman, D. and M.W. Riepe (1998) 'Aspects of Investor Psychology', *Journal of Portfolio Management* **24** (4) 52–65.

Karlsson, F. and S. Thoren (1997) 'Long-Term Stock Market Overreaction and the Contrarian Investment Strategy', Working Paper, Gothenburg School of Economics and Commercial Law.

Kryzanowski, L. and H. Zhang (1992) 'The Contrarian Strategy Does Not Work in Canadian Markets', *Journal of Financial and Quantitative Analysis* **27** (3) 383–395.

La Porta, R. (1996) 'Expectations and the Cross-Section of Stock Returns', *Journal of Finance* **51** (5) 1715–1742.

La Porta, R., J. Lakonishok, A. Shleifer, and R. Vishny (1997) 'Good News for Value Stocks: Further Evidence on Market Efficiency', *Journal of Finance* **52** (2) 859–874.

Lakonishok, J., A. Shleifer, and R. Vishny (1994) 'Contrarian Investment, Extrapolation, and Risk', *Journal of Finance* **49** (5) 1541–1578.

Lee, C.M.C. and B. Swanimathan (1998) 'Price Momentum and Trading Volume', Working Paper, Cornell University.

LeRoy, S.F. (1989) 'Efficient Capital Markets and Martingales', *Journal of Economic Literature* **27** (4) 1583–1621.

Lord, C., M.R. Lepper, and L. Ross (1979) 'Biased Assimilation and Attitude Polarization: The Effects of Prior Theories on Subsequently Considered Evidence', *Journal of Personality and Social Psychology*, 2098–2210.

Loughran, T. and J. Ritter (1996) 'Long-Term Market Overreaction: The Effect of Low-Priced Stocks', *Journal of Finance* **51** (5) 1959–1970.

Mackay, C. (1841) *Extraordinary Popular Delusions and the Madness of Crowds*, Farrar, Straus and Giroux, New York. Reprinted (1932).

Matsusaka, J.G. (1990) 'Takeover Motives During the Conglomerate Merger Wave', Working Paper, University of Chicago.

Mehra, R. and E. Prescott (1985) 'The Equity Premium: A Puzzle', *Journal of Monetary Economics*, 145–161.

Meyer, B. (1994) *Der Overreaction-Effekt am Deutschen Aktienmarkt*, Verlag Fritz Knapp, Frankfurt.

Moskowitz, T.J. and M. Grinblatt (1998) 'Momentum Investing: Industry Effects, Methodological Issues, and Profit Opportunities', Working Paper, University of Chicago.

Mutz, D.C. (1998) *Impersonal Influence*, Cambridge University Press, Cambridge, UK.

Nisbett, R. and L. Ross (1980) *Human Inference: Strategies and Shortcomings of Social Judgment*, Prentice-Hall, Englewood Cliffs, NJ.

Odean, T. (1998) 'Volume, Volatility, Price, and Profit When All Traders are Above Average', *Journal of Finance*, forthcoming.

O'Hanlon, J. and R. Whiddett (1991) 'Do UK Security Analysts Overreact?', *Accounting and Business Research* **22** 63–74.

Palomino, F. (1996) 'Noise Trading in Small Markets', *Journal of Finance* **51** (4) 1537–1550.

Poteshman, A.M. (1998) 'Does Investor Misreaction to New Information Increase in the Quantity of Previous Similar Information?', Working Paper, University of Chicago.

Pring, M.J. (1991) *Technical Analysis Explained* (3rd ed.), McGraw-Hill, New York.

Pring, M.J. (1993) *Investment Psychology Explained*, Wiley, New York.

Rabin, M. (1998) 'Psychology and Economics', *Journal of Economic Literature* **36** (1) 11–46.

Rayner A.C. and I.M.D. Little (1966) *Higgledy Piggledy Growth Again*, Basil Blackwell, Oxford.

Richards, A. (1997) 'Winner-Loser Reversals in National Stock Market Indices: Can They Be Explained?', *Journal of Finance* **52** (5) 2129–2144.

Ritter, J. (1991) 'The Long-Performance of Initial Public Offerings', *Journal of Finance* **46** (1) 3–27.

Rouwenhorst, K.G. (1998) 'International Momentum Strategies', *Journal of Finance* **53** (1) 267–284.

Schiereck, D., W.F.M. De Bondt, and M. Weber (1999) 'Contrarian and Momentum Strategies in Germany', Working Paper, University of Mannheim.

Schiereck, D. and M. Weber (1995)

Shefrin, H. and M. Statman (1994) 'Behavioral Capital Asset Pricing Theory', *Journal of Financial and Quantitative Analysis* **29** (3) 323–349.

Shefrin, H. and M. Statman (1997) 'Comparing Expectations About Stock Returns To Realized Returns', Working Paper, Leavey School of Business, Santa Clara University.

Shiller, R.J. (1989) *Market Volatility*, MIT Press, Cambridge, MA.

Shiller, R.J. (1990) 'Speculative Prices and Popular Models', *Journal of Economic Perspectives* **4** (2) 55–65.

Shleifer, A. and R. Vishny (1997) 'The Limits to Arbitrage', *Journal of Finance* **52** (1) 35–55.

Shleifer, A. and L.H. Summers (1990) 'The Noise Trader Approach to Finance', *Journal of Economic Perspectives* **4** (2) 19–33.

Siegel, J.J. (1992) 'The Equity Premium: Stock and Bond Returns Since 1802', *Financial Analysts Journal* **48** (1) 28–38.

Siegel, J.J. (1994) *Stocks for the Long Run*, Irwin, New York.

Simon, H.A. (1983) *Reason in Human Affairs*, Stanford University Press, Stanford.

Slovic, P. (1972) 'Psychological Study of Human Judgment: Implications for Investment Decision-Making', *Journal of Finance* **27** (4) 779–799.

Soros, G. (1998) *The Crisis of Global Capitalism*, Public Affairs, New York.

Teh, L.L. and W.F.M. De Bondt (1997) 'Herding Behavior and Stock Returns: An Exploratory Investigation', *Swiss Journal of Economics and Statistics*, 293–324.

Tversky, A. and D. Kahneman (1974) 'Judgment Under Uncertainty: Heuristics and Biases', *Science* **185** 1124–1131.

Tversky, A. and D. Kahneman (1986) 'Rational Choice and the Framing of Decisions', *Journal of Business* **59** (4) 67–94.

Vriezen, M.C. (1996) *Mean Reversion and Trending Among Stocks and Stock Markets*, Thesis Publishers, Amsterdam, The Netherlands.

A View of the Current Status of the Size Anomaly

Jonathan B. Berk

It is in many ways fitting that a conference devoted to security market anomalies should take place at the Isaac Newton Institute. The paradigm in physics for which Isaac Newton is responsible is arguably the most influential paradigm ever conceived. Most of the technological advances since that time would not have been possible without the paradigm. Yet, there have always been puzzles that, at the time of their discovery, appeared to be inconsistent with this paradigm. For instance, in the sixty odd years that it took to reconcile the orbit of the moon with Newton's inverse square law, some scientists suggested that the law needed to be modified. Ultimately, like countless other puzzles, the orbit of the Moon was shown to be predicted perfectly by the paradigm. The paradigm is, nevertheless, wrong. The reason we know this is that not every puzzle could be reconciled with the theory – a few were true anomalies, events that could only be explained by a better paradigm. As useful as Newton's laws are, they cannot explain the motion of Mercury's perihelion while Einstein's theory of general relativity can. Without the existence of this and one or two other anomalies, it is likely that Newton's laws would still be the currently accepted paradigm (and this Institute would not have hosted the Hawking–Penrose lectures on black holes.)

Although the case of Newton's law is only one example, there are good reasons for believing that puzzles and anomalies play a critical role in the advancement of knowledge. Thomas Kuhn (1970) in his seminal work on the subject, convincingly argues that the advancement of knowledge begins with the development of a paradigm. Puzzles that, at least initially, are not obviously explained by the paradigm are then identified. The field then advances by demonstrating why these puzzles are consistent with the paradigm. Sometimes, this effort fails and it becomes clear (usually only *ex post*), that these puzzles cannot be explained by the prevailing paradigm. A new paradigm that can explain these puzzles then replaces the old paradigm. The puzzles that prompted the development of the new paradigm are then known as anomalies.

In the last decade or so a number of empirical regularities in financial economics have been identified that have become known as 'anomalies'. Given the critical role anomalies play in the advancement of knowledge, identifying them is a significant milestone. In light of this, the question of whether these empirical regularities should legitimately be labelled as anomalies is an important one. That is, are these regularities so incongruous that a completely new asset pricing paradigm is required to explain them, or are they merely puzzles to be solved, under the existing paradigm, in the course of what Kuhn describes as normal science? In this chapter I will critically examine this question in the context of what perhaps is the most notorious empirical regularity – the size effect.

I will present the view that the so called 'size effect' (the relation between realized returns and variables such as market value, book-to-market, earnings-to-price, etc.) should not even be classed as a puzzle, let alone an anomaly. Furthermore, while I will argue that a derivative of this research, the development of sized-based factor models, might be useful in helping to explain the cross-sectional variation of expected returns, such 'models' provide no information on the underlying economic cause of this variation. Finally, I will argue that the size variables nevertheless have a useful role to play in tests of asset pricing theories.

The rest of this chapter is organized as follows. In the next section I briefly review the current empirical research on the size-related anomalies and then try to define precisely what asset pricing paradigm researchers have in mind when they refer to this evidence as an anomaly. In Section 2, I pose the question: Is the size effect an anomaly? I argue that is is not. I then examine the existing empirical evidence in light of this argument in Section 3. Following that, in Section 4, I examine the importance of size based factor models. Section 5 concludes the chapter.

1 The Size Effect

The cross-sectional relation between market value and return (the 'size anomaly') was discovered by Banz (1981). He not only found that market value is an excellent predictor of expected return, but that it is better in this regard than the CAPM itself. Reinganum (1981) then established that another effect, the predictive power of the earnings-to-price ratio (E/P), was in fact related to the size anomaly. Indeed, Reinganum showed that once size is controlled for, the E/P anomaly disappears.[1] Stattman (1980) and Rosenberg, Reid and Lanstein (1985) document another anomaly, that the ratio of book equity to market equity (book-to-market) is a predictor of average return. Other studies include Chan, Hamao and Lakonishok (1991) who show that these anomalies are also present in the Japanese stock market and Keim (1983) and Keim and Stambaugh (1984) who document important seasonals in the empirical relation.[2] Recently, both book-to-market and market

---

[1]Banz (1983), however, argues that the P/E ratio still has additional explanatory power once size and book-to-market have been controlled for.

[2]The reader who is interested in a more comprehensive survey is directed to one of the many excellent texts that document the varies stock market anomalies (e.g., Dimson (1988) or Ziemba (1994).)

value have also been shown to provide predictive power in the time-series of expected returns (see Pontiff and Schall (1998) and Kothari and Shanken (1997)).

The recent interest in the anomaly was sparked by Fama and French (1992) who systematically redocument the relation between market value (book-to-market) and return. The paper then convincingly argues that the other related effects as well as beta (from the single beta model) are all subsumed by book-to-market and market value. In Fama and French (1993) and Fama and French (1995), the authors form portfolios based on these two variables and show that their portfolios can be used to predict return in much the same way as the market portfolio is used in the CAPM. As a result they interpret these portfolios as factor mimicking portfolios and advocate using them to price assets. The authors attribute the predictive power of these portfolios to their ability to capture risk. Whether or not such an inference follows from their empirical work is the focus of Section 4.

Although the size related regularities are widely referred to as anomalies, the precise paradigm for which they present an anomaly is not so clear. Authors such as Lee, Shleifer and Thaler (1991) seem to have one of the central tenants in financial economics in mind – that the expected return of an asset is determined solely by the asset's riskiness. Other authors, such as Fama and French (1995), interpret the anomaly more narrowly. They take specific asset pricing models such as the CAPM or APT as the paradigm. In the next section I will examine the implications of the size related regularities on both of these definitions of the underlying paradigm.

2 Why the Size Effect could never be an Anomaly

To be classed as an anomaly the size effect must have the feature that it cannot be explained by the paradigm under consideration. Yet, as we will see, under both the narrow and broad definition of the paradigm, market value (or book-to-market) *must be* inversely related to return and so there is no sense in which the discovery of such a relation should be regarded as an anomaly.

Consider a one period economy in which expected value of every firm's end-of-period cashflow is the same. In order to ensure that stock prices can be compared easily, also assume that each firm issues only one share of stock. Now consider populating the world with risk neutral agents. Then it follows that stock prices in this world will be equal and so the market value of all firms will be the same. Now consider the same world populated with risk averse agents. Since the riskiness of a firm's cashflow is likely to vary across firms, each firm's stock price will, in general, be different. Given that all firms have the same expected cashflow, riskier firms with lower market values will, by definition, have higher expected returns. In the cross-section, market value will be inversely related to expected return. Finally, consider the possibility that stock prices are not influenced by the risk preferences of agents. For the sake of argument, assume that green men on Mars pick stock prices. Since all stocks have the same expected cashflow, by definition, the stocks that these green men assign lower (higher) prices will have higher (lower) expected returns. Thus, even in this world, market value is cross-sectionally inversely related to return. Clearly the only condition under which the size effect will not be observed in such a world is if all assets have the same price. *Any* theory that successfully explains any cross-sectional variation in expected returns *must* also predict an inverse relation between expected return and market value. In such a world, the size effect could never be an anomaly.

One shortcoming of the above argument is the assumption that all firms have the same expected cashflow. However, as I have shown elsewhere (Berk (1995)), so long as expected returns are not positively correlated to expected cashflows, the above result extends to an economy in which expected cashflows are not equal. This argument therefore shows that the only asset pricing paradigms for which the size effect could be an anomaly are ones which require a *positive* cross-sectional correlation between expected returns and expected cashflows. Since I know of no asset pricing theory that has this requirement, the size effect cannot be inconsistent with any paradigm that I am aware of. In particular it is neither inconsistent with the hypothesis that risk and return are related nor is it inconsistent with a specific asset pricing model such as the CAPM or APT.

Finally, consider what would happen if, instead of investigating the relation between expected return and market value, the relation between expected return and the ratio of expected cashflows to market value is investigated. In a single period model, another name for the ratio of expected cashflows to market value is the expected return, so this ratio is tautologically perfectly correlated to expected return! This observation requires no auxiliary assumptions whatsoever. Unfortunately, expected cashflows are unobservable and so no such study has been undertaken. However, something quite similar has been done.

The book value of equity measures depreciated past investment. Since the amount invested is likely to be highly correlated with the expected cashflows of the investment, one would expect the book value of equity and expected cashflow to be highly correlated. Thus book equity can be used as a proxy for the expected cashflow.[3] Therefore, the ratio of book equity to market equity is, in principle, a proxy for the ratio of expected cashflows to market value

---

[3]Berk, Green and Naik (1999) provide a formal model in which value of book equity is a perfect proxy for the expected cashflow.

and should therefore be a better measure of expected return than market equity alone. In light of the above argument it is not surprising then, that both Fama and French (1992) and Jegadeesh (1992) find that the logarithm of book-to-market equity is a much better predictor of return than the logarithm of market equity alone.

3   What about the Empirical Evidence?

Since its discovery, the observed inverse relation between market value and return has generally been interpreted as evidence that small firms have higher expected returns.[4] One hypothesis that has been proposed to explain this is that firm size proxies for exposure to a specific risk factor. For example, Chen (1988, p. 183) argues that because 'small firms tend to be marginal firms, they fluctuate more with business cycles, and thus, have higher risk exposure to the changing risk premium'. Since small firms have greater exposure, on average, they are riskier, and so there is an inverse relation between the size of firms and returns in the cross-section. Furthermore, if the empirical asset pricing model being tested does not correctly price this factor, then a measure of firm size will add explanatory power to the asset pricing model. The idea that small firms are riskier was first suggested in response to the empirical evidence (in the context of the price-earnings effect) by Ball (1978) who conjectured that earnings proxied for risk.[5] This hypothesis has since become the conventional explanation for the empirical relation between market value and average return, or 'size effect'[6].

The argument made in the previous section suggests that the traditional interpretation of the empirical relation between market value and average return might be flawed. Rather than evidence of a 'size' effect, the relation might be due solely to the endogenous inverse relation between the market value and discount rate of firms. The object of this section is to test whether this argument can completely explain the size effect. I therefore take, as the null hypothesis, that there is no relation between the size of a firm and

---

[4]Banz (1978) himself interpreted the size effect in this way suggesting, in his dissertation, that the effect resulted from the increased costs of diversification associated with smaller stocks.

[5]Other researchers who provide similar empirically based explanations for the relation between firm size and return include Banz (1978), Roll (1981), Blume and Stambaugh (1983), Schultz (1983), Schwert (1983), Stoll and Whaley (1983), Chan, Chen and Hsieh (1985), Amihud and Mendelson (1986), Keim (1988), Ball and Kothari (1989), Chan, Hamao and Lakonishok (1991), and Fama and French (1992).

[6]Consequently, Fama and French (1993) have recently run time-series regressions of returns on mimicking portfolios for size because they suggest that 'size is associated with a common risk factor that might explain the negative relation between size and average return' (p. 8). They interpret their regression slopes as the equilibrium risk premium of the common risk factor associated with size.

its expected return. The conventional interpretation of the size effect, that firm size is inversely related to expected return, is taken as the alternative hypothesis. Under the null hypothesis, non-market related measures of firm size will not be correlated to average return; while under the alternative, they will be. Consequently, the null hypothesis can be tested by repeating the empirical tests that identified the size effect in the first place with measures of firm size other than market value.

In the standard test of the relation between firm size and return, researchers test whether γ_s is equal to zero in the following relation:

$$E[R_i] = f_i + \gamma_s \, \mathrm{SIZE}_i \qquad (3.1)$$

where,

$$
\begin{aligned}
E[R_i] \;&=\; \text{expected return of stock } i, \\
f_i \;&=\; \text{expected return specified by an asset pricing model (if it is} \\
&\qquad \text{included in the test) of stock } i, \\
\mathrm{SIZE}_i \;&=\; \text{a measure of the size of firm } i, \\
\gamma_s \;&=\; \text{coefficient relating firm size to expected return.}
\end{aligned}
$$

Using a variety of empirical procedures, and the logarithm of market value as the measure of firm size, researchers have convincingly demonstrated that γ_s is significantly different from zero.

So long as $f_i \neq E[R_i]$ (i.e., the asset pricing model does not capture all relevant risk factors), in Berk (1995) I show that if market value is used as the measure of firm size, then even if expected return and firm size are unrelated, $\gamma_s < 0$.[7] Therefore the empirical results relating market value and return cannot be used as evidence that the size of a firm is related to its expected return. To show that size and return are related, it is necessary to show that γ_s is different from zero even when one employs measures of firm size that are not market related.

In this section I first repeat, using four non-market size measures, the form of the test that does not include a specific asset pricing model: Equation (3.1) is tested with $f_i \equiv a$, where a is any constant. The relation between firm size and risk adjusted return is then studied. This entails repeating existing studies in which (3.1) is tested with $f_i = \hat{E}_i$, where $\hat{E}_i$ is the expected return predicted by the single beta model.

The data set consists of all NYSE firms for which data exists in the period July 1966 – June 1987. This data set is chosen for two reasons. First, it closely resembles the data set in which Fama and French (1992) detected a very large

---

[7]The actual assumptions required are that expected return, $E[R_i]$, and the prediction of the asset pricing model, f_i, be uncorrelated to expected cashflow.

size effect. The object is to determine whether the size effect results, in part, from a relation between firm size and return. Thus it is essential that the data set used is one in which the size effect has already been empirically demonstrated to exist. Secondly, restricting attention to NYSE stocks avoids clouding the inferences of the tests by including new firms. New firms are on average small, and one could argue, riskier.[8] If this is true, then they would have higher expected returns. But in this case, it is the firm's short life rather than its size that is associated with its higher expected return. Since a very strong size effect has been identified on the NYSE alone, and since few firms listed on the NYSE are new firms, limiting attention to firms on this stock exchange provides a more powerful test of the null hypothesis.

All return and market value data are obtained from the monthly CRSP return and price files. Four non-market measures of the size of the firm are used: (1) the book value of assets (BVA); (2) the book value of all, un-depreciated, property, plant and equipment (PPE); (3) the total value of annual sales; and (4) the total number of employees. These data are obtained from the COMPUSTAT annual industrial files of income statement and balance-sheet data.[9]

The above four measures of the size of the firm are natural choices because there is no obvious reason why any of them should contain an adjustment for risk. The two accounting measures (BVA and PPE) might neglect an important component of firm size, the value of intangible assets. The intangible assets of a firm are defined as the difference between the market capitalization and the book value of assets. Therefore, the risk premium is part of the intangible assets of the firm. Yet there are other intangibles besides the risk premium (e.g. patents, copyrights, etc.) that can be thought of as part of the size of the firm but would not be measured by accounting variables. It is hard to see why these assets would have a significantly different relation to the return of the firm than the tangible assets, but if they do, the accounting measures would misstate the relation between the size of the firm and return. For this reason non-accounting measures of firm size are included (i.e., annual sales and number of employees).

Despite their differences, all five measures are, cross-sectionally, highly correlated. (See Table 1 which reports the time-series averages of these correlations.) The correlations of the other four size variables with market value are lower than their correlations with each other.

A selection bias in this data set merits comment. The data set was intentionally selected because a strong relation existed between market value and

---

[8]The idea that new firms are necessarily riskier is by no means obvious. For instance Clarkson, Guedes and Thompson (1993) provide conditions when new firms will not be riskier.

[9]The COMPUSTAT identification numbers are 6,7,12 and 29 respectively.

Table 1: Average Cross-Sectional Correlation Coefficients and Standard Deviations of the Five Size Measures: 1966–1986

Each average correlation coefficient is the average of the monthly cross-sectional correlation coefficients of the five size variables. The average standard deviation (the last row in the table) is the average of the monthly cross-sectional standard deviations of the five size variables. The market value of equity (MV) is measured in June of year t. Book value of assets (BVA), value of property, plant and equipment (PPE), annual sales (sales) and number of employees (employees) are the reported numbers for firms with fiscal year ends from June of year $t-1$ to May of year t. With the exception of the number of employees, all units are billions of dollars. Number of employees is quoted in thousands.

	MV	BVA	PPE	Sales	Employees
MV	1				
BVA	.732	1			
PPE	.721	.938	1		
Sales	.692	.822	.767	1	
Employees	.683	.759	.703	.778	1
Standard Deviation	2.377	3.988	3.563	3.782	41.39

average return. As such, the analysis is subject to the kind of data snooping bias discussed by Denton (1985), Black (1992), Foster and Smith (1997), Lo and MacKinlay (1992) and others. As these authors point out, if classical statistical significance tests are used, this bias increases the probability of rejecting the null in the sample when it holds in the population.

The actual matching of the accounting and market data is accomplished as follows. The COMPUSTAT file in year t contains data for fiscal year ends from June of year t to May of year $t+1$. In order to guarantee that the relevant explanatory variables are known before the returns they are supposed to explain, accounting data in the year t COMPUSTAT file is matched with the market value at the end of June in year $t+1$. These variables are then used to explain returns from July of year $t+1$ through June of year $t+2$.[10] As a result one year of data is lost leaving a total of 20 years (or 240 months) of observations.

The one-month lag between the last possible fiscal year end (May) and the first return observation (July) is necessary because firms report their balance

---

[10]Thus, the accounting numbers could be at most a year behind the market value number. This disparity is not, however, a cause for concern. Using market values from June of year $t-1$ (i.e., ensuring that market value is never more current than any accounting number) only marginally changes the results and does not change any inference in the paper.

sheet data after their fiscal year ends. Some might argue that this lag is not long enough, since a small minority of firms do not report within even the legally mandated 90 day period. However, less than 10% of firms have fiscal year ends between February and May and so only a tiny minority of firms' accounting data would be unavailable as late as July. The cost in potential explanatory power was therefore not deemed worth the benefits obtained by increasing this lag.

3.1 The Relation Between Size and Return

Can any part of the inverse relation between the market value and return of stocks listed on the NYSE be attributed to a relation between firm size and expected return? To answer this question a standard Fama–Macbeth regression test of the null hypothesis that $\gamma_s = 0$ in (3.1) (when f_i is taken to be constant across assets) is undertaken.

Table 2 contains the time series average slopes (in percent) of the month by month Fama–Macbeth (FM) univariate cross-sectional regressions of the monthly return of all firms on the logarithm of each of the five size measures separately. The associated t-statistic is the standard FM t-statistic (the average slope divided by its time series standard error). The p-value represents the probability, under a normal distribution, of observing a t-value greater (in absolute magnitude) than the value actually observed. Since the dependent variables are all directly observable, there is no error-in-variables problem, and so, given the length of the sample period, the t-statistics should be normally distributed under the null hypothesis. This was confirmed by bootstrapping the distribution.

Table 2 verifies that market value is a significant predictor of future return. The results for the other four size measures differ, however. Because of the data snooping bias discussed earlier, classical p-values overstate the significance of the other size variables.[11] Normally the test would need to be adjusted to take this bias into account. However, since the null hypothesis cannot be rejected at even the classical cutoffs, adjusting for the data snooping bias is unnecessary, since it cannot change the inference.[12]

The null hypothesis that $\gamma_s = 0$ cannot be rejected based on the evidence from the above FM regression test. However, one might be concerned that

---

[11]The bias follows because a strong relation between market value and return was known to exist in the data sample prior to the study. Since the the non-market related measures are known to be highly correlated with market value, under the null hypothesis, this bias increased the probability of detecting an effect in the sample when none exists in the population (see Lo and MacKinlay (1990) for a discussion of this bias in the context of the size effect).

[12]There is no evidence of heteroscedasticity in any of the cross-sectional regressions. As such, the results are unaffected when the residual covariance matrix is estimated using White's heteroscedastic-consistent estimator.

Table 2: Average Slopes (in %) and *t*-statistics from Monthly Regressions of Stock Returns on Five Size Measures: July 1967 – June 1987
The regressions are month by month cross-sectional Fama–Macbeth (FM) univariate regressions of the monthly return of all firms on the logarithm of each of the five size measures separately. The coefficient is the average of the monthly regression slopes and the *t*-statistic is this average divided by its time series standard error. The *p*-values are two tailed and are calculated under the assumption that the *t*-statistic is normally distributed. The market value of equity (MV) is measured in June of year *t*. Book value of assets (BVA), value of property, plant and equipment (PPE), annual sales (sales) and number of employees (employees) are the reported numbers for firms with fiscal year ends from June of year $t-1$ to May of year *t*. The returns are measured monthly from July of year *t* to June of year $t+1$. Thus each monthly return, from July of year *t* to June of year $t+1$ is regressed onto size measures that are recorded on or before June of year *t*.

	MV	BVA	PPE	Sales	Employees
Coefficient (%)	−0.170	−0.099	−0.082	−0.069	−0.061
t-statistic	−2.50	−1.83	−1.52	−1.47	−1.68
p-value	.0124	.0671	.129	.142	.0922

Table 3: Average Cross-Sectional Correlation Coefficients of the Residuals of the Logarithm of Market Value Regressed onto the Logarithm of the Four Size Measures: 1967–1986
Each average correlation coefficient is the average of the monthly cross-sectional correlation coefficients of the residuals. The coefficients are listed under the size variable from which the residual is calculated. The market value of equity (MV) is measured in June of year *t*. Book value of assets (BVA), value of property, plant and equipment (PPE), annual sales (sales) and number of employees (employees) are the reported numbers for firms with fiscal year ends from June of year $t-1$ to May of year *t*.

	BVA	PPE	Sales	Employees
PPE	.832			
Sales	.775	.710		
Employees	.647	.638	.883	
All	.960	.847	.881	.762

Table 4: Average Slopes (in %) and t-statistics from Monthly Univariate Regressions of Stock Returns onto Orthogonalized Market Value Measures: July 1967 – June 1987

Each year the logarithm of every firm's market value is cross-sectionally regressed on the logarithm of each of the four size measures individually producing four residuals. A fifth residual is created by cross-sectionally regressing the logarithm of every firm's market value onto the logarithm of *all* four size measures together. Over the next year, the return of every firm in a given month is then cross-sectionally regressed onto these five residuals individually. The coefficient of the residual regressions is the average of these monthly regression slopes and the t-statistic is this average divided by its time series standard error. The results are reported under the size measure from which the residuals are calculated. The p-values are two tailed, and are calculated under the assumption that the t-statistic is normally distributed. The market value of equity (MV) is measured in June of year t. Book value of assets (BVA), value of property, plant and equipment (PPE), annual sales (sales) and number of employees (employees) are the reported numbers for firms with fiscal year ends from June of year $t-1$ to May of year t. The returns are measured monthly from July of year t to June of year $t+1$. Thus each monthly return, from July of year t to June of year $t+1$ is regressed onto orthogonalized size measures that are recorded on or before June of year t.

	Part of Market Value Orthogonal to:				
	BVA	PPE	Sales	Employees	All
Coefficient (%)	−0.244	−0.237	−0.295	−0.240	−0.280
t-statistic	−2.65	−2.90	−3.29	−2.67	−2.71
p-value	.0081	.0037	.0010	.0076	.0068

the above test might not be powerful enough to reject the null. One way to address this concern is to compare the relation between market value and return when firm size is controlled for and when it is not. Under the alternative, if firm size accounts for any part of return, the predictive power of market value should decrease when firm size is controlled for.

To control for firm size, the logarithm of market value is annually cross-sectionally regressed onto the logarithm of each size measure individually. Log market value is also annually cross-sectionally regressed onto the log of all four size variables together. The residuals of these regressions control for firm size because they are the part of market value that is orthogonal to each size measure respectively. They are consequently highly correlated (see Table 3).

The relation between these residuals and average return is then compared to the relation between market value and average return. Under the alternative this relation should diminish (γ_s should be closer to zero) when these residuals, rather than market value, are used as the measures of firm size. The above FM univariate cross-sectional regression test is therefore repeated using each residual individually as the independent variable. The coefficients of these regressions are reported in Table 4, and can be found under the size variable from which the residual is calculated. The coefficients are all negative and significantly different from zero. More importantly, the coefficient of every residual is *larger* (in absolute value) than the coefficient of the market value variable itself.

Finally, we can also measure the marginal contribution, over market value, of the other size variables by repeating the above procedure using the residuals of the cross-sectional regression of each size variable on market value. The results of this test are reported in Table 5. Controlling for market value, γ_s actually reverses sign, that is larger size is associated with higher not lower return. In addition, the difference in the coefficient estimates in Tables 2 and 5 is significant.[13]

A positive relation between firm size measures and return controlling for risk is consistent with the hypothesis that firm size and risk are unrelated. To see why, recall that the market value of a firm is influenced by two effects: larger firms have larger market values and firms with high discount rates have lower market values. Consider two firms of different sizes (or expected cashflows) with the same market value. This implies that the smaller (larger) firm must have the lower (higher) discount rate. Consequently, *once market value is controlled for*, the larger (smaller) the firm the higher (lower) the discount rate.

These results imply that there is no evidence to support the hypothesis that there is a cross-sectional relation between firm size and expected return. In particular, in a data set in which market value has strong cross-sectional relation to average return, the above study fails to find a similar significant relation between average return and other, non-market, measures of firm size. I also show that the relation between market value and average return is unaffected when non-market measures of firm size are used to control the size of the firm.

---

[13] A two-tailed *t*-test is used to test whether the coefficient estimate of the FM regression of return on a particular size measure is different to the coefficient estimate of the FM regression of return on the residuals of the same size measure on market value (that is, the *t*-statistic is defined as the mean of the monthly differences between the two slopes divided by the standard error of these differences). The *t*-statistics are $2.59, 2.92, 3.34$ and 2.62 with corresponding *p*-values of $0.0095, 0.0035, 0.0008$ and 0.0087 for BVA, PPE, sales and employees respectively.

Table 5: Average Slopes (in %) and t-statistics from Monthly Univariate Regressions of Stock Returns onto Size Measures Orthogonal to Market Value: July 1967 – June 1987

Each year the logarithm of each of the four size measures is individually cross-sectionally regressed on the logarithm of every firm's market value producing four residuals. Over the next year, the return of every firm in a given month is then cross-sectionally regressed onto these four residuals individually. The coefficient of the residual regressions is the average of these monthly regression slopes and the t-statistic is this average divided by its time series standard error. The results are reported under the size measure from which the residuals are calculated. The p-values are two tailed, and are calculated under the assumption that the t-statistic is normally distributed. The market value of equity (MV) is measured in June of year t. Book value of assets (BVA), value of property, plant and equipment (PPE), annual sales (sales) and number of employees (employees) are the reported numbers for firms with fiscal year ends from June of year $t - 1$ to May of year t. The returns are measured monthly from July of year t to June of year $t + 1$. Thus each monthly return, from July of year t to June of year $t + 1$ uses orthogonalized size measures that are recorded on or before June of year t.

	Orthogonalized to Market Value:			
	BVA	PPE	Sales	Employees
Coefficient (%)	0.091	0.079	0.163	0.093
t-statistic	1.39	1.32	3.16	1.76
p-value	.165	.186	.00157	.0782

3.2   The Relation Between Size and Beta

One might argue that the empirical literature on size is not only regarded as anomalous because it finds that market value is an inverse predictor of return, but also because when the single beta model and market value are tested simultaneously, market value has additional explanatory power.[14] Although Fama and French (1992) interpret this evidence as 'a shot strait at the heart of the CAPM' (p. 438), this result by itself cannot be used to reject the CAPM (see Berk (1995)). So long as there is any misspecification in the empirical test, it is possible to reject the CAPM in favour of market value even if the CAPM itself holds perfectly. The same result is not necessarily true of the other size measures. If the size of the firm is unrelated to its return, then even if the CAPM does not hold perfectly, the size of the firm should have no additional explanatory power in a joint test of it and the CAPM. The purpose of this section is to empirically verify this fact.

---

[14]See, e.g., Jegadeesh (1992) or Fama and French (1992)

Formally, the null hypothesis is that, in (3.1), $\gamma_s = 0$ when $f_i = r + \beta_i \bar{R}_m$, where β_i is the CAPM beta of stock i, r is the riskless rate of interest and $\bar{R}_m$ is the expected return of the market proxy. The alternative hypothesis is that firm size has additional explanatory power, that is, that $\gamma_s \neq 0$. Using market value as the measure of firm size many researchers have been able to reject this null. Perhaps the most convincing evidence can be found in Gibbons, Ross and Shanken (1989). The reason why this evidence is particularly strong is that the empirical procedure they use does not require measuring β_i and so their test does not contain an error-in-variables bias. In this section I therefore repeat their procedure, in a data set in which a strong relation between market value and risk adjusted return is known to exist, using other measures of firm size.

The methodology that Gibbons, Ross and Shanken (1989) use is based on the time series test of the CAPM originally developed by Black, Jensen and Scholes (1972) and extended to include a test of the size effect by Brown, Kleidon and Marsh (1983). Unfortunately, as Gibbons, Ross and Shanken (1989) show in their paper, over the time period in this paper, there is no significant relation between market value and risk adjusted return on the NYSE.[15] Most researchers would consider finding no relation between the non-market size measures and risk adjusted return in a data set in which no relation between market value and risk adjusted return was known to exist as uninteresting. It therefore makes little sense to use this sample in this study. It turns out, however, that if the SP500 stocks are removed from the sample, a strong relation between market value and risk adjusted return does exist. Consequently, all the empirical tests in this section will be restricted to the NYSE stocks that are followed by COMPUSTAT but are not in the SP500 index.

In June of each year, each size measure is individually used to sort the sample of stocks into 10 size-ranked portfolios. The excess return of each decile over the next 12 months is then recorded.[16] A 240 month time series of monthly average excess returns for each decile is then formed by concatenating these returns over the 20 year period July 1967 – June 1987. Each time series is then separately regressed onto the time series of monthly excess returns of the CRSP value weighted index. The null hypothesis that $\gamma_s = 0$ is equivalent to the condition that the intercepts of the decile regressions are all jointly zero.

Table 6 contains the regression intercepts for each decile. They are also graphed in Figure 1.

---

[15] Gibbons, Ross and Shanken (1989) divided the time period in this paper into two subperiods and report the results for each subperiod. The comparable W-statistic for the whole time period is 0.069 which corresponds to a p-value of 0.11.

[16] The excess return of each stock is calculated by subtracting the 1 month riskless rate in the CRSP bond file from the monthly return of each asset.

Table 6: The Mean-Variance Efficiency of the CRSP Value Weighted Index:
Univariate Test
Using each size measure as well as book-to-market, the firms are (separately) sorted
into 10 sized based deciles in June of each year. That is, each June, stocks are sorted
into deciles based on the observed market value of equity (MV) in June of year t
and on the reported values between June of year $t-1$ and May of year t, of the book
value of assets (BVA), the value of property, plant and equipment (PPE), annual
sales (sales) and the number of employees (emp). The monthly excess return (i.e.,
the realized return minus the CRSP one month riskless rate) of each decile from
July of year t to June of year $t+1$ is then recorded. A 20 year time series of
monthly returns for each decile is then formed by concatenating the 12 monthly
returns for each decile in each year over the period July 1967 – June 1987. Each
of these time series are then separately regressed onto the time series of monthly
excess returns of the CRSP value weighted index. The intercept (in %) of each
regression and its associated t-statistic are recorded in the table.

Size	MV		BVA		PPE		Sales		Emp	
Dec	$\hat{\alpha}_{im}$	t	$\hat{\alpha}_{im}$	t	$\hat{\alpha}_{im}$	t	$\hat{\alpha}_{im}$	t	$\hat{\alpha}_{im}$	t
lg	−0.19	−2.12	0.17	1.26	0.13	1.03	0.10	0.78	0.16	1.01
9	−0.11	−1.10	0.06	0.40	0.17	1.32	0.10	0.64	0.10	0.61
8	−0.06	−0.45	0.05	0.34	0.17	1.08	0.20	1.20	0.14	0.74
7	0.02	0.15	0.19	1.16	0.22	1.21	0.21	1.20	0.17	0.88
6	0.21	1.25	0.16	0.90	0.24	1.29	0.19	0.97	0.21	1.10
5	0.12	0.65	0.19	0.91	0.22	1.09	0.13	0.68	0.14	0.73
4	0.42	1.91	0.33	1.64	0.19	0.92	0.23	1.16	0.17	0.88
3	0.27	1.13	0.16	0.69	0.31	1.33	0.27	1.32	0.33	1.83
2	0.56	2.10	0.40	1.73	0.24	1.02	0.42	1.83	0.46	2.58
sm	0.83	2.45	0.43	1.69	0.36	1.33	0.31	1.26	0.40	2.09

The estimates increase approximately monotonically as the market value
of the portfolio decreases. Furthermore, of the ten t-statistics of the portfolios
sorted by market value, three are above 1.96. The corresponding results for
the other four size measures are not as convincing. Of the 40 t-statistics of all
the portfolios sorted by the other four size measures only two are above 1.96.
However, as Gibbons, Ross and Shanken (1989) point out, the t-statistics do
not provide conclusive evidence on whether all the intercepts together are
significantly different from zero.

Table 7 reports the results of the F-test that the intercepts are jointly
significantly different from zero. The results show that when either market
value or book-to-market is used to form the portfolios, the intercepts are

Figure 1: **Intercepts of the Univariate Test for Efficiency of the CRSP Value Weighted Index.**

Using each size measure, the firms are (separately) sorted into 10 sized based deciles in June of each year. The average excess return (i.e., the realized return minus the CRSP one month riskless rate) of each decile over the next 12 months is then recorded. A time series of monthly returns for each decile is then formed by combining these annual returns over the 20 year period July 1967 – June 1987. These time series are then separately regressed onto the time series of monthly excess returns of the CRSP value weighted index. The graph shows the intercept terms of each regression.

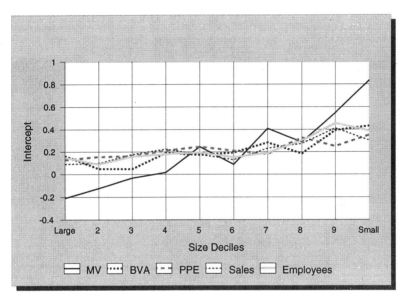

significantly different from zero. However, the other four size measures have no detectable explanatory power whatsoever. None of the other effects are significant (BVA had the smallest $p = 0.08$) at even classical statistical cutoffs.

In a data set in which a strong relation between market value and return exists, there is no evidence that the size of the firm has any additional explanatory power over and above the single beta model. In short, even when the proxy portfolio is shown to be inefficient, there is no evidence that firm size has additional explanatory power.

In a data set in which market value provides significant additional explanatory power over and above the single beta model, the results in this sub-section demonstrate that other, non-market, measures of firm size fail to provide similar explanatory power. Taken together, these results are evidence against the hypothesis that firm size proxies for exposure to specific

Table 7: The Mean Variance Efficiency of the CRSP Value Weighted Index: Multivariate Test

Using each size measure, the firms are (separately) sorted into 10 sized based deciles in June of each year. That is, each June stocks are sorted into deciles based on the observed market value of equity (MV) in June of year t and on the reported values between June of year $t-1$ and May of year t, of the book value of assets (BVA), the book value of common equity, the value of property, plant and equipment (PPE), annual sales (sales) and the number of employees (employees). The monthly excess return (i.e., the realized return minus the CRSP one month riskless rate) of each decile from July of year t to June of year $t+1$ is then recorded. A 20 year time series of monthly returns for each decile is then formed by concatenating the 12 monthly returns for each decile in each year over the period July 1967 – June 1987. Each of these time series are then separately regressed onto the time series of monthly excess returns of the CRSP value weighted index. The intercepts are then used to calculate the Gibbons, Ross and Shanken W statistic. Under the null hypothesis, $\frac{T(T-N-1)}{N(T-2)}W$ has a central F distribution with degrees of freedom N and $T-N-1$. Thus the p-values represent the probability, under a F distribution with degrees of freedom 10 and 229, of observing a (transformed) W value greater than the observation actually observed.

	MV	BVA	PPE	Sales	Employees
W	0.149	0.074	0.013	0.029	0.066
p-value	.00031	.082	.980	.755	.129

risk factors. This raises the question of what these arguments imply for the recent empirically based asset pricing models that rely on a market value and book-to-market factors. This is the subject of the next section.

4   What do we learn from Size-Based Factor Models?

The fact that the relation between market value (book-to-market) and return should not be regarded as anomalous does not imply that it is not potentially useful. Consequently, Fama and French (1995) have used portfolios constructed on the basis of market value and book-to-market to construct a 'factor model' of asset returns. The authors argue, based on the ability of these portfolios to explain cross-sectional variation in asset returns, that they collectively span the risk factors in the economy. In fact a portfolio that explains cross-sectional differences in asset returns need not necessarily capture risk in any form.

There are at least two reasons why a portfolio might explain cross-sectional differences in expected returns that are not risk related. The first is explained in a recent paper by Ferson (1995). In this paper, the author shows how an arbitrary set of firm specific attributes that are assumed to be related to expected returns but unrelated to any economy wide risk factor can nevertheless be used to form portfolios that will explain cross-sectional differences in expected returns. The second reason follows from the fact that the existence of a mean-variance efficient frontier does not depend on there being a risk-return trade-off in the economy. By the Roll (1977) critique, a mean-variance efficient portfolio always exists. This portfolio will always explain cross-sectional variation in expected return, yet its existence does not in any way depend on whether this cross-sectional variation is risk based.

Of course, the mere existence of frontier portfolios does not necessarily explain why the particular portfolios Fama and French identify work as well as they do. However, mean-variance efficient portfolios are not the only portfolios that can explain cross-sectional differences in expected return. As explained in Green (1986) even portfolios that are not mean-variance efficient will still explain cross-sectional differences in asset returns. Like the Roll critique this result does not rely on a presumed relation between risk and return. Indeed, in an economy in which risk is assumed to be completely unrelated to return it is possible to derive an example in which *any* portfolio (that does not require short selling) will explain cross-sectional variation in asset returns. To see why, consider the following economy.

Take a multiperiod economy that consists of a set of firms, I, each of which is a claim to an uncertain and perpetual dividend stream, c_i^t, where $i \in I$ and $t \geq \tau_i$. $\tau_i \in \mathcal{T}$ is the date of the initial cashflow of the firm. Thereafter the cashflows of the firms follow a lognormal random walk:

$$c_i^{t+1} = c_i^t \exp\left[\mu_i - \frac{1}{2}\sigma_i^2 + \sigma_i \epsilon_i(t+1)\right] \tag{4.1}$$

with $c_i^{\tau_i} = C_i$ and where the sequence $\{\epsilon_i(t), t \geq \tau_i\}$ is *iid* and $N(0,1)$. For any two firms, i and j, let the correlation between $\epsilon_i(t)$ and $\epsilon_j(t)$ be denoted ρ_{ij}. Assume that C_i, τ_i, σ_i, μ_i and ρ_{ij} are all non-negative and have (cross-sectional) distributions that are independent of one another. That is, each realization represents a one time (non-negative) draw from the cross-sectional distribution of that characteristic that is independent of everything else in the model.[17] For expositional clarity I will distinguish cross-sectional moments (as opposed to time-series moments) with a superscript 'c'. For example, if x_i is an arbitrary characteristic of firm i, then the cross-sectional variance (across firms) of this characteristic is denoted $\text{var}^c(x_i)$.

---

[17]More formally, if $f_x(\cdot)$ is the cross-sectional density function of firm characteristic x, then the joint density function of any two characteristics, say C_i and σ_i, is $f_{C_i}(\cdot)f_{\sigma_i}(\cdot)$.

The above assumptions are certainly plausible if not standard. Most models do not restrict the cross-sectional distribution of variables such as the initial cashflow, average cashflow growth and the standard deviation of cashflows. There is no obvious reason why the size of the initial cashflow should be related to the variance, covariance or subsequent growth of the cashflows. Nor should date of the first cashflow matter.

For any $t > \tau_i$, if $\sigma_i(t) \equiv \sigma_i\sqrt{t - \tau_i}$ and $\mu_i(t) \equiv \mu_i\,(t - \tau_i)$, then from (4.1) we get an expression for c_i^t as a function of C_i:

$$c_i^t = C_i \exp\left[\mu_i(t) - \frac{1}{2}\sigma_i^2(t) + \sigma_i(t)\psi_i(t)\right] \tag{4.2}$$

where $\psi_i(t) \equiv \frac{1}{\sqrt{t-\tau_i}}\sum_{\tau=\tau_i+1}^{t}\epsilon_i(\tau)$ is $N(0,1)$ and depends only on information between τ_i and t.

The firms are traded on a spot market at each time t. Since I am constructing an economy in which risk and return are unrelated I will assume that the dividend stream plays no part in determining the firm's price, p_i^t, on these spot markets. Instead, at each time t, prices are determined randomly, that is, each firm i's price on every spot market is drawn independently of everything else in the model. The total return is then given by

$$R_i^t \equiv \frac{c_i^{t+1} + p_i^{t+1}}{p_i^t}.$$

If the mean of the price distribution is denoted p, then the expected return is:

$$E_t[R_i^t] \equiv \frac{\bar{c}_i^{t+1} + p}{p_i^t}.$$

This economy represents perhaps starkest alternative to the kind of economy in which the current asset pricing paradigms are applied. Not only are prices unrelated to risk, they are unrelated to every aspect of the firm. Green men on Mars might as well be pricing assets in this economy. Yet, if an econometrician were to calculate the beta of any stock with any arbitrary portfolio that requires no short sales then she would observe a positive relation between these betas and expected returns. To demonstrate why, take any portfolio in this economy that contains no short positions, denote it m, and consider pricing all other stocks off stock m. By definition,

$$R_m^t = \sum_{k \in I} \alpha_k R_k^t.$$

where the portfolio weights, α_k sum to 1 and are non-negative. In the single beta model, the beta (at time t) of stock i is defined to be

$$\beta_i^t \equiv \frac{\text{cov}\left(R_m^t, R_i^t\right)}{\text{var}\left(R_m^t\right)} = \frac{\text{cov}\left(\sum_{k \in I}\alpha_k R_k^t, R_i^t\right)}{\text{var}\left(R_m^t\right)} = \sum_{k \in I}\alpha_k\frac{\text{cov}\left(R_k^t, R_i^t\right)}{\text{var}\left(R_m^t\right)}$$

$$
= \sum_{k \in I} \alpha_k \frac{\operatorname{cov}\left((c_i^{t+1} + p_i^{t+1}), (c_k^{t+1} + p_k^{t+1})\right)}{\operatorname{var}(R_m^t)\, p_i^t p_k^t} = \sum_{k \in I} \alpha_k \frac{\operatorname{cov}\left(c_i^{t+1}, c_k^{t+1}\right)}{\operatorname{var}(R_m^t)\, p_i^t p_k^t}
$$

$$
= \sum_{k \in I} \alpha_k \frac{c_i^t e^{\mu_i} c_k^t e^{\mu_k} \operatorname{cov}\left(e^{-\frac{1}{2}\sigma_i^2 + \sigma_i \epsilon_i (t+1)}, e^{-\frac{1}{2}\sigma_k^2 + \sigma_k \epsilon_k (t+1)}\right)}{\operatorname{var}(R_m^t)\, p_i^t p_k^t}
$$

$$
= \sum_{k \in I} \alpha_k \frac{c_i^t e^{\mu_i} c_k^t e^{\mu_k}\left(e^{\sigma_i \sigma_k \rho_{ik}} - 1\right)}{\operatorname{var}(R_m^t)\, p_i^t p_k^t}
$$

$$
= \sum_{k \in I} K_k \frac{c_i^t e^{\mu_i}\left(e^{\sigma_i \sigma_k \rho_{ik}} - 1\right)}{p_i^t}, \tag{4.3}
$$

where $K_k \equiv \alpha_k c_k^t e^{\mu_k}/\operatorname{var}(R_m^t)\, p_k^t > 0$. Next, compute the *unconditional* cross-sectional correlation of the time-t beta and the time-t expected return:

$$
\operatorname{cov}^c\left(E_t[R_i^t], \beta_i^t\right) = \sum_{k \in I} K_k \operatorname{cov}^c\left(\frac{\bar{c}_i^{t+1} + p}{p_i^t}, \frac{c_i^t e^{\mu_i}\left(e^{\sigma_i \sigma_k \rho_{ik}} - 1\right)}{p_i^t}\right)
$$

$$
= \sum_{k \in I} K_k \operatorname{cov}^c\left(\frac{c_i^t e^{\mu_i}}{p_i^t}, \frac{c_i^t e^{\mu_i}\left(e^{\sigma_i \sigma_k \rho_{ik}} - 1\right)}{p_i^t}\right) \tag{4.4}
$$

$$
+ K_k \operatorname{cov}^c\left(\frac{p}{p_i^t}, \frac{c_i^t e^{\mu_i}\left(e^{\sigma_i \sigma_k \rho_{ik}} - 1\right)}{p_i^t}\right)
$$

$$
= \sum_{k \in I} K_k \left(E^c\left[\left(\frac{c_i^t e^{\mu_i}}{p_i^t}\right)^2 \left(e^{\sigma_i \sigma_k \rho_{ik}} - 1\right)\right]\right.
$$

$$
\left. - E^c\left[\frac{c_i^t e^{\mu_i}}{p_i^t}\right] E^c\left[\frac{c_i^t e^{\mu_i}}{p_i^t}\left(e^{\sigma_i \sigma_k \rho_{ik}} - 1\right)\right]\right)
$$

$$
+ p K_k \left(E^c\left[\frac{c_i^t e^{\mu_i}\left(e^{\sigma_i \sigma_k \rho_{ik}} - 1\right)}{(p_i^t)^2}\right]\right.
$$

$$
\left. - E^c\left[\frac{1}{p_i^t}\right] E^c\left[\frac{c_i^t e^{\mu_i}\left(e^{\sigma_i \sigma_k \rho_{ik}} - 1\right)}{p_i^t}\right]\right). \tag{4.5}
$$

The terms in parentheses in (4.5) are positive. To see why, note that by the independence of p_i^t,

$$
E^c\left[\frac{c_i^t e^{\mu_i}\left(e^{\sigma_i \sigma_k \rho_{ik}} - 1\right)}{(p_i^t)^2}\right] - E^c\left[\frac{1}{p_i^t}\right] E^c\left[\frac{c_i^t e^{\mu_i}\left(e^{\sigma_i \sigma_k \rho_{ik}} - 1\right)}{p_i^t}\right]
$$

$$
= E^c\left[c_i^t e^{\mu_i}\left(e^{\sigma_i \sigma_k \rho_{ik}} - 1\right)\right] E^c\left[\frac{1}{p_i^t}\right]^2 - \left(E^c\left[\frac{1}{p_i^t}\right]\right)^2 E^c\left[c_i^t e^{\mu_i}\left(e^{\sigma_i \sigma_k \rho_{ik}} - 1\right)\right]
$$

$$
= E^c\left[c_i^t e^{\mu_i}\left(e^{\sigma_i \sigma_k \rho_{ik}} - 1\right)\right] \operatorname{var}^c\left(\frac{1}{p_i^t}\right) > 0, \tag{4.6}
$$

where the inequality follows from the fact that c_i^t, μ_i and ρ_{ik} are all strictly positive. Next, note that

$$E^c \left[\left(\frac{c_i^t e^{\mu_i}}{p_i^t} \right)^2 \left(e^{\sigma_i \sigma_k \rho_{ik}} - 1 \right) \right]$$

$$= E^c \left[\left(\frac{C_i \exp\left[\mu_i(t+1) - \frac{1}{2}\sigma_i^2(t) + \sigma_i(t)\psi_i(t) \right]}{p_i^t} \right)^2 \left(e^{\sigma_i \sigma_k \rho_{ik}} - 1 \right) \right]$$

$$= E^c \left[\frac{C_i e^{\mu_i(t+1)}}{p_i^t} \right]^2 \quad \times$$

$$\qquad E^c \left[E^c \left[\exp\left[-\sigma_i^2(t) + 2\sigma_i(t)\psi_i(t) \right] \left(e^{\sigma_i \sigma_k \rho_{ik}} - 1 \right) \Big| \sigma_i, \tau_i, \rho_{ik} \right] \right]$$

$$= E^c \left[\frac{C_i e^{\mu_i(t+1)}}{p_i^t} \right]^2 \quad \times$$

$$\qquad E^c \left[\left(e^{\sigma_i \sigma_k \rho_{ik}} - 1 \right) E^c \left[\exp\left[-\sigma_i^2(t) + 2\sigma_i(t)\psi_i(t) \right] \Big| \sigma_i, \tau_i, \rho_{ik} \right] \right]$$

$$= E^c \left[\frac{C_i e^{\mu_i(t+1)}}{p_i^t} \right]^2 E^c \left[e^{\sigma_i^2(t)} \left(e^{\sigma_i \sigma_k \rho_{ik}} - 1 \right) \right]. \tag{4.7}$$

Similarly,

$$E^c \left[\frac{c_i^t e^{\mu_i}}{p_i^t} \right] E^c \left[\frac{c_i^t e^{\mu_i}}{p_i^t} \left(e^{\sigma_i \sigma_k \rho_{ik}} - 1 \right) \right]$$

$$= E^c \left[\frac{C_i \exp\left[\mu_i(t+1) - \frac{1}{2}\sigma_i^2(t) + \sigma_i(t)\psi_i(t) \right]}{p_i^t} \left(e^{\sigma_i \sigma_k \rho_{ik}} - 1 \right) \right] \quad \times$$

$$\qquad E^c \left[\frac{C_i \exp\left[\mu_i(t+1) - \frac{1}{2}\sigma_i^2(t) + \sigma_i(t)\psi_i(t) \right]}{p_i^t} \right]$$

$$= \left(E^c \left[\frac{C_i e^{\mu_i(t+1)}}{p_i^t} \right] \right)^2 E^c \left[\exp\left[-\frac{1}{2}\sigma_i^2(t) + \sigma_i(t)\psi_i(t) \right] \left(e^{\sigma_i \sigma_k \rho_{ik}} - 1 \right) \right] \quad \times$$

$$\qquad E^c \left[\exp\left[-\frac{1}{2}\sigma_i^2(t) + \sigma_i(t)\psi_i(t) \right] \right]$$

$$= \left(E^c \left[\frac{C_i e^{\mu_i(t+1)}}{p_i^t} \right] \right)^2 \quad \times$$

$$\qquad E^c \left[\left(e^{\sigma_i \sigma_k \rho_{ik}} - 1 \right) E^c \left[\exp\left[-\frac{1}{2}\sigma_i^2(t) + \sigma_i(t)\psi_i(t) \right] \Big| \sigma_i, \tau_i, \rho_{ik} \right] \right]$$

$$= \left(E^c \left[\frac{C_i e^{\mu_i(t+1)}}{p_i^t} \right] \right)^2 E^c \left[\left(e^{\sigma_i \sigma_k \rho_{ik}} - 1 \right) \right]. \tag{4.8}$$

Using (4.7) and (4.8), the first term in parentheses in (4.5) becomes,

$$E^c \left[\left(\frac{c_i^t e^{\mu_i}}{p_i^t} \right)^2 \left(e^{\sigma_i \sigma_k \rho_{ik}} - 1 \right) \right] - E^c \left[\frac{c_i^t e^{\mu_i}}{p_i^t} \right] E^c \left[\frac{c_i^t e^{\mu_i}}{p_i^t} \left(e^{\sigma_i \sigma_k \rho_{ik}} - 1 \right) \right]$$

$$= E^c \left[\frac{C_i e^{\mu_i(t+1)}}{p_i^t} \right]^2 E^c \left[e^{\sigma_i^2(t)} \left(e^{\sigma_i \sigma_k \rho_{ik}} - 1 \right) \right]$$

$$- \left(E^c \left[\frac{C_i e^{\mu_i(t+1)}}{p_i^t} \right] \right)^2 E^c \left[\left(e^{\sigma_i \sigma_k \rho_{ik}} - 1 \right) \right]$$

$$> E^c \left[\frac{C_i e^{\mu_i(t+1)}}{p_i^t} \right]^2 E^c \left[\left(e^{\sigma_i \sigma_k \rho_{ik}} - 1 \right) \right]$$

$$- \left(E^c \left[\frac{C_i e^{\mu_i(t+1)}}{p_i^t} \right] \right)^2 E^c \left[\left(e^{\sigma_i \sigma_k \rho_{ik}} - 1 \right) \right]$$

$$= \text{var}^c \left(\frac{C_i e^{\mu_i(t+1)}}{p_i^t} \right) E^c \left[\left(e^{\sigma_i \sigma_k \rho_{ik}} - 1 \right) \right] > 0. \tag{4.9}$$

Using (4.6) and (4.9) in (4.5) implies that for any stock i, $\text{cov}^c \left(E_t[R_i^t], \beta_i^t \right) > 0$.

In this economy in which risk has no role in stock prices, any portfolio that does not require short selling will appear to price stocks. Thus, the mere existence of an empirical factor model is not sufficient to conclude that cross-sectional differences in expected returns are risk based. Furthermore, the portfolios themselves need provide no economic insight on the underlying reasons for the cross-sectional variance in security returns.

5 Conclusion

The object of this chapter was to provide a perspective on three important issues in financial economics. First I have argued that the empirically observed size effect should not be considered an anomaly. Second, while asset pricing model that are motivated purely by the empirically observed relation between market value (book-to-market) and return might well do a good job capturing cross-sectional variation in expected returns, they provide no insight on the *economic causes* of this cross-sectional variation. In particular, the variation need not result from cross-sectional variation in firm riskiness. Finally, while the additional explanatory power of market value (book-to-market) over asset pricing models such as the CAPM or APT clearly indicate that, at least as far these empirical specifications are concerned, these models cannot be capturing all cross-sectional variation in expected returns, this result cannot be used, by itself, to reject the model. It is quite conceivable that the result could derive from errors in the empirical specification rather than shortcomings in the model under consideration.

References

Amihud, Y. and H. Mendelson (1986), 'Asset pricing and the bid-ask spread,' *Journal of Financial Economics*, **17**:223–250.

Ball, R. (1978), 'Anomalies in relationships between securities' yields and yield-surrogates,' *Journal of Financial Economics*, **6**:103–126.

Ball, R., and S.P. Kothari (1989), 'Nonstationary expected returns: Implications for tests of market efficiency and serial correlations in returns,' *Journal of Financial Economics*, **6**:103–126.

Banz, R.F. (1978), *Limited Diversification and Market Equilibrium: An Empirical Analysis*, Ph.D. Dissertation, University of Chicago, Chicago, Ill.

Banz, R.F. (1981), 'The relationship between return and market value of common stocks' *Journal of Financial Economics*, **9**:3–18.

Basu, S. (1983), 'The relationship between earnings yield, market value, and return for NYSE common stocks: Further evidence,' *Journal of Financial Economics*, **12**:129–156.

Berk, J.B. (1995) 'A Critique of Size Related Anomalies,' *Review of Financial Studies*, **8**:275–86.

Berk, Jonathan B. (1996), 'Sorting Out Sorts,' forthcoming, *Journal of Finance*.

Berk, Jonathan B., Richard C. Green and Vasant Naik (1999), 'Optimal Investment, Growth Options and Security Returns,' forthcoming, *Journal of Finance*.

Black, F. (1992), 'Beta and Return'. Presentation at The Berkeley Program in Finance: Are Betas Irrelevant? Evidence and Implications for Asset Management.

Black, F., M.C. Jensen and M. Scholes (1972), 'The Capital Asset Pricing Model: Some Empirical Tests.' In *Studies in the Theory of Capital Markets*, Michael C. Jensen (ed.), Praeger Publishers, New York, USA.

Blume, M. and R. Stambaugh (1983), 'Biases in computed returns: An application to the size related effect,' *Journal of Financial Economics*, **12**:387–404.

Brown, P., A.W. Kleidon and T.A. Marsh (1983), 'New evidence on the nature of size-related anomalies in stock prices,' *Journal of Financial Economics*, **12**:33–56.

Chan, K.C., N. Chen and D. Hsieh (1985), 'An exploratory investigation of the firm size effect,' *Journal of Finance*, **46**:1467–1484.

Chan, L. K., Y., Hamao and J. Lakonishok (1991), 'Fundamentals and stock returns in Japan,' *Journal of Finance*, **46**:1739–1789.

Chen, N. (1988), 'Equilibrium asset pricing models and the firm size effect'. In *Stock Market Anomalies*, Elroy Dimson (ed.), Cambridge University Press.

Chen, N., S.A. Ross and R. Roll (1983), 'Economic Forces and the Stock Market,' *Journal of Business*, **59**:383–403.

Clarkson, P., J. Guedes and R. Thompson (1993), 'Is Estimation Risk Priced? Is It Observable?' Cox School of Business, S.M.U., working paper.

Denton, F. (1985), 'Data Mining As An Industry,' *Review of Economics and Statistics*, **February**:124–127.

Dimson E. (ed.) (1988) *Stock Market Anomalies*, Cambridge University Press.

Fama, E.F., and K.R. French (1992), 'The Cross-Section of Expected Stock Returns,' *Journal of Finance*, **47**:427–466.

Fama, E.F., and K.R. French (1993), 'Common risk factors in the returns on stocks and bonds,' *Journal of Financial Economics*, **33**:3–56.

Fama, E.F., and K.R. French (1995), 'Size and Book-to-Market Factors in Earnings and Returns,' *Journal of Finance*, **50**:131–84.

Ferson, W.E. (1995), 'Warning: Attribute-sorted Portfolios Can be Hazardous to Your Research!'. In *Modern Finance Theory and Applications*, Saitou, S., K. Sawaki and K. Kubota (eds.), Center for Academic Societies, Osaka, Japan, 21–32.

Foster, F.D. and T. Smith (1997), 'Assessing Goodness-of-Fit of Asset Pricing Models: The Distribution of the Maximal R^2,' *Journal of Finance*, **52**(2):591–607.

Jegadeesh, N. (1992), 'Does Market Risk Really Explain the Size Effect,' *Journal of Financial and Quantitative Analysis*, **27**:337–352.

Green, Richard C. (1986), 'Benchmark Portfolio Inefficiency and Deviations from the Security Market Line,' *Journal of Finance*, **41**: 1051–68.

Gibbons, M.R., S.A. Ross and J. Shanken (1989), 'A Test of the Efficiency of a Given Portfolio,' *Econometrica*, **57**:1121–1152.

Keim, D.B. (1983), 'Size-related anomalies and stock return seasonality: Further empirical evidence,' *Journal of Financial Economics*, **12**:13–32.

Keim, Donald B. and Robert F. Stambaugh (1984), 'A further investigation of the weekend effect in stock returns,' *Journal of Finance*, **39**:819–840.

Keim, D.B. (1988), 'Stock market regularities: A synthesis of the evidence and explanations'. In *Stock Market Anomalies*, Elroy Dimson (ed.), Cambridge University Press.

Kothari, S. P. and J. Shanken (1997), 'Book-to-market, Dividend Yield, and Expected Market Returns: A Time-series Analysis,' *Journal of Financial Economics*, **44**(2):169–203.

Kuhn, T. (1970), *The Structure of Scientific Revolutions*, University of Chicago Press.

Lee, C.M.C., A. Shleifer and R.H. Thaler (1991), 'Investor Sentiment and the Closed-End Fund Puzzle,' *Journal of Finance*, **46**:75–110.

Lo, A.W. and A.C. MacKinlay (1990), 'Data-Snooping Biases in Tests of Financial Asset Pricing Models,' *Review of Financial Studies*, **3**:431–468.

Lo, A.W. and A.C. MacKinlay (1992), 'Maximizing Predictability in the Stock and Bond Markets,' MIT Working Paper.

Pontiff, J. and L. Schall (1998), 'Book-To-market as predictors of market returns,' *Journal of Financial Economics*, **49**:141–160.

Reinganum, M.R. (1981), 'Misspecification of capital asset pricing: Empirical anomalies based on earnings' yields and market values,' *Journal of Financial Economics*, **9**:19–46.

Rosenberg, B., K. Reid and R. Lanstein (1985), 'Persuasive evidence of market inefficiency,' *Journal of Portfolio Management*, **11**:9–17.

Roll, R. (1977), 'A Critique of the Asset Pricing Theories Tests; Part 1: On Past and Potential Testability,' *Journal of Financial Economics*, **4**:129–76.

Roll, R. (1981), 'A Possible Explanation of the Small Firm Effect,' *Journal of Finance*, **36**:879–88.

Schultz, P. (1983), 'Transaction costs and the small firm effect: A comment,' *Journal of Financial Economics*, **12**:81–88.

Schwert G.W. (1983), 'Size and stock returns, and other empirical regularities,' *Journal of Financial Economics*, **12**:3–12.

Stattman, D. (1980), 'Book values and stock returns,' *The Chicago MBA: A Journal of Selected Papers*, **4**:25–45.

Stoll H., and R. Whaley (1983), 'Transaction costs and the small firm effect,' *Journal of Financial Economics*, **12**:57–79.

Ziemba, W.T. (1994), 'World Wide Security Market Regularities,' *European Journal of Operational Research*, **74**:198–229.

The Demise of Size*

Elroy Dimson and Paul Marsh

Abstract

Many researchers have uncovered empirical regularities in stock market returns. If these regularities persist, investors can expect to achieve superior performance. Unfortunately, nature can be perverse. Once an apparent anomaly is publicised, only too often it disappears or goes into reverse. The latter seems to have happened to the small firm premium. After the UK size premium was documented and disseminated, a historical small-cap *premium* of 6% was followed by a small-cap *discount* of around 6%. This study presents evidence of and some explanations for the disappearance of the small firm premium.

1 Introduction

Stock market anomalies generate strong and conflicting opinions from students of finance. Consider the views of four leading scholars:

Robert Haugen (1999) writes that "In the course of the last 10 years, financial economists have been struggling to explain... the *huge, predictable premiums in the cross-section of equity returns*" (italics in original). Haugen cites numerous regularities that provide an opportunity for investors to achieve superior stock market returns by taking advantage of known anomalies.

Paul Samuelson (1989) is more cautious: "Out of the thousands of published and unpublished statistical testings of various forms of the [efficient markets] hypothesis, a few dozen representing a minuscule percentage have isolated profitable exceptions to the theory." Samuelson concludes that while a few geniuses can successfully exploit anomalies in the market, recognising those individuals *ex ante* is difficult or impossible.

Fischer Black (1993) wrote that "Most of the so-called anomalies that have plagued the literature on investments seem likely to be the result of datamining. The researcher who finds [a profit opportunity] writes it up, and we have a new anomaly. But it generally vanishes as soon as it's discovered".

---

*We thank Don Keim and Bill Ziemba for valuable comments and David Stolin for research assistance. An earlier version of this paper was published in *Journal of Portfolio Management*, 1999.

Black was asserting that the out-of-sample return from seeking to exploit a market anomaly is expected to be zero.

Finally, Richard Roll (1994) reveals that "Over the past decade, I have attempted to exploit many of the seemingly most promising 'inefficiencies' by actually trading significant amounts of money... Many of these effects are surprisingly strong in the reported empirical work, but I have never yet found one that worked in practice." If none of these opportunities makes a profit, while some inevitably make losses, the expected return from following an anomaly is negative.

Roll was not the first person to experience the perversity of financial markets. In the literary world, too, nature's perversity had been noted, for example, by Samuel Beckett (1938) in his first novel, *Murphy*. Beckett's belief that "if anything can go wrong, it will" is presented by Dickson (1980) as the origin of a rule that has subsequently been attributed to a variety of individuals carrying the names Murphy, Finnagle or Sod. *Murphy's Law*, sometimes summarised as "bread always falls with the buttered side down", suggests that the return from trying to profit from a market anomaly will usually be negative.

However, Murphy's Law is not as simple as this. Dickson (1978) illustrates the complexity of Murphy's Law with the story of the man whose bread fell and landed buttered side up. He ran to his rabbi to report this deviance from one of the basic rules of the universe. At first the rabbi would not believe him but finally became convinced that it had happened. However, he did not feel qualified to deal with the question and passed it along to one of the world's leading Talmudic scholars. After months of waiting, the scholar finally came up with an answer: "The bread must have been buttered on the wrong side." Murphy's Law implies true perversity, and an absence of predictable behaviour. Just as an anomaly cannot be relied upon to continue, it cannot be relied upon to go into reverse. When an investor has become convinced that reversal is highly probable, Murphy's Law implies that the anomaly will persist after all. This aspect of Murphy's Law is sometimes referred to as *Mrs Murphy's Corollary*, which is summarised by Müller (1995) as "you cannot successfully determine beforehand which side of the bread to butter."

Fama (1998) integrates Murphy's Law into his efficient markets framework by pointing out that there should be about as many cases of perversity as of persistence of an anomaly. While many anomalies disappear when they are measured in a different way, some are apparent regardless of the choice of methodology. It is therefore of value to look at the consistency over time of those anomalies for which the empirical evidence is the most robust.

In this study, we examine the premier stock market anomaly – the striking outperformance of smaller companies.[1] We evaluate their performance from

---

[1]The outperformance of smaller companies has been documented in numerous studies

the time at which the size premium was documented and publicised, to the present day.[2] Because it is potentially explained by risk and cost considerations, the small-firm regularity has been regarded as one of the most robust patterns in stock price behaviour, and least likely to be a manifestation of data mining. Roll (1994), for example, pointedly excludes from his list of trading opportunities those anomalies "that have potential explanations based on misestimation of risks or costs. In this latter category, I would place empirical findings that small firms earn more than large firms (beta adjusted)." Yet we find that this anomaly did not persist, that the outperformance of smaller companies vanished, and that the out-of-sample small-firm premium turned negative. In brief, small-cap investors experienced Murphy's Law.

Though our study focuses on the size effect in the United Kingdom, we also contrast the British with the American experience. In Section 2 we summarise the evidence that convinced investors in 1987 that UK small-caps can be expected to outperform their larger counterparts. In Section 3 we document the return achieved over the following decade by smaller companies and draw comparisons with the United States. Section 4 discusses some of the implications of our observations, and Section 5 concludes.

2 The Long-Term History of Smaller Companies

The UK small firm premium was comprehensively documented by Dimson and Marsh (1987), and was brought to prominence by the launch in early 1987 of the Hoare Govett Smaller Companies Index.[3] The HGSC Index covers the smallest tenth by equity capitalisation of the UK market. A companion index, the Hoare Govett 1000 (HG1000) Index, covers companies that are sufficiently small to fall within the bottom 2% by value of the market (excluding closed-end funds). These indices, compiled daily at the London Business School, are described in detail in Dimson and Marsh (1998). The indices are rebalanced annually,[4] and total returns include gross dividends reinvested in the index.

---

since it was first identified by Banz (1981) and Reinganum (1981) (see Hawawini and Keim, 1999). It has been highlighted as an investment strategy in books such as Haugen (1999), Haugen and Lakonishok (1988), Klein and Lederman (1993), Mott *et al.* (1997), O'Shaughnessy (1998a, 1998b) and Siegel (1998).

[2]See Booth and Keim (1999) for a similar before-and-after analysis of the US size effect, although with a focus on the US small-firm January seasonal.

[3]Earlier, less comprehensive, studies had suggested the existence of a UK size premium, but these articles had little impact on the investment community. The previous evidence on UK small-cap performance includes Marsh (1979), Dimson and Marsh (1984, 1986).

[4]We minimise any potential bias by weighting index constituents according to their market capitalisation, identifying index composition using data prior to the year-end, and rebalancing each index at the last trading day of the year. The rationale for this is described

The cumulative performance of smaller companies is compared below to the All Share Index, which covers the vast majority by value of the entire UK equity market. For this purpose, we employ the FTSE All Share Index from its 1962 inception onwards, and a reconstructed index that follows the same design principles for the period 1955–62. Index returns are based on the London Share Price Database (LSPD), which provides monthly prices, capital changes and dividends for all UK registered, sterling denominated shares traded on the London Stock Exchange. For the period from 1975 onwards, there is 100% coverage of all relevant equities. For the twenty-year period beginning in 1955, the LSPD comprises a very large, fully representative sample of all companies and all new issues (see Dimson and Marsh, 1999). The fully-representative nature of the LSPD ensures that non-surviving, as well as recent, companies are included in the index backhistories. The availability of dividend information ensures that we produce an accurate estimate of total returns, including gross income reinvested at the ex-dividend date for each stock.

Table 1 presents a summary of the annual returns achieved by small and micro-capitalisation companies, as represented by the HGSC and HG1000 indices, and the All Share. The upper panel presents the long term statistics, as published in the 1987 launch document for the Hoare Govett index (see Dimson and Marsh, 1987). The lower panel provides corresponding statistics for the entire history to date. It is clear that, at least up to the January 1987 launch of the HGSC, smaller companies outperformed their larger counterparts, and the very smallest (micro) companies performed best of all. An investment at the start of 1955 in the HGSC Index would have given an annualised return of 20.2% by the date of the launch, as compared to 14.2% if the same money had been invested in the All Share Index. Over the same interval, the return on the very smallest companies, as represented by the HG1000 Index, would have been 23.5%. Up to the index launch date, and indeed to the present time, the long-term cumulative performance of smaller companies has been markedly superior to the market as a whole.[5]

The pre-launch history of the HGSC recorded strong outperformance by smaller companies. The annualised rate of return on the smaller companies index exceeded the return on the All Share by six percentage points per year. Since the All Share covers some 98% of the market by value, it already benefits from the performance of smaller companies. Relative to large (non-HGSC)

---

in Blume and Stambaugh (1983), Roll (1983a) and Campbell, Lo and MacKinlay (1997).

[5]Note that, although returns were highest from the smallest companies, they cannot be bought in the same quantities as large-cap stocks. For UK stocks, the velocity (proportion of equity market capitalisation traded each year) is similar for small and large companies (see Dimson and Marsh, 1997). It follows that a transaction of average size in every constituent of the HG1000 would represent an investment which is tiny (i.e., about one-fiftieth of the value) compared to a transaction of average size in, say, every constituent of the All Share Index.

Table 1. Distribution of Annual Percentage Returns Since 1955.

Period	Index	Geometric Mean	Arithmetic Mean	Standard Deviation
1955–86	All Share	14.2	18.3	33.4
	HGSC	20.2	24.5	32.9
	HG1000	23.5	27.0	30.6
	HGSC – All Share	**6.0**	**6.1**	**14.0**
	HG1000 – All Share	**9.2**	**8.7**	**20.7**
1955–97	All Share	14.4	17.7	29.5
	HGSC	17.8	21.3	29.9
	HG1000	19.8	23.1	29.3
	HGSC – All Share	**3.4**	**3.6**	**14.2**
	HG1000 – All Share	**5.4**	**5.5**	**20.8**

companies, the return differential in favour of low-cap strategies was even higher than is suggested by comparison with the All Share.

The pre-launch backhistory also demonstrated that the volatility of small-cap returns was no higher than that of the All Share. Since the HGSC represents a universe with a far lower aggregate value than the All Share, this is all the more notable. Indeed, with only one (trivial) exception, over all periods of five years duration between 1955 and 1986, the annualised standard deviation of monthly returns was systematically lower for the HGSC than for the All Share (see Dimson and Marsh, 1998).

For investors, a further attraction of smaller UK companies was that they proved to have a beta, relative to the All Share, that was low. Whether estimated using the entire backhistory of annual returns, or employing the Dimson (1979) estimator with 60 monthly returns, the beta of the HGSC was below one. This is consistent with the low level of beta noted for small companies in Japan (Ziemba and Schwartz, 1991), and also for other countries (see the references in Hawawini and Keim, 1999). Risk-adjusted returns, calculated as in Banz (1981), were therefore even larger than the unadjusted returns reported in Table 1.

Comprehensive evidence on the UK small firm premium therefore confirmed findings for the US by Banz (1981) and Reinganum (1981), as well as research that was to appear on other markets over periods of varying length, ending in the 1980s.[6] In the United States, Schwert's (1983) special

---

[6]Amongst the latter are Brown, Keim, Kleidon and Marsh (1983) for Australia; Hawaw-

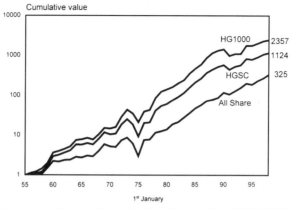

Figure 1. Cumulative Return from Smaller Companies, 1955–1998. Source:
Dimson and Marsh (1998).

issue of the *Journal of Financial Economics* on stock market anomalies had
been reported widely, and Dimensional Fund Advisors, amongst others, had
promptly offered investors comprehensive, passive exposure to the small company effect. Just as the size effect had achieved substantial visibility amongst
investment professionals in the US, the same thing happened in the UK.

Figure 1 shows the cumulative performance of a £1 investment in the All
Share and in the two small company indices. The dramatic outperformance
of small-cap stocks up to 1987 attracted substantial media attention, and after publication of the long-run history of smaller company returns in Britain
(Dimson and Marsh, 1987) there were over 200 follow-up articles in the UK
press. By the end of 1988, at least 30 open- and closed-end funds had been
launched to exploit the perceived outperformance of the HGSC, and numerous institutions developed a strategy of investing in smaller companies as a
distinct asset class. We therefore refer to 1987–88 as the UK 'launch period'.
In the next section, we document the size premium both during this launch
period, and more importantly, over the post-launch period from end 1988
onwards.

ini, Michel and Corhay (1989) for Belgium; Berges, McConnell and Schlarbaum (1984) for
Canada; Wahlroos and Berglund (1986) for Finland; Louvet and Taramsco (1991) for
France; Stehle (1992) for Germany; Coghlan (1988) and McKillop and Hutchinson (1988)
for Ireland; Ziemba (1991) and Comolli and Ziemba (1999) for Japan; Van den Burgh and
Wessels (1983) for the Netherlands; Gillan (1990) for New Zealand; Rubio (1986) for Spain;
Cornioley and Pasquier-Dorthe (1991) for Switzerland; Ma and Shaw (1990) for Taiwan;
and the earlier work by Dimson and Marsh (1984, 1986) for the UK. Most of these studies
are reviewed in Dimson (1988) and in Hawawini and Keim (1999).

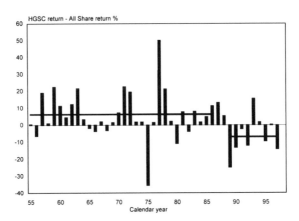

Figure 2. Annual Size Premium, 1955–97. Source: Dimson and Marsh (1998).

3   The Recent Performance of Smaller Companies

Over the period from 1955 up to the 1987 launch of the HGSC, smaller companies had outperformed the All Share by an annual average of six percentage points. This outperformance is illustrated in Figure 2 by the horizontal line that identifies the arithmetic average size premium over the period 1955–86. But although smaller companies continued to outperform the All Share during the 1987–88 index launch period, their subsequent performance was extremely disappointing.

Over the post-launch period subsequent to 1988, the HGSC underperformed the All Share. The lower horizontal line in Figure 2 shows how the arithmetic average return premium on the HGSC turned negative. Over the period 1989–97, it was lower than *minus* six percentage points per year. For the very smallest companies, the contrast was even greater: the HG1000 index, whose return premium had averaged between eight and nine percentage points during the 1955–86 period, provided an annual return over the post-1988 period that was some eight percentage points lower than the All Share.

The reversal in the performance of small-cap stocks is summarised in Table 2, which shows the arithmetic average returns for the indices over the period prior to 1987, and subsequent to 1988. During the first period, coinciding with the returns publicised when the HGSC was launched, the total return on the HGSC averaged 24.5%, as compared to 18.3% for the All Share. Smaller companies provided a return premium of just over six percentage points.

Over the following decade, the performance of the All Share was almost identical to its historical level, namely 17.1% per year. But over this period,

Table 2. Reversal of the UK Size Effect, 1955–97.

Index	1955–86		1989–97	
	Arithmetic mean return	*t*-value of size premium	Arithmetic mean return	*t*-value of size premium
All Share	18.3	na	17.1	na
HGSC	24.5	na	10.6	na
HGSC – All Share	**6.1**	**2.46*****	**−6.5**	**−1.67***
HG1000	27.0	na	9.2	na
HG1000 – All Share	**8.7**	**2.37****	**−7.9**	**−1.29**

*** Significant at the 99% level
** Significant at the 97.5% level
* Significant at the 90% level

the HGSC provided an average return of less than 11%, a return *discount* of more than six percentage points annually. For the smallest companies, as represented by the HG1000, the reversal was even more extreme.

This reversal is not only large in economic terms, but it is surprising from the perspective of statistical significance. Over the 32-year backhistory, the standard deviation of the annual HGSC return premium is 14.0%, as shown in Table 1. The *t*-value for the annual premium of 6.1% is therefore $6.1\sqrt{32}/14.0 = 2.46$ which is significantly above zero at the 99% confidence level (see Table 2). The *t*-statistic for the nine-year post-launch period, based on the standard deviation of 11.7%, is equal to $-6.5\sqrt{9}/11.7 = -1.67$ which is significantly below zero at the 90% confidence level, and significantly below the historical average at the 99% confidence level. Whereas the pre-launch HGSC return premium appeared statistically significant, the subsequent decade's underperformance by smaller companies also appeared significant, but now in the opposite direction.

We have also looked at the performance of smaller companies in the United States, and the story is similar. The publication and dissemination of the Banz (1981) and Reinganum (1981) research led to considerable investment interest in US small-caps. This spurred the launch of significant new small-cap investment vehicles led by Dimensional Fund Advisors, who raised several billion dollars within two years of their December 1981 launch. As in the UK, this honeymoon period lasted for approximately two years, until the end of 1983, and during this period, US small-caps continued to outperform. For comparability with the UK, we refer to 1982–83 as the US 'launch period'. Over the period from the start of 1955 (chosen for comparability with the UK research) until the end of 1983 (the end of the 'launch period'), American small-caps outperformed large-caps by more than four percentage points per year. As in the UK, however, after the launch period, smaller companies in the US also underperformed their larger brethren with a return that was on average several percentage points lower than large-caps.

The dramatic reversal in the fortunes of smaller companies can be seen in Table 3, which records the difference in annualised returns between small-cap and large-cap stocks (the 'small-cap premium') and between micro-cap and large-cap stocks (the 'micro-cap premium').[7] While the entire long-term record from 1955 to 1997 favours smaller companies, this has comprised two subperiods. From 1955 up to the end of the 'launch period' in each country, small-cap and micro-cap stocks performed very well in relative terms.[8] But from the end of the launch period, through to the end of 1997, they underperformed. The reversal in the small firm premium is striking, and calls for an explanation.

4   Explaining the Reversal

There is an extensive literature on the small firm effect, much of which is summarised in Dimson (1988), Dimson and Marsh (1989), Keim and Ziemba (1999) and Hawawini and Keim (1999). Most of the research, however, focuses on documenting and explaining a positive premium in the return on small companies. If we are to explain the reversal in small-cap performance, then we must look for factors that explain not only high, but also low subsequent returns from smaller companies.

In the early 1980s, the puzzle was why small companies had performed so well, and the need for a model to explain low returns from small-cap companies was not at all obvious. Schwert (1983), for example, wrote that "to successfully explain the 'size effect', new theory must be developed". He visualised a "search for the variable or combination of variables that will make the 'size effect' go away". Yet within a few years, it became clear that this

---

[7]Small-caps are defined in the UK as HGSC constituents and in the US as deciles 6–8 of the New York Stock Exchange. Micro-cap stocks are defined in the UK as the constituents of the HG1000 and in the US as deciles 9–10 of the NYSE. Premia are measured relative to a large-cap index: in the UK we use the All Share, and in the US we use deciles 1–5 of the NYSE. The returns data for the US indices are taken from Ibbotson Associates (1998).

[8]Our UK statistics run from 1955 onwards, and for comparability we report US results beginning in 1955. If the US premia are calculated from 1926 onwards as in Ibbotson Associates (1998), the small-cap (i.e. deciles 6–8) premium is 1.6% (1926–97) and 2.6% (1926–end of launch period), and the micro-cap premium (i.e. deciles 9–10) is 1.5% (1926–97) and 4.7% (1926–end of launch period).

Table 3. Small-cap and Micro-cap Premia in the UK and US, 1955–97.

Period	Interval for measuring premium	Small-cap premium*		Micro-cap premium*	
		UK	US	UK	US
Entire period	1955–97	3.4	2.1	5.4	1.5
Subperiods	1955–end of launch period**	5.9	4.1	9.1	5.7
	End of launch period–1997	–5.6	–2.4	–7.8	–7.2

*Relative to All Share Index (UK) or NYSE deciles 1–5 (US)
**End of index launch period is 1988 (UK) or 1983 (US)

search had been unsuccessful. By the end of the decade, Dimson and Marsh (1989) concluded that "Even if we mine the share price databases mercilessly, we are unlikely to be able to explain the very high returns achieved, for example, by the HGSC index. In the long run, this suggests that smaller companies can be expected to continue to beat the market as a whole by a modest margin, but by rather less than the high margin of outperformance achieved in recent years".

Relative to the returns backhistory, these predictions of reduced returns from small-cap stocks were pessimistic. But in the light of the subsequent performance of smaller companies, they were not pessimistic enough. There is now a need to re-examine some of the arguments that had been used to explain the superior returns from smaller companies, and link them to subsequent empirical evidence of underperformance, as well as the more recent literature.

4.1 Backhistory Biases

One possibility for the reversal of the UK small firm premium is that the backhistory was biased, so that the premium estimated from the 1955–86 period was overstated or perhaps even non-existent. Four potential sources of bias are the survival of markets or asset categories, survivorship bias in the database, bid-ask bounce and omission of transaction costs. It seems appropriate to consider these potential sources of inaccuracy in the HGSC returns backhistory, before considering alternative explanations for the changing fortunes of smaller companies.

Goetzmann and Jorion (1999) point out that expected returns are overstated if they are inferred from an asset category like US equities which, with hindsight, is known to have survived.[9] The historically measured UK equity premium may therefore exceed future expectations because the UK market has over the long haul been one of the best performing of the world's stock markets. This reflects the fact that the UK market has 'survived' rather than suffering extreme discontinuities from wars or revolutions which afflicted countries such as Germany, Japan and Russia. In our research, however, smaller companies are defined as a prespecified proportion of the value of the equity market. Defined this way, the small-cap asset category is guaranteed to survive (at least, relative to large-caps), and survival does not taint the 1955–86 backhistory.

Another possibility is identified by Elton, Gruber and Blake (1996) who find that annualised returns are markedly higher for a security that continues to trade, than for one that ceases to trade.[10] As pointed out earlier, the HGSC backhistory is based on all companies that had a listing at the start of each year, and index returns include the subsequent performance of every company in the HGSC. No company is ever omitted at a particular date because it was subsequently delisted or suspended. Given that the London Share Price Database is fully representative in its coverage,[11] our backhistory is not subject to look-ahead bias.

Portfolio returns can also be biased as a result of failing to consider the bid-ask spread in security prices. Canina, Michaely, Thaler and Womack (1998) warn about the dangers of using equally-weighted index returns or frequently-rebalanced portfolio returns to estimate long run premia. This bias has been known since the papers by Blume and Stambaugh (1983) and Roll (1983a), and does not arise in the context of our research, which uses market capitalisation-weighted indices that are rebalanced annually.

This leaves the question of the transactions costs associated with investing in smaller companies. Some researchers (e.g., Stoll and Whaley, 1983,

---

[9]Goetzmann and Jorion's view that *ex post* US equity returns have been anomalously high has been expressed vigorously by Rietz (1988) and Brown, Goetzmann and Ross (1995). To those who understand the deeper implications of Murphy's Law, it is no surprise that over the period since these papers appeared, the US equity risk premium has been even larger than the high levels observed prior to publication of these papers. Over the period 1926–87, the annualised equity premium is estimated by Ibbotson Associates (1998) to be 6.2%; the 1988–97 average was larger, at 9.3%, while the 1995–97 average was larger still at 14.1%.

[10]Elton, Gruber and Blake consider mutual funds, where survival is strongly linked to superior performance. For companies, non-survivors comprise those firms that experience financial distress and also those that are taken over at a premium. Stolin (1999) examines the impact of company survival within the London Share Price Database.

[11]As explained earlier, the LSPD covers every eligible stock from 1975 and a one-in-three random sample for the period 1955–74. Stolin (1998) shows that the survivorship of the one-in-three sample is indistinguishable from the survivorship of the omitted two-thirds of companies.

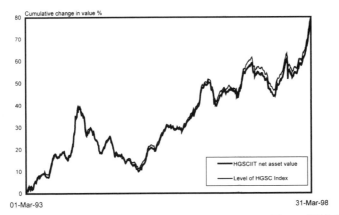

Figure 3. Net Asset Value of the HGSC Index Investment Trust, 1993–98.

and Knez and Ready, 1996) have argued that larger costs are incurred in managing a portfolio of small companies, as compared to large ones. This is undoubtedly true if we compare the costs of running segregated large-cap portfolios with those of small-cap portfolios, given that smaller stocks typically have higher dealing spreads. From the perspective of an investor with an initial portfolio of large-cap stocks, however, the relevant comparison is between restricting investment to large-caps or allowing a portfolio of small companies to accumulate.

Restricting investment to large-caps would require disposal of holdings in companies whose market capitalisation becomes small. For these investors, there would be no incremental cost in allowing a portfolio of small companies to accumulate. Costs arise through rebalancing back into large-cap stocks. The cost of running a stand-alone small-cap portfolio would hence be irrelevant. For many investors, transaction costs would actually have been larger as a result of limiting investment to large companies.

Fortunately, there is also empirical evidence on the costs of running a passive portfolio to track an index of smaller companies. Figure 3 illustrates the tracking error of a passive portfolio run by a listed closed-end fund, the HGSC Index Investment Trust, that invests in a diversified portfolio of HGSC constituents. Over its five-year life the net asset value of this fund has had a tracking error, the annualised standard deviation of which is just 1.46%. Over the five years, the fund has underperformed the HGSC Index by 0.37% per year. Since the portfolio incurs dealing costs while the HGSC Index does not, this underperformance provides us with estimates of the annual transactions costs associated with running a diversified portfolio of smaller companies.

The experience of this portfolio shows that not only is it possible to replicate the performance of the HGSC Index, but that the transactions costs involved in doing so, even when considered on a stand-alone basis, have been a modest 37 basis points per year. It therefore seems unlikely that the historical performance of the smaller companies index was illusory.[12] Over the period 1955–86, a well-diversified portfolio of smaller companies would indeed have beaten the All Share by close to six percentage points per year.

4.2 Risk Factors

Given the integrity of the backhistories for various countries, including the UK, the academic literature has examined whether smaller companies achieved higher returns through chance (Black, 1986), because of their exposure to various risk factors. Factors that might be relevant are the systematic risk of smaller companies, exposure to other pervasive risk factors, and agency problems that have an impact on asset pricing.

In the United States, smaller companies have an above-average beta relative to the market as a whole. Roll (1981) suggested that the risk of smaller companies might nevertheless be underestimated due to thin trading in their shares. Using the Dimson (1979) estimator for beta, Roll and others showed that when risk is estimated properly, the magnitude of the small-firm premium decreases but does not disappear. Hawawini and Keim (1999) show that in most countries outside the US, the beta of small companies, even after the Dimson adjustment, remains below one. Given that this is the case, the small firm premium cannot be explained away through misestimation of smaller companies' CAPM betas.

For explaining the relative performance of smaller companies, Ross' (1976) arbitrage pricing theory is potentially promising. In the APT, there are many different types of risk that may be disliked by investors, and which are therefore rewarded with a larger expected return. In early research, firm size appeared to be one of the APT factors associated with a higher long-run return in the stock market; see Chen (1988), Connor and Korajczyk (1988), Huberman and Kandel (1987), and Chan, Chen and Hsieh (1985). However, since these observations simply reflect hindsight about the historical size premium they are of limited interest for explaining the magnitude of small firm returns.

More recently, Fama and French (1992, 1993, 1995, 1996, 1998), Lakonishok, Shleifer and Vishny (1994) and others have examined premia in multifactor models of stock returns, and confirm earlier observations that equity capitalisation has historically been associated with a premium in common stock returns. Fama and French (1998) assert that the international evidence on the Fama–French model is less clear cut because the widely used MSCI database of large stocks "does not allow meaningful tests for a size effect, such

---

[12]Keim (1999) provides US evidence on Dimensional Fund Advisors' small-cap experience which reinforces this conclusion.

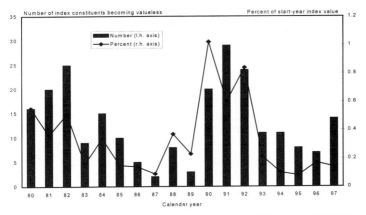

Figure 4. HGSC Index Constituents Becoming Valueless, 1980–97.

as that found by Banz (1981) in US returns." However, many studies have shown that the size effect is approximately linear in size decile, so that while omission of the smallest companies reduces statistical power, it is still possible to perform 'meaningful' tests, many of which are summarised in Hawawini and Keim (1999). Unfortunately, however, even if these tests tell us that low market capitalisation appears to be rewarded by a premium in returns, the tests do not tell us why this should be the case.

One possibility is that a distress factor premium can explain the size effect. He and Ng (1994) and Shumway (1996) suggest a measure of distress that is related to loss of listing on the stock exchange. While Fama and French (1995) believe that their book-to-market factor proxies for relative distress, it appears from Shumway's (1996) results that size is the better proxy. However, as shown by the bars in Figure 4, only a small number of HGSC index constituents (see the left-hand axis) become valueless each year. Even in the peak year of 1991, the 28 companies which became valueless represented only some 2% of the total population of index constituents. Additionally, those companies that become valueless start out with an initial market capitalisation that is on average well below the typical size of an index constituent. This is shown by the line-plot in Figure 4, which indicates the start-year index weighting of all the companies which subsequently became valueless in each year (see the right-hand axis). On this metric, the peak year was 1990, when bankruptcies accounted for just 1% of the start-year index value. Quite clearly, the companies that become valueless have relatively little impact on index performance.

It seems unlikely that the small firm premium can be attributed to a premium for the risk of default. The exposure of smaller companies to some sort of distress risk that is not related to delisting, such as low earnings (Fama and French, 1995), may be more relevant. However, there is unlikely to be sufficient variation in such risk attributes to explain a persistent discount, as well as the historical premium, in small company returns. We need to look for other potential explanations for the size effect.

4.3   Other Explanations

In the US, the size premium has been closely linked with the turn of the year. Keim (1983) was the first to discover a strong relationship between January returns and firm size, finding that nearly half of the size premium was due to January abnormal returns. This has been confirmed by many subsequent researchers (e.g., Roll, 1983b) who have shown the effect was confined to the last trading day of December and the first few trading days of January. Various explanations have been put forward to explain this, including year-end tax loss selling and a propensity to window-dress portfolios at the turn-of-the-year. These factors, coupled with difficulties in arbitraging price pressure at the end of the tax year, are said to produce stock exchange seasonalities that favour the performance of smaller companies.

Unfortunately, neither these explanations, nor the empirical findings they were seeking to explain, travel well. As Dimson and Marsh (1989, 1999) point out, in the UK the tax year for individuals starts on April 6, and many portfolio managers also report their results on an April–March cycle, while other corporations and institutional investors work to an end-December tax and reporting year-end. Arguably, any UK turn-of-the-year effect should be observable in April as well as, or instead of, January. However, in sharp contrast to the US, there is no evidence of a year-end effect in the UK size premium (i.e., HGSC returns minus All Share returns), regardless of whether we look at the end of the calendar year or the end of the tax year (see Dimson and Marsh, 1999).

Other explanations for the historical outperformance of smaller companies include a variety of agency arguments. Arbel and Strebel (1983) were the first to suggest that the relative neglect of smaller companies makes them problematic as portfolio holdings, thereby depressing price and enhancing expected returns. Maug and Naik (1995) explain why a focus on (large-company oriented) benchmarks leads investors to modify their investment strategy, again depressing prices and increasing returns for non-benchmark, and hence, lower capitalisation securities.

There are at least two problems with these agency-based explanations for the size premium. First, these explanations give rise to premia for small companies that are small, and nowhere near their historical level. And second, as shown in Dimson and Marsh (1989), many of these seasonality- and agency-based explanations are country specific, and do not explain why the small firm premium should have been present in so many different markets.

4.4   The Last Decade

The biggest challenge to theories of the size premium, however, is recent experience. Many explanations appear consistent with the historical US data on the size premium and with the backhistory for the HGSC index; indeed, they were largely assembled with hindsight about the previous performance of smaller companies. But if these theories are to have real value, they ought to help us understand why the most recent decade has witnessed underperformance by smaller companies. We need to explain recent underperformance in terms of factors such as transactions costs, systematic risk, factor exposure, default risk, analyst neglect, returns seasonality or the impact of performance benchmarks.

We have looked at these and other explanations for the recent performance of smaller companies. To give some idea of the difficulty of explaining returns through these theories, consider the role of systematic risk. If the beta of the HGSC Index were nowadays lower than it once was, then the expected return on smaller companies would be lower, which is consistent with recent history. But the risk of smaller companies has in fact risen, not fallen. Measured relative to the All Share Index, the beta of the HGSC was 0.69 for the sixty months ending on the index launch date in New Year 1987. By the end of the launch period, in December 1988, the estimated beta had risen to 0.86. If beta had jumped up sharply when the HGSC was launched, then prices would have fallen in order to give rise to a higher expected return. But in reality, small-cap prices continued to outperform throughout the two year index launch period.

Over the nine year post-launch period, the beta of the HGSC rose further from 0.86 at the start of the period to 0.98 for the sixty months ending New Year 1998. It is possible, therefore, that this further increase in beta could help explain the HGSC's underperformance during the post-launch period. Suppose, for example, that the overall increase of $0.98 - 0.86 = 0.12$ in the HGSC's estimated beta during the post-launch period reflected the true change in beta, and that this occurred as a result of a change averaging $0.12/9 = 0.0133$ over each of the nine years. With an equity risk premium of, say, 7%, an increase in beta of 0.0133 equates to an increase in the real required rate of return of around 0.093%. With a riskless real interest rate of 3% and an average beta of 0.92 over this period, the average annual expected return on the HGSC, based on the CAPM, would be $3 + (0.92 \times 7) = 9.44\%$. If a constant real growth rate of $g = 5\%$ were already factored into stock prices, then, applying the Gordon Model, the impact on the HGSC could be a decline of some 2% per annum (i.e., $(0.0944 - g)/(0.0944 + .0093 - g) - 1.0$).

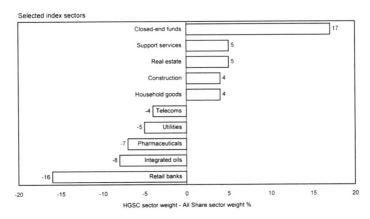

Figure 5. HGSC Sector Weights Relative to All Share, 1 January 1998.

In reality, however, an assumed growth rate of $g = 5\%$ seems implausibly high given historical levels of real dividend growth. For more realistic estimates, the expected impact on the HGSC would be less than the 2% indicated in the above example. It is thus difficult to explain more than a small proportion of the recent underperformance of smaller companies in terms of a re-rating to compensate for higher risk. More generally, over the entire period from 1955 to date, we have little hope of explaining the large fluctuations in the size premium just by looking at changes in beta and the risk premium. The existing hypotheses thus do relatively little to help explain UK small firm performance, either during the underperformance of the last decade, or during the outperformance of the preceding three decades.

4.4.1   The Impact of Sector Exposure

A more promising explanation would seem to be that, over this period, the fundamentals of the smaller company universe have changed in a way not envisaged at the time of the HGSC launch in the late 1980s. In particular, UK smaller companies may have underperformed over the last decade because they are concentrated in sectors that have themselves underperformed. Certainly, as Figure 5 shows, the HGSC and All Share indexes have very different sector weightings for a number of important sectors. The HGSC is severely under-represented in sectors such as retail banks, integrated oil companies, pharmaceuticals and utilities, while it is over-represented in closed-end funds, support services, real estate and construction.

In many cases, the low (or high) exposure of the HGSC to certain sectors is a reflection of the natural size of businesses within that sector. For example, the low weighting in integrated oil or power companies reflects the fact that

Table 4. Impact of Sector Weightings After Launch Period of the HGSC.

Index return %	1989	1990	1991	1992	1993	1994	1995	1996	1997	Mean
HGSC	10.8	−23.1	18.5	8.5	44.0	−4.1	14.4	17.0	9.5	10.6
All Share	35.6	−9.8	20.8	20.5	28.4	−5.8	23.9	16.7	23.6	17.1
HGSC − All Share	**−24.8**	**−13.3**	**−2.3**	**−12.0**	**15.6**	**1.7**	**−9.5**	**0.3**	**−14.0**	**−6.5**
Reweighted All Share	28.6	−16.9	16.8	19.1	37.7	−7.1	17.4	17.0	12.5	13.9
HGSC − r/w All Share	**−17.8**	**−6.2**	**1.7**	**−10.6**	**6.3**	**3.0**	**−3.0**	**0.0**	**−3.0**	**−3.3**

these businesses tend naturally to be large because of economies of scale. Similarly, in retail banking, size provides security to a bank's customers as well as economies of scale, and in the UK this makes it natural for retail banks to be tilted heavily towards the large-cap universe, and virtually absent from the HGSC. In contrast, there is no comparable commercial pressure in the UK real estate market that would force all property and construction companies to be large. Similarly, there are fewer economies of scale in assembling and managing closed-end funds, and it is natural to find that this sector is quite heavily represented within the HGSC.

Whatever the explanation for these sector differences, it is clear that a bet on smaller companies is also a bet on relative sector performance. Furthermore, in recent years, this latter bet has contributed to the poor performance of the small-cap universe.

Table 4 demonstrates the impact of sector weightings on index performance. The top two rows show the year-by-year performance of the HGSC and the All Share indexes during the post-launch period from 1989–97. The third row of Table 4 shows the annual small-cap premium, i.e. the return difference between the HGSC and the All Share. In six out of the nine years, the 'premium' was negative, while in a further two years it was positive but close to zero. The final column shows that over this nine-year period the arithmetic average annual small-cap premium was −6.5% as presented earlier in Table 2.

The fourth row of Table 4 shows the return that would have been achieved on the All Share Index had each industrial sector been held in its HGSC weighting rather than its All Share weighting.[13] In seven out of nine years, this reweighting led to a decrease in the All Share return. The final row shows the revised size premium estimate, relative to the reweighted All Share Index. The last column of this row shows that the superimposition of HGSC weightings on the All Share Index would have halved the UK size 'premium'

---

[13]The reweighted All Share Index return R, is given by $R = \sum R_j X_j$ where R_j is the return on All Share Index sector j and X_j is the HGSC weighting in sector j.

from -6.5% to -3.3%. Clearly sector weightings serve to explain a sizeable portion of the HGSC's poor relative performance over this period.

4.4.2 Relative Dividend Growth

A second aspect of fundamental performance is related to the underlying business performance of small, as compared to large, companies. While performance could be monitored in terms of earnings growth, cash flow or a variety of measures, dividends provide the most direct measure of cash flow to shareholders. We therefore examine the relative dividends of HGSC, as compared to non-HGSC companies.

In 1955, the prospective dividend for the HGSC was 3.6% higher than for non-HGSC companies. Over the period 1955–88, these dividends grew at an annualised rate that was 1.9% larger than for non-HGSC companies. For the duration of the backhistory and launch period, the HGSC premium of 6.0% was therefore supported by a difference in dividend levels and dividend growth that totalled 5.5% (i.e. $(3.6 + 1.9)\%$) more than non-HGSC companies.

The left-hand side of Figure 6 summarises this performance record over the period 1955–88. The first bar denotes the return differential between small-cap and large-cap companies, while the second bar indicates the magnitude of the dividend yield plus *ex post* growth differential between small-caps and large-caps. The stock market relative performance is close to the fundamental, dividend-based relative performance of small-cap companies.

Over the subsequent period, the right-hand side of Figure 6 shows that the HGSC underperformed large-cap companies by well over seven percentage points per year. At the start of 1989, the prospective dividend for the HGSC was 1.2% lower than for non-HGSC companies. Over the period 1989–97 dividends grew at an annualised rate that was 3.4% lower than for non-HGSC companies. During the post-launch period, the HGSC 'premium' of -7.7% was accompanied by a difference in dividend levels and dividend growth that totalled -4.6%. Again, stock market performance reflects fundamental performance, although there appears to be some expectation of worse that is still to come.

With a larger dividend yield for the HGSC than for large companies in 1955 (and, indeed, a larger yield every year until the late 1970s), small-caps were priced in the expectation of lower subsequent dividend growth.[14] The outcome, however, was a long period of superior growth up to 1988. At that point, small-caps were yielding markedly less than large-caps, having adjusted to an expectation of higher dividend growth. Unfortunately, as Figure 6 shows, the following nine years witnessed lower growth from small stocks.

---

[14]We assume that the 3.6% annual yield differential in favour of smaller companies was too large to represent an expected small firm premium.

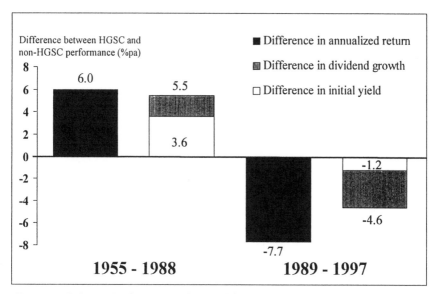

Figure 6. Relative Performance of the HGSC and Non-HGSC Indices, 1955–97.

We can now return to the fundamental question: was the 1955–88 stock market return on small-caps 'too high', or the 1989–97 return 'too low'? If in 1955 investors had had perfect foresight about the long-term dividend growth rates of large and small companies, then the Gordon model (see, e.g., Brealey and Myers, 1997) would have indicated a long-term HGSC premium of 5.5%.[15] If in 1989 there had been a reappraisal, and investors had forecast accurately the subsequent dividend growth rates, the Gordon model would have indicated an HGSC 'premium' of −4.6%.[16] *Ex post* stock market performance was close to what would have been expected with perfect foresight about fundamental performance. Over the long haul, fundamentals thus appear to have driven the relative share prices of UK smaller companies.

We have also replicated our analysis using data for the United States,

---

[15]We assume that there is no variation from year to year in forecast dividend growth rates, and that our 34-year horizon is close to the infinite horizon of the Gordon model. We do not mean to imply that the assumptions of the Gordon model are strictly correct. With perfect foresight about fundamental performance, investors would promptly have driven stock prices to a level that reflects anticipated dividend flows. The *ex post* stock market performance of the HGSC would thus have been realised at the start of our measurement interval. Since the superior fundamental performance of small-caps materialised gradually over time, it is clear that investors' expectations in fact altered over the years, and the HGSC responded accordingly.

[16]We again assume constant dividend growth rates, and also that investors expected the dividend growth to persist at the same level after 1997.

and the results are similar. Taking the 'launch period' for small-cap investing in the US as the two years ending in 1983, we noted in Table 3 that
US small-caps (NYSE deciles 6–8) had historically outperformed large-caps
(NYSE deciles 1–5) by an average of 4.1% per year, while from 1984 to the
present day they underperformed by 2.4% per year (see Ibbotson Associates,
1998). In the earlier period, the outperformance by US smaller companies
was accompanied by an initial dividend yield plus subsequent growth rate
that totalled 3.5% higher than on large-caps. During the subsequent 1984–97
period the underperformance of smaller companies was accompanied by an
inferior dividend yield plus growth, that was 2.6% lower than on large-caps.

We do not mean to suggest that the pricing of small-cap stocks is inex-
tricably linked, in the short term, to dividend payments. On the contrary,
index levels are largely determined by the expectations surrounding future
cash flows. Nor should we focus solely on dividends, since investors can also
receive cash flows through buybacks, cash acquisitions, and so on. But at least
in the UK, where stock repurchases have hitherto been infrequent, dividend
growth rates provide a clear indication of the long-term performance of the index. It appears that HGSC Index returns have in the long term reflected, and
in the short term partly anticipated, the fundamentals for smaller companies.

This, of course, just replaces one question with another. If we conclude
that the stock market performance of smaller companies over the last decade
reflects investor disappointment in smaller company fundamentals, the new
puzzle which emerges is why the fundamental performance of small-caps has
been so much worse than large-caps over this period. We can only speculate here. A partial explanation is that economies of scale in exploiting new
technology, for example in the fields of information processing and pharmaceutical development, were not fully anticipated, and these economies accrued
to large firms. Another partial explanation is that increases in market power,
as global businesses found new ways of capitalising on their brands, patents
and core competencies, also favoured big firms. The 1990s were a period when
technology and market power gave rise to a large company premium. Rather
than small-caps underperforming over this period, perhaps we should think
of large-caps outperforming investors' expectations.

5   Conclusion

In this paper, we have documented the long-term performance of smaller
companies, as compared to larger capitalisation equities in the UK. Their level
of outperformance over the period from 1955 to date has been substantial.
But this long-term record masks a change around the time of the launch
of the HGSC Index, and the accompanying dissemination of information on
the historical size premium. Up to and including the launch period for the

HGSC, smaller companies had outperformed larger companies by some six percentage points per year. Over the subsequent decade, smaller companies underperformed by a similar margin.

We have investigated a number of potential explanations for the disappointing performance of smaller companies. It is difficult to attribute the changing fortunes of smaller companies to risk factors that are commonly cited as explaining the small firm premium. The small firm premium, perhaps the best documented of all stock market anomalies, simply went into reverse. Large companies, instead of being left behind, stormed ahead of smaller companies, a pattern that was replicated in the US, and also in other countries.

Two points are worth stressing. First, we are not suggesting that the reversal of the small-cap premium was a consequence of its discovery and dissemination. Instead, our evidence suggests that the reversal resulted from a change in fundamentals, not just a change in sentiment. Second, the size premium may have disappeared and gone into reverse, but the size effect lives on. The terms size premium and size effect are often used interchangeably in the literature, but adopting a more precise definition, the term 'size effect' should strictly refer to the tendency for small-cap stocks to perform *differently* from large-caps. In both the UK and US, the volatility of the size premium has continued to be as great in the 'post-launch' period as in the pre-launch backhistory. However, the size effect has manifested itself over the last decade through a negative mean size premium, rather than a positive mean.

The size effect thus continues to be a crucial consideration in areas such as asset allocation, benchmarking, performance measurement and attribution, and the design of event studies (see Dimson and Marsh, 1986). Investors and researchers who fail to take account of the size effect thus still run the risk of reaching seriously flawed conclusions. Even Murphy's Law is unlikely to afflict the size effect, as opposed to the size premium.

The size premium is not the only regularity that appears to have evaporated or reversed. Over the period 1932–79 the US equity market provided a premium for beta risk of approximately $R_m - R_f$, the value predicted by the capital asset pricing model; but after 1979 the compensation for beta risk has been far lower than the CAPM indicates.[17] Basu's (1977) price-to-earnings anomaly is shown by Loughran (1997), using another measure of value, to be entirely absent for a large majority by value of the entire stock market. The contrarian strategy of buying the previous week's winners and selling losers (see Lehman, 1990) is challenged by Conrad, Gultekin and Kaul (1997) who show that these profit opportunities are not available to investors who have to pay the bid-ask spread when they trade. Cross' (1973) day-of-the-week effect is shown by Dubois and Louvet (1996) to have disappeared in the US, and it

---

[17]For devotees of Murphy's Law, it is interesting to record that 1979 was the launch year of London Business School's beta service (Dimson and Marsh, 1997).

also now behaves inconsistently across countries (Agrawal and Tandon, 1994) and over time (Wang, Li and Erickson, 1997). While the turn-of-the-month effect documented by Ariel (1987) appears to persist out-of-sample, the days on which the effect occurs are now different (Hensel, Sick and Ziemba, 1999). The Rozeff and Kinney (1976) turn-of-the-year effect is one which, according to Keim (1989), Bhardwaj and Brooks (1992), and Booth and Keim (1999) cannot be used to earn abnormal returns. Leinweber (1995) reports that while the Value Line paper portfolio had an annualised return of 26%, the real Value Line Fund returned only 16%, a substantial implementation shortfall for the Value Line anomaly. This list could be extended further.

Since we wrote this article, Murphy's Law has continued to be applicable to the stock market. During 1998, the HGSC Index underperformed the All-Share Index by 17.8%. Investors who became enthusiastic about smaller companies in the UK did so just in time to gain exposure before a period of unexpectedly poor fundamental performance. But, just as many of these investors were losing the last of their enthusiasm and reweighting their portfolios towards large-caps, the small-cap reversal seems itself to have gone into reverse. In the first nine months of 1999, the HGSC Index has outperformed the All-Share Index by 22.6%.

For investors seeking to take advantage of stock market anomalies, bread often falls with the buttered side down. But not even this can be relied upon, since Murphy's Law embraces the notion that the Law itself is not infallible. Several anomalies do appear to have persisted after their discovery and dissemination, notably Ball and Brown's (1968) post-earnings announcement drift. The reason why not all anomalies go into reverse is presumably because of *Murphy's Second Law* (Dickson, 1980): "Nothing is ever as simple as it seems".

References

Agrawal, Anup and Kishore Tandon (1994). 'Anomalies or Illusions? Evidence from Stock Markets in Eighteen Countries', *Journal of International Money and Finance* **13** 83–106.

Arbel, Avner and Paul Strebel (1983). ' Pay Attention to Neglected Firms!', *Journal of Portfolio Management* **9** (Winter) 37–42.

Ariel, Robert (1987). 'A Monthly Effect in Stock Returns', *Journal of Financial Economics* **18** 161–174.

Ball, Ray and Phillip Brown (1968). 'An Empirical Examination of Accounting Income Numbers', *Journal of Accounting Research* **6** (2) 159–178.

Banz, Rolf (1981). 'The Relationship Between Return and Market Value of Common Stocks', *Journal of Financial Economics* **9** 3–18.

Basu, Sanjoy (1977). 'Investment Performance of Common Stocks in Relation to their Price-Earnings Ratios: A Test of the Efficient Market Hypothesis', *Journal of Finance* **32** 663–682.

Beckett, Samuel (1938). *Murphy*. G. Routledge and Co. London.

Berges, Angel, John McConnell and Gary Schlarbaum (1984). 'The Turn of the Year in Canada', *Journal of Finance* **39** 185–192.

Bhardwaj Ravinder and LeRoy Brooks (1992). 'The January Anomaly: Effects of Low Share Price, Transaction Costs, and Bid-Ask Bias', *Journal of Finance* **47** 553–575.

Black, Fisher (1986). 'Noise', *Journal of Finance* **41** 529–543.

Black, Fischer (1993). 'Beta and Return', *Journal of Portfolio Management* **20** 8–18.

Blume, Marshall and Robert Stambaugh (1983). 'Bias in Computed Returns: An Application to the Size Effect', *Journal of Financial Economics* **12** 387–404.

Booth, David and Donald Keim (1999). 'Is There Still a January Effect?', this volume, 169–178.

Brealey, Richard and Stewart Myers (1997). *Principles of Corporate Finance*. Fifth edition, McGraw Hill.

Brown Phillip, Donald Keim, Alan Kleidon and Terry Marsh (1983). 'Stock Return Seasonalities and the Tax-Loss Selling Hypothesis: Analysis of the Arguments and Australian Evidence', *Journal of Financial Economics* **12** 105–128.

Brown, Stephen, William Goetzmann and Stephen Ross (1995). 'Survival', *Journal of Finance* **50** 853–873.

Campbell, John, Andrew Lo and Craig MacKinlay (1997). *The Econometrics of Financial Markets*. Princeton University Press.

Canina, Linda, Roni Michaely, Richard Thaler and Kent Womack (1998). 'Caveat Compounder: A Warning about Using the Daily CRSP Equal-Weighted Index to Compute Long-Run Excess Returns', *Journal of Finance* **53** 403–416.

Chan, KC, Chen Nai-fu and David Hsieh. (1985). 'An Exploratory Investigation of the Firm Size Effect', *Journal of Financial Economics* **14** 451–471.

Chen, Nai-fu (1988). 'Equilibrium Asset Pricing Models and the Firm Size Effect'. In *Stock Market Anomalies*, Elroy Dimson (ed.), Cambridge University Press.

Coghlan, Helen (1988). 'Small Firms versus Large on the Irish Stock Exchange: An Analysis of the Performances 1977–86', *Irish Business and Administrative Research* **9** 10–20.

Comolli, Luis and William Ziemba (1999). 'Japanese Security Market Regularities, 1990–1994', this volume, 458–491.

Connor, Gregory and Robert Korajczyk (1987). 'Estimating Pervasive Economic Factors with Missing Observations'. Working paper **34**, Department of Finance, Northwestern University.

Conrad, Jennifer, Mustafa Gultekin and Gautam Kaul (1997). 'Profitability of Short-Term Contrarian Strategies: Implications for Market Efficiency', *Journal of Business & Economic Statistics* **15** 379–386.

Cornioley, Claude and Jacques Pasquier-Dorthe (1991). 'CAPM, Risk Premium Seasonality and the Size Anomaly: The Swiss Case', *Finance* **12** 23–44.

Cross, Frank (1973). 'The Behavior of Stock Prices on Fridays and Mondays', *Financial Analysts Journal* (Nov.–Dec.) 67–69.

Dickson, Paul (1978). *The Official Rules.* Arrow Books, London.

Dickson, Paul (1980). *The Official Explanations.* Arrow Books, London.

Dimson, Elroy (1979). 'Risk Measurement When Shares Are Subject to Infrequent Trading', *Journal of Financial Economics* **17** 197–226.

Dimson, Elroy (ed.) (1988). *Stock Market Anomalies.* Cambridge University Press.

Dimson, Elroy and Paul Marsh (1984). 'Hedging the Market: The Performance of the FTSE 100 Share Index', *Journal of the Institute of Actuaries* **111**(2) 403–430.

Dimson, Elroy and Paul Marsh (1986). 'Event Study Methodologies and the Size Effect', *Journal of Financial Economics* **17** 113–142.

Dimson, Elroy and Paul Marsh (1987). *The Hoare Govett Smaller Companies Index for the UK.* Hoare Govett Limited, January.

Dimson, Elroy and Paul Marsh (1989). 'The Smaller Companies Puzzle', *The Investment Analyst* **91** 16–24.

Dimson, Elroy and Paul Marsh (1997). *Risk Measurement Service.* London Business School.

Dimson, Elroy and Paul Marsh (1998). 'The Hoare Govett Smaller Companies Index 1955–1997', ABN-AMRO Hoare Govett, January.

Dimson, Elroy and Paul Marsh (1999). 'Financial Market Returns 1955–1999', Working paper, London Business School.

Dubois, Michel and Pascal Louvet (1996). 'The Day-of-the-Week Effect: The International Evidence', *Journal of Banking & Finance* **20** 1463–1484.

Elton, Edwin, Martin Gruber and Christopher Blake (1996). 'Survivorship Bias and Mutual Fund Performance', *Review of Financial Studies* **9** 1097–1120.

Fama, Eugene (1998). 'Market Efficiency, Long-Term Returns, and Behavioral Finance', *Journal of Financial Economics* **49** 283–386.

Fama, Eugene and Kenneth French (1992). 'The Cross-section of Expected Returns', *Journal of Finance* **47** 427–465.

Fama, Eugene and Kenneth French (1993). 'Common Risk Factors in the Returns on Stocks and Bonds', *Journal of Financial Economics* **33** 3–56.

Fama, Eugene and Kenneth French (1995). 'Size and Book-to-Market Factors in Earnings and Returns', *Journal of Finance* **50** 131–155.

Fama, Eugene and Kenneth French (1996). 'Multifactor Explanations of Asset Pricing Anomalies', *Journal of Finance* **51** 55–87.

Fama, Eugene and Kenneth French (1998). 'Value Versus Growth: the International Evidence', *Journal of Finance* **53** 1975–1999.

Gillan, S (1990). 'An Investigation into CAPM Anomalies in New Zealand: the Small Firm and Price Earnings Ratio Effects', *Asia Pacific Journal of Management* **7** (Special issue) 63–78.

Goetzmann, William and Philippe Jorion (1999). 'Global Stock Markets in the Twentieth Century', *Journal of Finance* **54** (3) 953–980.

Haugen, Robert (1999). *The New Finance: The Case Against Efficient Markets.* Second edition, Prentice Hall.

Haugen, Robert and Nardin Baker (1996). 'Commonality in the Determinants of Expected Stock Returns', *Journal of Financial Economics* **41** 401–439.

Haugen, Robert and Josef Lakonishok (1988). *The Incredible January Effect: The Stock Market's Unsolved Mystery.* Dow-Jones–Irwin, Homewood, Il.

Hawawini, Gabriel, Pierre Michel and Albert Corhay (1989). 'A Look at the Validity of the Capital Asset Pricing Model in Light of Equity Market Anomalies: The Case of Belgian Common Stocks'. In *A Reappraisal of the Efficiency of Financial Markets*, Rui Guimarães, Brian Kingsman and Stephen Taylor (eds.) NATO ASI Series, Springer-Verlag.

Hawawini, Gabriel and Donald Keim (1999). 'The Cross-Section of Common Stock Returns: A Synthesis of the Evidence and Explanations', this volume, 3–43.

He, Jia and Lilian Ng (1994). 'Economic Forces, Fundamental Variables, and Equity Returns', *Journal of Business* **67** 599–609.

Hensel, Chris, Gordon Sick and William Ziemba (1999). 'A Long Term Examination of the Turn-of-the-Month Effect in the S&P 500', this volume, 218–246.

Huberman, Gur and Shmuel Kandel (1987). 'Mean Variance Spanning', *Journal of Finance* **42** 873–888.

Ibbotson Associates (1998). *Stocks, Bonds, Bills and Inflation: 1998 Yearbook.* Ibbotson Associates, Chicago.

Keim, Donald (1983). 'Size-Related Anomalies and Stock Return Seasonality: Further Empirical Evidence', *Journal of Financial Economics* **12** 473–490.

Keim, Donald (1989). 'Trading Patterns, Bid-Ask Spreads, and Estimated Security Returns: The Case of Common Stocks at Calendar Turning Points', *Journal of Financial Economics* **25** 75–98.

Keim, Donald (1999). 'An Analysis of Mutual Fund Design: The Case of Investing in Small-Cap Stocks', *Journal of Financial Economics* **51** (2) 173–194.

Keim, Donald and William Ziemba (eds.) (1999). *Security Market Imperfections in World Wide Equity Markets.* Cambridge University Press.

Klein, Robert and Jess Lederman (1993). *Small Cap Stocks: Investment and Portfolio Strategies for the International Investor.* Probus Publishing, Chicago.

Knez, Peter and Mark Ready (1996). 'Estimating the Profits from Trading Strategies', *Review of Financial Studies* **9** 1121–1163.

Lakonishok, Josef, Andrei Shleifer and Robert Vishny (1994). 'Contrarian Investment, Extrapolation and Risk', *Journal of Finance* **49** 1541–1578.

Lehmann, Bruce (1990). 'Fads, Martingales and Market Efficiency', *Quarterly Journal of Economics* **105** 1–28.

Leinweber, David (1995). 'Using Information from Trading in Trading and Portfolio Management', *Journal of Investing* **4** (2) 40–50.

Loughran, Tim (1997). 'Book-to-Market Across Firm Size, Exchange, and Seasonality: Is There an Effect?', *Journal of Financial and Quantitative Analysis* **32** 249–268.

Louvet, Pascal and Ollivier Taramasco (1991). 'The Day-of-the-Week Effect on the Paris Stock Exchange: A Transactional Effect', *Journal de la Société Statistique de Paris* **133** 50–76.

Ma Tai and T Y Shaw (1990). 'The Relationships Between Market Value, P/E Ratio, Trading Volume and the Stock Return of Taiwan Stock Exchange'. In *Pacific-Basin Capital Markets Research, Volume 1*, Ghon Rhee and Rosita Chang (eds. North Holland, Amsterdam.

Marsh, Paul (1979). 'Equity Rights Issues and the Efficiency of the UK Stock Market', *Journal of Finance* **34** 839–862.

Maug, Ernst and Narayan Naik (1995). 'Herding and Delegated Portfolio Management: The Impact of Relative Performance Evaluation on Asset Allocation'. Working Paper No. 223, Institute of Finance and Accounting, London Business School.

McKillop, Donal and Robert Hutchinson (1988). 'The Small Firm Effect, Seasonality and Risk in Irish Equities', *Irish Business and Administrative Research* **9** 21–29.

Mott, Claudia *et al.* (1997). *Investing in Small-Cap and Microcap Securities.* Association for Investment Management and Research. Virginia.

Müller, Didier (1995).   An Abridged Collection of Interdisciplinary Laws. http://www.robin.no/rcc/murphy.html.

O'Shaughnessy, James (1998a). *What Works on Wall Street.* Revised edition. McGraw Hill, NY.

O'Shaughnessy, James (1998b). *How to Retire Rich.* Broadway Books, NY.

Reiganum, Marc (1981). 'Misspecification of Capital Asset Pricing: Empirical Anomalies Based on Earnings Yields and Market Values', *Journal of Financial Economics* **9** 19–46.

Rietz, Thomas (1988). 'The Equity Risk Premium: A Solution', *Journal of Monetary Economics* **22** 117–131.

Roll, Richard (1981). 'A Possible Explanation of the Small Firm Effect', *Journal of Finance* **36** 879–888.

Roll, Richard (1983a). 'On Computing Mean Returns and the Small Firm Premium', *Journal of Financial Economics* **12** 371–386.

Roll, Richard (1983b). 'Vas Ist Das? The Turn-of-the-Year Effect and the Return Premia of Small Firms', *Journal of Portfolio Management* **9** 18–28.

Roll, Richard (1994). 'What Every CEO Should Know About Scientific Progress in Economics: What is Known and What Remains to be Resolved', *Financial Management* **23** (Summer) 69–75.

Ross, Stephen (1976). 'The Arbitrage Theory of Capital Asset Pricing', *Journal of Economic Theory* **13** 341–360.

Rozeff, Michael and William Kinney (1976). 'Capital Market Seasonality: The Case of Stock Returns', *Journal of Financial Economics* **3** 379–402.

Rubio, Gonzalo (1986). 'Size, Liquidity and Valuation'. Working Paper, Southern European Discussion Series (Universidad del Pas Vasco, Lejona).

Samuelson, Paul (1989). 'The Judgement of Economic Science as Rational Portfolio Management: Indexing, Timing and Long-Horizon Effects', *Journal of Portfolio Management* **16**(1) 4–12.

Schwert, William (1983). 'Size and Stock Returns and Other Empirical Regularities', *Journal of Financial Economics* **12** 3–12.

Shumway, Tyler (1996). 'Size, Overreaction, and Book-to-Market Effects as Default Premia'. Working paper, University of Michigan.

Siegel, Jeremy (1998). *Stocks for the Long Run*. Second edition. McGraw Hill, NY.

Stehle, Richard (1992). 'The Size Effect in the German Stock Market'. Paper presented at the 1992 Meeting of the European Finance Association, University of Augsburg, Germany, August 1992.

Stolin, David (1998). 'UK Share Delistings: A Survival Analysis', *Transaction of the 26th Annual Congress of Actuaries* **7** 73–105.

Stolin, David (1999). 'Share Survivorship and Return on the LSE'. Working paper, London Business School.

Stoll, Hans and Robert Whaley (1983). 'Transaction Costs and the Small Firm Effect', *Journal of Financial Economics* **12** 57–79.

Van den Bergh, Willem and Roberto Wessels (1985). 'Stock Market Seasonality and Taxes: An Examination of the Tax-Loss Selling Hypothesis', *Journal of Business Finance & Accounting* **12** 515–530.

Wahlroos Bjorn and Tom Berglund (1986). 'Risk, Return and Equilibrium Returns in a Small Stock Market', *Journal of Business Research* **14** 423–440.

Wang, Ko, Yuming Li and John Erickson (1997). 'A New Look at the Monday Effect', *Journal of Finance* **52** 2171–2186.

Ziemba, William (1991). 'Japanese Security Market Regularities: Monthly, Turn-of-the-Month and Year, Holiday and Golden Week Effects', *Japan and the World Economy* **3** 119–146.

Ziemba, William and Sandra Schwartz (1991). *Invest Japan*. Probus Publishing, Chicago.

Direct Evidence of Non-Trading of NYSE and AMEX Stocks*

Stephen R. Foerster and Donald B. Keim

Abstract

This paper documents the frequency of non-trading for NYSE and AMEX stocks based on information in the CRSP monthly and daily data files. We find a declining pattern of non-trading over the 1926–1990 period: 23.4% of NYSE stocks do not trade on an average (end-of-month) day during the 1926 to 1945 period, compared with 1.29% on average over all days during the 1973–1990 period. In the 1973–1990 period, non-trading averaged more than 15% for AMEX firms. We find that the average amount of non-trading is larger for smaller stocks, is lowest at the end of the year, and tends to be lowest at the beginning of the week and highest at the end of the week. We also find substantial heterogeneity in the amount of non-trading across the stocks within each size decile. For example, while 10% of the stocks in the smallest decile trade virtually every trade day, 10% of the stocks in that decile do not trade on 51% of the trade days during the year, and 1% do not trade on 76% of the trade days during the year. Finally, based on our findings, studies that used the pre-1991 CRSP files were exposed to errors related to the way CRSP stored information relating to the trade status of a security during the period 1964–1972.

1   Introduction

Empirical research in financial economics has long recognized that infrequent or nonsynchronous trading can result in measurement error in security returns that, in turn, biases the measurement of risk and autocorrelation of returns.[1]

---

*We thank Andrew Lo, Craig MacKinlay, Ananth Madhavan, Charles Mossman, Dave Porter, Matt Richardson, Jay Ritter and seminar participants at the Northern Finance Association Meetings, the University of Manitoba, and the University of Toronto for helpful comments and discussions. Financial support was provided by the Social Sciences and Humanities Research Council of Canada and the Plan for Excellence (Foerster) and by the Geewax–Terker Program in Financial Instruments at Wharton (Keim).

[1]Fama (1965) and Fisher (1966) examine infrequent trading and the measurement error in security returns. Scholes and Williams (1977), Dimson (1977), Cohen, Hawawini, Maier, Schwartz and Whitcomb (1983), Fowler and Rourke (1983), and Dimson and Marsh (1983) examine the impact of infrequent trading on risk measurement. Perry (1985), Atchison, Butler and Simonds (1987) and Lo and MacKinlay (1990a, 1990b) examine the impact of infrequent trading on autocorrelation of returns.

For example, Keim (1983) reports OLS beta estimates (based on daily data) of only 0.76 for a portfolio of small market-capitalization stocks, and Lo and MacKinlay (1990a) report a daily return first-order autocorrelation of 0.35 for a portfolio of small stocks. The conventional 'wisdom' is that non-trading frequency is inversely related to market capitalization, resulting in an inverse relation between non-trading-induced biases and market capitalization. However, few studies have directly examined the degree of non-trading in the data.[2] Given the increasing emphasis placed on potential biases in estimated moments due to market frictions like non-trading, examination of the actual contamination of the data due to non-trading is important.

This paper makes three contributions. First, we provide documentation of the extent to which non-trading exists among NYSE and AMEX firms for the 1926–1990 period, and we show how non–trading varies through time. Such documentation provides useful bounds for future research that relates to non-trading, both for the frequently-studied post-1962 period and earlier periods as well. Second, we examine cross-security patterns in non-trading. We also find a relation between non-trading and many well-known stock return patterns such as the size effect and day-of-the-week effect, building on Keim's (1989) earlier findings relating non-trading to the turn-of-the-year effect. Third, we provide an application of our non-trading estimates by reexamining previous models which relate non-trading to daily portfolio autocorrelations.

In this paper we define non-trading as the failure of a stock to trade over a particular period (usually one day) when the NYSE and AMEX are open for trading. We identify non-traded stocks by exploiting the convention used by the Center for Research in Security Prices (CRSP) to record the prices of non-traded securities. Specifically, when a stock does not trade CRSP records the price as the *negative* value of the average of the last bid and ask price. We find that, on average, over 23% of NYSE securities failed to trade on a given day during the period from 1926 to 1945; for the 1946–1963 period, this number drops to 7.7%; and over the last decade of the study, NYSE non-trading is approximately 1%. We also examine trading frequency by month and by day, by exchange, and for securities in deciles ranked by market value of equity. The evidence indicates that as many as 75% of the prices used to compute small stock portfolio returns were not actual transaction prices during the first ten years of the CRSP history of returns. Between 1973 and 1990, on average, 24% of NYSE and AMEX stocks in the smallest size decile failed to trade on a given day. We also find substantial heterogeneity in the amount of non-trading across the stocks within each size decile. For example,

---

[2] A few exceptions are Lakonishok and Smidt (1984) and Keim (1989) who examine non-trading around the turn-of-the year; and Lo and MacKinlay (1990a) who examine month-end non-trading between 1973 and 1987.

while 10% of the stocks in the smallest decile trade virtually every trade day, 10% of the stocks in that decile do not trade on at least 52% of the trade days during the year, and 1% do not trade on at least 76% of the trade days during the year.

Interestingly, we find an abnormally low number of negative prices recorded by CRSP during the 1964–1972 period. In several of these years no prices were recorded as negative values. This finding does not indicate an absence of non-trading since a random check of month-end bid and ask prices in the *Wall Street Journal* revealed numerous securities that did not trade during this period. According to the CRSP Stock File Guide, the source of daily price data during this period was Standard & Poor's Price Tape. After September 1972 the price data are provided by Interactive Data Corporation. The change in data source in 1972 coincides with the resumption of non-zero estimates of non-trading. The 1990 edition of the CRSP Stock File Guide warns users of the omission of the negative price convention for the 1964–1972 period and subsequent Guides indicate steps were being taken to rectify its recording of non-traded securities. As a check for potential differences, we compare the non-trading frequencies estimated with the 1990 CRSP file (and, thereby, implicit in all CRSP-related research conducted to that point) with the non-trading frequencies estimated from the 1993 CRSP file. We find that there are substantial differences, even for the period after 1972. Nonetheless, the results we report below are drawn from the 1990 CRSP file that contains the recording inaccuracies because it was the pre-1991 editions of the CRSP files that were used in much of the defining literature that relates to return autocorrelations and stock return patterns. Because of the problems with the data in the 1964–1972 period (inclusive), we exclude those years when computing the summary statistics reported below.

Our results provide useful bounds for research related to non-trading. As an example, we revisit the relation between non-trading and estimates of autocorrelation, and find mixed results. Our evidence casts some doubts on previous research that concludes non-trading cannot explain the positive *daily* return autocorrelations found in portfolios of common stocks during the 1973–1990 period. However, in the 1928–1940 period when non-trading was at its highest levels, the autocorrelations of the daily S&P Composite Index were at the lowest levels for the entire century.

The paper is organized as follows. Section 2 presents NYSE non-trading evidence from 1926 to 1990, relates non-trading to the size (market value of common equity) of the firm, and examines daily and monthly patterns of non-trading of NYSE and AMEX securities over the more recent 1973 to 1990 period. Section 3 uses the stylized facts from Section 1 to reexamine the relation between non-trading and estimates of portfolio autocorrelations. Section 4 concludes the paper.

2   Non-trading Evidence

In this section we document the amount of non-trading among NYSE and AMEX stocks. The data are obtained from the 1990 edition of the daily and monthly stock price files provided by CRSP. The monthly file contains prices for NYSE stocks for the period December 1925 to December 1990. The daily file contains prices for both NYSE and AMEX securities from July 1962 to December 1990. If no trade occurs on the actual trading day in the case of the daily files, or the last trading day of the month in the case of the monthly files, then the average of the bid and ask prices is entered as a negative value. We exploit this convention in the CRSP files to compute the percentage of stocks that did not trade during a particular interval. Specifically, we compute the percentage of total NYSE stocks that did not trade on the last day of each month during the 1926 to 1990 period. For the period from July 1962 to December 1990, we also compute for each day the percentage of total NYSE and AMEX stocks that did not trade. Using these series of monthly and daily percentages, we compute the average amount of non-trading for a particular year, month of the year, and day of the week.

2.1   A long perspective: 1926 to 1990

Table 1 contains evidence on non-trading frequencies based on information in the monthly CRSP files for the years 1926 to 1990. We report frequency distribution cutoffs (medians, 75th percentile and maximum value) for the entire sample, and for ten subgroups based on an annual year-end sort by market value of the outstanding common stock for the firms. The top row in Table 1 reports results for all NYSE securities. During the overall period, 14.2% of the stocks on the NYSE did not trade on an average end-of-month day. Within-year averages (i.e., the mean of the twelve month-end trading frequencies) range from 43.3% in 1932 to 0.5% in 1987. The median across all years is 7.9%. We examine three subperiods: the 'early' period of 1926–1945; the post-World War II period of 1946–1963; and the most recent period of usable data, 1973–1990. During the 1926–1945 subperiod, the mean (median) non-trading of NYSE securities on an average end-of-month day is 27.4 (23.4)%. This number drops to 10.8 (7.7)% between 1946 and 1963 and is 2.9 (0.9)% from 1973 to 1990, indicating a general decline in the frequency of non-trading.

The remainder of Table 1 examines the relation between non-trading and market capitalization of NYSE securities. To create the size categories, we use the 'year-end capitalization portfolio assignment' supplied by CRSP. Percentages of non-trading are computed at each month-end for each size category. Summary statistics are computed as described above. For the overall period in panel A, the smallest market value stocks have the highest mean (median)

Table 1. Frequency of non-trading measured by the average percent of negative
prices at month-end as recorded on the monthly CRSP file of NYSE stocks

If a stock does not trade on the last trading day of the month, the CRSP monthly
file reports the average of the bid and ask prices as a negative value. Portfolios of
NYSE and AMEX stocks are ranked by year-end market value of equity. The
sample period is 1926 to 1990. The sample excludes the years 1964–1972 because of
the abnormally low number of negative prices recorded by CRSP during that period.

	A. 1926–1990		B. 1926–1945			C. 1946–1963			D. 1973–1990		
All Firms	Mean 14.2%	Median 7.9%	Median 23.4%	75%	Max.	Median 7.7%	75%	Max.	Median 0.9%	75%	Max.
Size Decile											
Smallest	29.3	25.3	49.7	64.7	90.6	25.3	33.1	68.6	7.8	16.5	66.7
2	21.5	17.0	36.4	50.0	78.5	16.9	23.2	73.3	4.9	9.7	27.5
3	19.1	13.8	34.6	46.9	84.7	12.9	19.4	61.1	3.2	5.7	20.3
4	16.9	10.8	30.4	42.6	73.9	10.8	17.5	52.2	2.0	3.6	10.5
5	14.6	8.6	25.9	40.3	72.3	8.4	13.3	55.6	1.2	2.4	11.1
6	12.5	6.9	23.8	35.5	79.0	6.1	9.7	41.9	0.5	1.6	7.6
7	10.5	4.9	22.0	30.2	58.4	4.6	7.6	42.7	0.5	1.3	7.4
8	8.0	2.8	16.4	23.3	50.6	2.2	5.9	36.2	0.4	0.9	3.3
9	5.8	2.0	11.8	17.0	39.2	1.9	3.8	24.5	0.0	0.4	2.3
Largest	3.5	1.0	7.3	10.7	30.0	0.9	1.9	25.0	0.0	0.0	0.7

percentage of constituents that do not trade in an average month: 29.3 (25.3).
Non-trading declines monotonically through the groups to a minimum value
of 3.5 (1.0)% for the largest firms.[3] The 75th percentile and maximum non-
trading% also show monotonic patterns. Non-trading for each size group is
more prevalent in the early period (panel B) than the later periods (panels C
and D), and is especially pronounced for the smallest firms in the earliest years
of the sample period. The largest *within-year average* for the smallest stocks
is 74.8% (1932) versus 12.0% (1941) for the largest firms. With the exception
of the late 1940s, when as many as 40% of the smaller NYSE stocks did not
trade on an average month-end, the latter subperiods display substantially-
reduced levels of non-trading. This is evident in Figure 1 which charts the
within-year average non-trading percentage in each year for both the smallest
and largest size deciles. The 1964-72 period of missing non-trading data is
also clearly illustrated in Figure 1.

We also examine a month-by-month breakdown of non-trading for all securities and for the two extreme size categories. The greatest frequency of
non-trading occurs in the summer months of June, July and August. The
lowest level of non-trading occurs at the end of the year in November and,

---

[3]Stoll and Whaley (1990) report a monotonic relation between infrequent *intraday* trading and size during the 1982 to 1986 period for NYSE stocks.

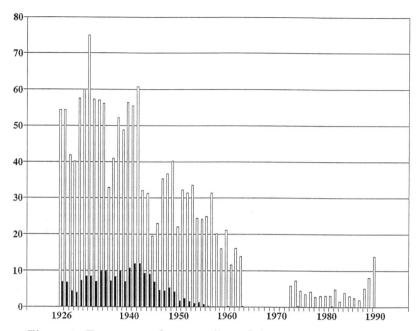

Figure 1. Frequency of non-trading of the smallest and largest portfolios of NYSE stocks. The smallest portfolio is represented by the white bars and the largest by the black. If a stock does not trade on the last day of the month, the CRSP monthly file reports the average of the bid and ask prices as a negative value. Month-end frequencies of non-trading are averaged to create within-year frequencies. The sample period is 1926–1990.

especially, December. These year-end results are consistent with Lakonishok and Smidt's (1984) observation (using data from 1970 to 1981) of increased trading on the last day of the year, as well as Keim's (1989) evidence (for data from 1972 to 1987). Examining the size-based results, we find a much larger decrease in the mean (median) non-trading from November to December for the smallest firms (27.0 (23.1)% to 17.2 (13.6)%) than the largest firms (3.1 (0.2)% to 2.7 (0.5)%). This tendency is robust across subperiods and is consistent with the hypothesis of increasing levels of tax-loss selling of smaller stocks as the end of the tax year approaches.[4] This tendency is also consistent with 'window dressing' by portfolio mangers – increased trading activity including the selling of poor performers and the buying of good performers prior to the release of year-end portfolio contents.

---

[4]We computed F-tests of the equality of non-trading across months for the overall period (all stocks). We tested each month against *all* other eleven months. Only December is significantly different from the other eleven months.

In summary, our analysis of non-trading using the CRSP monthly data finds an inverse relation between firm size and degree of non-trading. The amount of non-trading is substantial, particularly for smaller firms in the earlier sample years; for example, 74.8% of the prices used to compute small firm portfolio returns in some years are not based on actual transaction prices. However, non-trading has generally declined over the last 70 years for both small and large stocks.

2.2   A closer look: 1973 to 1990

Use of the CRSP daily stock price file permits analysis of intra-week and intra-month patterns of non-trading, as well as analysis of non-trading for American Stock Exchange (AMEX) stocks. Table 2 reports non-trading results based on the day-by-day percentage of negative prices relative to the entire sample, reported separately by day, month and year for the period 1973 to 1990. The overall mean percentage of non-traded securities over this time period is 6.05% for the combined sample of NYSE and AMEX securities, 1.39% among NYSE securities, and 15.86% among AMEX securities (the medians are similar to the means). Figure 2 graphs daily non-trading percentages from 1973 to 1990 for NYSE stocks (upper panel) and AMEX stocks (lower panel). In the 1970s non-trading of NYSE (AMEX) stocks was as high as 8 (40)% on some days. The lowest levels of non-trading occurred during several periods during the 1980s including the October 1987 market crash. Year-by-year average non-trading is roughly ten times higher among AMEX versus NYSE securities. Over the entire 1973–1990 period, average non-trading is 24.8% for the smallest decile firms and 0.2% for the largest decile firms.

An interesting day-of-the week pattern emerges for both NYSE and AMEX stocks: non-trading increases monotonically through the week. In conjunction with Keim's (1989) and Porter's (1992) finding of a tendency for Friday's price to close at an ask and Monday's price to close at a bid, these findings may partially account for the tendency for Monday's measured returns to be negative. For example, since closing prices on Monday are more likely to be transaction prices (versus means of the bid and ask) than on any other day-of-the-week, there is a greater probability that average Monday returns will reflect movement within the bid-ask spread resulting from such systematic investor trading patterns, in this case from the ask to the bid, which will bias the average return downward.

The month with the lowest incidence of non-trading for both NYSE and AMEX securities is December. The average non-trading percentage for the entire year is 6.05%, versus 4.17% for an average day in December. This confirms the inference drawn from the month-end non-trading results in the previous section. There also tends to be a higher incidence of non-trading

Table 2. Frequency of non-trading measured by the daily mean percent of
negative prices as recorded on the daily CRSP file of NYSE and AMEX stocks

If a stock does not trade on a particular day, the CRSP daily file reports the average
of the bid and ask prices as a negative value. Portfolios of NYSE and AMEX stocks
are ranked by year-end market value of equity. The sample period is 1973 to 1990.

		All Stocks	NYSE Only	AMEX Only	Smallest Decile	Largest Decile
Day	Monday	5.74	1.28	15.12	23.44	0.18
	Tuesday	5.86	1.31	15.43	24.41	0.14
	Wednesday	6.04	1.38	15.85	24.83	0.15
	Thursday	6.26	1.46	16.37	25.59	0.15
	Friday	6.32	1.50	16.50	25.65	0.16
Month	January	5.67	1.22	15.09	24.50	0.15
	February	5.77	1.31	15.18	24.60	0.14
	March	5.69	1.29	14.98	23.78	0.14
	April	5.94	1.32	15.66	25.15	0.16
	May	6.20	1.41	16.32	25.90	0.15
	June	6.48	1.53	16.92	26.67	0.19
	July	6.47	1.50	16.95	27.24	0.18
	August	6.75	1.63	17.45	27.47	0.19
	September	6.68	1.59	17.35	26.71	0.17
	October	6.66	1.55	17.40	27.09	0.16
	November	6.00	1.37	15.77	23.37	0.14
	December	4.17	0.90	11.10	14.76	0.14
Year	1973	6.61	1.88	15.19	25.60	0.38
	1974	11.16	3.50	25.96	38.05	0.61
	1975	10.84	3.04	26.20	38.40	0.34
	1976	7.83	1.95	19.91	30.63	0.50
	1977	7.87	1.93	20.35	32.10	0.22
	1978	5.82	1.45	15.46	23.44	0.11
	1979	5.41	1.25	14.92	22.99	0.02
	1980	4.56	0.98	12.87	21.77	0.02
	1981	4.59	1.10	12.90	21.97	0.02
	1982	5.27	1.35	14.14	24.85	0.07
	1983	2.75	0.57	7.68	12.67	0.10
	1984	4.35	0.95	12.28	19.80	0.20
	1985	3.88	0.60	11.46	18.21	0.10
	1986	3.21	0.44	9.62	16.89	0.05
	1987	3.36	0.47	9.79	15.24	0.00
	1988	6.91	1.06	18.61	25.89	0.04
	1989	6.42	0.98	17.07	24.74	0.03
	1990	7.98	1.46	21.08	33.01	0.00
Overall		6.05	1.39	15.86	24.79	0.16

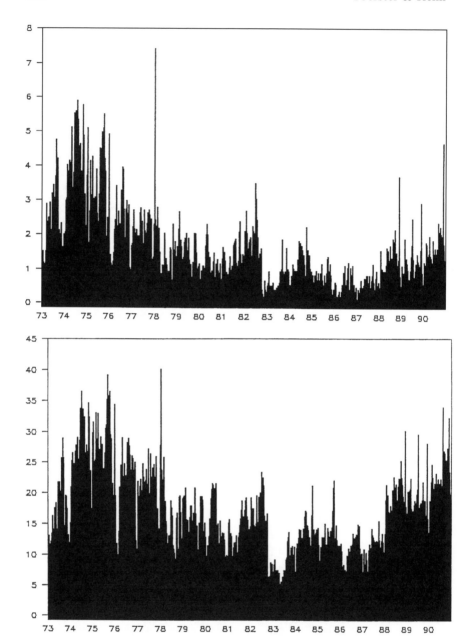

Figure 2. Frequency of non-trading of NYSE and AMES stocks. The
upper panel presents NYSE frequencies and the lower, AMEX frequencies. If a
stock does not trade on a particular day, the CRSP daily file reports the average
of the bid and the ask prices as a negative value. The sample period is 1973–1990.

during the summer and fall months. Average non-trading in each month from November to April is less than 6.0% and in each month from May to October is greater than 6.2%.

Recall that the monthly non-trading results reported in the previous section are based on the last trading day of the month. Using the CRSP daily file, we find that, in most cases, the incidence of non–trading (i.e., the occurrence of negative prices) for the NYSE firms is statistically the same for the last trading day of the month as for the other days. One exception is December where there is a statistically lower incidence of negative prices on the last trading day than on the other trading days in the month. In other words, there is a statistically greater tendency for the last trading price in December to be a transaction price than for other trading days in December. Additionally, the last trading day in December is one of the five lowest non-trading days in ten of the years during the 1973–90 period.

We also examine the likelihood of a stock going without a trade for up to five consecutive days. We group firms into size decile portfolios. Overall, 2.24% of all stocks went untraded for two consecutive days; only 0.42% ever go untraded for five consecutive days. For the smallest size decile, over 10% went untraded for two consecutive days, and over 2% remained without a trade for five consecutive days. These results are quite similar to those of Keim's (1989) based on the turn-of-the-year. Based on the 1973–1990 period, we find the probability of a conditional on a *non-trade* on the previous day is, on average, 37.6%, considerably greater than the unconditional probability of 6.0%. For the smallest (largest) size portfolio, the conditional probability is 45.7 (30.2)%. The probability of a stock *not trading* conditional on a *trade* on the previous day is, on average, 4.6% and declines monotonically from the smallest portfolio (18.1%) to the largest portfolio (less than 0.02%).

2.3 Heterogeneity of non-trading among stocks with similar market capitalizations

The above sections examine average levels of non-trading. Although averages are useful in summarizing cross-portfolio and cross-time differences in the amount of non-trading, they do so at the expense of lost information about potential differences in non-trading among securities within a portfolio. To the extent that individual security non-trading is not homogeneous among otherwise-similar securities, summarization by means may be misleading and might result in potentially erroneous inferences about the impact of non-trading on measured portfolio returns.

To examine the degree of heterogeneity in non-trading, Table 3 reports the fractiles of the distribution of percent non-trading for NYSE and AMEX stocks within each decile of size. We define percent non-trading as the ratio of

Table 3. Distribution of the number of non-traded days as a percentage of total
trading days during the year, within size deciles (1973–1990)

The non-trading percentage is measured for each NYSE and AMEX stock within
each year between 1973 and 1990 as the ratio of the number of non-traded days to
the total number of trading days during that year. Averages and distribution
fractile are measured over the entire 1973–1990 period.

Portfolio	Mean	Max	99%	95%	90%	75%	50%	25%	10%
Smallest	31.0	91.2	75.9	60.2	51.0	36.7	20.0	7.4	0.8
2	18.7	83.4	62.7	47.1	37.1	22.5	9.1	2.0	0.0
3	11.5	86.9	55.9	36.7	27.5	12.3	2.9	0.4	0.0
4	8.3	96.4	50.5	28.6	18.0	5.7	0.8	0.0	0.0
5	5.7	95.0	43.8	18.7	9.9	2.0	0.0	0.0	0.0
6	9.9	96.4	56.5	12.3	5.7	0.8	0.0	0.0	0.0
7	6.5	94.9	47.1	9.1	3.8	0.4	0.0	0.0	0.0
8	2.9	83.7	32.9	3.8	1.0	0.0	0.0	0.0	0.0
9	1.0	82.1	18.0	1.0	0.8	0.0	0.0	0.0	0.0
Largest	0.2	48.2	1.0	0.0	0.0	0.0	0.0	0.0	0.0

the number of days a stock did not trade during a calendar year to the total
number of trading days for that stock in that year. We compute this ratio
for each stock for each year. Using the values of non-trading for all stocks
that appeared in a portfolio across all years from 1973 to 1990, we report
distribution fractiles and means for each portfolio. The means, reported in
the leftmost column in Table 3, indicate the percentage of days within an
average year that the 'representative' stock in a portfolio did not trade. For
example, the average stock in the smallest size decile did not trade on 31% of
the trading days during an average year. The mean non-trading percentage
declines almost monotonically with market capitalization to a value of 0.2%
for the stocks in the decile of largest market cap.

Interestingly, the distribution fractiles indicate a substantial degree of heterogeneity in the amount of non-trading, particularly in the smaller market
cap deciles. For example, while 10% of the stocks in the smallest decile trade
virtually every day, 10% of the stocks in that decile do not trade on 51% of
the trade days during the year, and 1% do not trade on 76% of the trade
days during the year. Moreover, this cross-sectional heterogeneity does not
vary much during the sample period. Figures 3a and 3b plot the year-by-year
distributional fractiles for the two smallest market cap deciles. It is apparent

from the figures that the non-trading heterogeneity reported in Table 3 is not due to extreme levels of non-trading in only a small number of years.

2.4   Non-trading estimates from the new CRSP files

As mentioned previously, CRSP has made an effort to clean the historical record of non-traded securities in their data files, particularly for the period after 1962. As a result, the non-trading frequencies computed from the newer versions of the file are sometimes quite different from those computed with the earlier files. An important consequence of this revision is that previous research results that are sensitive to non-trading issues and conducted using the earlier files cannot be replicated with the revised files.

So, how significant are the differences in estimated non-trading frequencies between the earlier files and the revised files? To answer this question, we examine the average daily non-trading frequencies for the smallest decile of NYSE and AMEX stocks for each year from 1962 to 1990 using the 1993 CRSP daily file. The results are graphed in Figure 4 along with the small-stock non-trading frequencies computed with the 1990 CRSP file. The answer is that the changes are quite significant. For the 1964–1972 period during which non-trading frequencies had been recorded as zero, the frequencies now range from 2.5% in 1967 to 17% in 1964. Interestingly, the non-trading frequencies also change for most of the other years during this period, often by orders of magnitude. Particularly large revisions occur during 1973–1975. For example, the former estimate of non-trading of 5% in 1975 increases to 32% in the revised file. The estimated frequency differences are, of course, much smaller for larger stocks. However, the important point is that for small stocks, where issues of non-trading-induced biases are most important, the non-trading frequencies are dramatically higher in the revised file.

3 An Application of the Non-trading Evidence

While our non-trading findings are interesting in their own right, the evidence in Section 2 is useful for future research because it provides reasonable bounds for levels of nontrading. As an example, in this section we use the stylized facts from Section 2 to reexamine the relation between stock return autocorrelations and non-trading. Most previous research in this area relies primarily on *conjectures* about the extent of non-trading. Our contribution is to demonstrate the potential impact of non-trading on the measurement of return autocorrelations using *actual* non-trading levels.

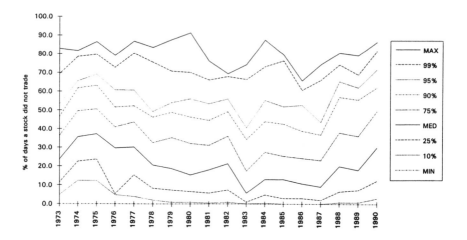

Figure 3a. Fractiles of the distribution of the number of non-traded days as a percentage of total trading days for the decile of smallest NYSE and AMEX stocks, measured within each year from 1973–1990. The non-trading precentage is measured for each stock in the smallest market cap decile at the beginning of each year as the ratio of the number of non-traded days to the total number of trading days during the year.

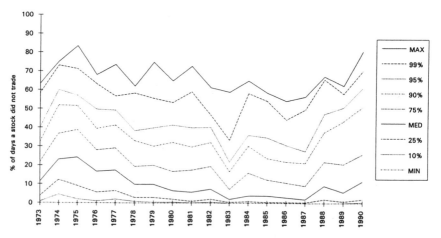

Figure 3b. Fractiles of the distribution of the number of non-traded days as a percentage of total trading days for the second (next to smallest) decile of NYSE and AMEX stocks, measured within each year from 1973–1990. The non-trading precentage is measured for each stock in the second decile at the beginning of each year as the ratio of the number of non-traded days to the total number of trading days during the year.

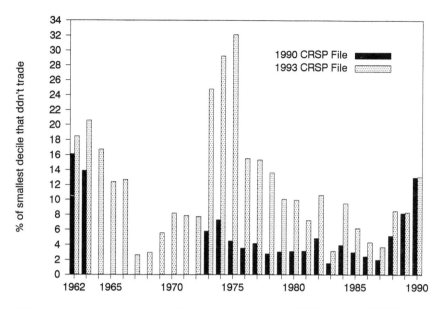

Figure 4. Average non-trading frequencies for NYSE and AMEX – a comparison of the differences between the 1990 and 1993 CRSP daily files. The figure reports average daily non-trading frequencies for the stocks in the smallest size decile of NYSE and AMEX stocks for the 1962–1990 period. If a stock does not trade on a particular day, the CRSP file reports the average of the closing bid and ask prices as a negative value. Daily non-trading frequencies are averaged to create within-year frequencies. The light bars represent estimates of non-trading frequencies computed from the 1993 CRSP daily file, the dark bars from the 1990 CRSP files.

3.1   Daily portfolio autocorrelations implied by the Lo-MacKinlay non-trading model

A hypothesis, long conjectured by researchers, is that nonsynchronous or infrequent trading can result in significant small stock portfolio autocorrelations (e.g., Fisher (1966)). However, return-generating models that condition on non-trading have not been entirely successful in explaining autocorrelation patterns in actual returns. For example, Atchison, Butler and Simonds (1987) compare theoretical and observed daily index autocorrelations between 1978 and 1981 based on transaction-frequency data between January and March 1980. They find that a Scholes and Williams (1977) model of nonsynchronous trading can explain only a small portion of the observed index autocorrelation over their period of study.

Lo and MacKinlay (1990a, 1990b) develop a general model of non-trading and examine the magnitude of index autocorrelations implied by the model.[5] They derive expressions for moments of observed returns as a function of the non-trading process. In each period t there is a probability p_i that security i will not trade, and this probability is assumed to be *iid*. If a stock does not trade, then its observed return, R_{it}^o, is simply zero. If it trades the next period, then its observed return is simply the sum of R_{it}^o and R_{it+1}^o. Within this framework, Lo and MacKinlay show that the autocorrelation for *portfolio* a can be calculated as:

$$\text{Corr}\left[R_{at-n}^o(q), R_{at}^o(q)\right] = \frac{(1 - p_a^q)^2 p_a^{nq-q+1}}{q(1 - p_a^2) - 2p_a(1 - p_a^q)}, \qquad (3.1)$$

where q represents the time aggregation interval (e.g., $q = 1$ for daily returns, $q = 5$ for weekly returns, etc.), n represents the number of lags and p_a represents a common (*homogeneous*) non-trading probability for portfolio a securities. For daily returns, where $q = 1$, equation (3.1) reduces to:

$$\text{Corr}\left[R_{at-n}^o, R_{at}^o\right] = p_a^n. \qquad (3.2)$$

Thus, the *first-order* (i.e., $n = 1$) autocorrelation for *daily* portfolio returns is simply the average non-trading probability for securities in that portfolio. With time aggregation (i.e., $q > 1$), even relatively large probabilities of daily non-trading will have a small impact on weekly and monthly autocorrelations.

We reexamine the impact of non-trading on *daily* portfolio autocorrelations by computing implied values for portfolio autocorrelations as a function of the level of non-trading during the 1973–1990 period. We report results in Table 4 using equation (3.2) and our estimates of average non-trading for size portfolio deciles from Table 3. The model-implied portfolio autocorrelations based on equation (3.2) are reported in the rightmost column of Table 4. For comparison with the theoretically-implied values, the estimated daily autocorrelations for value-weighted NYSE and AMEX size deciles for the 1973 to 1990 period are in the first column. Comparing the two columns, the daily autocorrelations implied by the model for the 'representative' stocks in each portfolio uniformly underestimate the observed autocorrelations over the sample period, although the model-implied autocorrelation for the smallest-stock portfolio is quite close to the empirical estimate.[6]

---

[5] As Lo and MacKinlay (1990a) indicate, the Scholes and Williams (1977) model requires each security to trade within a fixed time interval. It is worth noting that the Scholes and Williams empirical application attempts to eliminate securities that did not trade on a particular day, but their sample includes much of the 1964 to 1972 period during which CRSP omitted the information necessary to identify non-traded securities.

[6] The impact of non-trading on *weekly* and *monthly* autocorrelations is not presented in the table, but can be inferred from equation (3.1). For the smallest portfolio, the implied weekly and monthly first-order autocorrelations are 0.059 and 0.012, respectively, much lower than actual levels reported by Lo and MacKinlay (1990a) of 0.46 (weekly) and 0.20 (monthly) for their smallest portfolio.

Table 4. Actual and model-implied daily autocorrelations for NYSE
and AMEX stocks for the 1973–1990 period

	Daily Autocorrelation	
Market Cap Decile	Implied[1]	Actual
Smallest	0.379	0.310
2	0.349	0.187
3	0.330	0.115
4	0.336	0.083
5	0.314	0.057
6	0.315	0.099
7	0.316	0.065
8	0.287	0.026
9	0.268	0.010
Largest	0.169	0.002

[1] 1st order autocorrelation implied by the
Lo–MacKinlay non-trading model and the
non-trading values from Table 3

3.2 Heterogeneity of non-trading and the Boudoukh, Richardson and Whitelaw (1994) Model

The substantial heterogeneity of nontrading evident in Table 3 within each
of the size deciles is inconsistent with the Lo–MacKinlay assumption of homogeneous non-trading among stocks in a portfolio. Boudoukh, Richardson
and Whitelaw (1994) present an alternative to the Lo–MacKinlay model that
incorporates non-trading heterogeneity and demonstrate that such differences
in the distribution of non–trading among securities in a portfolio can have a
substantial influence on that portfolio's autocorrelation. Boudoukh, Richardson and Whitelaw (1994) use the evidence on non-trading presented here in
Table 3 to show that the assumption of homogeneity underestimates the relation between return autocorrelation and non-trading. They conclude (p.558)
that "large spurious autocorrelations are principally driven by severe nontrading for some stocks coupled with frequent trading in others" i.e., the pattern
documented in Table 3.

3.3 Daily portfolio return autocorrelations and non-trading: a longer perspective

Most studies of short-horizon return autocorrelations in US markets examine the period after 1962, the starting date for the CRSP daily data files. Is it possible that the relation between non-trading and autocorrelation reported above is specific to the recent sample period? To address this question we examine the autocorrelation of the daily returns of the S&P Composite Index which is available back 1928.[7] The daily first-order autocorrelation for the S&P Composite during the commonly-studied July 1962 to December 1990 period is 0.135. However, the estimated first-order daily autocorrelation for the January 1928 to June 1962 period is 0.038, approximately one-fourth as large as in the later subperiod. The estimate of the S&P daily return autocorrelation for the entire 1928–1990 period is 0.062. This is consistent with Schwert's (1990, Table 8) estimate of 0.05 for the first-order serial correlation of a daily stock index series for the period 1885 to 1987.

The time series behavior of the S&P daily return autocorrelation for the 1928-90 period is illustrated in Figure 5. The figure displays: (1) the within-year estimates of autocorrelations based on daily returns for the S&P and; (2) the within-year non-trading estimates for NYSE stocks compiled from the monthly data described in Section 2. The early period is interesting. From the 1920s through the early 1940s, a period when non-trading was at its highest levels during the century, the autocorrelations of the daily returns on the S&P were at their lowest.[8] Whereas there is no relation between the S&P autocorrelation and the within-year non-trading in the 1962–1990 period ($\rho = 0.010$, excluding 1964–1972), there is a significant *negative* relation for the 1928–1962 period ($\rho = -0.429$). These general observations suggest that for the pre-1960 time period non-trading is not a cause of spurious daily portfolio autocorrelation.

The decline of the daily S&P autocorrelations through the 1970s and 1980s is also interesting. This recent decline in autocorrelations is evident across most stocks. For comparison, in Figure 6 we show daily return autocorrelations computed within each quarter for the period 1973–1990 for three portfolios – an equal-weighted portfolio of stocks in the smallest quintile of market capitalization, an equal-weighted portfolio of stocks in the largest quintile, and an equal-weighted portfolio of the 30 stocks in the Dow Jones Industrial Average. We include the Dow Jones portfolio because it is virtually free of non-trading on a daily basis. Figures 6a to 6c plot the time series

---

[7]The pre-1962 segment of this series was compiled by Rob Stambaugh and is the same index used in Keim and Stambaugh (1984).

[8]The S&P Composite contained only 90 stocks during the period from 1928 to February 1957. It is unlikely, though, that this reduction in the number of securities is responsible for the low autocorrelation in the early period.

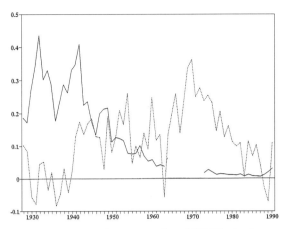

Figure 5. Frequency of non-trading NYSE stocks and daily autocorrelations for the S&P Index, measured within-years. The solid line represents the frequency of non-trading and the dotted line represents the daily autocorrelations. If a stock does not trade on the last day of the month, the CRSP monthly file reports the average of the bid and ask prices as a negative value. Month-end frequencies of non-trading are averaged to create within-year frequencies. The sample period is 1928–1990. The sample for non-trading frequencies excludes the years 1964–72 because of the abnormally low number of negative prices recorded by CRSP during that period.

of within-quarter autocorrelations for the smallest quintile (6a), the largest quintile (6b), and the Dow Jones portfolio (6c). The figures show a general decline in the within-quarter autocorrelations in all three portfolios. However, it is difficult to equate such a general decline in autocorrelations to the modest declines in non-trading over the period.

An alternative explanation for the decline in autocorrelations in the 1970s and 1980s involves the relative proportions of institutional and individual investors in the market. The idea is that if institutional investors tend to trade more frequently at the level of portfolios of stocks rather than individual stocks (e.g., program trading, indexing), then buys and sells across securities contained in their portfolios will have greater synchronousness. Further, to the extent that such investors are index funds and use 'market at close' orders to permit closer alignment with their benchmark portfolios, these firms might influence the distribution of daily closing prices by causing clustering of prices at the bid or ask quotes. For example, if one of these firms were in the midst of a buying program, a larger percentage of transactions would occur closer to the market maker's ask price. A natural outcome is an increased tendency for *portfolio* bid-ask bounce. That is, the bid-ask bounce that is evident in individual security returns, but which is normally elimi-

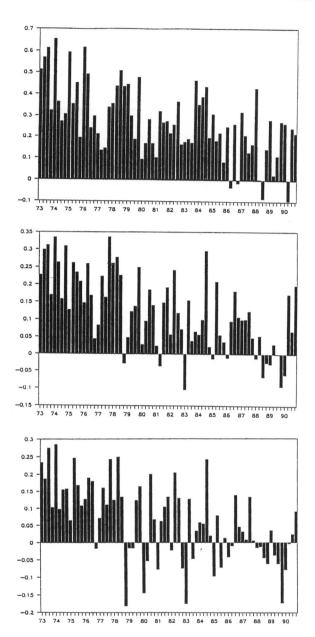

Figure 6. Daily portfolio autocorrelations, measured within-quarters.
The top panel presents autocorrelations for the quintile of smallest NYSE and
AMEX stocks. The middle panel presents autocorrelations for the quintile of
largest NYSE and AMEX stocks. The bottom panel presents autocorrelations for
an equal-weighted portfolio of the Dow Jones 30 stocks. The sample period is
1973–1990.

nated in diversified portfolios, will not be eliminated by diversification and will be evident in portfolio returns. The resulting negative autocorrelation component will tend to offset the positive autocorrelation component induced by positive cross-autocovariances (Lo and MacKinlay (1990b)). Other things equal, we should observe a reduction in portfolio autocorrelations since the introduction of such trading practices, especially for portfolios that are heavily concentrated in the stocks comprising the indexes that the institutional portfolios mimic (e.g., S&P 500, Russell 2000). Thus, the 1970s and 1980s, which display a pattern of declining index autocorrelation, may constitute an institutionally-dominated regime that is distinct from earlier periods.[9] An increased institutional presence is indeed evident in Figure 7 which plots annual small-stock non-trading frequencies against the percentage of total stock market value that is held in institutionally-traded portfolios. The data on institutional holdings are from the Federal Reserve Board's Flow of Funds data. The percentage of the market held in institutional portfolios increased from 33% in 1973 to 53% in 1990. Of course, this undoubtedly understates the proportion of total *trading volume* represented by institutional traders.

4   Concluding Remarks

We document a declining pattern of non-trading over the 1926 to 1990 period. Frequency of non-trading is directly related to firm size. Non-trading occurs less frequently in December relative to other months, and this is particularly true for smaller firms. This finding is consistent with both the tax-loss selling and window dressing hypotheses. Since 1973, non-trading has averaged about 1.3% for NYSE firms, 15.1% for AMEX firms and 24.1% for NYSE and AMEX firms in the smallest size decile portfolio. Non-trading is shown to increase monotonically from Monday through Friday. Our analysis excludes the years 1964 to 1972 (inclusive) because of the abnormally low number of negative prices recorded by CRSP during that period on the pre-1991 versions of the files. Subsequent versions of the CRSP files (both daily and monthly files) have cleaned these errors. Consequently, any non–trading-related research that used the pre-1991 versions of the files was exposed to the reporting inaccuracies documented above and cannot be replicated with the current CRSP files.

    We also find substantial heterogeneity in the amount of non-trading across the stocks within each size decile. For example, while 10% of the stocks in the smallest decile trade virtually every trade day, 10% of the stocks in that

---

[9]See Froot and Perold (1995) who also discuss similar arguments. Brenner, Subrahmanyam and Uno (1990) and Kishimoto (1990) find a similar decline in the daily autocorrelations of the Nikkei Stock Average and the Topix index after September 1988 when futures contracts on these indexes were introduced.

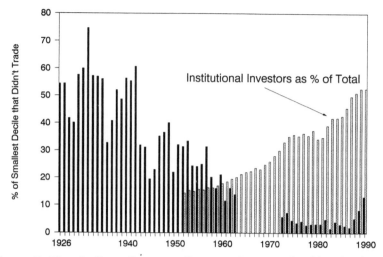

Figure 7. The decline of non-trading vs. the growth of institutional investing. The figure reports average non-trading frequencies for the smallest decile of NYSE based on the CRSP monthly file (dark bars) and the percentage of total US stock market value that is held in institutionally-traded portfolios (light bars). If a stock does not trade on the last day of the month, the CRSP file reports the average of the bid and ask price as a negative value. Month-end frequencies are averaged to create within-year frequencies. The data on institutional holdings is from the Federal Reserve Board's flow of funds data.

decile do not trade on 51% of the trade days during the year, and 1% do not trade on 76% of the trade days during the year. Further, this heterogeneity is prevalent in every year from 1973 to 1990, indicating that a large percentage of NYSE and AMEX stocks consistently exhibit substantial non-trading. These findings have important implications for inferences about return autocorrelations drawn from models of non-trading that assume homogeneous non–trading among similar (market-capitalization) stocks. The impact of non-trading on daily portfolio autocorrelation measures may have been understated in the literature. Our application highlights the need for more research on the relation between non-trading and return autocorrelations.

Finally, our results provide useful bounds for future studies relating to non-trading. We would expect to find a greater amount of non-trading on markets like NASDAQ that contain a large proportion of smaller-cap stocks. While we have documented a relation between non-trading and firm size, size may only be a proxy for other variables closely related to non-trading, such as the amount (and timing) of new information being released about particular firms and the subsequent trading in these stocks. Further research will yield additional insights into these relationships.

References

Atchison, M., K. Butler and R. Simonds (1987) 'Nonsynchronous security trading and market index autocorrelation', *Journal of Finance* **42** 111–118.

Blume, M. and R. Stambaugh (1983) 'Biases in computed returns: An application to the size effect', *Journal of Financial Economics* **12** 387–404.

Boudoukh, J., M.P. Richardson and R.F. Whitelaw (1994) 'A tale of three schools: Insights on autocorrelations of short-horizon returns', *The Review of Financial Studies* **7** 539–573.

Brenner, M., M.G. Subrahmanyam and J. Uno (1990) 'The volatility of the Japanese stock indices: evidence from the cash and futures market', working paper, NYU.

Cohen, K., G. Hawawini, S. Maier, R. Schwartz and D. Whitcomb (1983), 'Frictions in the trading process and the estimation of systematic risk', *Journal of Financial Economics* **12** 263–278.

Dimson, E. (1979) 'Risk measurement when shares are subject to infrequent trading', *Journal of Financial Economics* **7** 197–226.

Dimson, E. and P. Marsh (1983) 'The stability of UK risk measures and the problem of thin trading', *Journal of Finance* **38** 753– 783.

Fowler, D. and C. Rourke (1983) 'Risk measurement when shares are subject to infrequent trading: Comment', *Journal of Financial Economics* **12** 279–283.

Fama, E. (1965) 'Tomorrow on the New York Stock Exchange', *Journal of Business* **38** 285–299.

Fisher, L. (1966) 'Some new stock market indices', *Journal of Business* **39** 191–225.

Froot, K. and A. Perold (1995) 'New trading practices and short-run market efficiency', *Journal of Futures Markets* **15** 731–765.

Keim, D. (1983) 'Size-related anomalies and stock return seasonality: Further empirical evidence', *Journal of Financial Economics* **12** 13–32.

Keim, D. (1989) 'Trading patterns, bid-ask spreads and estimated security returns: The case of common stocks at calendar turning points', *Journal of Financial Economics* **25** 75–97.

Keim, D. and R. Stambaugh (1984) 'A further investigation of the weekend effect in stock returns', *Journal of Finance* **39** 819–835.

Kishimoto, K. (1990) 'A new approach for testing the randomness of heteroscedastic time-series data', working paper, University of Tsukuba.

Lakonishok, J. and S. Smidt (1984) 'Volume and turn-of-the-year behavior', *Journal of Financial Economics* **13** 435–456.

Lo, A., and A.C. MacKinlay (1990a) 'An econometric analysis of nonsynchronous-trading', *Journal of Econometrics* **45** 181–211.

Lo, A. and A.C. MacKinlay (1990b) 'When are contrarian profits due to stock market overreaction?', *The Review of Financial Studies* **3** 175–205.

Porter, David C. (1992) 'The probability of a trade at the ask: an examination of interday and intraday behavior', *Journal of Financial and Quantitative Analysis* **27** 209–228.

Perry, P. (1985) 'Portfolio serial correlation and nonsynchronous trading', *Journal of Financial and Quantitative Analysis* **20** 517–523.

Scholes, M. and J. Williams (1977) 'Estimating betas from nonsynchronous data', *Journal of Financial Economics* **5** 309–327.

Schwert, W. (1990) 'Indexes of U.S. stock prices', *Journal of Business* **63** 399–426.

Stoll, H. and R. Whaley (1990) 'Stock market structure and volatility', *The Review of Financial Studies* **3** 37–71.

Part II
Seasonal Patterns in Stock Returns
and Other Puzzles

Is There Still a January Effect?*

David G. Booth and Donald B. Keim

1 Introduction

The 'January effect' refers to the tendency for the size premium to be significantly larger in January than in the other eleven months.[1] Studies of the January effect (Keim 1983, Roll 1983) and the size effect (Banz, 1981 and Reinganum 1981) were widely circulated in both academic and practitioner circles by early 1982. For example, a conference on the small firm effect and associated phenomenon, organized by Mark Reinganum at the University of Southern California in April 1982, attracted a broad cross section of academics and practitioners. Coinciding with the increased interest in the size and January effects was the startup of Dimensional Fund Advisors' small cap mutual fund, named the '9–10 Fund' because its composition is determined by the two size deciles (the 9th and 10th) containing the smallest stocks.

The 9–10 Fund was the first passively-managed mutual fund designed to provide access to small-cap stocks, and has become a widely-accepted benchmark of small-cap performance in that it represents realized returns from a live portfolio.[2] As such, the 9–10 Fund returns include trading costs. Ability to measure the January and size effects with actual portfolio returns is important because trading costs for these illiquid stocks can be large enough to offset the premia observed in the simulated returns.

The 15-year out-of-sample period, plus the availability of an equally-long series of live portfolio returns, justifies another look at the January effect. In this paper, therefore, we revisit the January seasonal pattern in the size premium by addressing two often-asked questions:

(1) Is the January effect evident in the post-1981 period when measured with decile portfolio returns as in the early research? and

(2) Is the January effect evident in the live portfolio returns?

The answer is yes to the first question, no to the second. We show that the absence of a January size premium in the live portfolio returns is due to

*We thank Truman Clark, Ken French, Rex Sinquefield and Bill Ziemba for helpful comments. Any errors are our own.

[1]Hawawini and Keim (1999) discuss the evidence relating to the size premium.

[2]The small-cap index published by Ibbotson Associates for the post-1981 period is the DFA 9–10 Fund.

the exclusion of the lowest-price and lowest-market cap stocks from the 9–10 Fund. It is precisely these very illiquid stocks that tend to amplify the effect in the 9th and 10th decile portfolio returns.

2 Historical Perspective – The Evidence Prior to 1982

Rozeff and Kinney (1976) provided the first academic evidence that January returns may differ from the returns in other months, finding that equal-weighted indexes containing all the stocks listed on the NYSE displayed significantly higher returns in January than in the other eleven months over the period 1904–1974. Keim (1983) documented that the magnitude of the size effect varied by month of the year. He found that fifty percent of the annual size premium was concentrated in the month of January. Subsequent research by Blume and Stambaugh (1983) demonstrated that, after correcting for an upward bias in average returns for small stocks that was common to the experimental design in the early studies on the size effect, the size effect is evident only in January.

In this section we present the evidence on the size and January effects that was available at the beginning of 1982. Following early research, we define company size by reference to NYSE capitalization deciles. 'Decile 10' is a capitalization-weighted small-cap portfolio of the smallest 10% of NYSE companies (plus AMEX and NASDAQ stocks of similar size). 'Decile 1' is a corresponding large-cap portfolio.

Table 1 reports average values for the size and January effects for the 1926–81 period, as well as for ten-year subperiods. The size premium is computed as in the early research – the average difference in returns between Decile 10 and Decile 1. The column labeled 'All Months' contains the average monthly size premium measured over all 12 months. The 'January' column reports the January size premium, and the column labeled 'Feb–Dec' contains the average monthly size premium outside of Janaury. The t-values in the right-most column test the statistical significance of the difference between the January and the non-January size premium.

The size effect, 1926–81. The size premium ('All months') is significant when estimated for the entire period (81 basis points per month, $t = 2.38$). It is significant in the 1966–81 subperiod and marginally significant in the 1936–45 subperiod.

The January effect, 1926–81. The January effect is significant, both statistically and economically, in the overall period (10.8% per month, $t = 7.11$)

Table 1. Summary Statistics for the Size and January Effects, 1926–81.
The summary statistics below reflect the stylized facts for the Size and January effects prior to inception of Dimensional Fund Advisors and the publication of the academic research on the effects. Reported are means (*T*-Statistics) of the difference in monthly percentage returns between Decile 10 (Small-Cap) and Decile 1 (Large-Cap) stocks for the period 1926–1981.

Time Period	All Months	January	NonJan (Feb to Dec)	t(Jan − NonJan)
1926–1981	0.81	10.80	−0.10	9.32
	(2.38)	(7.11)	(−0.31)	
1926–1935	1.19	16.48	−0.20	3.54
	(0.88)	(2.96)	(−0.15)	
1936–1945	1.94	16.56	0.62	4.27
	(1.76)	(3.19)	(0.60)	
1946–1955	−0.19	5.34	−0.69	5.69
	(-0.58)	(6.35)	(−2.23)	
1956–1965	0.24	6.55	−0.33	7.98
	(0.83)	(5.80)	(−1.38)	
1966–1981	0.84	9.72	0.03	7.56
	(2.10)	(6.31)	(0.08)	

and in each subperiod. Based on the results in the third column of Table 1, the size effect is insignificant outside of January.

The January effect is also evident in many international stock markets. Worldwide evidence of January seasonality in the size premium is summarized in Figure 1 which is based on the survey of international markets by Hawawini and Keim (1985) (see their table 9). The monthly size premia reported in the table are defined as the difference in return between the smallest and largest market-cap stocks in the respective samples, reported separately for January and for the rest of the year (as in Table 1). In all countries except France and the United Kingdom the size premium is significantly larger during January than during the rest of the year, although it generally remains positive after January.

3   The 1982–95 Evidence for the Decile 10 Small-Cap Stocks

To draw comparisons with the original research, we measure the January effect for the 1982-95 period using the same methods and data as in Table

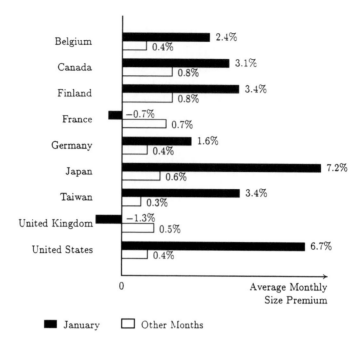

Figure 1. International Evidence on the Monthly Size Premium. Size premium = $\overline{R}_{\text{small}} - \overline{R}_{\text{large}}$. *Sources. Belgium*, Hawawini *et al.* (1989); *Finland*, Berglund (1985); *France*, Hamon (1986); *Germany*, Stehle (1992); *Japan*, Ziemba (1991); *Taiwan*, Ma *et al.*; *United Kingdom*, Levis (1985).

1. The top row in Table 2 contains the updated size and January effects, computed as the average difference in monthly returns between Decile 10 and Decile 1. Unlike the earlier period, the average size effect is negative for the 14 years from January 1982 to December 1995, a reflection of the prolonged inverted size effect from mid-1983 through the end of 1990. Although negative, the size premia is insignificantly different from zero during this period.

Like the earlier period, the average January size premium is significantly positive in the 1982–95 period (4.48%, $T = 2.83$). In addition, the January size premium is greater than the non-January premium. The difference between the January and non-January premium is 5.24% ($T = 5.09$). Although the 1982–95 January premium has a lower mean and higher variability than the earlier periods, the premium for the 1982–95 period is not significantly different from the post-war 1946–81 premium ($T = -1.89$), a result that is graphically illustrated in Figure 2. In sum, the January size premium, when measured as in the early research, still persists in the out-of-sample period that followed the first appearance of the research.

Table 2. The Post-1981 Size and January Effects for Various Small-Cap Portfolios and Indexes.

Reported are means (*T*-Statistic) of the difference in monthly percentage returns between several small-cap portfolios and Decile 1 (Large-Cap) Stocks (January 1982–December 1995).

	All Months	January	NonJan (Feb to Dec)	t(Jan − NonJan)
Decile 10–Decile 1	−0.33	4.48	−0.76	5.09
	(−1.08)	(2.83)	(−2.77)	
CRSP910–Decile 1	−0.26	3.36	−0.59	4.25
	(−0.98)	(2.51)	(−2.34)	
LIVE910–Decile 1	−0.07	2.00	−0.26	2.68
	(−0.30)	(1.68)	(−1.13)	
SIM910–Decile 1	−0.07	2.23	−0.28	3.18
	(−0.32)	(2.13)	(−1.28)	
CRSP–LIVE	-0.19	1.37	-0.33	6.60
	(−2.40)	(3.80)	(−4.71)	
CRSP–SIM	−0.19	1.13	−0.31	5.62
	(−2.47)	(3.08)	(−4.47)	
SIM–LIVE	0.00	0.23	−0.02	1.34
	(0.01)	(0.82)	(−0.41)	

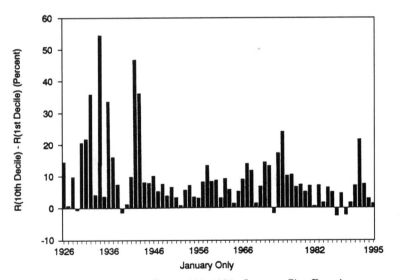

Figure 2. The January Effect, 1926–1995. January Size Premium = R(10th) − R(1st).

4   Trading Costs, Realizable Returns and the January Size Premium

4.1   An Alternative Small-Cap Benchmark: The CRSP 9–10 Index.

From a practical perspective, some might argue that 10th decile stocks are not the best benchmark of small-stock performance because those stocks are extremely illiquid. Due to trading cost considerations, a more practical investable benchmark of small stocks is the 9–10 Index constructed by the Center for Research in Security Prices (*CRSP910*). *CRSP910* contains the smallest 20% of NYSE stocks, the 9th and 10th deciles. The second row in Table 2 reports the average value of the size premium measured as the difference in returns between *CRSP910* and Decile 1. The results are similar to those for the size premium measured with the 10th decile only. The main difference is that the average January size premium, although still significant, is reduced by 25% from 4.48% to 3.36%.

Keep in mind that these results, like the previous results for the Decile 10 stocks, do not include trading costs. The Decile 10 portfolio and *CRSP910* will, in practice, have a high turnover rate and incur large trading costs, possibly offsetting the higher returns. Stoll and Whaley (1983) warned that the size effect may be illusory because trading costs might eliminate the observed premium. Keim and Madhavan (1997) analyzed the recent trade costs of 21 institutional investors. They estimate that *one-way* trade costs for buy transactions of small-cap stocks that reside approximately in the 9th and 10th market-cap deciles range from 0.35% for the smallest transactions to 2.35% for the largest transactions in NYSE-AMEX stocks. The comparable range for NASDAQ stocks is 0.92% to 3.34%. Estimates of this magnitude, coupled with significant turnover in the portfolio, could easily confirm Stoll and Whaley's warning.

4.2   Is the January Effect Realizable? Evidence from the DFA 9–10 Fund

The DFA 9–10 Fund (*LIVE910*) is constructed primarily from stocks in *CRSP910*. An indexed-type portfolio, the *LIVE910* should theoretically exhibit return behavior similar to *CRSP910*, except that *LIVE910* returns include transactions costs and management fees. The average size premium measured with *LIVE910* returns is shown in the third row in Table 2. Like the Decile 10 and *CRSP910* size premia, the non-January *LIVE910* size premium is negative in the 1982–95 period, but not significantly different from zero (-0.26%, $T = -1.13$). When measured over all months, the size premium for the live portfolio is a negative seven basis points per month ($T = -0.30$) for the 1982–95 period.

In contrast to results for *CRSP910*, the January size premium for *LIVE910* is substantially smaller at 2.00% for the 1982–95 period, and insignificant ($T = 1.68$). The *CRSP910* size premium exceeds the *LIVE910* size premium in January by 137 basis points ($T = 3.80$), but it is smaller on average by 33 basis points ($T = -4.71$) in the remaining months (see row 5 in Table 2). So the results in Table 2 for *LIVE910* show a much-attenuated January size premium when compared to *CRSP910*, even though the live portfolio is constructed primarily from those same 9th and 10th decile stocks. At first glance, the difference would appear to be attributable to the costs associated with trading the live portfolio. However, the difference is difficult to assign to trade costs because the influence of those costs should be reasonably constant through the year. As is evident in Table 2, however, the *LIVE910* size premium and the *CRSP910* size premium behave very differently at different times of the year. Indeed, the non-January *LIVE910* premium is significantly greater than the non-January *CRSP910* premium, a result that cannot be attributed to trade costs.

5 Why Does the Seasonal Pattern Appear Weaker in Live Returns?

If not trading costs, what can explain the attenuation of a January effect in *LIVE910*? A possible difference between *LIVE910* and *CRSP910* concerns the rules imposed by DFA's portfolio managers that restrict the composition of *LIVE910*, causing it to differ from *CRSP910*. To examine the influence of these rules, we use a value-weighted portfolio that is designed to simulate the DFA 9–10 investment strategy (*SIM910*).[3] *SIM910* is designed to reflect those DFA rules that are likely to cause the largest deviations in portfolio holdings when compared to the Index. Accordingly, the simulated portfolio incorporates the following rules:

(1) An 8th decile hold range. The simulated portfolio contains the stocks in the 9th and 10th deciles as determined by the quarterly sort updates, plus those stocks that had previously been members of the 9th and 10th deciles but migrated to the 8th decile. Stocks that exit the 8th decile into the 7th decile are 'sold' and do not reenter the portfolio unless their market cap drops below the 9th decile breakpoint. Note that 8th decile stocks that arrived in that decile from the 7th decile are not included in the portfolio.

---

[3]See Keim (1999) for details on the construction of this index.

(2) IPOs are included in the portfolio only after they accumulate one year of trading history.

(3) Stocks with share prices less than $2.00 or with market capitalization less than $10 million are excluded.

The constraints imposed by these rules result in a simulated live portfolio that is very different from *CRSP910*. For example, the exclusion of the extra-low price and extra-small market cap stocks effectively eliminates the influence of the very illiquid stocks that tend to reside in the 10th decile. *A priori*, this exclusion should have a big impact on the measured January size premium since much of that premium has been shown by Blume and Stambaugh (1983), Keim (1989) and others to be caused by illiquid, low-price stocks. Inclusion of the 8th decile stocks (as a result of the hold range) should also tend to dampen the magnitude of the January size premium.

Indeed, these rules create large differences in the average size premium between *SIM910* and *CRSP910* – the average monthly difference between the *CRSP910* and *SIM910* size premia is –0.19% ($T = -2.47$) across all months, 1.13% ($T = 3.08$) in January, and –0.31% ($T = -4.47$) in the remaining months (see '*CRSP – SIM*' in row 6 of Table 2). These differences are similar to those that exist between *LIVE910* and *CRSP910* (see '*CRSP – LIVE*' in row 5 of Table 2), indicating that the simulated series is capturing the important characteristics of the live portfolio. Consistent with this, the difference in the size premium between *SIM910* and *LIVE910* is insignificant, indicating that the attenuation of the January size premium for *LIVE910* is attributable to the rules governing the composition of the live portfolio that tend to exclude those (illiquid) stocks for which the January size premium is largest and including (8th decile) stocks for which the effect is attenuated. The results appear to confirm that the attenuation is not due to the trading costs embedded in the *LIVE910* returns.

6 Conclusions

Enough time has gone by since the first publication of the January effect to warrant an update. The 1982–95 experience confirms many of the original research findings that covered the 1926–81 period. In particular, for Decile 10 stocks the size premium is significant in January but insignificant in the remaining months.

The 9–10 Fund returns demonstrate that small-stock return behavior can be captured in a live portfolio. However, there is not a significant January effect in the live returns. Using a simulated portfolio designed to capture the important characteristics of the live portfolio, we are able to show that this attenuation of a January size premium is due to the exclusion of the lowest-price and lowest-market cap stocks from the 9–10 Fund. It is precisely these very illiquid stocks that tend to amplify the January effect in the Decile 10 returns.

Thus, the January effect is alive and well, but difficult to capture. That is not to say knowledge of the seasonal has no value. For example, investors with new commitments to small stocks who are going to trade anyway, and have some flexibility in the timing of the transaction, should do so before the end of the year. Likewise, redemptions are best deferred till late January. However, short-term round-trip transactions in common stocks intended to capture the January size premium are unlikely to be profitable because of transactions costs.

References

Banz, R.W. (1981) 'The relationship between returns and market value of common stocks', *Journal of Financial Economics* **9** 3–18.

Berglund, T. (1985) *Anomalies in stock returns in a thin security market: the case of the Helsinki Stock Exchange.* Doctoral Thesis, The Swedish School of Economics and Business Administration, Helsinki, Finland.

Blume, M.E. and Stambaugh, R.F. (1983) 'Biases in computed returns: an application to the size effect', *Journal of Financial Economics* **12** 387–404.

Hamon, J. (1986): 'The Seasonal Character of Monthly Returns on the Paris Bourse', *Finance* **7** 57–74, (in French).

Hawawini, G. and Keim. D.B. (1995) 'On the predictability of common stock returns: world-wide evidence'. In *Finance*, R. Jarrow *et al.* (eds.), Elsevier.

Hawawini, G. and Keim. D.B. (1999) 'The cross section of common stock returns: a review of the evidence and some new findings', this volume, 3–43.

Hawawini, G., Michel, P., and Corhay, A. (1989) 'A look at the validity of the capital asset pricing model in light of equity market anomalies: the case of Belgian common stocks'. In *A Reappraisal of the Efficiency of Financial Markets*, R.C. Guimaraes and B.G. Kingsman and S. Taylor (eds.), Springer Verlag.

Keim, D.B. (1983) 'Size-related anomalies and stock return seasonality: further empirical evidence,' *Journal of Financial Economics* **12** 13–32.

Keim, D.B. (1989) 'Trading patterns, bid-ask spreads, and estimated security returns: the case of common stocks at the turn of the year', *Journal of Financial Economics* **25** 75–98.

Keim, D.B. (1999) 'An analysis of mutual fund design: the case of investing in small-cap stocks', *Journal of Financial Economics* **51** 173–194.

Keim, D.B. and Madhavan, A. (1997) 'Execution costs and investment style: an inter-exchange analysis of institutional equity trades,' *Journal of Financial Economics* **46** 265–292.

Levis, M. (1985) 'Are small firms big performers?', *The Investment Analyst* **76** (April) 21–27.

Ma T., and Shaw, T.Y. (1990) 'The relationships between market value, P/E ratio, trading volume and the stock return of the Taiwan Stock Exchange'. In *Pacific-Basin Capital Markets Research*, Volume I, S.G. Rhee and R.P. Chang (eds.), North Holland, 313–335.

Reinganum, M. (1981) 'A misspecification of capital asset pricing: empirical anomalies based on earnings yields and market values', *Journal of Financial Economics* **12**, 89–104

Roll, R. (1983) 'The turn of the year effect and the return premia of small firms', *Journal of Portfolio Management* **9** (Winter) 18–28.

Rozeff, M. and Kinney, W. (1976) 'Capital market seasonality: the case of stock returns', *Journal of Financial Economics* **3** 379–402.

Stehle, R. (1992) 'The size effect in the German Stock Market,' Unpublished manuscript, University of Augsburg.

Stoll, H.R., and Whaley, R.E. (1983) 'Trading costs and the small firm effect,' *Journal of Financial Economics* **12** 57–79.

Ziemba, W.T. (1991) 'Japanese security market regularities. Monthly, turn-of-the-month and year, holiday and Golden Week effects,' *Japan and the World Economy* **3**(2) 119–146.

Anticipation of the January Small Firm Effect in the US Futures Markets[*]

Chris R. Hensel and William T. Ziemba

Abstract

This paper studies small and large cap equity returns in the US Small caps have outperformed in the calendar month of January (particularly in the first half) since at least 1904. We study this 'January effect' in the futures markets for all sixteen turn-of-the-years that futures have been traded, 1982/83 through 1997/98. Futures, which have low transactions costs, should be a way to exploit this effect. The evidence is that: (a) there has been anticipation of the turn-of-the-year (TOY) effect in the second half of December; (b) there was a small cap advantage in this December period; (c) there was a small cap advantage in the first half of January in the 1980s and early 1990s, but not in recent years; and d) there was a small cap disadvantage in the second half of January. The conclusion is that there now is no January small cap effect except during the second half of December in the futures markets and that it has been possible to exploit the effect in the futures markets.

1 Introduction

Rozeff and Kinney (1976) are generally credited with the discovery that US small cap stock returns in January were higher than in other months.[1] Using equally weighted data from 1904–74, the average return in January was 3.5% versus 0.5% in other months; the equal weighting means that it was small cap stocks that out-performed. The so-called January turn-of-the-year (TOY) effect refers to the phenomenon that small cap stocks outperform large cap stocks in January. Keim (1983) and Roll (1983b) were the first to rigorously document this phenomenon which Fama (1991) refers to as the 'premier anomaly'. Keim and Roll showed that most of the TOY effect was concentrated in the last trading day of December and the first four trading days of January.

Fundamental questions in applied financial economics and statistics are:

---

[*] This research was supported by the Program in Financial Modeling at the University of British Columbia and the Frank Russell Company. We wish to thank Donald Keim for helpful comments on a previous draft.
    [1]However, Wachtel (1942) first discussed the January effect with a brief study. Banz (1981) and Reinganum (1981) showed that small cap stocks, when risk adjusted by the CAPM, outperformed large caps over the whole year.

(a) Do small capitalized stocks have higher annual returns than large cap stocks and if so is the advantage because of increased risk?

(b) Is there really a January small firm effect?

(c) If there is a January small firm effect, how can we use this effect to add risk-adjusted value to portfolio performance or how can we directly exploit it?

(d) How can one tell if in a particular data subset where there is no January small firm effect whether or not the effect no longer exists or is it a small sample phenomenon?

Many research papers and books have discussed these points[2] although the vast majority of research is concerned with (a) and/or (b). The papers by Berk(1999), Booth and Keim (1999), Dimson and Marsh (1999) and Hensel and Ziemba (1999) in this volume are part of this literature. Berk (1999), addressing (a), argues that while a small firm effect may exist it can be explained by increased risk and that other measures of size related to physical attributes may be more appropriate measures of size rather than market capitalization.

Booth and Keim (1999) argue that the January effect does exist but it is very hard to capture in real portfolios. For the example for the DFA 9–10 decile fund, the primary difficulty is that actual portfolio management uses rules that eliminate investment in very small micro-cap stocks and initial public offerings that have had high returns. Hensel and Ziemba (1999) argue that over the whole year small cap returns, relative to large cap returns, have been higher when there are Democrats in the White House. The Ibbotson data, see e.g. Riepe (1997) and Ziemba (1994), suggests that most or all of the small cap advantage from 1926–97 occurred in the month of January.

Dimson and Marsh (1999) argue that there was a significant small cap effect in the United Kingdom, but that the effect reversed in the early 1990s and there has since been a significant small cap disadvantage. Black (1986, 1993), Fama (1998), Roll (1994) and Samuelson (1989) argue that while anomalies may exist for short periods of time it is difficult to exploit them and there may well be as many negative as positive outcomes from anomaly trading. Various researchers such as those above and others such as Haugen and Jorion (1996) who argue that the 'January effect is still there after all these years'

---

[2]See e.g. Agrawal and Tandon (1994), Berges et al. (1984), Bhardwaj and Brooks (1992), Blume and Siegel (1992), Blume and Stambaugh (1983), Brown et al. (1983), Chan et al. (1985), Constantinides (1984, 1988), Dimson (1988), Fama and French (1995), Gultekin and Gultekin (1983), Haugen (1995), Keim (1989, 1999), Lakonishok and Smidt (1988), Loughran (1996), Reinganum (1983), Ritter (1988), Ritter and Chopra (1989), Roll (1983a), Rogalski and Tiniç (1986), Schwert (1983), Siegel (1988), Thaler (1992), Tiniç and West (1984) and Ziemba (1989, 1994).

use different sets of data. In Section 2 we present some of this data for the period 1926–98. In Section 3 we investigate the day by day cash and futures returns of the Value Line small cap and S&P500 large cap indices during December and January for the sixteen TOYs for which there have been futures trading. One would expect that given a low cost way of exploiting a pattern it would be exploited. This was the case and leads to futures anticipation that violates theoretical futures pricing relative to cash. Section 4 discusses a particular trading strategy that attempts to capture profits from the small cap advantage at the TOY by entering the market with a small minus large cap futures spread in mid December and exiting in mid January. The effect is exploitable. It is strongest and most reliable in the second half of December and in recent years did not occur in the first half of January. Concluding remarks appear in Section 5.

2   The historical record: large and small cap equity returns, 1926–1998

We examined December and January monthly small minus large return differentials using CRSP data. Small stocks are represented by the smallest capitalization quintile (CRSP 9th & 10th deciles) and large stocks are represented by the largest capitalization quintile (CRSP 1st & 2nd deciles). Figure 1 indicates that small stocks have consistently outperformed large stocks in January. This was especially true up through 1982 when research on the January effect was becoming known. Post-1982, the results for the January effect have been less consistent with large caps outperforming in 5 of the 16 Januarys (previously large caps had only outperformed in 3 out of 57 Januarys). Still over the January 1983 through 1998 period, the average monthly outperformance by small stocks was 2.47%.

Given the less consistent performance of the January effect post-1982, we examined the December returns to see if the effect might have moved into December in the monthly returns. Figure 2, again using CRSP data, shows the small minus large cap stock returns for December 1927 through December 1997. On average, large cap stocks have outperformed small stocks in December during the entire 1927–97 period and the post-1982 period. Based on the monthly CRSP data, we see little evidence of the January effect moving into December in the cash market. There is even some evidence that there has been a reverse January effect in the second half of the 1990s. This is analogous to the results for United Kingdom stocks in Dimson and Marsh (1999) which show that over the whole twelve months large cap stocks outperformed small cap stocks in the 1990s. We now turn to the subject of this paper: the attempt to exploit the January turn-of-the-year small cap effect in the futures markets.

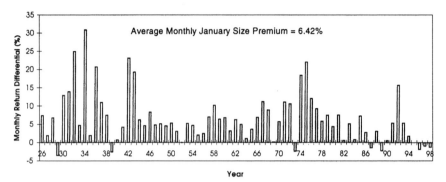

Figure 1. January Small Cap (CRSP 910) minus Large Cap (CRSP 12)
Returns, 1926–98

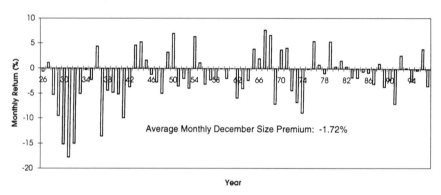

Figure 2. December Small Cap (CRSP910) minus Large Cap (CRSP12)
Returns, 1926–97

3   Value line and S&P500 returns for futures and cash in December and January, 1982–98

The evidence suggests that small stocks outperform large stocks at the turn of the year. Yet, transactions costs, particularly bid-ask spreads and price presures, take away most, if not all, of the potential gains. See e.g., Stoll and Whaley (1983) and Keim (1989) on these effects in January. Jones and Singh (1997) show that call options on stocks with low prior returns have increased implied volatility during the TOY. They conclude that there is some anticipation of the January seasonal but it survives in the cash stock market because of transaction costs. However, transaction costs to trade index futures are a tenth or less those for the corresponding basket of securities, and even more important, there is much less market impact. Hence, it may be profitable to buy long positions in small stock index futures and sell short positions in large stock index futures. This pair of positions is known as a spread trade. The strategy must anticipate the effect in the market place, in particular, the price impact of buying and selling futures contracts. Stock index futures began trading in the US in May 1982. The Value Line / S&P500 index spread is a way to measure and possibly capture any advantage small stocks may have over large cap stocks.

The Value Line small cap index is currently a price weighted arithmetic average of about 1,650 securities.[3] [4] Futures trade on the Kansas City Board of Trade and futures options trade on the Philadelphia Stock Exchange. The index was geometrically weighted until January 1988. Microsoft and the smallest market cap company in the index are treated equally in the weighting. The Value Line contains few of the smallest decile stocks on the NYSE, but it is arguably the best[5] available trading index for the purposes of capturing the effect. Both the S&P500 and Value Line futures have four contract months, with contracts expiring in March, June, September and December. Usually three of these contracts will trade at any moment in time with declining liquidity the longer the time to expiry. The contracts expire on the third Friday of the contract month; this was know as 'triple witching' day, since options, futures, and options on futures all expired at that time. On occasion this day has been marked with tremendous volatility largely because program traders must unwind (sell) their positions before expiry; see e.g., Stoll and Whaley (1987). Beginning with the June 1987 S&P contract, new rules went into effect: trading now ends on Thursday's close

---

[3]In what follows we argue that there has been some anticipation of the TOY effect in December and January in the futures markets. Given that the Value Line cannot easily or effectively be index arbitraged due to the number and liquidity of the stocks involved there is little in the way of 'moving the cash effect into December via futures anticipation.'

[4]From May 1982 to January 1988 the Value Line index was geometrically weighted. Hence because of the arithmetic-geometric inequality, the geometric mean index value will always lag the arithmetic mean. Hence the theoretical futures pricing equation is more complex than the usual cost of carry formula. Eytan and Harpaz (1986) derive the correct formula; see Modest and Sundaresan (1983) for general geometric mean futures pricing formulas. The correct formula is $F_T = S_t(1 + r - d - \frac{1}{2}\sigma)^{T-t}$, where F_T is the futures price at T, S_t is the spot price at t, r is the risk free rate of interest, d is the dividend yield and σ is the average unique risk of the stocks in the index. See Ritter (1996) for a discussion of the dangers of using the wrong pricing equation for these futures if the futures are held for a long time. Since the average unique risk is about 10% per year, the downward drift is about 0.4% per month. This is a significant but small amount for the trade discussed in Section 4, which is held for about one month, and does not greatly affect the results discussed there.

[5]The Russell 2000 index, which is a value weighted index of the smallest 2000 stocks in the Russell 3000 universe, is a similar index in terms of average capitalization. The discussion that follows is concerned only with the Value Line which has a much longer data history. However, the conclusions using the Russell 2000 which trades on the Chicago Mercantile Exchange are very similar.

but settlement is based on Friday's opening prices. The Value Line contract's expiry as well as those of stock options remains at Friday's close. The change to double witching greatly decreased the volatility on the expiring Fridays; see e.g., Stoll and Whaley (1991).

The S&P500 futures contract, which is value weighted and hence emphasizes the larger capitalized stocks, is traded on the Chicago Mercantile Exchange.

A 1:1 weighting in the spread attempts to mitigate the risk of the overall position, since the standard deviation of the S&P500 and Value Line index futures were about 1.82% and 1.74% per day, respectively, with a daily correlation of 0.967, according to Donaldson, Lufkin, and Jenrette's Futures Service (based on data up to 1993). The dollar value of the 1:1 weighting is, however, not identical for the futures contracts and indeed in some years was 5–20% different. Hence, investors may equalize these exposures with slightly different weightings over time. The S&P500 is concentrated in stocks with capitalization in the $25 billion plus range while the Value Line has roughly equal weights among stocks with capitalization under $100 million to $200 billion plus.

For maximum liquidity and the smallest bid-ask spreads, especially since the Value Line / S&P500 futures spread trade must be made over two exchanges, it is best to use the March contract. Also, since the trade should be completed during January, there is no need to consider the June contract. The S&P500 contract is highly liquid, trading about 150,000 contracts a day for the lead month with an open interest of about 350,000 contracts. The Value Line is much less liquid, trading less than several hundred contracts a day with open interest of 500–1500 contracts during the period when it is most in demand (November to March). We discuss this trade, via a specific decision rule, in Section 4. Now we turn to the day by day, S&P500 and Value Line futures and cash results for 1982–98.[6]

Figure 3(a)–(p) show the cash index and March futures spreads (Value Line minus S&P500) day by day in December and January for the sixteen turn-of-the-years 1982/83 through 1997/98. The spread between the futures difference and the cash difference represents the futures anticipation.[7] Table 1

---

[6]Keim and Smirlock (1987) investigated the Value Line futures and cash indices along with those of the S&P500 for the period April 1982–December 1986. They found that in this period the S&P500 tended to have most of its gains during the day whereas the Value Line had all of its gains at night. Hence, there was a positive size (small minus large cap) effect at night and the reverse effect during the day. Furthermore, many of the seasonal patterns in the cash market were in the futures markets such as the turn-of-the-month and the turn-of-the-year effects. The Value Line minus S&P500 futures spread was positive during the turn-of-the-year and much larger than in other months. Finally, hourly futures variances were, respectively, 56 and 62 times as high during trading hours than at night for the Value Line and S&P500 at the turn-of-the-year.

[7]The S&P has a higher dividend rate than the Value Line which is a positive for the

Table 1. Anticipation in the futures markets of the January turn-of-the-year small cap advantage in December and January for 1982/83 to 1997/98 (+ =anticipation; − =reverse)

	Second Half of December to Turn	Turn (−1 to +4)	Rest of First Half of January	Second Half of January
82/83	−,0	+	+	−
83/84	0, −	+	+	0
84/85	+	+	+	+
85/86	+	+	0	0
86/87	−	−	−	−
87/88	−	−	−	−
88/89	+	+	−	0
89/90	+	+	−	−
90/91	0	0	0	0
91/92	+	+	0	0
92/93	0	0	0	0
93/94	+	+	−	−
95/96	+	+	−	+
96/97	0	0	−	−
97/98	+	−	0	0

attempts to summarize the anticipation of lack of it in the second half of December, during the turn-of-the-year (trading days −1 to +4), the rest of the first half of January and the second half of January for the sixteen turn-of-the years 1982/83 to 1997/98. The year by year patterns vary greatly. However, there seems to be a shift from anticipation in the first half of January in the early years to the second half of December in the later years. There is definitely no small cap anticipation past the turn-of-the-year in the rest of the first half or the second half of January. These periods have negative anticipation except in the first three years 1982/83 to 1984/85. The strongest futures anticipation in recent years (1988/89 to 1997/98) is from mid December to the −1 trading day of January (the last trading day of December) or a few days into January. In most years, the spread anticipation in January is negative to anticipate the S&P500 outperforming the Value Line.[8]

Value Line minus S&P500 futures spread during the years 1988–98; see the cost of carry formula in Footnote 4 (eliminating σ). During 1982–88 when the Value Line was geometrically weighted, the theoretical difference of the futures spread depends on $d_{S\&P} − d_{VL} − \frac{1}{2}\sigma$, which was negative with values around −0.4% per month. Hence, one must consider this when viewing the graphs, which are presented here as a record for future researchers as well as for use in this paper.

[8]Roll (1983b) showed that the −1 trading day had the largest gain in the cash market

(a) 1997/98

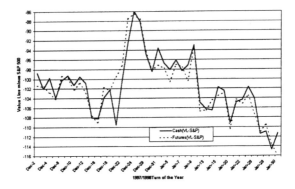

(b) 1996/97

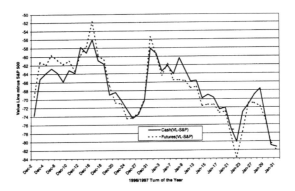

(c) 1995/96

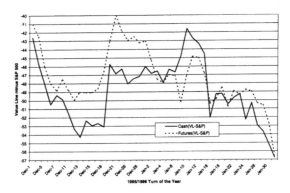

Figure 3. Value Line minus S&P500 1:1 Spread for Cash and March Futures
for the Sixteen Turn-of-the-Years 1982/83 through 1997/98

(d) 1994/95

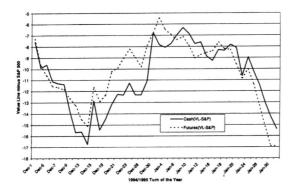

(e) 1993/94

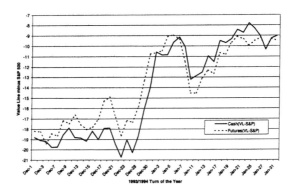

(f) 1992/93

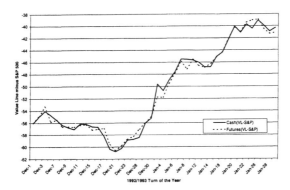

Figure 3. *cont.*

(g) 1991/92

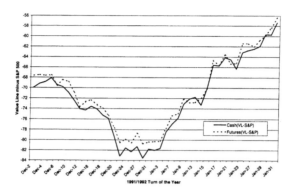

(h) 1990/91

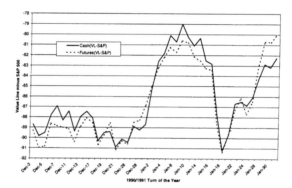

(i) 1989/90

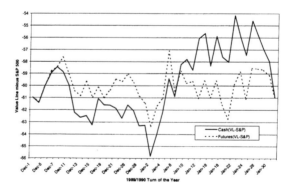

Figure 3. *cont.*

(j) 1988/89

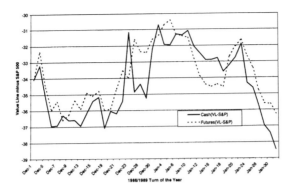

(k) 1987/88

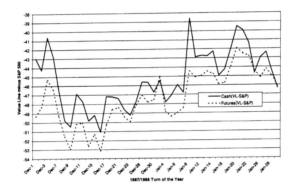

(l) 1986/87

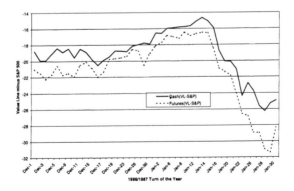

Figure 3. *cont.*

(m) 1985/86

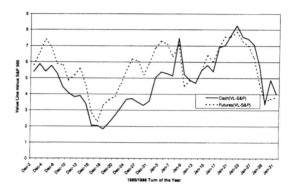

(n) 1984/85

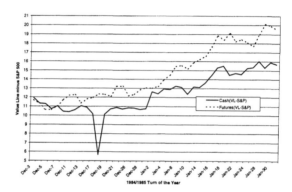

(o) 1983/84

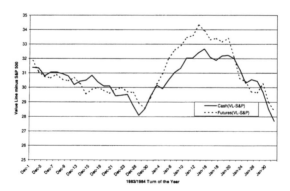

Figure 3. *cont.*

(p) 1982/83

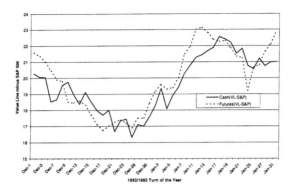

Figure 3. *cont.*

Figures 4a–d display the average Value Line / S&P500 index spreads day by day during December and January for the entire December 1982 to January 1998 period and three sub-periods of December 1982 to January 1990, December 1990 to January 1998 and December 1994 to January 1998, respectively. The notation −5 of January means the 5th from the last trading day of December, etc. The results show that: (1) the small caps had an advantage in the second half of December through the sample period; (2) the small caps had an advantage in the first half of January up to 1994 and since then the large caps have outperformed; and (3) the large caps outperform in the second half of January (except 1990–93).

Figures 5a–d display, using the same scale, the daily March futures spreads as presented in Figure 4 for the cash market. This supplements Table 1. The entire period (Figures 4a and 5a) shows a small cap advantage during the second half of December to mid January, about trading days −7 to +5 in both the cash and futures markets. Figure 5b shows that during 1982–90 the small cap futures outperformed large cap futures during both the second half of December and first half of January; large caps outperformed in the second half of January. However, as shown in Figure 5c, the small cap outperformance in the 1990s was solely in the second half of December. In recent years, 1994/95 to 1997/8 (Figure 6d), the large caps have outperformed in the first half of January.

---

of the small minus large cap spreads. In subsequent years this day has had high gains in the cash Value Line/S&P500 spread, but losses in the futures spread as the anticipation of the small firm effect dissipates.

(a) December 1982 through January 1998

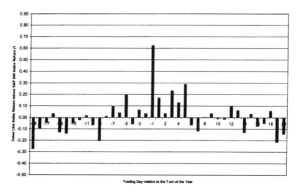

(b) December 1982 through January 1990

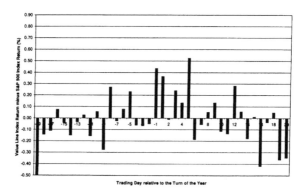

Figure 4: Average Index Differences—Value Line minus S&P500

4   Results from a turn-of-the-year futures trading strategy

Daily data on the Value Line and S&P500 cash and futures contracts for
December and January for the years 1982/83 through 1997/98 are shown in
Figures 4 and 5. These allow one to test expost, the results of particular
turn-of-the-year (TOY) trading rules. In their 1987 paper, Clark and Ziemba
suggested the following rule.

> We have found that the following rule works well: buy the spread
> on the first closing uptick, starting on December 15 and definitely
> by the 17th, and sell on January 15. Waiting until (−1) now seems
> to be too late: possibly finance professors and their colleagues, as

(c) December 1990 through January 1998

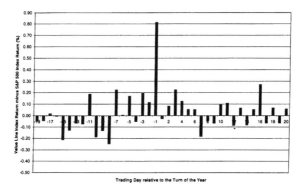

(d) December 1994 through January 1998

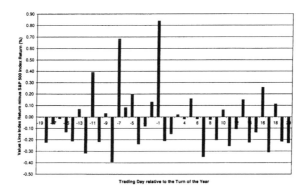

Figure 4: *cont.*

well as other students of the turn-of-the-year/January effect who
are in on the strategy, move the VL index. There seems to be a
bidding up of the March VL future price relative to the spot price.
Clark and Ziemba (1987), page 805

Their idea in 1987 was that the January small firm effect existed and occurred during the first two weeks of January in the cash market (as argued by
Ritter, 1988; see also comments by Ziemba, 1988), but that futures anticipation would move the effect in the futures markets into December. Hence, an
entrance into the Value Line/S&P500 futures spread trade in mid-December
and an exit in mid-January should capture the effect if it actually existed.
Investigation of the cash spot prices for the year 1976/77 to 1982/83 and
futures prices for 1983/84 to 1985/86 in their 1987 paper, as reproduced in

(a) December 1982 through January 1998

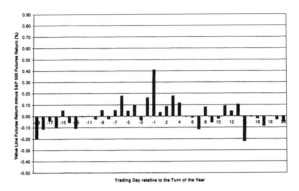

(b) December 1982 through January 1990

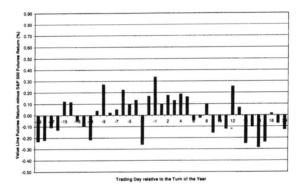

Figure 5: Average Futures Differences—March Value Line minus March S&P500

Table 2, shows the results of this trading rule.

The results showed that there was a small cap advantage in the December 15 to January 15 period with most of the excess in the January 1–15 period. Large caps outperformed small caps in the remainder of January. Clark, a futures broker, and Ziemba, a researcher and trader, used this trading rule on actual trades. Despite some volatility and year by year differences, the trades were successful. Table 4 has the cash results for 1982/83 as they were attempting to explain their actions. They did not trade in 1982/83, the first year of futures trading. This is not a long enough data for any reliable statistical tests.[9] Also the risk of the variable (VL futures – S&P500 futures)

[9]Of course, one has the years of data in Table 1 which suggests an advantage over long periods.

(c) December 1990 through January 1998

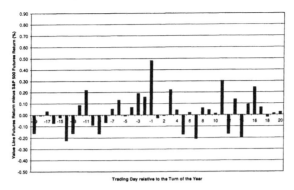

(d) December 1994 through January 1998

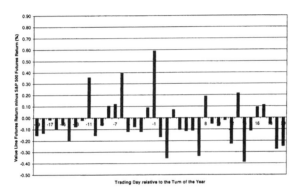

Figure 5: *cont.*

becoming negative, in the December 15 to January 15 period, is different than
the path independent quantity (VL futures – S&P500 futures) on January 15
minus (VL futures – S&P500 futures) on December 15. Hence because of
high volatility, trading problems could arise. Clark and Ziemba suggested a
fractional Kelly log utility strategy for this which is an attempt to balance risk
and return over a multiperiod (about 20 day) horizon for a trade to be made
in successive years; see the graphs in their paper and MacLean, Ziemba and
Blazenko (1992). Ziemba (1994) updated these results to the 1992/93 TOY
and discussed the 1993/94 TOY. Table 3 details this. The average differences
between the Value Line and S&P500 prices at the various dates are for cash
(the seven TOYs 1976/77 to 1982/83) and futures (the ten TOYs 1983/84 to
1992/93), respectively. This trade is known in the academic and investment
spheres; see e.g. the Wall Street journal article by Angrist (1990) where the

Table 2. Results from VL/S&P500 Spread Trades on Various Buy/Sell
Dates for the Ten Turn-of-the-Years (1977–86). *Source*: Clark and Ziemba
(1987)

Turn of the Year	Difference Dec 15 to (−1)	Difference (−1) to Jan 15	Difference Jan 15 to end Jan	Trade Gain Dec 15 to Jan 15
Spot Prices				
1976/77	0.67	3.15	−1.90	3.82
1977/78	−1.25	3.09	0.57	1.84
1978/79	3.74	−1.09	1.96	2.65
1979/80	−1.13	4.95	−0.64	3.82
1980/81	2.61	−0.55	0.02	2.06
1981/82	−0.87	0.99	−3.04	0.12
1982/83	−0.97	5.55	−2.51	4.58
March Futures				
1983/84	−1.43	2.69	−4.40	1.26
1984/85	1.05	2.69	−4.40	1.26
1985/86	3.15	0.25	−2.18	3.40
Average	0.56	2.28	−0.92	2.84

contrasting views of several traders including Ziemba are discussed.

These results were similar to the earlier results with most of the advantage coming in the first half of January and a disadvantage in the remainder of January with some anticipation in the second half of December. So far our trader had profits in all the years.

We have updated the futures results up to 1997/98 in Table 4. There are some minor differences in the values in Tables 2–4, because of different data sources. Table 4 suggests that there still is a turn-of-the-year effect in the futures markets in the second half of December. Only in three of the sixteen years is the VL/S&P500 difference negative in this period and the largest difference was in 1997/98. However, the average difference was 3.18 with standard deviation of 4.10 and t-statistic of 0.78 so the hypothesis that this quantity is actually negative cannot be rejected. There does seem to be a major change in the first two weeks of January. The period 1982/83 to 1992/93 indicates an advantage for the small caps. However, the large caps had the advantage in the 1994/95 to 1997/98 TOYs.

The second half of January continued to have a large cap advantage. Our trader observed in the 1995/96 TOY[10] that this trade was becoming less reliable, the open interest was declining, his share of the open interest was

---

[10]Dimson and Marsh (1999) discuss reasons why the small firm effect in England has failed to materialize in recent years. One additional very important reason why the US TOY situation has changed is the enormous growth of S&P500 index funds. In a strong

Table 3. Results from VL/S&P500 Spread Trades on Various Buy/Sell
Dates for the Seven Turn-of-the-Years 1986/87 to 1992/93 and Summary
Results for the Cash Markets from 1976/77 to 1982/83 and Futures Markets
from 1983/84 to 1992/93. *Source*: Ziemba (1994).

Turn of the Year	Difference Dec 15 to (–1)	Difference (–1) to Jan 15	Difference Jan 15 to end Jan	Trade Gain Dec 15 to Jan 15
March Futures				
1986/87	2.75	1.68	–14.98	4.43
1987/88	3.80	1.65	0.25	5.45
1988/89	4.20	–2.60	–2.05	1.60
1989/90	–0.50	1.85	0.90	1.35
1990/91	1.35	3.95	3.20	5.30
1991/92	–8.10	15.75	8.25	7.65
1992/93	5.45	6.55	4.80	12.00
Averages				
Cash (76/–82/3)	0.40	2.30	–0.79	2.70
Futures (83/4–92/3)	1.17	3.55	–0.30	4.72
Combined (76/7–92/3)	0.79	3.04	–0.52	3.83

rising and holding the position into January was not profitable. Despite the
small gain in the December 15, 1995 to January 15, 1996 period it was clear
that this trade was now very risky with increasing volatility and decreasing
liquidity; see Thomas (1996) for a history of the Value Line contract and
its declining volume and open interest. Our trader was in this market in
1996/97, with a modest position, and unwound the position in early January
with a profit; however, this was by selling Value Line futures high and buying
S&P500 futures cheap. This used additional risk and was not in the script
for the original Clark and Ziemba rule. Using that rule would have lead to
losses in 1996/97. Our TOY trader did not participate in 1997/98; although
it was possible to do the trade successfully in the second half of December
and the Clark and Ziemba rule was profitable; 1998/99, not shown here, was
similar.

---

bull market such as 1995 to mid-1998 the investment in such funds was very large and
strategies that short the S&P500 were very risky.

Table 4. Results from VL/S&P500 March Futures Spread Trades on Various
Buy/Sell Dates for the 16 Turn-of-the-Years 1982/83 to 1997/98

Turn of the Year	Difference Dec 15 to (–1)	Difference (–1) to Jan 15	Difference Jan 15 to end Jan	Trade Gain Dec 15 to Jan 15
1982/1983	1.50	3.05	0.65	4.55
1983/1984	–0.70	4.00	–4.90	3.30
1984/1985	1.10	3.55	2.90	4.65
1985/1986	3.15	0.45	–2.60	3.60
1986/1987	2.75	–0.30	–9.85	2.45
1987/1988	8.15	–0.90	–0.25	7.25
1988/1989	3.50	–2.95	–1.70	0.55
1989/1990	–0.50	1.85	–1.45	1.35
1990/1991	1.70	3.60	3.20	5.30
1991/1992	–7.15	10.20	13.80	3.05
1992/1993	5.45	6.55	4.05	12.00
1993/1994	4.65	0.00	1.15	4.65
1994/1995	6.15	–1.65	–8.50	4.50
1995/1996	6.00	–3.75	–9.70	2.25
1996/1997	4.35	–15.90	–10.75	–11.55
1997/1998	10.70	–6.50	–12.30	4.20
Average	3.18	0.08	–2.27	3.26

5   Final remarks

This paper provides some statistical evidence concerning the first three of
the four fundamental questions regarding the TOY phenomenon discussed in
the introduction. The data suggest that the TOY effect has changed greatly
in the 1990s particularly since 1994: whether this is a fundamental change
or a small sample problem remains to be seen. Dimson and Marsh (1999)
find similar evidence in England and offer some possible reasons for this that
relate to changing fundamentals such as the economies of scale in information processing and pharmaceutical developments whose advantages largely
accrued to large firms. Added to that is the large growth in S&P500 index
funds in the US which favors the large caps. Evidence beginning with Tiniç
and West (1984) indicates that there seems to be higher risk in January and
possibly other months, see e.g. Berk (1999), for the small caps. We find
that the TOY effect still exists in the second half of December in the futures
markets but not in the futures or cash market during January. Moreover,
there is a strong reverse effect in the second half of January favoring large
cap stocks. It seems possible to use these effects to add to portfolio value.
However, the small cap futures markets are very thin so this cannot be done

on a large scale. Moreover, the results, while positive, have little in the way of traditional statistical significance due to the small sample sizes and the volatility of the time series. Whether or not in future years the TOY effect will return to its behavior prior to the 1990s is unclear.

References

Angrist, S.W. (1990) 'Futures offer cheap play on small stocks' annual rally', *Wall Street Journal*, December 13, 1.

Agrawal, A. and Tandon, K. (1994) 'Anomalies or Illusions? Evidence from stock markets in eighteen countries', *Journal of International Money and Finance* **13** 83–106.

Banz, R.W. (1981) 'The relationship between return and market value of common stocks', *Journal of Financial Economics* **9** (1) 3–18.

Berges, A., McConnell, J. and Schlarbaum, G. (1984) 'The turn of the year in Canada', *Journal of Finance* **39** 184–192

Berk, J. (1999) 'The Current Status of the Size Effect', this volume, 90–115,

Bhardwaj, R. and Brooks, L. (1992) 'The January anomaly: effects of low share price, transactions costs, and bid-ask bias', *Journal of Finance* **47** 553–575.

Black, F. (1986) 'Noise', *Journal of Finance* **41** 529–543.

Black, F. (1993) 'Beta and return', *Journal of Portfolio Management* **20** 8–18.

Blume, M.E. and Siegel, J.J. (1992) 'The theory of security pricing and market structure', *Journal of Financial Markets, Institutions and Instruments* **1** (1) 1–58.

Blume, M.E. and Stambaugh, R.F. (1983) 'Biases in computed returns: in computed returns: an application to the size effect', *Journal of Financial Economics* **12** (3) 387–404.

Booth, D. and Keim, D. (1999) 'Is there still a January effect?', this volume, 169–178

Brown, P., Keim, D., Kleidon, A. and Marsh, T. (1983) 'Stocks return seasonalities and the tax-loss selling hypothesis: analysis of the arguments and Australian evidence', *Journal of Financial Economics* **12** 105–128.

Chan, K.C., Chen, N.-F., and Hsieh, D.A. (1985) 'An exploratory investigation of the size effect', *Journal of Financial Economics* **14** (3) 451–471.

Clark, R. and Ziemba, W.T. (1987) 'Playing the turn-of-the-year effect with index futures', *Operations Research* **35** (6) 799–813.

Constantinides, G.M. (1984) 'Optimal stock trading with personal taxes', *Journal of Financial Economics* **13** 65–89.

Constantinides, G.M. (1988) 'Comments on *Stock return seasonality*', in *Stock Market Anomalies*, E. Dimson, ed., Cambridge University Press, 123–128.

Dimson, E. (ed.) (1988) *Stock Market Anomalies*, Cambridge University Press.

Dimson, E. and Marsh, P. (1999) 'The demise of size', this volume, 116–143.

Eytan, T.H. and Harpaz, G. (1986) 'The pricing of futures and options contracts on the Value Line Index', *Journal of Finance* **41** (September) 843–855.

Fama, E.F. (1991) 'Efficient capital markets: II', *Journal of Finance* **46** (5) 1575–1617.

Fama, E.F. (1998) 'Market efficiency, long-term returns and behavioral finance', *Journal of Financial Economics* **49** (3) 283–306.

Fama, E.F., and French, K. (1995) 'Size and book-to-market factors in earnings and returns', *Journal of Finance* **51** 131–155,

Gultekin, M.N. and Gultekin, N.B. (1983) 'Stock market seasonality: international evidence', *Journal of Financial Economics* **12** 461–81.

Haugen, R.A. (1995) *The new finance: the case against efficient markets*, Prentice Hall.

Haugen, R.A. and Jorion, P. (1996) 'The January effect: still there after all these years', *Financial Analysts Journal*, January–February, 7–31.

Hensel, C.R. and Ziemba, W.T. (1999) 'How does Clinton stand up to history: US stock market returns and presidential party affiliations, 1928-1997', this volume, 203–217.

Jones, S.L. and Singh, M.K. (1997) 'The distribution of stock returns implied in their options at the turn-of-the-year: a test of seasonal volatility', *Journal of Business* **70** (2) 281–311.

Kamara, A. (1996) 'The effects of the evolution of the US equity markets on the January-size seasonal'. Report, University of Washington.

Keim, D.B. (1983) 'Size-related anomalies and stock market seasonality: further empirical evidence', *Journal of Financial Economics* **12** (1) 13–32.

Keim, D.B. (1989) 'Trading patterns, bid-ask spreads, and estimated security returns: the case of common stock returns at the turn of the year', *Journal of Financial Economics* **25** 75–98.

Keim, D.B. (1999) 'An analysis of mutual fund design; the case of investing in small-cap stocks', *Journal of Financial Economics* **51** (2) 173–194.

Keim, D.B. and Smirlock, M. (1987) 'The behavior of intraday stock index futures prices', *Advances in Futures and Options Research* **2** 143–166.

Lakonishok, J. and Smidt, S. (1988) 'Are seasonal anomalies real? A ninety-year perspective', *Review of Financial Studies* **1** (4) 403–426.

Loughran, T. (1996) 'Book-to-market across firm size, exchanges and seasonality: is there an effect?' Report, University of Iowa.

MacLean, L.C., Ziemba, W.T., and Blazenko, G. (1992) 'Growth versus security in dynamic investment analysis', *Management Science* **38** 1562–1585.

Modest, D.M and Sundaresan. M. (1983) 'The relationship between spot and futures prices in stock index futures markets', *Journal of Futures Markets* **3** 15-41.

Reinganum, M.R. (1981) 'A mis-specification of capital asset pricing: empirical anomalies based on earnings yields and market values', *Journal of Financial Economics* **9** (1) 19–46.

Reinganum, M.R. (1983) 'The anomalous stock market behavior of small firms in January: emperical tests for tax-loss selling effects', *Journal of Financial Economics* **12** 89-104.

Riepe, M.W. (1997) 'Is publicity killing the January effect?' Report, Ibbotson Associates, Chicago.

Ritter, J.R. (1988) 'The buying and selling behavior of individual investors at the turn-of-the-year', *Journal of Finance* **43** (3) 701–717.

Ritter, J.R. (1996) 'How I helped make Fischer Black wealthier', *Financial Management* **25** (Winter) 104–107.

Ritter, J.R. and Chopra, N. (1989) 'Portfolio rebalancing and the turn-of-the-year effect', *Journal of Finance* **44** (1) 149–166.

Rogalski, R.J. and Tiniç, S.M. (1986) 'The January size effect: anomaly or risk measurement?', *Financial Analysts Journal* **42** (November–December) 63–70.

Roll, R. (1983a) 'On computing mean returns and the small firm premium', *Journal of Financial Economics* **12** 371–386.

Roll, R. (1983b) 'Vas ist das? The turn-of-the-year effect and the return premia of small firms', *Journal of Portfolio Management* **9** (1) 18–28.

Roll, R. (1994) 'What every CEO should know about scientific progress in economics: what is known and what remains to be resolved', *Financial Management* **23** (Summer) 69–75.

Rozeff, M. and Kinney, W. (1976) 'Capital market seasonality: the case of stock returns', *Journal of Financial Economics* **3** 379–402.

Samuelson, P. (1989) 'The judgment of economic science as rational portfolio management; indexing, timing and long-horizon effects,' *Journal of Portfolio Management* **16** (1) 4–12.

Schwert, W. (1983) 'Size and stock returns and other empirical regularities', *Journal of Financial Economics* **12** 3–12.

Siegel, J. (1998) *Stocks for the Long Run*, 2nd edition, McGraw Hill.

Stoll, H.R. and Whaley, R.E. (1983) 'Trading costs and the small firm effect', *Journal of Financial Economics* **12** 57–79.

Stoll, H.R. and Whaley, R.E. (1987) 'Program trading and expiration day effects', *Financial Analysts Journal* **43** 16–28.

Stoll, H.R. and Whaley, R.E. (1991) 'Expiration day effects: what has changed?', *Financial Analysts Journal* **47** 58–72.

Thaler, R. (1992) *The Winner's Curse*, The Free Press.

Thomas, S. (1996) 'The saga of the first stock index futures contract: was it a case of the market using the wrong model and not learning?' Report, Weatherhead School of Management, Case Western Reserve University.

Tiniç, S.M. and West, R.W. (1984) 'Risk and return: January and the rest of the year', *Journal of Financial Economics* **13** 561–574.

Wachtel, S. (1942) 'Certain observations on seasonal movements in stock prices', *Journal of Business* (April) 184–193.

Ziemba, W.T. (1988) 'Comment on *J. Ritter: The Buying and Selling Behavior of Individual Investors at the Turn-of-the-Year*', *Journal of Finance* **43** (3) 717–719.

Ziemba, W.T. (1994) 'World wide security market anomalies', *European Journal of Operational Research* **74** (April), 198–229.

How does Clinton Stand up to History: US Stock Market Returns and Presidential Party Affiliations, 1928–97[*]

Chris R. Hensel and William T. Ziemba

Abstract

This paper analyzes the returns from large and small capitalization equities, various bond indices and cash following elections won by Democratic and Republican presidents from 1928–92 and during Clinton's two Democratic administrations from 1993–97. From 1928 to 1992, small cap stocks had significantly higher returns during Democratic administrations than with Republicans because of gains rather than losses in the April–December period. This phenomenon was not a manifestation of the January small-firm effect, rather it is a significant small cap effect outside January during Democratic presidencies. Large cap stocks had statistically identical returns under both administrations. For both Democratic and Republican administrations, small and large cap stock returns were significantly higher during the last two years of the presidential term than during the first two years. Corporate, long term government and intermediate term government bonds and cash had significantly higher returns during Republican administrations than under Democratic rule.

From 1937 to 1993, the simple investment strategies of investing in small cap stocks during Democratic administrations and either intermediate-term government bonds or large cap stocks during Republican administrations, produced higher mean returns – with higher standard deviations – than did investing in large cap stocks throughout the period. The cumulative wealth from these politically correct investment strategies greatly exceeded that from other strategies – small and large cap stocks, all types of bonds, cash and a 60/40 stock/bond mix.

Clinton's first term produced results consistent with past democratic administrations. Small cap equities outperformed large cap, small cap returns

---

[*]This paper is an update of Hensel and Ziemba (1995) to include the Clinton years, 1993–97. We would like to thank the Frank Russell Company and the Social Sciences and Humanities Research Council of Canada for financial support of this research and Gayle Nolte for computational assistance.

outside of January were positive, and the small and large cap equity returns in the last two years exceeded those of the first two. However, in the first fourteen months of his second term, large cap equities outperformed small cap equities and returns from the equity strategies greatly exceeded those from bonds, cash and the 60/40 mix.

1 Introduction

There is some previous research on the effect of presidential elections on US stock returns. Herbst and Slinkman (1984), using data from 1926–77, found a 48-month political/economic cycle during which returns were higher than average; this cycle peaked in November of presidential election years. Riley and Luksetich (1980) and Hobbs and Riley (1984) showed that, from 1900–80, positive short-term effects followed Republican victories and negative returns followed Democratic wins. Huang (1985), using data from 1832–1979 and for various subperiods, found higher stock returns in the last two years of political terms than in the first two. This finding is consistent with the hypothesis that political reelection campaigns create policies that stimulate the economy and are positive for stock returns.

Using data from 1928 to 1997, we investigate several questions concerning US stock, bond and cash returns such as: Do small and large capitalization stock returns differ between Democratic and Republican administrations? Do corporate bond, intermediate and long-term government bonds and Treasury bill returns differ between the two administrations? Do the returns of various assets in the second half of each four-year administration differ from those in the first half? Were Clinton's administrations analogous to past Democratic administrations?

This paper expands on the previous literature in several ways. First, we examine small-cap stock and various cash and bond returns whereas the previous literature focused on large cap stocks. The results indicate a significant small cap effect during Democratic presidencies. Small cap stocks had higher returns during Democratic than Republican administrations but there has also been a small cap minus large cap advantage outside the month of January for the Democrats. The turn-of-the-year small firm effect, in which small cap stock returns significantly exceed those for large cap stocks in January, under both Republican and Democratic administrations, occurred during these 70 years. (See Hensel and Ziemba (1999) for an analysis of the turn-of-the-year small cap effect in the 1980s and 1990s.) This advantage was slightly higher for Democrats, but the difference is not significant. Moreover, bond and cash returns were significantly higher during Republican compared with Democratic administrations. The results also confirm and extend previous findings that equity returns have been higher in the second half compared with the first half of presidential terms. This finding is documented for small and large cap stocks during both Democratic and Republican administrations. Finally, two simple investment strategies based on these findings yielded superior portfolio performance than did common alternatives during the sample period. The results cast doubt on the long run wisdom of the common 60/40 stock-bond strategy since all 100% equity strategies investigated had much higher wealth at the end of the sample period. Indeed the 1942–97 returns were twenty-four times higher with the strategy small caps with Democrats and large caps with Republicans than the 60/40 mix.

2 US Stock Returns after Presidential Elections

This analysis is based on monthly return data for the 70-year period from January 1928 to December 1997. The large cap data are from the S&P500 Index. Since March 1957, the Index has consisted of 500 large stocks weighted by market value (price times number of shares outstanding); previously it consisted of 90 large stocks. The small cap stocks are the bottom 20% (by capitalization) of companies on the New York Stock Exchange. The data from Ibbotson Associates are monthly, continuously compounded total returns including dividends. The return distributions are close enough to being normally distributed that statistical tests based on normality are valid.

Figure 1 shows the average monthly return differences of large and small cap stocks for election months and the following 13 months (to the end of the following year) minus the 1928–97 average for each month. Large and small cap stocks had higher mean returns during November of the election year and March, April, May and July of the following year. In all other months, small and large-cap stocks had lower than average or average mean returns following an election.

Table 1 presents the aggregate information from Figure 1. Both large and small cap stocks had lower than average mean returns during election years. Small stock returns were 551 basis point a year lower than the average; however, this result was not statistically significant.

The 1928–97 period encompassed 18 presidential elections. The end of 1997 was into Clinton's second term. There were 33 years of Republican and 37 years of Democratic administrations during this period. Table 2 lists and compares the first year, first two years, last two years and whole term mean returns under Democratic and Republican administrations from January 1929 to December 1997 and for January 1937 to December 1997, a period that excludes one term for each party during the 1929 crash, subsequent depression and recovery period. Each term is considered separately, so two-term presidents have double entries. The *t*-values shown in Table 2 test the hypothesis

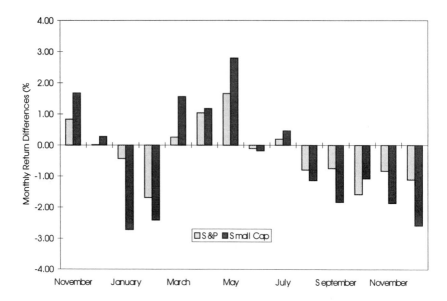

Figure 1. Stock monthly return differences: presidential election months and the subsequent 13 months minus monthly averages for 1928–97

Table 1. Annual average equity returns for presidential election months and the subsequent 13 months, 1928–97, minus annualized monthly averages for 1928–97*

Return Period	Large Cap	Small Cap
Election + Next 13 Months	8.12	6.51
1928–97 Annual Average	10.12	12.02
Annual Difference	−2.00	−5.51

*Monthly means were annualized by multiplying by 12.

that, during each of these periods, returns did not differ between Democratic and Republican administrations.

From 1929 to 1997, the mean returns for small stocks were statistically higher during the Democratic presidential terms than during the Republican terms. The data confirm the advantage of small cap over large cap stocks under Democratic administrations. Small cap stocks returned, on average, 20.15% a year under Democrats compared with 1.94% under Republicans for the 1929–97 period. This difference, 18.21%, was highly significant. The first year return differences for this period were even higher, averaging 33.51%.

The right-hand panel of Table 2 presents the return results after eliminating the 1929 crash, the Depression and the subsequent period of stock price volatility. Removing these eight years (1929–36) from the study eliminates one Democratic and one Republican administration from the data. The small stock advantage under Democrats was still large (an average of 7.55% per year) but was no longer statistically significant. The large cap (S&P500) returns during Democratic rule were statistically indistinguishable from the returns under Republican administrations.

For Democratic and Republican administrations, the mean small and large cap stock returns were much higher in the last two years compared with the first two years of presidential terms for both of the time periods presented in Table 2. For example, small cap stocks returned 24.65% during the last two years compared with 15.90% during the first two years for Democrats and 10.18% compared with –6.29% for Republicans from 1929 to 1992. Returns on large cap stocks increased to 17.40 from 8.09% for Democrats and to 9.06 from 3.77% for Republicans for the same period. This result is consistent with the hypothesis that incumbents embark on favorable economic policies in the last two years of their administrations to increase their reelection chances and that the financial markets view these policies favorably.

The advantage of small stocks over large stocks under Democratic administrations was not a manifestation of the January small stock effect. Instead Table 3 and Figure 2 show the relative advantage of small over large cap stocks under Democrats compared with that under Republicans was attributable to having fewer small stock losses, as well as higher mean small stock returns, in the April–December period. Under Democrats, mean returns were positive in each of these months, except October, and the small minus large differential was positive during 10 of the 12 months; under Republicans, the small minus large differential was negative during 9 of the 12 months.

The small cap advantage also occurred in the months following Democrat Clinton's election. From November 1992 to December 1993 the small cap index rose 36.9% versus 14.9% for the S&P500. This domination continued until the second election. Small caps returned 1.58% per month versus 1.31% per month for the S&P500 from November 1992 to October 1996. However, large cap S&P500 returns began exceeding small cap returns in 1994 and this continued through 1997. The January 1994 to December 1996 returns were small cap 1.36% per month versus 1.50% per month for the S&P500. From November 1996 to December 1997 small caps returned 1.81% per month and the S&P500 2.44% per month. There was a phenomenal growth in S&P500 index funds and much foreign investment in large cap stocks during this period. While small caps had very large returns, those of the S&P500 were even higher.

We investigated how inflation varied with political regimes. The results for

Table 2. Average annual returns for the first year and four years
of Democratic and Republican presidencies[*]

	January 1937 to December 1997		January 1929 to December 1997	
	S&P500 TR	US Small Stk TR	S&P500 TR	US Small Stk TR
Democrat				
Avg. 1st Yr	6.58	11.32	10.24	19.06
Avg. 1st 2Yrs	6.14	11.85	8.09	15.90
Avg. Last 2Yrs	16.13	24.11	17.40	24.65
Avg. Term	10.81	16.71	12.62	20.15
Std. Dev. Term	16.35	27.76	18.26	30.69
# Years	36.00	36.00	37.00	37.00
Republican				
Avg. 1st Yr	1.87	−6.22	0.54	−14.45
Avg. 1st 2Yrs	6.98	1.39	3.77	−6.29
Avg. Last 2Yrs	15.03	16.95	9.06	10.18
Avg. Term	11.00	9.17	6.42	1.94
Std. Dev. Term	15.12	19.89	21.17	27.81
# Years	28.0	28.0	32.0	32.0
Diff. 1st Yr	4.72	17.54	9.71	33.51
Diff. 1st 2Yrs	−0.84	10.46	4.32	22.19
Diff. Last 2Yrs	1.10	7.16	8.33	14.47
Diff. Term	−0.19	7.55	6.20	18.21
1st year t-values (Ho:Diff=0)	0.67	1.39	1.15	**2.58**
First 2-years t-values (Ho:Diff=0)	-0.14	1.13	0.69	**2.39**
Last 2-years t-values (Ho:Diff=0)	0.20	0.69	1.20	1.41
Term t-values (Ho:Diff=0)	−0.05	1.04	1.29	**2.57**

[*] In this and subsequent tables, statistically significant differences at the 5%
level (2-tail) are shown in bold.

the 1929–97 period, using the Ibbotson inflation index, indicate that inflation
was significantly higher under Democrats, but this difference was contained in
the 1929–36 period. Excluding this early period, inflation was slightly higher,
on average, under Democrats but not statistically different from inflation under Republicans. Inflation rates differed across the years of the presidential

Table 3. Average monthly small- and large-cap stock returns
during Democratic and Republican presidencies, January 1929–
December 1997

| | Democratic Administrations | | | Republican Administrations | | |
	S&P500 Total Return	US Small Cap Total Return	Small Cap minus Large Cap	S&P500 Total Return	US Small Cap Total Return	Small Cap minus Large Cap
Jan.	1.72	6.45	4.72	1.65	5.93	4.28
Feb.	−0.38	0.74	1.11	1.59	2.78	1.19
Mar.	−0.58	−0.91	−0.34	0.96	1.21	0.25
Apr.	2.25	2.58	0.33	−0.24	−1.82	−1.57
May	1.07	1.40	0.33	−0.50	−1.52	−1.02
June	1.57	1.71	0.14	0.78	−0.40	−1.18
July	1.95	2.81	0.86	1.69	1.11	−0.58
Aug.	1.17	1.65	0.47	1.73	1.25	−0.47
Sept.	0.40	0.78	0.38	−2.87	−3.31	−0.45
Oct.	0.42	−0.24	−0.67	−0.40	−2.66	−2.26
Nov.	1.44	1.61	0.17	0.44	−0.53	−0.97
Dec.	1.56	1.58	0.02	1.59	−0.09	−1.68

terms. For example, for the 1937–97 period, in the first year of the presidential term, inflation under the Democrats was significantly lower than it was under the Republicans. An analysis of the first and second two years of administrations during this same period indicated that inflation was higher under Democrats but the difference was not statistically significant.

3   US Bond Returns after Presidential Elections

The bond data are also from Ibbotson Associates and consist of monthly, continuously compounded total returns for long term corporate bonds, long term (20-year) government bonds, intermediate (5-year) government bonds, and cash (90-day T-bills).

Figure 3 illustrates average return differences for bonds during election months and the subsequent 13 months (1929–97) minus each month's 1928–97 average return. Corporate, long term government, and intermediate government bond returns were all higher than the monthly average in the year following an election only in May, October, and November. Both government bonds also exceeded the average in some other months.

Throughout the 14-month period from the election month through the following year, corporate, intermediate government and long-term government bonds returned less than the monthly average, while cash returned more than

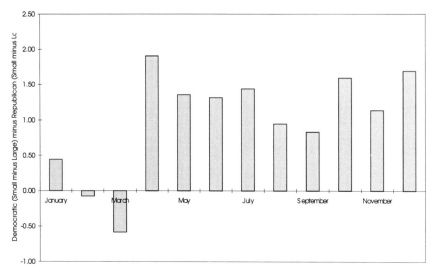

Figure 2. Cap size effects and Presidential Party, 1929–97. Democratic (small minus large) minus Republican (small minus large)

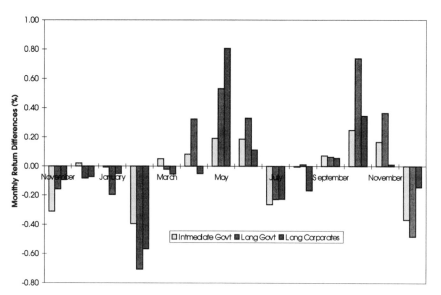

Figure 3. Bond monthly return differences: Presidential election months and the subsequent 13 months minus monthly average for 1928–97

Table 4. Annual average bond returns for presidential election months and the subsequent 13 months, 1928–97, minus annualized monthly averages, 1928–97.

Return Period	Long-term Government, %	Intermediate Government, %	Long Corporate, %	US 30-day T-bill, %
Election years	5.01	5.13	5.53	3.71
1928–97 average	5.73	5.04	5.76	3.98
Annual difference	0.72	−0.09	0.23	0.28

the average (Table 4).

As Table 5 indicates, the performance of fixed income investments differed significantly between Democratic and Republican administrations. All fixed income and cash returns were significantly higher during Republican than during Democratic administrations during the two study periods. The high significance of the cash difference stems from the low standard deviation over terms. The performance of fixed income investments differed very little between the first two years and the last two years of presidential terms.

The distribution of Democratic and Republican administrations during the 1929–97 period played a part in the significance of the fixed income and cash returns. As Table 6 indicates, the cash returns for the first four Democratic administrations in this period (1933–48) were very low (0.20%, 0.08%, 0.25% and 0.50% annually). This result largely explains why the term cash-return differences in Table 5 are so significant (t-value $= -12.31$ for 1929–97). Democratic administrations were in power for three of the four terms during the 1941–56 period, when government bonds had low returns. Bond returns in the 1961–68 period (both Democratic terms) and 1977–80 period (Democratic) were also low.

4   Some Simple Presidential Investment Strategies

Two presidential party based investment strategies suggest themselves. The first is equity only and invests in small caps with Democrats and large caps with Republicans; the second, a simple alternating stock-bond investment strategy, invests in small cap stocks during Democratic administrations and intermediate government bonds during Republican administrations. The test period was January 1937 through December 1997.

Figure 4 illustrates the growth of a $1 investment in these two strategies as well as in large cap and small cap stocks and in a benchmark of 60% large

Table 5. Average annual returns for the first year and four years of Democratic and Republican presidencies, in %

	January 1937 to December 1997				January 1929 to December 1997			
	US LT Gvt TR	US IT Gvt TR	US LT Corp TR	US 30-Day T-Bill TR	US LT Gvt TR	US IT Gvt TR	US LT Corp TR	US 30-Day T-Bill TR
Democrat								
Avg. 1st Yr	5.56	3.28	4.80	2.30	5.00	3.13	5.30	2.10
Avg. 1st 2 Yrs	3.84	3.11	3.90	2.27	3.63	3.04	4.22	2.16
Avg. Last 2 Yrs	2.22	3.30	2.72	3.06	2.36	3.49	3.08	2.90
Avg. Term	2.73	3.04	2.94	2.80	3.01	3.26	3.67	2.52
Std. Dev. Term	6.43	3.81	5.53	0.82	6.20	3.73	5.55	0.81
# Years	36.00	36.00	36.00	36.00	37.00	37.00	37.00	37.00
Republican								
Avg. 1st Yr	7.13	7.79	6.36	6.74	6.66	7.54	5.97	6.48
Avg. 1st 2 Yrs	9.15	8.99	8.67	6.22	8.50	8.64	8.27	5.88
Avg. Last 2 Yrs	6.62	6.65	7.72	4.74	6.43	6.20	7.28	4.28
Avg. Term	7.89	7.82	8.19	5.48	7.46	7.42	7.77	5.08
Std. Dev. Term	9.56	5.21	8.49	0.88	9.20	5.00	8.13	0.90
# Years	28.0	28.0	28.0	28.0	32.0	32.0	32.0	32.0
Diff. 1st Yr (Ho:Diff=0)	−1.57	−4.51	−1.56	−4.44	−1.66	−4.41	−0.66	−4.37
Diff. 1st 2 Yrs	−5.31	−5.89	−4.77	−3.96	−4.87	−5.60	−4.05	−3.72
Diff. Last 2 Yrs	−4.41	−3.35	−4.99	−1.68	−4.07	−2.70	−4.20	−1.37
Diff. Term	−5.16	−4.78	−5.25	−2.68	−4.45	−4.16	−4.11	−2.56
1st year t-values (Ho:Diff=0)	−0.34	−1.91	−0.39	−10.06	−0.35	**−1.79**	−0.16	**−11.33**
1st 2 years t-values (Ho:Diff=0)	**−1.70**	**−3.54**	**−1.77**	**−12.36**	−1.61	**−3.51**	−1.54	**−12.67**
Last 2 years t-values (Ho:Diff=0)	−1.59	**−2.03**	**−1.96**	**−5.42**	−1.58	−1.76	−1.79	**−4.79**
Term t-values (Ho:Diff=0)	**−2.44**	**−4.03**	**−2.81**	**−11.77**	**−2.31**	**−3.85**	**−2.40**	**−12.31**

Table 6. Presidents, political parties, and fixed income returns, 1929–97

| Term | President | Party | Annualized Average Monthly Return | | Difference |
			Intermediate Gvt Bonds	Cash	Intermediate Bonds minus Cash
1929–32	Hoover	Republican	4.61	2.26	2.35
1933–36	Roosevelt	Democratic	5.05	0.20	4.85
1937–40	Roosevelt	Democratic	3.73	0.08	3.65
1941–44	Roosevelt	Democratic	1.74	0.25	1.49
1945–48	Roosevelt/Truman	Democratic	1.48	0.50	0.99
1949–52	Truman	Democratic	1.24	1.35	−0.11
1953–56	Eisenhower	Republican	1.19	1.66	−0.48
1957–60	Eisenhower	Republican	4.24	2.54	1.70
1961–64	Kennedy/Johnson	Democratic	3.21	2.84	0.37
1965–68	Johnson	Democratic	2.76	4.43	−1.67
1969–72	Nixon	Republican	7.06	5.19	1.87
1973–76	Nixon/Ford	Republican	7.42	6.25	1.17
1977–80	Carter	Democratic	3.17	8.11	−4.94
1981–84	Reagan	Republican	13.71	10.39	3.32
1985–88	Reagan	Republican	10.35	6.22	4.12
1989–92	Bush	Republican	10.77	6.11	4.66
1993–96	Clinton	Democratic	5.74	4.30	1.44
1997	Clinton	Democratic	8.05	5.12	2.93

Table 7. 1997 value of $1 initial investment

Date	Large Cap (S&P)	Small Cap	Presidential (SC/Int)	Presidential (SC/LC)	60/40 Benchmark
Jan. '37–Dec. '97	346.1	453.2	527.9	963.2	140.5
Jan. '42– Dec. '97	639.0	2044.1	2380.9	4343.8	180.9

cap stocks and 40% intermediate government bonds (rebalanced monthly). The 60/40 portfolio is a common investment strategy and provides a benchmark for comparison with the stock-bond strategy. Transaction costs were not included, but they would have a minor effect on the results because the higher return presidential strategies trade at most every four years. These investment strategies, as shown in Figure 4 all lost money until the early 1940s.

The cumulative wealth relatives for Figure 4 are shown in Table 7 along with the results achieved if the investments had been made in January 1942 just prior to when they started performing better. The first line of Table 7 shows, for each strategy the amount that a dollar invested in 1937 would be worth at the end of 1997. Cumulative returns for both presidential strategies

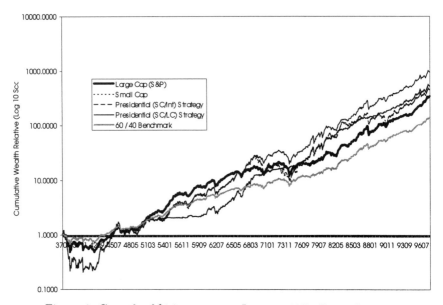

Figure 4. Growth of $1 investment, January 1937–December 1997

exceeded those for all the other investment strategies. The presidential stock strategy ended with more than double the small cap value (the highest returning single equity category). The advantage of the presidential strategies relative to the large cap strategy was even greater if the initial investment was made in January 1942.

Throughout both study periods, these simple presidential party based investment strategies produced higher cumulative wealth than either stock investment and especially the 60/40 benchmark. Figure 4 and Table 7 make the case for the presidential strategies appear strong, because of the long time period. However, Table 8 indicates that those strategies carried high risk: their standard deviations were almost as high as those of small cap stocks from 1937 to 1997.

The first line of Table 8 shows that, for the 1937-97 period, the standard deviations of small stocks and presidential strategies (stock and stock-bond) were 24.5%, 22.8%, and 20.7%, respectively. The presidential stock strategy had the highest arithmetic mean return (14.1%). Small stocks had a higher mean return than the stock-bond strategy, 13.3% versus 12.6%, but the cumulative wealth was less than under the stock-bond strategy because of small stocks' higher volatility.

Table 8 also shows the results for 60-, 50-, 40-, 30-, 20-, 10-, 5- and 3-year

Table 8. Average returns and standard deviations for different investment strategies for different investment horizons

Dates	# years	Large Cap Stocks Ave. Ret.	Large Cap Stocks Std. Dev.	Small Cap Stocks Ave. Ret.	Small Cap Stocks Std. Dev.	Presidential (SC/Int) Ave. Ret.	Presidential (SC/Int) Std. Dev.	Presidential (SC/LC) Ave. Ret.	Presidential (SC/LC) Std. Dev.	60/40 Benchmark Ave. Ret.	60/40 Benchmark Std. Dev.
Jan. '37–Dec. '97	61.0	10.9	15.8	13.3	24.5	12.6	20.7	14.1	22.8	8.6	10.0
Jan. '38–Dec. '97	60.0	11.8	15.5	14.9	23.7	14.3	19.8	15.8	22.0	9.2	9.8
Jan. '48–Dec. '97	50.0	12.3	14.0	13.9	19.2	13.1	12.7	14.9	16.5	9.7	9.0
Jan. '58–Dec. '97	40.0	11.6	14.2	14.7	20.2	14.6	13.0	15.6	17.0	9.7	9.3
Jan. '68–Dec. '97	30.0	11.4	15.1	12.7	21.5	14.3	12.4	14.3	17.5	10.1	9.9
Jan. '78–Dec. '97	20.0	15.4	14.7	16.3	19.5	15.9	13.8	17.2	17.9	12.9	9.9
Jan. '88–Dec. '97	10.0	16.6	11.9	15.2	14.9	13.7	9.9	16.2	13.2	13.2	8.2
Jan. '93–Dec. '97	5.0	18.4	10.6	17.7	13.3	17.7	13.3	17.7	13.3	13.5	7.4
Jan. '95–Dec. '97	3.0	27.1	11.2	22.1	15.1	22.1	15.1	22.1	15.1	19.7	7.6
Jan. '93–Dec. '93	1.0	9.5	6.1	19.0	9.4	19.0	9.4	19.0	9.4	10.0	4.3
Jan. '93–Dec. '94	2.0	5.4	8.5	11.1	9.8	11.1	9.8	11.1	9.8	4.3	6.5
Jan. '95–Dec. '96	2.0	26.3	8.4	22.9	14.2	22.9	14.2	22.9	14.2	19.3	5.9
Jan. '97–Dec. '97	1.0	28.8	15.8	20.5	17.6	20.5	17.6	20.5	17.6	20.5	10.5

intervals ending December 1997. In all intervals (except the 3 years 1995–97), the presidential stock strategy had a standard deviation greater than that of large cap stocks, but the returns were also always greater. The risk of this strategy was lower and the return was higher than those of small cap stocks. For all the investment horizons examined (except the last 5 years), the presidential stock-bond strategy had lower (or equal) risk than large cap and small cap stocks. For the 50-, 40-, 30-, and 20-year periods, the presidential stock-bond strategy had higher mean returns than large cap stocks, with lower risk; that is in a mean-variance sense, this strategy dominated the large-cap strategy for these periods. For the 30-year period, the presidential stock-bond strategy also had higher mean returns and lower risk than small cap stocks. The 60/40 benchmark exhibited lower risk than the presidential strategies, but the generally lower risk was always accompanied by lower returns. The total returns shown in Table 7 show the impact of many years of compounding and the advantage of being in the higher return assets. The lower standard deviation of the 60/40 mix is little consolation for total returns only 1/24th of the presidential equity strategy.

5   Concluding Remarks

An interesting finding of this study was the much higher small-stock returns during Democratic administrations as compared with Republican administrations. This finding is consistent with the hypothesis that Democrats devise economic policies that favor small companies and consequently, their stock prices. The 33.51 percentage point difference between small stock performance in Democratic and Republican administrations in the first year in office and the 18.21% difference for the full four-year term are very large.

This political party effect is different from the well-known January small firm effect which was present for Republicans as well as Democrats. In January, returns were higher for small cap stocks than for large cap stocks during Republican as well as Democratic rule. This phenomenon is well documented; see Hensel and Ziemba (1999) for an analysis of the 1980s and 1990s. We found in addition a substantial small stock/large stock differential outside of January during Democratic rule (see Table 3). Large stock returns were statistically indistinguishable between Democrats and Republicans, but bond and cash returns were significantly higher during Republican than during Democratic administrations. We also confirmed and updated Huang's finding that large cap stocks have had higher returns in the last two years of presidential terms; this finding applies regardless of political party and for both small and large cap stocks.

A study of the differences in economic policies that lead to the divergence of investment results according to which political party is in office would be interesting. Clearly, candidates seeking reelection are likely to favor economic policies that are particularly attractive to the public; and those policies are consistent with higher stock prices. Cash returns did not differ significantly between the first and second two-year periods of Democratic and Republican presidential terms.

Two simple presidential investment strategies performed well over the sample period. The strategy of investing in small cap stocks during Democratic administrations and large cap stocks during Republican administrations produced greater cumulative wealth than other investment strategies during the 1937–92 period. The alternating stock-bond strategy of investing in small cap stocks under Democrats and intermediate bonds under Republicans produced the second highest cumulative wealth. Both of these presidential party based strategies had higher standard deviations than large cap stocks alone during the 1937–97 period. Clinton's first administration had returns for small and large cap stocks, bonds, and cash consistent with the past. However, in the first fourteen months of his second administration large cap stocks have produced higher returns than small cap stocks.

References

Hensel, Chris R. and William T. Ziemba (1995) 'United States investment returns during democratic and republican administrations, 1928–1993', *Financial Analysts Journal* (March–April), 61–69.

Hensel, Chris R. and William T. Ziemba (1999) 'Anticipation in the January effect in the US futures markets', this volume, 218–246.

Herbst, Anthony F. and Craig W. Slinkman (1984) 'Political-economic cycles in the US stock market', *Financial Analysts Journal* **40**(2) 38–44.

Hobbs, Gerald R. and William B. Riley (1984) 'Profiting from a presidential election', *Financial Analysts Journal* **40**(2) 46–52.

Huang, Roger D. (1985) 'Common stock returns and presidential elections', *Financial Analysts Journal* **41**(2) 58–65.

Riley, William B. Jr. and William A. Luksetich (1980) 'The market prefers Republicans: myth or reality?', *Journal of Financial and Quantitative Analysis* **15**(3) 541–59.

A Long Term Examination of the Turn-of-the-Month Effect in the S&P500*

Chris R. Hensel, Gordon A. Sick and William T. Ziemba

Abstract

We present evidence on the turn-of-the-month effect using data on the S&P500 value-weighted cash index from 1928–1993, with an update to the end of 1996. Following Ariel (1987), the turn-of-the-month is trading days –1 to +4, the first-half-of-the-month days –1 to +9, and the rest-of-the-month days +10 to –2. The results are stable across decades and indicate that the returns in these sixty-five years were significantly positive in the turn-of-the-month and first-half-of-the-month and significantly negative in the rest-of-the-month. The results update, supplement and expand on the Lakonishok and Smidt (1988) study of the Dow Jones Industrials from 1897-1986 with the more representative S&P500 Index. We also investigate the seasonality across months and find substantial differences between months during these 65 years. There were strong seasonality effects with significantly above-average returns at the turn-of-the-month in January, March, May and July. The rest-of-the-month had negative returns in all but four months. May, September and November had statistically significant losses in the rest-of-the-month. The S&P500 returns during the three and a half year update period are above average, namely 1.4% per month. There were high returns during the turn, first half and rest of the month, but no statistical significance to the results because of the small sample. The paper also investigates the largest gains and losses across the three time periods and strategies being in the S&P500 in favorable periods and in cash otherwise. These strategies were successful in the 65 year sample but not in the update period.

*This research was partially supported by the Social Sciences and Humanities Research Council of Canada and the Frank Russell Company. Without implicating them, we would like to thank Gayle Nolte for computational assistance and Lynn Nansen-Dale for editing assistance.

Introduction

Investment advisors such as Merrill (1966), Fosback (1976) and Hirsch (1986) have argued that US stocks have substantial rises at the turn-of-the-month. Ariel (1987) has documented this for small-and large-capitalization stocks for the 19 years from 1963–1981. His data consisted of the equal- and value-weighted indexes of all NYSE stocks from the Center for Research on Security Prices (CRSP) tape. The turn-of-the-month (called TOM) is defined to be the last trading day of the previous month (–1) and the first four trading days of the new month (+1 to +4). Ariel's research showed there were very high returns during TOM. The rest of the market gains during 1963–1981 occurred in the second-week-of-the-month (days +5 to +9). The first-half-of-the-month (called FH, namely, trading days –1 to +9 or TOM + second-week) had all the gains. The second-half-of-the-month (called ROM, for the rest-of-the-month, which is trading days +10 to –2) had negative returns. Hence, investment in the first-half-of-the month provided more than all the year's stock market gains. Some additional support for Ariel's findings for the period from 1982–1988 appears in Cinar and Vu (1991), Keim and Smirlock (1987), and Linn and Lockwood (1988).

Roll (1983) had previously documented that there were significant positive returns from 1963–1978 in January during TOM and, in particular, that the small capitalization stocks significantly outperformed the large capitalization stocks on these days. Lakonishok and Smidt (1988) investigated various seasonal anomalies using a 90-year data set on the Dow Jones Industrial Average from 1897–1986. The DJIA is a large capitalization price-weighted index of 30 major NYSE stocks. The index rose 0.475% during the four-day period of –1 to +3 each month, whereas the average gain for a four-day period was 0.061%. The average gain per month over these 90 years was 0.349%. Hence, aside from these four days at the turn of the month, the DJIA had negative returns. Lakonishok and Smidt's test of whether or not all the gains in the market are in the first-half was inconclusive. However, this may have been because of their decision to include the –1 day in the second-half of the previous month. Ariel's definition of the turn-of-the-month to include the –1 day of the month coincides with the hypothesis that the high returns at the turn-of-the-month accrue because of considerable cash flows coming into the stock market beginning on that day. Many salaries, dividends, principal payments and debt interest are payable on the –1 and +1 days of the month. There are also institutional corporate and pension fund purchases during the turn-of-the-month. These cash flows vary by month and lead to higher average returns in January, which has the highest cash flow. This is also consistent with Roll's (1983) conclusion that more trades occur at the ask than at the bid starting on day –1.

Ogden (1987, 1990) discusses empirical support for the hypothesis about flow of funds into the stock market hypotheses from cash flows and monetary action of agencies such as the Federal Reserve System. Ogden reported that according to Moody's Manuals, 70% of the interest and principal payments on corporate debt (90% on municipal debt) are payable on the first or last days of the month. In addition, data in Standard & Poor's Stock Guide indicates that about 45% of dividends on common stock (65% on preferred stock) are payable on the first or last day of the month. The payable dates for interest, principal and dividend payments on corporate and municipal securities have, according to Moody's Manuals, been consistently on these dates throughout the twentieth century. Ogden's research provides empirical support for the hypothesis that stock prices tend to rise at the beginning of months that are preceded by months in which aggregate, economywide liquid profits are large. In particular year-end bonuses, large Christmas sales (which, according to Schwadel (1988), typically amount to 35% to 55% of annual retail sales) and other cash flows make December the highest economic activity month and seem to provide part of the reason for January's high TOM returns. Cadsby and Ratner (1991) provided further support for the turn-of-the-month cash flow hypothesis using data from Canada, the United Kingdom, Australia, Switzerland and West Germany. Cadsby and Ratner found a significant turn-of-the-month effect on trading days –1 to +4 in these countries.

Using data from 1949 to 1988 on the Nikkei stock average, Ziemba (1991) found that the turn-of-the-month in Japan is similar to that in the US. However, the dates change, because of cash flow and institutional differences, with the TOM being trading days –5 to +2, with +3 to +7 being the remainder of the first-half-of-the-month. Each of the days –5 to +2 had significantly positive returns. Moreover, all the gains in the stock market were in the FH. The securities firms have some information regarding these trading patterns and plan their monthly sales effort during the –5 to +7 period. The high returns in Japan during this period seem to be based on cash flow and institutional trading. In Japan most salaries are paid from the 20th to the 25th of the month, with the majority on the 25th. Security firms generally invested for their own accounts based on their capitalization, which in this period resulted in buying commencing on the –3 day with Japan's three-day settlement. Individual investors bought mutual funds with their pay, which they received on calendar days 15 to 25 of the month; the funds were then invested in stocks during the turn-of-the-month. Employee stock-holding plans and mutual funds also received money in this period to invest during the turn-of-the-month. An update of these results in the highly negative return period 1990–1994 by Comolli and Ziemba (1999) appears in this volume.

Since the early studies on the day-of-the-week effect by Cross (1973), the January effect by Rozeff and Kinney (1976) and the small-firm effect by Banz

(1981) and Reinganum (1981), there has been an intense investigation of seasonal anomalies in US and other security markets. For example, discussion of the day-of-the-week effect appears in French (1980), Gibbons and Hess (1981), Jaffe and Westerfield (1985), Ball and Bowers (1986), Pettengill and Jordan (1988), Kato, Schwartz and Ziemba (1989), Lakonishok and Maberly (1990), Solnik (1990) and Ziemba (1993). Holiday effects are analyzed by Pettengill (1989), Ariel (1990), Cadsby and Ratner (1991), Ziemba (1991) and Cervera and Keim (1999). Monthly effects are studied by Gultekin and Gultekin (1983), Brown, Kleidon and Marsh (1983), Jacobs and Levy (1988), Hawawini (1991), Jaffe and Westerfied (1989) and Cadsby (1992). Harris (1986) studied time-of-day effects. Besides those mentioned above, January and size effects are also analyzed by Clark, McConnell and Singh (1992), Keim (1983, 1989); Roll (1983); Ritter (1988); Ritter and Chopra (1989); Jaffe, Keim and Westerfield (1989); and Booth and Keim (1999). Recent surveys of this literature appear in Hawawini and Keim (1995, 1999) and Ziemba (1994).

This research and other studies bring up the issue of data mining. Merton (1985), Black (1986, 1992) and Lo and MacKinlay (1990) in particular discuss the dangers of finding what appear to be genuine anomalies but in fact may be simply random data variations. This is especially true because even with strong seasonality effects the average daily magnitudes of these anomalies are typically less than a two-way transaction cost, which is about 0.46% for NYSE stocks according to Berkowitz, Logue and Noser (1988), or the typical bid-ask spread, which at an eighth is 0.31% of the average-priced $40 NYSE stock. Perhaps the best remedy against data snooping is new data and convincing reasons for the effects. The reasons for the turn-of-the-month effect are largely cash-flow and institutionally based, as discussed above. Another factor seems to be behavioral. For example, bad news such as that relating to earnings announcements is delayed and announced late in the month, while good news is released promptly at the beginning of the month (see Penman 1987), and the market does not fully anticipate this information.

The following explanations for the turn-of-the-month have also been advanced: inventory adjustments of different traders (Rock 1989 and Ritter 1988); the timing of trades by informed and uninformed traders (Admati and Pfleiderer 1988) and specialists' strategies in response to informed traders (Admati and Pfleiderer 1989); seasonal tax-induced trading (Lakonishok and Smidt 1986); and window dressing induced by periodic evaluation of portfolio managers (Haugen and Lakonishok 1988 and Ritter and Chopra 1989). Hensel, Sick and Ziemba (1994) duplicated the Ariel (1987) tests on the turn- and first-half-of-the-month effects using data from 1982–1992 on the S&P500 and Value Line Composite cash and futures indexes that measured large and small capitalization stocks, respectively. They found strong effects in this period, which does not overlap with Ariel's 1963–1981 data. They also found partial anticipation in the futures markets on the three trading days before the turn-of-the-month but the cash effect was still present and similar to Ariel's results. Ziemba (1989) also showed that the futures market on the SIMEX in Singapore anticipated but did not change the effect in the Japanese cash market during 1986–1988, when there was essentially no derivative securities trading in Japan.

In this paper we investigate the turn-of-the-month effect in a 65-year sample from 1928–1993 for the S&P500, which is arguably the most common US stock index. It is widely used as a benchmark for active investing and a target for passive index portfolios. The paper significantly expands the data period studied by Ariel (1987) and uses a portfolio that is closer to the overall market basket than the paper of Lakonishok and Smidt (1988). Monthly variations and large gain and loss effects are investigated. Some of the results in the paper appeared in Hensel and Ziemba (1996); see also Gonzales (1995).We also update the results to 1996 in a postscript to the paper.

In Section 1 we investigate the monthly return patterns on the S&P500 by trading day of the month. The trading days during TOM had very high returns, with three of these five days having returns significantly above average. Moreover, all of the rest-of-the-month trading days from +9 to –3 had returns below average. The results are consistent across decades from the 1920s. Seasonality effects are also investigated and indicate that January, February, April, July, October and December all had high returns at the turn-of-the-month. The percentages of positive daily returns were studied for the 65-year period by decade and month. The evidence is that the higher mean returns during TOM and FH and lower mean returns during ROM were at least partly a result of a higher percentage of positive return days.

In Section 2 we investigate the days with the 100 largest percentage gains and 100 largest percentage losses over these 65 years. The first-half-of-the-month had a higher percentage of the highest return days and a lower percentage of the lowest return days than expected. The rest-of-the-month had less of the high-return days and more of the low-return days than expected. This suggests, however, that the higher positive mean returns during the FH is partly due to a higher mean return conditional on a positive return. However, these differences are not statistically significant.

Section 3 investigates more fully the turn-of-the-month, first-half-of-the-month and rest-of-the-month returns. The average return during the turn-of-the-month for the 65 years was significantly above average and equaled seven times a typical day's return. The first-half returns were also significantly above average. However, during the rest-of-the-month, the returns were significantly below average and also significantly negative. With trading day –1 in the first-half-of-the-month, all the gains in the S&P500 were in the first-half-of-the-month. Strategies that were long the S&P during the turn-of-the-month or first-half and then in cash (Treasury bills) otherwise, were mean-variance superior to the passive buy-and-hold S&P strategy. There were significant seasonality effects in the S&P500 returns. In every month the returns during the turn-of-the-month were higher than average. In four months these returns were significantly above average. In all but four months there were negative returns in the rest-of-the-month. This section also investigates the returns of small stocks, various types of bonds and cash and their correlations with the S&P500 returns during various months. Wealth relatives are also computed that show the compounded effects of monthly as well as intramonth seasonality during the TOM, FH and ROM subperiods. The TOM and FH plus-cash strategies have relatively low correlation with large and small capitalization stocks and mean-variance dominate other strategies.

Section 4 presents an update on these results from July 1993 to the end of December 1996. The three and a half year period from June 1993 to December 1996 was an era of high S&P500 returns averaging 1.4% per month. While the turn and first half of the month had high returns, there were also high returns in the rest of the month. Because of the small sample, none of the returns are significantly different from each other or from zero. The TOM and FH plus cash strategies lagged the S&P500. In this latter period there is a possible shift of the TOM period in trading days −4 to +3 because of futures anticipation.

Concluding remarks appear in Section 5.

1   The Monthly Return Patterns in the S&P500

The data used in this study consists of the daily closing prices of the S&P500 stock index for the 65-year period from February 1928 to June 1993. This data was supplied by Data Resources Incorporated. The S&P500 is a market value-weighted index of large capitalization US stocks. From March 1957 this has consisted of the 500 largest stocks weighted by market value (price times number of shares outstanding). Prior to then it consisted of the 90 largest stocks. This index is referred to as the S&P composite or S&P500 as we call it here. Index futures contracts on the S&P500 have been trading since 1982 on the Chicago Mercantile Exchange, and futures options are traded at the Chicago Board of Trade. Hensel, Sick and Ziemba (1994) have studied the turn-of-the-month and monthly return patterns of the S&P500 and the Value Line Composite small stock index during the period of futures trading from 1982–1992. This paper takes a deeper look at these monthly patterns in the S&P500 during the longer time period back to 1928. We extend the

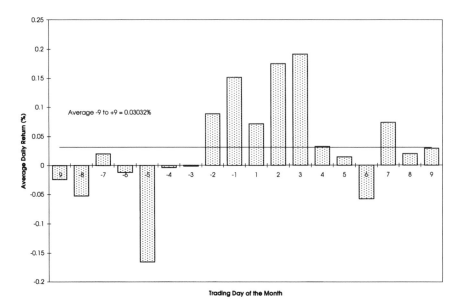

Figure 1. Average Daily Returns in the S&P500 Cash Market by Trading Day of
the Month (February 1928 to June 1993).

Hensel, Sick and Ziemba study and those by Ariel (1987) on small and large
capitalization cash market data from 1963–1981 and Lakonishok and Smidt
(1988) on the Dow Jones Industrials from 1897–1986. This paper supplements
and updates the latter paper with the S&P500, which is arguably a superior
index since it has more securities, (500 versus 30), is value weighted and
closer to the market portfolio, and is the most common index used in passive
portfolio management to represent the market.

Figure 1 shows the average return pattern by trading day. The average
return on trading days –9 to +9 over the 65-year sample was 0.0303% per
day.[1] Days –9 to –1 are during the previous month and +1 to +9 during the
current month. For example, day –9 of January is a day in December that
is 9 trading days before the start of January. Significantly higher returns
occurred on trading days –2 to +3. The return on these five days averaged
0.136% or more than four times as much. Hence, the bulk of the monthly
returns occurred at the turn of the month. To avoid data-snooping biases,
researchers such as Roll (1983), Ariel (1987), Ogden (1990) and Hensel, Sick
and Ziemba (1994) have generally referred to the five-day period –1 to +4 as
the turn of the month (TOM). Figure 1 shows that this period does indeed

---

[1]In this section we compare the return patterns using the trading days –9 to +9. Comparisons using all trading days appear in Section 3.

have mean returns significantly above average. However, over the 65-year period studied here, the five-day period –2 to +3 had even higher returns.

Figure 1 indicates that all the monthly gains occurred in the FH, on trading days –1 to +9. The ROM, trading days +10 to –2, had negligible returns. The first row of Table 1a details these mean returns by trading day of the month. The last three columns have the average daily returns in the –9 to +9 period, the turn-of-the-month (–1 to +4) and the first-half-of-the-month (–1 to +9). The mean daily returns over the 65 years were higher in TOM at 0.1236% than in the period –9 to +9, at 0.0303%, with the first-half having average daily returns of 0.0695%. The first row of Table 1b gives the t values for the hypothesis that the mean daily returns on the various trading days –9 to +9 were above the mean of the –9 to +9 trading-day returns, which was 0.0303% per day. Trading days –1, +2 and +3 had significantly higher returns than average. Trading days –2 and +1 also had high average returns. All of the trading days from –9 to –3 had returns below average, with two of these (–8 and –5) significantly negative.

Tables 1a and b present the mean returns and t statistics, respectively, for the hypothesis that the daily returns were greater than the –9 to +9 average by trading day of the month, by decade from the 1930s to the 1980s and by month. Investigation of the t statistics shows that trading days –1, +2 and +3 have significantly higher mean returns than average in most decades and for the entire 65-year period. Every decade had high returns on –1 and +2, and only in the 1980s were the returns on +3 negative. The latter may be the anticipation of the turn-of-the-month effect in the S&P500 futures market and the associated index arbitrage; see Hensel, Sick and Ziemba (1994). The t statistics in Table 1b are not independent of each other because the full period is not independent of the subperiod returns on particular days. Hence, these t statistics should be viewed as flags to draw attention to the magnitudes of some of the numbers and not as true t statistics. Tables 1a and b also investigate the monthly effect. January returns for the S&P500 at the turn-of-the-month were high (see also Figure 2) but not as high as for the small stocks; see Table 5 for comparison. January, March, May and July all had high returns at the turn-of-the-month.

Figures 3a and 3b show the rolling 60-month returns and standard deviations of the contrasting trading days –1 and –5. For the high-return day –1, there are consistently positive returns with relatively constant standard deviation. For the low-return day –5, there are consistently negative returns and somewhat higher and more variable deviations.

Table 2 details the percentage of positive daily returns for the 65-year period by decade and by month. The results show that the higher mean returns during TOM and FH and the lower mean return during ROM are partly effects of a higher percentage of positive return days. We now investigate

Table 1a. Average Daily Returns in the S&P500 Cash Market by Decade and Month for Monthly Trading Days −9 to +9 (February 1928 to June 1993).*

	−9	−8	−7	−6	−5	−4	−3	−2	−1	+1	+2	+3	+4	+5	+6	+7	+8	+9	−9 to 9	TOM	FH
1928-1993	−0.02	−0.05	0.02	−0.01	−0.17	0.00	0.00	0.09	0.15	0.07	0.17	0.19	0.03	0.01	−0.06	0.07	0.02	0.03	0.03	0.12	0.07
1928-1939	0.08	−0.21	0.12	−0.01	−0.44	−0.14	−0.19	0.08	0.10	−0.01	0.17	0.40	0.24	0.13	−0.25	0.18	0.16	0.07	0.03	0.18	0.12
1940-1949	−0.12	−0.03	0.00	0.00	−0.07	−0.07	−0.02	0.25	0.16	0.09	0.21	0.19	−0.14	0.05	0.03	0.01	−0.16	−0.06	0.02	0.10	0.04
1950-1959	0.03	−0.02	0.10	0.05	−0.17	0.11	0.10	0.08	0.16	0.20	0.30	0.18	0.02	0.01	−0.20	−0.02	0.06	0.05	0.06	0.17	0.08
1960-1969	−0.08	−0.10	−0.04	−0.05	−0.20	−0.05	−0.03	0.07	0.15	0.04	0.14	0.13	0.10	0.02	0.02	0.11	0.02	0.02	0.01	0.11	0.07
1970-1979	−0.12	−0.08	−0.03	−0.08	−0.03	0.00	0.08	−0.05	0.09	0.02	0.03	0.21	0.03	0.00	0.01	0.05	−0.01	0.00	0.01	0.07	0.04
1980-1993	0.03	0.10	−0.03	0.00	−0.07	0.11	0.06	0.09	0.23	0.10	0.19	0.04	−0.06	−0.10	0.04	0.08	0.02	0.08	0.05	0.10	0.06
1928-1993	**−9**	**−8**	**−7**	**−6**	**−5**	**−4**	**−3**	**−2**	**−1**	**+1**	**+2**	**+3**	**+4**	**+5**	**+6**	**+7**	**+8**	**+9**	**−9 to 9**	**TOM**	**FH**
January	−0.03	−0.15	0.10	0.13	−0.06	0.07	0.05	0.14	0.32	0.01	0.56	−0.01	0.19	−0.17	0.10	0.07	0.08	−0.10	0.07	0.21	0.10
February	0.15	−0.13	−0.10	0.08	−0.28	−0.12	0.09	0.01	0.24	0.07	0.15	−0.16	0.07	−0.09	−0.26	0.12	0.00	0.00	−0.01	0.08	0.02
March	−0.05	0.03	−0.03	−0.18	−0.21	0.00	−0.06	0.09	0.16	0.20	0.22	0.14	0.22	−0.04	−0.12	0.17	−0.10	−0.12	0.02	0.19	0.08
April	0.19	−0.10	−0.04	−0.19	−0.22	0.03	−0.14	−0.13	−0.16	0.08	0.15	0.12	0.10	−0.06	−0.02	0.20	0.24	−0.06	0.00	0.05	0.06
May	−0.17	−0.11	0.04	−0.06	−0.11	0.08	−0.07	0.22	0.31	−0.03	0.20	0.32	0.04	0.25	−0.21	0.11	−0.19	0.04	0.04	0.17	0.08
June	−0.04	0.00	0.18	0.02	−0.21	0.01	−0.15	0.08	−0.04	−0.02	0.14	0.31	0.27	−0.02	−0.18	−0.07	0.16	0.15	0.03	0.13	0.07
July	−0.07	−0.10	−0.13	0.23	−0.17	0.14	−0.05	0.10	0.14	0.28	0.32	0.39	−0.01	0.02	0.21	0.08	−0.05	0.31	0.09	0.23	0.17
August	0.03	0.05	0.14	0.07	−0.06	−0.05	−0.01	0.01	0.15	0.11	0.06	0.21	−0.19	0.13	0.03	0.22	−0.08	0.06	0.05	0.06	0.07
September	−0.09	−0.01	0.24	−0.35	−0.25	−0.19	0.06	−0.22	0.25	−0.17	0.17	0.33	−0.09	−0.31	−0.32	0.11	0.02	−0.22	−0.05	0.10	−0.02
October	−0.15	0.19	−0.11	−0.02	−0.49	−0.12	−0.10	0.30	0.00	0.06	0.26	0.18	−0.23	0.47	−0.07	−0.13	−0.26	0.40	0.01	0.04	0.06
November	−0.03	−0.20	−0.14	0.15	−0.03	0.00	−0.03	−0.07	0.21	0.25	−0.17	0.34	−0.06	−0.10	0.04	0.08	0.41	−0.05	0.04	0.11	0.09
December	0.03	−0.13	0.13	−0.01	−0.03	0.10	0.33	0.52	0.27	−0.02	0.10	0.15	0.11	0.12	0.04	−0.09	−0.04	−0.09	0.08	0.12	0.06

*In this and succeeding tables, t values significant at the 5% level are in bold. Months are tested against the −9 to +9 mean for the entire period. Specific periods are tested against the mean for the same period.

Table 1b. t Values for (a) and Figures 1 and 2 for the Hypothesis that the Mean Daily Returns are Greater than the Overall Mean for the Same Time Period (for days −9 to +9) by Trading Day of the Month for the S&P500 Cash Market by Decades and Month (February 1928 to June 1993)

	−9	−8	−7	−6	−5	−4	−3	−2	−1	+1	+2	+3	+4	+5	+6	+7	+8	+9	−9 to 9	TOM	FH
1928–1993	−1.21	**−1.99**	−0.29	−1.12	**−4.82**	−0.86	−0.78	1.52	**3.31**	1.07	**3.66**	**4.04**	0.04	−0.38	**−2.42**	1.05	−0.27	−0.04	0.00	**5.15**	**3.00**
1928–1939	0.26	−1.41	0.69	−0.21	**−3.11**	−1.02	−1.33	0.35	0.50	−0.29	0.86	**2.33**	1.08	0.59	−1.91	0.93	0.88	0.25	0.00	**2.10**	1.77
1940–1949	−1.51	−0.74	−0.22	−0.25	−1.18	−0.97	−0.57	**3.07**	**2.22**	0.86	**2.52**	**2.52**	−1.88	0.37	0.12	−0.09	**−2.01**	−0.93	0.00	**2.57**	0.74
1950–1959	−0.38	−1.15	0.65	−0.11	**−2.50**	0.88	0.72	0.37	1.77	**2.18**	**4.46**	1.68	−0.72	−0.92	**−4.01**	−1.21	−0.04	−0.19	0.00	**4.21**	0.76
1960–1969	−2.07	**−2.07**	−1.04	−1.13	**−3.88**	−1.14	−0.48	0.84	**2.21**	0.34	**2.00**	**2.28**	1.50	0.11	0.14	1.71	0.16	0.02	0.00	**3.34**	**2.75**
1970–1979	−1.63	−1.05	−0.50	−1.20	−0.43	−0.12	0.88	−0.69	1.09	0.12	0.35	**2.38**	0.35	−0.11	0.10	0.52	−0.21	−0.12	0.00	1.67	1.28
1980–1993	−0.24	0.48	−1.10	−0.69	−1.42	0.93	0.16	0.59	**2.53**	0.67	**2.06**	−0.10	−1.43	**−2.01**	−0.20	0.41	−0.47	0.38	0.00	1.40	0.47
1928–1993	**−9**	**−8**	**−7**	**−6**	**−5**	**−4**	**−3**	**−2**	**−1**	**+1**	**+2**	**+3**	**+4**	**+5**	**+6**	**+7**	**+8**	**+9**	**−9 to 9**	**TOM**	**FH**
January	−0.59	−1.31	0.91	0.96	−0.59	0.34	0.08	1.06	**3.66**	−0.13	**3.33**	−0.39	0.86	−1.35	0.55	0.42	0.59	−1.09	1.28	**3.03**	1.78
February	0.70	−1.66	−1.00	0.61	**−2.13**	−1.13	0.58	−0.12	**2.31**	0.40	1.26	−1.55	0.46	−1.19	**−2.44**	0.59	−0.20	−0.33	−1.26	1.02	−0.36
March	−0.69	0.03	−0.62	−1.60	**−2.08**	−0.24	−0.63	0.45	1.49	1.85	1.58	1.01	0.73	−0.61	−1.10	1.30	−1.33	−1.22	−0.34	**2.04**	1.16
April	0.81	−0.92	−0.63	**−2.48**	**−2.09**	0.05	−1.38	−1.23	−1.26	0.29	1.02	0.91	0.45	−0.74	−0.19	1.41	1.35	−0.70	−0.77	0.32	0.54
May	−1.48	−0.89	0.22	−0.82	−1.06	0.42	−0.44	1.57	1.80	−0.55	1.49	**2.70**	0.02	1.40	**−2.22**	0.74	**−2.11**	−0.07	0.21	**2.34**	1.27
June	−0.68	−0.16	1.28	−0.14	−1.50	−0.16	−1.68	0.32	−0.43	−0.31	0.78	1.53	1.44	−0.60	−1.63	−0.60	0.93	0.71	0.04	1.89	0.97
July	−0.56	−0.74	−1.62	1.28	−1.22	0.75	−0.61	0.60	1.11	**2.11**	**2.55**	**2.83**	−0.33	−0.06	1.85	0.43	−0.80	**2.02**	1.71	**3.58**	**3.96**
August	−0.02	0.12	0.87	0.36	−0.65	−0.70	−0.40	−0.26	**0.95**	0.69	0.21	1.04	−1.89	0.68	0.00	1.28	−0.91	0.26	0.43	0.61	0.74
September	−0.48	−0.28	1.15	**−2.19**	−1.47	−1.58	0.24	−1.79	**2.04**	−1.24	0.98	1.53	−1.31	**−2.00**	**−3.18**	0.67	−0.08	−1.39	**−2.40**	1.12	−0.93
October	−1.01	0.72	−0.80	−0.35	**−2.83**	−0.64	−0.58	1.17	−0.21	0.15	1.92	0.83	−1.21	1.85	−0.62	−0.92	−1.68	**2.18**	−0.62	0.18	0.62
November	−0.36	−1.68	−1.27	0.89	−0.10	−0.18	−0.34	−0.84	1.17	1.48	−0.92	**2.07**	−0.46	−0.71	0.10	0.24	1.78	−0.61	0.21	1.17	1.08
December	0.14	−1.35	1.02	−0.42	−0.51	0.55	**2.44**	**4.50**	1.64	−0.39	0.47	0.94	0.78	0.68	0.17	−0.88	−0.56	−1.45	**2.08**	1.51	0.66

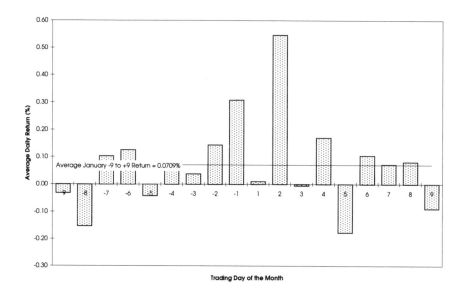

Figure 2. Average Daily Returns at the Turn-of-the-Year for the S&P500 Cash Market by Trading Day (February 1928 to June 1993).

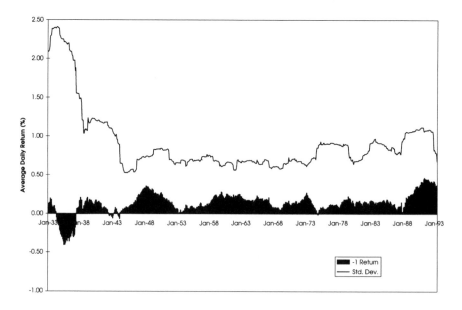

Figure 3a. Rolling 60 Month Returns and Standard Deviations for Trading Day −1 (February 1928 to June 1993).

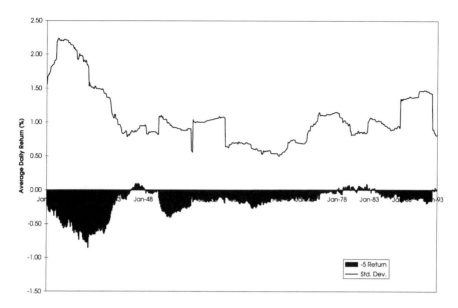

Figure 3b. Rolling 60 Month Returns and Standard Deviations for Trading Day
−5 (February 1928 to June 1993).

the days with the highest percentage of positive returns and largest returns
to investigate whether the mean return, conditional on a positive return, is
higher during TOM.

2 Days with the Largest Gains and Losses

The evidence is that all of the gains in the S&P500 from 1928 to 1993 were
at the turn and during the first-half-of-the-month. The returns in the rest-
of-the-month were nonpositive. There are several ways to investigate the
distribution of returns. One measure is the frequency distribution of the
largest percentage gains and losses for various times of the month. Tables
3a and 3b summarize the results from the 100 largest gains and losses (plus
ties). It is illustrative to investigate the recent period 1950–1993 as well as
the entire 1928–1993 sample.

In the later 43-year period the turn-of-the-month had 23.7% of the total
trading days. Of these there were 28.0% of the highest return days and
19.0% of the lowest return days. The second week (SW) had less (20.0%)
than its expected share (23.7%) of high-return days and more (26.0%) than
its expected share of low-return days (23.7%). The rest-of-the-month had
slightly less (52.0%) of the high-return days and slightly more (55.0%) of the

Table 2. Percentage of Positive Daily Returns in the S&P500 Cash Market by Decade and Month for Monthly Trading Days 9 to +9 (February 1928 to June 1993) Average −9 to +9 = 52%.

	−9	−8	−7	−6	−5	−4	−3	−2	−1	+1	+2	+3	+4	+5	+6	+7	+8	+9	−9 to 9	TOM	FH
1928–1993	47%	48%	52%	49%	45%	52%	52%	55%	58%	55%	60%	57%	51%	51%	50%	52%	51%	51%	58%	64%	64%
1928–1939	51%	48%	53%	53%	41%	48%	48%	54%	54%	48%	57%	55%	52%	50%	41%	53%	51%	50%	61%	64%	68%
1940–1949	43%	49%	55%	48%	47%	46%	45%	59%	60%	60%	60%	56%	39%	57%	58%	50%	46%	46%	57%	64%	62%
1950–1959	50%	51%	54%	52%	45%	60%	61%	58%	61%	67%	68%	66%	51%	53%	38%	50%	53%	53%	63%	77%	65%
1960–1969	40%	44%	48%	48%	43%	53%	50%	50%	60%	50%	65%	63%	63%	50%	58%	58%	55%	52%	58%	68%	69%
1970–1979	49%	48%	44%	46%	46%	48%	51%	49%	53%	56%	53%	55%	55%	53%	53%	53%	53%	45%	54%	59%	60%
1980–1993	48%	48%	50%	44%	50%	56%	57%	57%	63%	52%	57%	52%	46%	44%	52%	49%	52%	59%	57%	57%	60%

1928–1993	−9	−8	−7	−6	−5	−4	−3	−2	−1	+1	+2	+3	+4	+5	+6	+7	+8	+9	−9 to 9	TOM	FH
January	52%	45%	61%	58%	48%	50%	45%	61%	70%	48%	77%	47%	56%	44%	50%	52%	58%	50%	59%	69%	66%
February	48%	38%	48%	54%	42%	42%	55%	54%	63%	57%	60%	48%	49%	51%	37%	49%	54%	46%	55%	65%	57%
March	51%	48%	49%	42%	29%	52%	54%	46%	58%	65%	57%	55%	52%	48%	58%	58%	40%	57%	63%	68%	68%
April	48%	52%	49%	35%	51%	57%	48%	45%	43%	60%	55%	51%	60%	54%	52%	62%	55%	48%	52%	58%	60%
May	45%	43%	57%	45%	46%	57%	46%	62%	57%	48%	71%	72%	48%	52%	43%	46%	46%	48%	60%	66%	72%
June	40%	52%	55%	52%	43%	48%	49%	55%	58%	45%	52%	55%	55%	52%	49%	42%	63%	59%	58%	62%	65%
July	48%	51%	46%	49%	40%	60%	52%	58%	58%	65%	63%	63%	54%	57%	58%	51%	43%	67%	71%	75%	75%
August	48%	51%	54%	55%	40%	43%	60%	52%	65%	52%	50%	55%	42%	55%	44%	52%	42%	55%	48%	58%	59%
September	42%	53%	56%	39%	50%	42%	56%	47%	66%	61%	56%	58%	44%	50%	39%	56%	63%	41%	52%	63%	53%
October	44%	55%	48%	44%	33%	58%	52%	52%	48%	50%	67%	56%	53%	56%	52%	38%	42%	56%	58%	61%	64%
November	58%	41%	52%	61%	55%	58%	45%	47%	55%	63%	53%	67%	45%	48%	53%	67%	59%	41%	53%	67%	66%
December	44%	45%	50%	50%	52%	56%	59%	73%	61%	48%	58%	61%	53%	48%	56%	52%	50%	50%	67%	64%	64%

low-returns days than the expected (52.7%). The standard deviation of the largest gains and losses was much lower in the TOM than in the SW or ROM. The largest gains and losses during the entire sample shown are dominated by the high volatility period around the 1929 crash and the subsequent depression era. The average daily gains or losses of these extreme days are much larger than during the latter periods. Five of the six comparisons are as expected. However, the observed frequency of the largest gains during TOM (21.0%) is less than the expected frequency. Investigation of the 1960–1993 period yields results similar to the recent 43-year period with a larger percentage (31.0%) of the highest gains in TOM versus the expected (23.8%).

To summarize, the evidence is consistent with the statement that the higher average returns during the FH in the 1928–1993 sample were partly composed of a higher frequency of very large returns and a lower frequency of very low returns. However, the results are not strong enough to have statistical significance. Thus, the largest 100 gains and losses have at most a minor effect on the results elsewhere in the paper that all the gains were in the FH during 1928–1993. This is consistent with the hypothesis that large gains and losses are the result of financial news shocks that occur randomly in time.

The cumulative wealth effects of investment during various time periods magnify the effects. The higher average returns during TOM in the 1928–1993 sample were partly composed of this higher frequency of very large returns and lower frequency of very low returns.

3   Investigation of the Turn-of-the-Month, First-Half and Rest-of-the-Month Returns

Table 4 provides the average daily mean returns by month during the turn-of-the-month trading days (–1 to +4), the first-half (–1 to +9) and the rest-of-the-month (+10 to –2) for the S&P500 from 1928–1993. The average return per day for the 65 years was 0.0186%. The mean returns during the turn-of-the-month were nearly seven times as high at 0.124%, which has a t value of 5.94 for the hypothesis that the turn-of-the-month days have returns above the mean. The first-half returns were also significantly higher than average. The rest-of-the-month had mean returns significantly below the overall average and also significantly negative.[2]

---

[2]These results also show the effect of approximately four days per month not included in the –9 to +9 calculations in Section 2 and Tables 1a and 1b. The mean for –9 to +9 was 0.0303% versus 0.0186% for all days. Had we used 0.0186% for the mean in Table 2, the significant daily t values during the positive parts of the month, particularly during TOM, would be even higher.

Table 3a. Summary Statistics of the 100 Largest Gains and Losses in the S&P500, January 1950 to June 1993.

	Typical Year	Expected Frequency	Largest Gains			Largest Losses		
			Average Daily Returns	Average Daily St. Dev.	Observed Frequency	Average Daily Returns	Average Daily St. Dev.	Observed Frequency
TOM	60	23.7	2.79	0.49	28.0	-2.75	0.56	19.0
SW	60	23.7	2.84	0.67	20.0	-2.95	1.04	26.0
ROM	133	52.7	3.12	1.11	52.0	-3.55	2.98	55.0
All days	253	100.0	2.97	0.90	100.0	-3.24	2.30	100.0

Table 3b. Summary Statistics of the 100 Largest Gains and Losses in the S&P500, February 1928 to June 1993.

	Typical Year	Expected Frequency	Average Daily Returns	Largest Gains Average Daily St. Dev.	Observed Frequency	Average Daily Returns	Largest Losses Average Daily St. Dev.	Observed Frequency
TOM	60	22.4	6.07	1.82	21.0	−5.71	1.73	18.0
SW	60	22.4	5.63	1.76	28.0	−5.36	0.90	20.0
ROM	146	55.2	5.68	2.55	51.0	−6.21	2.79	62.0
All days	267	100.0	5.72	1.97	100.0	−5.95	2.36	100.0

Table 4. Average Daily Returns and t-values for the Hypothesis that the Mean Returns are Greater than Zero and the Overall Mean Return, respectively, in the S&P500 Cash Market by Month, During the Turn, First-Half, Rest-of-the-Month and Whole-Month (February 1928 to June 1993).

	Average Daily Returns (%)				t-values (Ho: return = 0)				t-values (Ho: ret diff = 0)			
	TOM −1 to +4	FH −1 to +9	ROM +10to−2	All Days −1 to−1	TOM −1 to +4	FH −1 to +9	ROM +10to−2	All Days −1 to−1	TOM −1 to +4	FH −1 to +9	ROM +10to−2	All Days −1 to−1
S&P500 Index									Avg Daily S&P Index Ret = 0.0186			
January	0.2061	0.1025	0.0359	0.0651	**3.42**	**2.54**	1.15	**2.62**	**3.11**	**2.08**	0.56	1.87
February	0.0807	0.0170	−0.0214	−0.0024	1.70	0.49	−0.51	−0.09	1.31	−0.04	−0.96	−0.77
March	0.1876	0.0768	−0.0212	0.0208	**2.89**	1.86	−0.65	0.81	**2.61**	1.41	−1.22	0.09
April	0.0503	0.0566	−0.0169	0.0161	0.91	1.38	−0.45	0.58	0.57	0.93	−0.95	−0.09
May	0.1653	0.0819	−0.0836	−0.0107	**3.20**	**2.14**	**−2.10**	−0.38	**2.84**	1.65	**−2.56**	−1.04
June	0.1287	0.0669	0.0033	0.0315	1.81	1.41	0.08	1.03	1.55	1.02	−0.38	0.42
July	0.2258	0.1697	−0.0050	0.0738	**4.32**	**4.67**	−0.12	**2.67**	**3.96**	**4.16**	−0.59	**2.00**
August	0.0645	0.0672	0.0129	0.0364	1.12	1.63	0.35	1.33	0.80	1.18	−0.15	0.65
September	0.0976	−0.0175	−0.0978	−0.0605	1.50	−0.37	**−1.96**	−1.76	1.22	−0.77	**−2.33**	**−2.29**
October	0.0445	0.0632	−0.0787	−0.0178	0.65	1.19	−1.31	−0.43	0.38	0.84	−1.62	−0.88
November	0.1108	0.1038	−0.0821	0.0071	1.44	1.87	**−1.96**	0.21	1.20	1.53	**−2.41**	−0.33
December	0.1217	0.0564	0.0599	0.0584	**2.11**	1.42	1.71	**2.22**	1.79	0.95	1.18	1.52
All Months	0.1236	0.0703	−0.0235	0.0186	**6.99**	**5.61**	**−1.98**	**2.15**	**5.94**	**4.13**	**3.71**	0.00
All except Jan	0.1162	0.0674	−0.0292	0.0143	**6.28**	**5.12**	**−2.30**	1.56	**5.28**	**−3.54**	**−3.76**	−0.47

There was significant seasonality in the S&P500 returns. January, March, May and July had mean returns during the turn-of-the-month significantly above the overall average. In every month the mean returns during TOM were higher than average. January and July also had significantly higher mean returns than average during the first-half-of-the-month. The mean returns in the first-half were positive in all months except September. The rest-of-the-month had negative mean returns in all months except January, June, August and December. The rest-of-the-month for September and October – which are widely cited in the media as a time for stock market losses – did have large negative mean returns, but only September is significant. The ROM of September, however, had mean returns significantly below average. May and November have significantly negative mean returns during ROM. Again the reader is cautioned that the number of t statistics in Table 4 plus subperiod dependencies may lead to overstatement of the significance of results for particular months.

A comparison of the S&P500 returns during TOM, the FH and ROM by month and decade, with the returns from other investment strategies, appears in Table 5. The monthly total return data on the other investment strategies was supplied by Ibbotson Associates. The first three columns of Table 5 are the periods –9 to +9, TOM and the FH. In the bottom panel of Table 5 the first three columns adjust these periods so that the correlations that appear in the second panel constitute the same month.[3] For example, TOM (+1 to +4, –1) consists of the first four trading days of the given month plus the last. The FH and –9 to +9 are also modified. The next six columns are Ibbotson Associates' returns for the large capitalization (S&P composite), small capitalization (bottom 20% of companies, capitalization weighted), high-yield corporate bonds, long-term (20 years) and intermediate-term (five years) government bonds, and cash measured by the 90-day T-bill return. The next to the last column has returns from the strategies that invested in the S&P500 during the turn-of-the-month and then in cash for the remainder of the month. The final column is the same strategy for the first-half. It is assumed that these strategies are in cash 80% and 60% of the month, respectively. These two strategies have higher mean returns and lower standard deviations than during TOM, FH, or the whole investment period. The first panel provides return results by decade rather than by month. Although there was some variability, there were consistently high returns during TOM and the FH in all the decades. The strategies of being long in the S&P500 during TOM or the FH and then in cash had consistently high mean returns that were mean-variance superior to the buy-and-hold S&P500 strategy. Monthly effects, such as the 6.38% mean return in January

---

[3]These correlations utilize +1 to +r and –1 of the next month for TOM, etc., so that the comparisons are all within the same month.

for small stocks, appear in the second panel.

The last panel of Table 5 displays the correlations between the various investment returns from February 1928 to June 1993. The TOM and FH returns are for days within the same month -namely, trading days +1 and +4 and −1, and +1 to +9 and −1, respectively. The TOM- and FH-plus-cash strategies have a relatively low correlation with large and small capitalization stocks – namely, 0.46 and 0.38 for TOM and 0.67 and 0.57 for the FH, respectively. Hence these strategies may be considered as separate asset classes in mean-variance or related asset allocation studies just as small or large capitalization or foreign stocks might be. Wealth relatives for the various investment strategies by month, decade and for the entire 1928–1993 period are shown in Table 6. The results show the growth of one dollar over the 65 years of the sample. These values show the compounded effects of monthly as well as intramonth seasonality during the TOM, FH and ROM subperiods. In interpreting these results, one must consider the number of trading days in each period. TOM had five days, FH had 10 days, ROM had about 10–12 days, and the whole month had about 20–22 days. Glaringly apparent is the $63.40 total wealth of an initial $1 invested in small capitalization stocks in January. The total growth in the small capitalization index ($1,483.63) was higher than the TOM-plus-cash and FH-plus-cash strategies that returned $758.36 and $1,290.97, respectively. These strategies dominated the others in wealth relatives. Adjustment for risk via standard deviation points to the superiority (high returns and relatively low standard deviations) of these later strategies during the 65-year sample period.

Suppose the value of the small capitalization portfolio and S&P500 portfolio follow a joint geometric random walk with a constant drift, correlation and variances throughout the month. Also suppose that investing in cash yields no return variance. That is, interest rates are constant. Let f denote the fraction of days of the month for which the trading is invested in stock. For the TOM-plus-cash strategy, f is approximately 0.2, and for the FH-plus-cash strategy, f is approximately 0.4. Let the correlation between the returns of the fully invested S&P500 portfolio and either the small capitalization or large capitalization portfolio be ρ.[4] Then the correlation between the returns of the TOM- and FH-plus-cash trading strategies with the returns of S&P portfolio or the small capitalization portfolio equals $\sqrt{f}\rho$. The correlations of the fully invested portfolio with the S&P500 portfolio is $\rho = 1$, so we would expect the TOM-plus-cash and FH-plus-cash strategies to have correlations of approximately $\sqrt{0.2} = 0.45$ and $\sqrt{0.4} = 0.63$ with the large-capitalization

---

[4]To see this, note, for example, that the full-month return is the TOM return plus noise. Thus, the variance of the traded portfolio is f times the variance of the full-month portfolio. The covariance between the full-month return on the S&P or small capitalization portfolio and the traded portfolio is f times the product of ρ with the product of the variances of the full-month portfolios. The result follows from the standard formula for correlation.

Table 5. Returns in Percent During Various Monthly or Part-of-Month Positions for the S&P500 and Other Investment Strategies (February 1928 to June 1993).

-9 to +9	TOM (-1 to +4)	FH (-1 to +9)	LgCap	SmCap	CorpBnd	LTgvtBnd	ITgvtBnd	Cash	TOM(1 to +4) + 0.8Cash	FH(-1 to +9) + 0.6Cash
Feb1928–Jun1993	0.55	0.62	0.70	0.79	0.96	0.45	0.40	0.43	0.28	0.84	0.92
1928–1939	0.47	0.89	1.18	0.19	-0.21	0.51	0.36	0.36	0.10	0.96	1.26
1940–1949	0.32	0.51	0.38	0.73	1.57	0.22	0.27	0.15	0.03	0.54	0.41
1950–1959	1.05	0.86	0.75	1.47	1.30	0.08	-0.01	0.11	0.15	0.98	0.87
1960–1969	0.26	0.55	0.74	0.63	1.20	0.14	0.12	0.29	0.32	0.80	0.99
1970–1979	0.11	0.37	0.42	0.47	0.91	0.50	0.45	0.56	0.51	0.78	0.82
1980–1993	0.93	0.51	0.63	1.22	1.15	1.02	1.00	0.93	0.53	0.94	1.06

1928–1993	TOM (-1 to +4)	FH (-1 to +9)	LgCap	SmCap	CorpBnd	LTgvtBnd	ITgvtBnd	Cash	TOM(1 to +4) + 0.8Cash	FH(-1 to +9) +0.6Cash
January	1.28	1.03	1.03	1.56	6.38	0.83	0.18	0.40	0.28	1.26	1.25
February	-0.12	0.40	0.17	0.50	1.67	0.15	0.37	0.34	0.27	0.62	0.39
March	0.37	0.94	0.77	0.38	0.24	0.24	0.36	0.43	0.28	1.16	0.99
April	-0.04	0.25	0.57	1.02	0.56	0.12	0.30	0.38	0.26	0.46	0.78
May	0.68	0.83	0.82	0.17	-0.19	0.44	0.27	0.42	0.28	1.05	1.04
June	0.57	0.64	0.67	1.13	0.61	0.55	0.78	0.41	0.29	0.87	0.90
July	1.65	1.13	1.71	1.82	2.07	0.33	0.21	0.32	0.28	1.35	1.93
August	0.84	0.32	0.67	1.57	1.32	0.32	0.10	0.19	0.28	0.55	0.90
September	-0.99	0.49	-0.18	-1.32	-1.32	0.38	0.16	0.42	0.29	0.72	0.05
October	0.16	0.22	0.63	0.02	-1.33	0.81	0.90	0.73	0.31	0.47	0.88
November	0.67	0.55	0.94	1.05	0.85	0.59	0.73	0.66	0.28	0.78	1.17
December	1.50	0.61	0.56	1.61	0.70	0.63	0.41	0.44	0.29	0.84	0.80

Table 6. Summary Statistics for Various Monthly or Part of Month Strategies

Feb28–Jun93	−9 to +9	TOM (−1 to +4)	FH (−1 to +9)	LgCap	SmCap	CorpBnd	LTgvtBnd	ITgvtBnd	Cash	TOM(1 to +4) + 0.8Cash	FH(−1 to +9) + 0.6Cash
Monthly Average	0.55	0.62	0.70	0.79	0.96	0.45	0.40	0.43	0.28	0.84	0.92
Monthly Std. Dev.	4.99	2.54	3.65	5.80	8.71	2.01	2.20	1.27	0.24	2.54	3.64
Monthly Maximum	35.16	20.82	23.80	35.46	55.08	13.27	14.18	11.32	0.99	20.85	23.82
Monthly Minimum	−27.68	−13.98	−24.30	−35.28	−45.79	−9.32	−8.78	−6.62	0.00	−13.89	−24.21
Monthly Skewness	−0.08	0.13	−0.45	−0.42	−0.04	0.73	0.74	1.00	0.68	0.09	−0.50
Monthly Kurtosis	7.33	8.50	6.76	7.99	7.57	6.78	5.27	10.02	−0.45	8.45	6.83
Yearly Average	6.55	7.42	8.35	9.50	11.53	5.38	4.78	5.13	3.39	10.13	11.06
Yearly Std. Dev.	17.28	8.80	12.64	20.11	30.18	6.95	7.62	4.39	0.83	8.79	12.62

Correlations	+1 to +9, −9 to −1	TOM(+1 to +4,−1)	FH(+1 to +9,−1)	LgCap	SmCap	CorpBnd	LTgvtBnd	ITgvtBnd	Cash	TOM(+1 to +4, −1) + 0.8Cash	FH(+1 to +9, −1) + 0.6Cash
+1 to +9, −9 to −1	1.00										
TOM(+1 to +4,−1)	0.59	1.00									
FH(+1 to +9,−1)	0.79	0.69	1.00								
LgCap	0.79	0.69	0.67	1.00							
SmCap	0.52	0.46	0.57	0.86	1.00						
CorpBd	0.45	0.38	0.14	0.22	0.18	1.00					
LTgvtBd	0.14	0.09	0.06	0.17	0.11	0.84	1.00				
ITgvtBd	0.07	0.03	0.03	0.12	0.08	0.77	0.84	1.00			
Cash	0.07	0.01	−0.05	0.00	−0.02	0.03	0.05	0.13	1.00		
TOM(+1 to +4, −1) + 0.8Cash	−0.02	−0.06	0.69	0.46	0.38	0.09	0.03	0.02	0.02	1.00	
FH(+1 to +9, −1) + 0.6Cash	0.59	1.00	0.69	0.67	0.57	0.15	0.06	0.03	−0.01	0.69	1.00

stocks. This is consistent with the data, suggesting that the variance of the S&P500 portfolio is the same during the TOM or FH as it is throughout the rest of the month. Similarly, by Table 5, the correlation between the S&P500 and the small capitalization portfolio is $\rho = 0.86$ over the whole month. This implies correlations between the small capitalization portfolio and the TOM-plus-cash and FH-plus-cash strategies of 0.39 and 0.54, respectively. These are roughly consistent with the empirical evidence. However, the theoretical correlations are less than the empirical correlations for the TOM-plus-cash strategies, which suggests that the variance rate of the S&P500 portfolio increases slightly during the turn-of-the-month.

4 Postscript: Update July 1993–Dec. 1996

This three and a half year period had very high S&P returns averaging 1.4% per month, well above the mean of the previous 65 years. The average daily returns and t values for the hypothesis that the mean daily returns are greater than the overall mean appear in Table 7. These returns are displayed in Figure 4. Returns were positive in all but four days, during TOM, FH, −9 to +9 and trading days −4 to +3. The latter period reflects possible shifts in returns at the turn-of-the-month because of futures anticipation (see Hensel, Sick and Ziemba, 1994); few of the days have returns with t values close to significance. The −4 to +3 returns were double the average day, and positive on all seven days and above those in TOM and FH.

Table 8 investigates the returns from various strategies. Because of the high equity returns, the TOM and FH plus cash strategies, and even the −4 to +3 plus cash strategies, lagged the S&P500. This is shown in the monthly returns, Sharpe ratios and growth of $1 values. The latter are shown in Figure 5.

5 Concluding Remarks

There was a substantial turn-of-the-month effect in US large capitalization stock prices as measured by the S&P500 during the 65-year period 1928–1993. To avoid data snooping biases such as those discussed by Merton (1985), Black (1986, 1992) and Lo and MacKinlay (1990), we defined the turn, first-half and rest-of-the-month as Ariel (1987) did in his study of the cash markets from 1963–1981. The turn is trading days −1 to +4, the first-half days −1 to +9 and second-half (rest-of-the-month) days +10 to −2. The results show that the mean returns in the cash market were significantly positive in the turn and first-half-of-the-month and significantly negative in the rest-of-the-month. Returns during the turn and the first-half were significantly above average, and those higher returns were nearly seven times as large as a typical

Table 7. Average Daily Returns in the S&P500 Cash Market for Monthly Trading Days −9 to +9, July 1993–December 1996 and t-statistics for hypothesis that mean daily returns are greater than the overall mean for the same time period.

	−9	−8	−7	−6	−5	−4	−3	−2	−1	+1	+2	+3	+4	+5	+6	+7	+8	+9	−9 to 9	TOM	FH	−4 to +3
Average daily returns	−0.23	−0.09	0.06	0.06	0.10	0.18	0.08	0.00	0.01	0.08	0.15	0.09	−0.07	0.00	−0.01	0.02	0.04	0.19	0.04	0.05	0.05	0.03
t-statistic	−2.01	−1.92	0.17	0.21	.81	1.71	0.54	−0.35	−0.32	0.46	0.98	0.65	−0.90	−0.42	−0.38	−0.15	0.00	1.73	0.00	0.30	0.41	1.52

Table 8. Monthly Returns and Standard Deviations in Percent and Sharpe Ratios and Growth of $1 of Various Monthly or Part-of-Month Positions for the S&P500 and Other Investment Strategies, June 1993 to December 1996.

	−9 to +9	TOM (−1 to +4)	FH (−1 to +9)	LgCap	SmCap	Cash	TOM(1 to +4) + 0.8Cash	FH(−1 to +9) + 0.6Cash	−4 to +3 + 0.7Cash
Monthly Returns	0.64	0.26	0.49	1.40	1.47	0.38	0.56	0.72	0.85
Standard Deviation	2.47	1.71	2.18	2.67	3.58	0.09	1.74	2.20	1.44
Sharpe Ratio	0.11	−0.07	0.05	0.38	0.31	0.00	0.10	0.16	0.33
Growth of $1	1.29	1.11	1.22	1.77	1.80	1.17	1.26	1.34	1.42

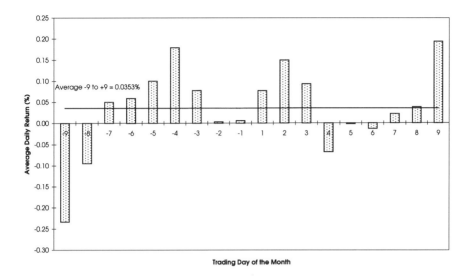

Figure 4. Average Daily Returns in the S&P500 Cash Market by Trading Day
(July 1993 to December 1996).

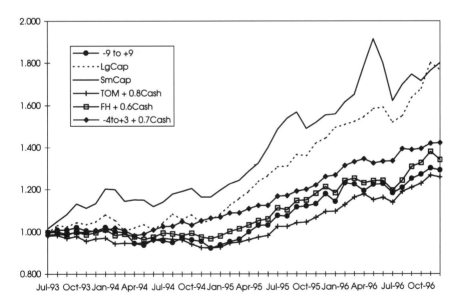

Figure 5. Growth of $1 Investment for Various Investment Strategies (July 1993
to December 1996)

day's returns. The returns on these days, particularly on days -1 and $+2$, were consistently high in every decade since the 1920s.

Monthly seasonality effects were strong, with significantly above-average returns on the S&P500 at the turn-of-the-month in January, March, May and July. January and July also had significantly higher returns than average during the first-half-of-the-month. January returns for the S&P500 were high but not as relatively high as for small capitalization stocks. April, June, July, August, November and December all had returns on the S&P500 that exceeded January's or were statistically indistinguishable. The rest-of-the-month had negative returns in all months except January, June, August and December. The popular press's widely discussed negative returns in September and particularly October are borne out by the data. But the losses were all in ROM. May and November also had large significant losses in ROM.

There were more positive return days during TOM and FH than during ROM. In addition, the mean return on the positive days was higher during the FH and lower on the negative days. This occurred for all the trading days as well as the 100 highest and lowest return days. The higher returns during TOM and ROM were earned through a shifting of the probability distribution of returns as well as the mean returns. The higher returns were not caused by a few outliers.

The cumulative wealth effects of investment during various time periods magnify the effects. The results indicate that the total return from the S&P500 over this 65-year period was mostly received during the turn-of-the-month. The strategy of being long the S&P500 during the TOM or the FH and T-bills otherwise had very high total returns (exceeded only by small stocks), and mean-variance dominated all other strategies considered, including the small stocks.

The postscript period July 1993 to December 1996 had very high annualized returns compared to the preceding 65 year period, namely 18.2% versus 10.0%, respectively. As a result there were positive returns on most days during TOM, the FH and during ROM. The TOM and FH plus cash strategies lagged the S&P500. As argued in our earlier 1994 futures paper, which uses 1982–1992 data, there may be a shifting of the ROM effect to trading days -4 to $+3$ because of future anticipations. The results in this period follow this behavior with positive returns as all seven trading days from -4 to $+3$ and average returns more than double the -9 to $+9$ average. However, with the small sample essentially all the results from the three and a half year postscript period are not statistically significant.

References

Admati, Anat R. and Paul Pfleiderer (1988) 'A Theory of Intra-Day Patterns: Volume and Price Variability', *Review of Financial Studies* **1** 3–40.

Admati, Anat R. and Paul Pfleiderer (1989) 'Divide and Conquer: A Theory of Intra-Day and Day of the Week Mean Effects', *Review of Financial Studies* **2** 189–223.

Ariel, Robert A. (1987) 'A Monthly Effect in Stock Returns', *Journal of Financial Economics* **18** 161–174.

Ariel, Robert A. (1990) 'High Stock Returns Before Holidays: Existence and Evidence on Possible Causes', *Journal of Finance* **45** 1611–1626.

Ball, R. and J. Bowers (1986) 'Daily Seasonals in Equity and Fixed-Interest Returns: Australian Evidence and Tests of Plausible Hypotheses', Working Paper. Australian Graduate School of Management.

Banz, Rolf (1981) 'The Relationship Between Return and Market Value of Common Stocks', *Journal of Financial Economics* **9** 3–18.

Berkowitz, Stephen A., Dennis E. Logue and Eugene A. Noser (1988) 'The Total Cost of Transactions on the NYSE', *Journal of Finance* **43** 97–112.

Black, Fisher (1986) 'Noise', *Journal of Finance* **41** 529–543.

Black, Fisher (1992) 'Estimating Expected Returns', Working Paper, Goldman Sachs, New York.

Booth, Donald G. and Donald B. Keim (1998) 'Is there still a January effect?', this volume, 169–178.

Brown, Phillip, Allen W. Kleidon and Terry A. Marsh (1983) 'New Evidence on the Nature of Size-Related Anomalies in Stock Prices', *Journal of Financial Economics* **12** 33–56.

Cadsby, Charles B. and Michael Ratner (1992) 'Turn-of-the-Month and Pre-Holiday Effects on Stock Returns: Some International Evidence', *Journal of Banking and Finance* **16** 497–509

Cadsby, Charles B. (1992) 'The CAPM and the Calendar: Empirical Anomalies and the Risk-Return Relationship', *Management Science* **38** 1543–1561.

Cervera, Alonso and Donald B. Keim (1998) 'The International Evidence on the Holiday Effect', this volume, 512–531.

Clark. R.A., J.J. McConnell and M. Singh (1992) 'Seasonalities in NYSE Bid-Ask Spreads and Stock Returns in January', *Journal of Finance* **47** 1999–2014

Cinar, E. Miné and Joseph D. Vu (1991) 'Seasonal Effects in the Value Line and S&P500 Cash and Futures Returns', *Review of Futures Markets* **10** 283–291.

Comolli, Luis R. and William T. Ziemba (1998) 'Japanese Security Market Regularities, 1990–1994', this volume, 458–491.

Cross, F. (1973) 'The Behavior of Stock Prices on Fridays and Mondays', *Financial Analysts Journal* (November-December) 67–69.

Fosback, Norman (1976) *Stock Market Logic*, The Institute for Economic Research, Fort Lauderdale, Florida.

French, Kenneth R. (1980) 'Stock Returns and the Weekend Effect', *Journal of Financial Economics* **8** 55–70.

Gibbons, Michael R. and Patrick Hess (1981) 'Day of the Week Effects in Assets Returns', *Journal of Business* **54** 579–596.

Gonzales, Michael (1995) 'Buying Stock? Consider Turn-of-the-Month Effect', *Wall Street Journal*, November 7, C1–C2.

Gultekin, Mustafa N. and N. Bulent Gultekin (1983) 'Stock Market Seasonality: International Evidence', *Journal of Financial Economics* **12** 469–482.

Harris, Lawrence (1986) 'A Transaction Data Study of Weekly and Intradaily Patterns in Stock Returns', *Journal of Financial Economics* **16** 99–117.

Haugen, Robert A. and Josef Lakonishok (1988) *The Incredible January Effect*, Dow Jones-Irwin, Homewood, Illinois, 231–250.

Hawawini, Gabriel (1991) 'Stock Market Anomalies and the Pricing of Equity on the Tokyo Stock Exchange', *In Japanese Financial Market Research*, William T. Ziemba, Warren Bailey and Yasushi Hamao (eds.), North-Holland, 231–250.

Hawawini, Gabriel and Donald B. Keim (1995) 'On the Predictability of Common Stock Returns: World-wide Evidence'. In *Finance*, R. Jarrow, V. Maksimovic and W. T. Ziemba (eds.), North-Holland, 497–544.

Hawawini, Gabriel and Donald B. Keim (1998) 'The Cross-Section of Common Stock Returns: A Synthesis of the Evidence and Explanations', this volume, 3–43.

Hensel, Chris R., Gordon A. Sick and William T. Ziemba (1994) 'The Turn-of-the-Month Effect in the US Stock Index Futures Markets (1982–1992)', *Review of Futures Markets* **13)(3)**, 827–856.

Hensel, Chris R. and William T. Ziemba (1996) 'Investment Results from Exploiting Turn-of-the-Month Effects', *Journal of Portfolio Management* Spring, 17–23.

Jacobs, Bruce I. and Kenneth N. Levy (1988) 'Calendar Anomalies: Abnormal Returns at Calendar Turning Points', *Financial Analysts Journal* **44** 28–39.

Jaffe, Jeffrey and Randolph Westerfield (1985) 'The Week-end Effect in Common Stock Returns: Day of the Week and Turn of the Year Effects', *Journal of Financial and Quantitative Analysis* **20** 261–272.

Jaffe, Jeffrey and Randolph Westerfield (1989) 'Is There a Monthly Effect in Stock Market Returns? Evidence From Foreign Countries', *Journal of Banking and Finance* **13** 237–244.

Kato, Kiyoshi, Sandra L. Schwartz and William T. Ziemba (1989) 'Day of the Week Effects in Japanese Stocks'. In *Japanese Capital Markets*, E.J. Elton and M.J. Gruber (eds.), Harper and Row, 249–281.

Keim, Donald B. (1983) 'Size Related Anomalies and Stock Return Seasonality: Further Empirical Evidence', *Journal of Financial Economics* **12** 3–32.

Keim, Donald B. (1989) 'Trading Patterns, Bid-Ask Spreads, and Estimated Security Returns: The Case of Common Stocks at Calendar Turning Points', *Journal of Financial Economics* **25** 75–98.

Keim, Donald B. and Michael Smirlock (1987) 'The Behavior of Intraday Stock Index Futures Prices', *Advances in Futures and Options Research* **2** 143–166.

Lakonishok, Josef and Edwin Maberly (1990) 'The Weekend Effect: Trading Patterns of Individual and Institutional Investors', *Journal of Finance* **45** 231–243.

Lakonishok, Josef and Seymour Smidt (1986) 'Volume for Winners and Losers: Taxation and Other Motives for Stock Trading', *Journal of Finance* **41** 951–974.

Lakonishok, Josef and Seymour Smidt (1988) 'Are Seasonal Anomalies Real? A Ninety-Year Perspective', *Review of Financial Studies* **1** 403–425.

Linn, Scott C. and Larry J. Lockwood (1988) 'Short Term Stock Price Patterns; NYSE, AMEX, OTC', *Journal of Portfolio Management* Winter, 30–34.

Lo, Andrew W. and A. Craig MacKinlay (1990) 'Data-Snooping Biases in Tests of Financial Asset Pricing Models', *Review of Financial Studies* **3** 431–467.

Merrill, Arthur A. (1966) 'Behavior of Prices on Wall Street', *Journal of Finance*.

Merton, Robert C. (1985) 'On the Current State of Stock Market Rationality Hypothesis'. In *Macroeconomics and Finance*, R. Dornbusch and S. Fischer (eds.), MIT Press.

Ogden, Joseph P. (1987) 'The End of the Month as a Preferred Habitat: A Test of Operational Efficiency in the Money Market', *Journal of Financial and Quantitative Analysis* **22** 329–344.

Ogden, Joseph P. (1990) 'Turn-of-the-Month Evaluations of Liquid Profits and Stock Returns: A Common Explanation for the Monthly and January Effects', *Journal of Finance* **45** 1259–1272.

Penman, Stephen H. (1987) 'The Distribution of Earnings News Over Time and Seasonalities in Aggregate Stock Returns', *Journal of Financial Economics* **18** 199–228.

Pettengill, Glenn N. (1989) 'Holiday Closings and Security Returns', *Journal of Financial Research* **12** 57–67.

Pettengill, Glenn N. and Bradford D. Jordan (1988) 'A Comprehensive Examination of Volume Effects and Seasonality in Daily Security Returns', *Journal of Financial Research* **11** 57–70.

Reinganum, Mark R. (1981) 'A Misspecification of Capital Asset Pricing: Empirical Anomalies Based on Earnings Yields and Market Values', *Journal of Financial Economics* **9** 19–46.

Ritter, Jay R. (1988) 'The Buying and Selling Behavior of Individual Investors at the Turn of the Year', *Journal of Finance* **43** 701–717.

Ritter, Jay R. and Nissan Chopra (1989) 'Portfolio Rebalancing and the Turn of the Year Effect', *Journal of Finance* **44** 149–166.

Rock, Kevin (1989) 'The Specialist's Order Book: A Possible Explanation for the Year-End Anomaly', Report, Harvard University.

Roll, Richard (1983) 'Vas Ist Dat: The Turn of the Year Effect and the Return Premia of Small Firms', *Journal of Portfolio Management* **9** (Winter) 18–28.

Rozeff, Michael S. and William R. Kinney, Jr. (1976) 'Capital Market Seasonality: The Case of Stock Returns', *Journal of Financial Economics* **3** 379–402.

Solnik, Bruno (1990) 'The Distribution of Daily Stock Returns and Settlement Procedures: The Paris Bourse', *Journal of Finance* **45** 1601–1609.

Ziemba, William T. (1989) 'Seasonality Effects in Japanese Futures Markets'. In *Research on Pacific Basin Security Markets*, Ghon H. Rhee and Rosita P. Chang (eds.), North-Holland, 379–407.

Ziemba, William T. (1991) 'Japanese Security Market Regularities: Monthly, Turn-of-the-Month and Year, Holiday and Golden Week Effects', *Japan and the World Economy* **3** 119–146.

Ziemba, William T. (1993, Comment on 'Why a Weekend Effect?' *Journal of Portfolio Management* (Winter), 93–99.

Ziemba, William T. (1994) 'World Wide Anomalies in Security Markets', *European Journal of Operational Research* **74** 198–229.

The Closed-End Fund Puzzle: a Literature Review

Carolina Minio-Paluello

1 Introduction

Closed-end funds, referred to in the UK as investment trusts, are characterized by one of the most puzzling anomalies in finance: the closed-end fund discount. Empirical research has shown that shares in US funds are issued at up to a 10% premium to net asset value (NAV). This premium represents the underwriting fees and start-up costs. Subsequently, within a matter of months, the shares trade at a discount[1], which persists and fluctuates according to a mean-reverting pattern.

Upon termination (liquidation or 'open-ending') of the fund, share price rises and discounts disappear (Brauer (1984), Brickley and Schallheim (1985)). Several theories of the pricing of closed-end funds attempt to make sense of the discount within the framework provided by the efficient market hypothesis but none can account for the aforementioned peculiarities. Agency costs, such as managerial performance and the present value of management fees, cannot account for the fluctuations in discount. Similarly, the assertion that NAVs are incorrectly calculated (as a result of letter stocks and tax liabilities relating to unrealized capital gains) cannot explain the price rise at open-ending. Additionally, some published studies have shown that discount based trading strategies can prove profitable when high discount shares are bought and low discount shares are sold (Thompson (1978), Pontiff (1995)). Thus the issue of closed-end fund shares may well represent a possible violation of the efficient market hypothesis.

A theory that encompasses some aspects of the puzzle is the limited rationality model of Lee, Shleifer and Thaler (1991). The irrationality of individual investors, the most prominent holders of closed-end fund shares in the US, places an additional risk on the assets they trade. The misperceptions of these investors translate into optimistic or pessimistic overreactions. Furthermore, there is evidence that discounts are correlated with the prices of other securities (such as small stocks), affected by the same investor sentiment. However, the limited-rationality theory is inconsistent with empirical evidence of the

---

[1]For a sample of 64 closed-end funds that went public from 1985–1987, Weiss (1989) shows that within 24 weeks of trading, US stock funds trade at a significant average discount of 10%. The cumulative index-adjusted return reaches -23.2% after 120 days.

UK closed-end fund market largely dominated by institutional ownership. This paper reviews most of the theories attempting to explain the existence and behaviour of the discount on closed-end funds.

2 Theoretical Principles and Performance of Closed-End Fund Shares

Several standard theories of the pricing of closed-end funds attempt to explain the discount within the framework provided by the efficient market hypothesis. The existence of agency costs, tax liabilities and illiquid assets have been put forward as possible explanations for the discount, but none of these, even when considered together, can account for the closed-end fund anomaly.

2.1 Agency costs

Agency costs – management fees and management performance – are a possible explanation of the closed-end fund discount if expenses are considered too high or if future portfolio management is expected to be below par. There are, however, several problems with this theory. Positive agency costs imply that funds should never be issued at a premium as long as no-load open-end funds[2] charging comparable fees exist. Furthermore, the agency costs theory neither accounts for the wide cross-sectional and periodic fluctuations in the discounts, nor for why some closed-end funds trade at a premium. An additional drawback of this theory is that it cannot explain why rational investors buy into closed-end funds that are issued at a premium, since they are aware of the likelihood of the fund subsequently trading at a discount.

2.1.1 Management fees

The agency costs theory claims that the discount on closed-end funds is a consequence of investors anticipating possible managerial dissipation and capitalizing future management fees. However, Malkiel (1977) finds no correlation between US discounts and management expenses[3]. Furthermore, the theory predicts that when long term interest rates fall, the present value of future

---

[2]Open-end funds, referred to in the UK as unit trusts, are characterized by the continual selling and redeeming of their units at or near NAV and this at the request of any unitholder. Therefore, these trusts have a variable capitalization. The open-end fund units not being traded implies that their managers are not priced by the market. In the US, some open-end companies, known as no-load funds, sell their shares by mail. Since no salesperson is involved, there is no sales commission (load) and the shares are sold at NAV.

[3]Turnover is also suggested as a possible explanation for the discount. While limited portfolio reallocations may be required to maintain diversification, some closed-end fund managers do a large amount of unnecessary portfolio shifting. Malkiel (1977) finds no correlation between discounts and turnover.

management fees should rise and discounts increase. Lee, Shleifer and Thaler (1991), however, show that changes in discounts are not correlated with unanticipated changes in the term structure.

Ammer (1990) shows, with both a simplified static and dynamic version of a new analytical arbitrage based framework, that expenses and yields account for a level of the discount that is typical of UK investment trusts. However, this framework fails to explain most of the time-series, cross-sectional, and international variations in discount. Kumar and Noronha (1992) re-examine the role of expenses by developing a present value model. Using a larger sample and a different specification of the expense variable from the one used by Malkiel (1977), they find that discounts are related to expenses[4].

2.1.2   Management performance

The agency cost theory predicts that if the fund pays more than the 'fair value' for managerial expertise, its shares should sell at a discount or earn an abnormally low return on investment, and vice versa. Thompson (1978) does not support this hypothesis. He observes that over long time periods many funds sell at a discount and simultaneously earn, on a before-tax basis, greater rates of return than can be justified by the two-parameter CAPM[5].

---

[4]Malkiel (1977) defines the expense variable using: $\text{EXP1}_i = \text{expenses}_i/\text{NAV}_i$. Kumar and Noronha (1992) propose the alternative specification: $\text{EXP2}_i = \text{expenses}_i/(\text{dividends}_i + \text{expenses}_i)$

[5]The two-factor CAPM is an asset pricing model which implies that, in equilibrium, the expected return on an asset is

$$E(r_j) = E(r_z) + \beta_j\{E(r_M) - E(r_z)\}$$

where

$E(r_j)$ = expected return on asset j

$E(r_z)$ = expected return on a security that has zero-beta with the market (which is riskless in the portfolio M)

$E(r_M)$ = expected return on the market portfolio

β_j = systematic risk, $\text{Cov}(r_j, r_M)/\text{Var}(r_M)$.

Thompson (1978) estimates the systematic risk, β_j, by regressing past asset returns on past market returns. The benchmark returns are taken to be those given by the restricted borrowing and lending CAPM and defined as

$$\hat{r}_{j,t} = d_{0,t} + d_{1,t}\beta_{j,t}$$

where d_0, d_1 are market determined parameters which describe ex-post the average relationship between systematic risk and realized returns. The values are taken from Fama and MacBeth (1973). The residuals are

$$\hat{e}_{j,t} = r_{j,t} - (d_{0,t} + d_{1,t}\beta_{j,t})$$

and the time series average of the e is the estimated α, the abnormal performance.

This we shall term the performance theory of closed-end fund discounts.

Agency costs do not seem to explain much of the cross-sectional variation in the discount. Malkiel (1977) finds no significant relationship between future fund performance and discount levels[6]. Roenfeldt and Tuttle (1973) find, in a very small sample, marginal support for a contemporaneous relationship. Assuming rational expectations, the performance theory predicts that large discounts reflect poor future NAV performance. However, Lee, Shleifer, and Thaler (1990) do not support the hypothesis. They find that assets of funds trading at large discounts tend to outperform those with smaller discounts. Furthermore, Pontiff (1995) shows that the ability to predict future discounts based on current discounts is almost entirely attributable to the ability to forecast stock returns as opposed to NAV returns. But Chay (1992), calculating the net managerial performance[7], shows that funds selling at discounts underperform funds selling at premiums. Thus, his findings tend to support the hypothesis that discounts reflect market expectations of fund manager future performance. However, Minio-Paluello (1998) shows that also using a definition of managerial performance that adjusts for the fund's effective asset exposure, discounts do not seem to reflect future managerial performance.

Deaves and Krinsky (1994) suggest a possible reconciliation of the conflicting findings on closed-end funds. They investigate the puzzling evidence that managerial contribution and discounts are not negatively related. They show that it is possible to explain some of the findings without abandoning market efficiency. The model has its foundations in the principle of rationality amongst investors and shows that it is not necessary for the relationship between managerial contribution (which is simply the difference between managerial performance and managerial fees) and discounts to be monotone. They argue that it is possible to imagine that, as managerial contribution declines, the discount narrows if investors attach an increased probability to open-ending, which by definition moves the price towards the NAV[8].

---

[6]Malkiel assumes that either short- or long-run past performance might serve as a useful proxy for expected future performance. In order to measure the performance of managers NAVs are used rather than market prices of the fund shares. The intercept a from the Ordinary Least Squares regression of excess returns is used as a measure of risk-adjusted performance:

$$r_{j,t} - r_F = \alpha + \beta(r_{m,t} - r_F).$$

[7]Chay defines 'Net managerial performance' as gross managerial performance less expenses charged by managers.

[8]The Japan Fund Inc., the first major US fund to invest in Japan, traded as a closed-end fund until August 14, 1987 when it was open ended. From the beginning of 1986 to the end of 1987, the shareholders earned 188% return, which includes the 19 October 1987 crash (Ziemba and Schwartz (1991)).

2.1.3 Agency problems

Agency theory focuses on the relationship between the principal (the shareholder of the trust) and the agent (the manager). Agency problems emerge when conflicts of interests between agents and principals affect the operations of the company.

Draper (1989) investigates the UK fund management market and finds that UK investment trusts are rarely managed 'in-house' but rather contract out their management to groups of specialists. These lucrative contracts act as an incentive to managers to impede shareholder asset realization, as a result of open-ending. Consequently, shareholders may be forced to bear substantial costs because of the difficulty of displacing management and liquidating their holdings. To some extent, US data supports this hypothesis[9]. Evidence from the UK is less satisfactory. The very low levels of liquidation and open-ending costs revealed by Draper's study suggest that far more trusts could profitably have been liquidated with beneficial effects for the shareholders of the trust. However, he also shows that investment trust managers receive considerably higher fees on open-ending and consequently it would be in their interest to open-end.

Additional evidence supporting agency problems is suggested by Barclay, Holderness and Pontiff (1993). They find that there is a stable and significant cross-sectional relationship between discounts and the concentration of ownership. The greater the managerial stock ownership in the closed-end fund, the larger the discounts to NAV. The average discount for funds with blockholders is 14%, whereas the average discount for funds without blockholders is only 4%. The idea is that blockholders receive private benefits that do not accrue to other shareholders and, therefore, tend to veto open-ending proposals to preserve these benefits.

2.2 Miscalculation of NAV

Explanations of the discount, consistent with market efficiency and frictionless capital markets, emphasize the notion that NAV may be overestimated. Tax liabilities relating to unrealized capital gains and restricted stocks are considered as possible causes of this miscalculation. However, this theory is neither consistent with the existence of premiums to NAV nor with the empirical regularity of price rises at 'open-ending'[10].

---

[9]Informal discussions with US closed-end fund managers revealed the existence of attempts to resist any open-ending pressure from the shareholders.

[10]The term 'open-ending' (referred to in the UK as 'unitization') refers to a set of techniques that force a closed-end fund's share price to NAV: converting the fund to an open-ended fund, merging the fund with an open-ended fund, and liquidating the fund's assets and distributing the proceeds to the shareholders.

2.2.1   Tax liabilities related to unrealized capital appreciation

In the US the regulations governing closed-end funds and the tax system are
such that the funds must distribute 90% of realized gains to qualify for exclusion from corporation tax. Therefore, shareholders receive two streams of
dividends: the income dividend and the capital gains dividend. If a closed-
end fund is characterized by large unrealized capital gains, this implies that
shareholders will be liable for capital gains taxes and the theory suggests
that this might explain part of the discount on closed-end funds[11]. However,
Malkiel (1977) finds, under fairly generous assumptions, that tax liabilities
can account for a discount of no more than 6%. Furthermore, the tax liabilities theory implies that, upon open-ending, NAV should decrease. Lee,
Shleifer and Thaler (1990) show the opposite – upon liquidation prices rise
to the NAV.

Fredman and Scott (1991) argue that discounts may partially be caused
by capital gains liabilities and suggest that if portfolio performance has been
good and capital gains liabilities are large, then discounts follow suit. However, Pontiff (1995) provides evidence that past NAV returns, net of market
return, are more strongly related to current discounts than simple NAV returns, which is inconsistent with our capital gains arguments since capital
gains are computed using unadjusted returns.

The theory of capital gains tax liabilities predicts that when stocks do
well, closed-end funds should accrue unrealized capital gains, and the discount should in general widen, if turnover rates on fund assets are constant.
However, Lee, Shleifer and Thaler (1991) find that the correlation between
returns on the market and changes in discounts is about zero.

2.2.2   Restricted shares

Bookkeeping procedures can potentially lead to a fund manager either under-
or overestimating the fund's NAV. Reporting restricted shares[12] (letter stocks)
at the same price as publicly traded common stocks can overstate the NAV.
Malkiel (1977) finds, over the period 1969-74, a significant relationship between the discount and the variable measuring the proportion of the portfolio
in restricted stock. However, Lee, Shleifer and Thaler (1990) show that restricted holdings cannot explain much of the closed-end fund puzzle as most

---

[11]However, UK regulations are different. The trusts are not allowed to distribute any
capital gains and the shareholder's revenue is the income dividend with a tax credit at-
tached. Unless they sell their shares, they will not be liable to any capital gains tax.

[12]Restricted or 'letter' stocks are like common stock except that they must be held
for investment and cannot be sold for a prespecified period of time. These stocks are
unregistered and highly illiquid which implies that the market price of these stocks is not
a fair indication of their liquidation value.

of the funds hold little letter stock and still sell at a discount. More importantly, if restricted stocks were overvalued, the NAV should drop down upon open-ending to the fund's price. Instead, the evidence shows that the share price in fact rises.

2.2.3 Liquidity

Seltzer (1990) argues that discounts can be accounted for by the mispricing of illiquid securities in the portfolio. He suggests that these securities are likely to be overvalued because of the difficulty to determine their fair market value. The importance of liquidity in terms of explaining stock returns is demonstrated by Datar, Naik and Radcliffe (1998). The liquidity argument is a possible explanation for the discounts. However, investors might be willing to pay higher management fees for holding the liquid shares of investment trusts that invest in less liquid securities, such as small capitalization stocks. Therefore, the importance of illiquid assets is difficult to measure.

2.3 Other possible explanations for the discount

2.3.1 Sales effort

Pratt (1966) and Malkiel and Firstenberg (1978) suggest that closed-end investment companies, when compared to open-end funds, sell at a discount primarily because of a lack of sales effort and public understanding. Malkiel (1977) and Anderson (1984) support this hypothesis and argue that brokers prefer to sell different securities from closed-end fund shares because of lower commissions on the former[13]. Furthermore, Anderson (1984) shows that after the reduction of commission fees in 1975, which in all probability reduced sales efforts, US discounts increased.

2.3.2 Holdings of foreign stocks

Some closed-end funds, referred to as country funds, invest exclusively in foreign securities. The existence of restrictions on direct foreign investment is suggested as a possible explanation for their trading, at certain times, at a premium. Bonser-Neal, Brauer, Neal and Wheatley (1990) test whether a relationship exists between announcements of changes in investment restrictions and changes in the price to NAV ratios. Using weekly data from 1981 to 1989, they find that four out of five country funds examined experience a significant decrease in the ratios following the announcement of a liberalization of investment restrictions. However, German and Spanish funds, which invest

---

[13]Malkiel argues that 'investors usually do not buy investment funds', but it is the public who is sold fund shares by brokers or other salesmen.

in completely open markets, have sold at large premiums in the 1980s, sometimes at levels above 100%. Furthermore, Malkiel (1977) found no significant relationship between discounts and holdings of foreign stocks[14].

3 Tax-Timing Option Values

A number of papers have attempted to justify the discount of investment trusts by the effect of the tax liabilities associated with unrealized capital gains. But, Tax Code legislation may have an additional implication related to the tax-timing argument. Constantinides (1983, 1984) investigates the influence of taxes on security returns. The ownership of stocks confers upon the investor a timing option as taxes on capital gains and losses are levied based on realization and not accrual. He argues that the optimal tax-trading strategy is to realize capital losses immediately and defer capital gains until a forced liquidation. Constantinides also shows that, compared with a suboptimal policy of never voluntarily realizing capital losses, the optimal tax-trading strategy would generate a tax-timing option value that constitutes 3 to 19% of the position in the stock. Constantinides' results can be relevant for solving the discount anomaly if we consider Merton's (1973) option pricing theorem which states that for all options, including tax-timing options, a portfolio of options is more flexible than an option on the corresponding portfolio.

3.1 Models

Brickley, Manaster and Schallheim (1991) and Kim (1994) suggest an explanation of the discount on closed-end funds based on the above mentioned papers. Both find evidence consistent with the hypothesis that the investment trust discount is partly driven by the fact that, by holding shares of a closed-end fund, investors lose valuable tax-trading opportunities associated with the idiosyncratic movements of the individual security prices in the portfolio. Brickley *et al.* (1991) show that, cross-sectionally, discounts are positively correlated with the average variance of the constituent assets in the fund and that in time-series the value of the discounts varies countercyclically[15].

---

[14]As an interesting aside, Chang, Eun, and Kolodny (1995) investigate the potential for closed-end country funds to provide international diversification. They show that funds exhibit significant exposure to the US market and behave more like US securities than do their underlying assets. This evidence supports Bailey and Lim's (1992) findings that these funds are poor substitutes for direct holdings of foreign securities. However, Chang *et al.* argue that closed-end country funds provide US investors with substantial diversification benefits. In particular, emerging markets country funds such as Brazil, Mexico, and Taiwan are shown to play a unique role in expanding the investment opportunity set.

[15]Discounts appear to increase during stock market declines and decrease during stock market increases. These findings are consistent with Schwert's (1989) results showing that the variances of stocks tend to increase during business downturns.

Kim (1994) uses the state-preference framework to develop a one-period 'horizon' model for discounts on closed-end funds. The model predicts that high correlations among assets will result in low discounts[16] and that funds with more volatile securities will show greater discounts than funds with less volatile securities in their portfolios[17].

3.2 Empirical evidence

Evidence from the US closed-end fund market tend to support the tax-timing option argument. The 1986 Tax Reform eliminated the favourable tax treatment on capital gains by making capital gains income taxable as ordinary income and made the long- and short-term capital gains tax rates equal. The end of the 'restart option'[18] implied that the tax law became less disadvantageous to closed-end funds. Consistent with this prediction, the number of closed-end funds increased dramatically after 1986[19]. However, De Long, Bradford and Shleifer (1992) show that the discount on US closed-end funds progressively widened between 1985 and 1990. The UK investment trust market does not provide much stronger evidence supporting the tax-timing argument. The 'restart option' was effectively[20] eliminated in 1985 and the average discount progressively narrowed from 20% in 1985 to 5% in 1990. However, the lack of statistical tests and the dramatic increase of the discount thereafter suggest that the tax-timing argument cannot explain the closed-end fund anomaly by itself.

The tax-timing option and investor sentiment explanations are not mutually exclusive and both factors may contribute to the discount. The investor

---

[16]In the extreme case where all assets are perfectly correlated, there are no discounts as the value of a portfolio of options on underlying securities is equal to the value of an option on the portfolio composed of the same underlying securities. For example, bond funds should have lower discounts than diversified funds since changes in the general level of interest rates affect the price of various bonds in a similar way. From 1979 to 1988, the average discount on diversified funds and bond funds was respectively 6.9 and 4.9%.

[17]Brickley *et al.* (1991) find a similar result. However, they do not consider correlations between assets.

[18]Constantinides (1984) investigates the 'restart option' based on the fact that short-term capital gains and losses used to be taxed at a higher rate. The restart strategy consist of recognizing all losses short term and all gains long term. The theory suggests that taxable investors should realize long term gains in high variance stocks in order to realize potential future losses short term.

[19]In March 1983, 45 closed-end funds operated in the US, with total assets of $6.9 billion. In 1991, the number had jumped to 270, with total assets of $60 billion. Kim (1994) tests the hypothesis that the average number of closed-end funds before 1986 is equal to the average number after 1986. The means are respectively 55.2 and 152.7. The hypothesis of equal means is rejected.

[20]In the UK, the differential taxation between long term and short term capital gains was effectively eliminated in 1985. In 1982 the tax rates were made equal, but it was not until 1985 that indexation of short term capital gains was introduced.

sentiment hypothesis does not appear to explain the cross-sectional findings that relate the discount to the variance of constituent assets within a given year. Additionally, however, the tax-timing hypothesis cannot account for funds selling at a premium[21]. However, the introduction of more complicated capital structures such as zero-dividend securities which generate a tax-free capital gains and income producing securities that can be held by UK investors free of income tax within a Personal Equity Plan, is particularly interesting because it allows the tax-timing argument to be consistent with both premiums and discounts to NAV.

3.3 Relevance of the tax-timing option

Seyhun and Skinner (1994) show the relevance of the tax-timing option in terms of the extent to which investors' transactions are motivated by Tax Code incentives. The results tend to indicate that investors' trades are consistent with simple tax-reduction strategies such as realizing losses short term and deferring gains, but not with the restart option suggested by Constantinides. Seyhun and Skinner estimate that in a given year only a small fraction (5 to 7%) of investors trade to reduce their tax payments and that the large majority (90%) buy and hold stocks. Overall, their results show that taxes are important to investors, but not to the extent that they continually adjust their portfolios to minimize the present value of their net tax payments.

4 Limited Rationality – Investor Sentiment

The failure of the standard theories to explain the anomalous behaviour of the discount on closed-end funds casts doubts on the rationality of the market. Furthermore, premiums seem to occur at times of great investor enthusiasm about stocks in general, such as the late 1920s or mid 1980s, or times of investor enthusiasm about particular securities, such as country stocks. In addition to high volatility, some country funds have also experienced violent fluctuations that cannot be related to the state of the foreign market. An anecdotal example is the behaviour of the discount in the US of the Germany fund. During the winter of 1989–90, after the fall of the Berlin Wall, the typical discount of 10% turned into a premium of 100%. This dramatic rise was attributed to speculations about investment opportunities in Germany. For a long time after this episode, the Germany fund traded at a premium. What made the behaviour most puzzling is that it seemed to have carried a cross-border contagion. Other country funds (Austria, First Iberia, Italy, Swiss,

---

[21]The tax option argument fails to account for the coexistence of open-end and closed-end funds, as the former are equally subject to the tax-timing option penalty, but sell at NAV.

Malaysia, Thai and Taiwan) experienced, over the same period, dramatic but short-lived increases (decreases) in the premium (discount).

Zweig (1973) is the first to have suggested that discounts might reflect the expectations of individual investors. Weiss (1989) supports[22] this conjecture and shows that individual investors, as opposed to institutional investors, own a larger stake in closed-end funds. Lee, Shleifer, and Thaler (1991) build on this evidence and speculate that the discount movements reflect the differential sentiment of individual investors since these investors hold and trade a preponderance of closed-end fund shares but are not as important an ownership group in the assets of the funds' portfolio.

4.1 Investor sentiment model

De Long, Shleifer, Summers, Waldmann (1990) and Lee, Shleifer, and Thaler (1991) have explored one possible explanation of the closed-end fund puzzle based on a model of noise traders, an argument clearly inconsistent with the efficient market hypothesis. The argument is that discounts on closed-end funds reflect changes in investor sentiment rather than changes in each fund management. They suggest the existence of two kinds of investors, the rational and the irrational (noise) traders. The former have unbiased expectations whereas the latter make systematic forecasting errors. Two important assumptions are made: rational investors are risk averse and have finite horizons. The intuition driving this model is that the fluctuations in the noise trader sentiment are unpredictable. This new source of risk deters rational investors from attempting aggressive arbitrage strategies. The evidence that funds, on average, sell at a discount does not rely on the average pessimism of noise traders, but stems completely from the risk aversion of rational investors that are willing to buy closed-end fund shares only if they are compensated for the noise trader risk, which means buying the fund at a discount.

Like fundamental risk, noise trader risk will be priced at equilibrium because the fluctuations in the same noise trader sentiment affect many assets and are correlated across noise traders, which implies that the risk that these fluctuations create cannot be diversified. As a result, assets subject to noise trader risk will earn a higher expected return than assets not subjected to such risk. Therefore, relative to their fundamental values, these assets will be underpriced. The returns earned by Thompson's (1978) portfolio strategies are earned at the expense of being exposed to the investor sentiment. Pontiff (1995) provides weak evidence supporting the hypothesis that funds with

---

[22]Weiss (1989) compares and contrasts the relative level of institutional ownership for closed-end fund IPOs and a control sample of 59 equity IPOs. Institutional ownership of equity is significantly higher for the control sample of IPOs than for closed-end funds, respectively 21.82% and 3.50% after the first quarter following the offering (the disparity of levels persists throughout the first three quarters).

larger discounts are exposed to greater investor sentiment risk.

The implications derived from the investor sentiment theory are supported by US empirical findings: discounts on various funds move together[23], new funds get started when seasoned funds sell at a premium or at a small discount[24], and discounts are correlated with prices of other securities, such as small stocks, that are affected by the same investor sentiment. Furthermore, Weiss Hanley and Seyhun's (1994) show that, because of the possibility of shifts in investor sentiment[25], short sellers earn significant abnormal returns only during the IPO period where price declines are fairly certain.

4.2 Discount and small firms effect

Evidence of individual investors specializing in holding small stocks justifies the Lee, Shleifer and Thaler (1991) conjecture that the investor clientele argument explains both the behaviour of the discount on closed-end funds and the 'small stock effect'[26]. Their results support the theory because discounts tend to narrow when small stocks perform well and vice versa. However, Chen, Kan and Miller (1993) challenge the sentiment theory by questioning the link between discounts and small firms. They argue that the time period tested and not institutional ownership matters for the results previously found. Chopra, Lee, Shleifer and Thaler (1993) respond by providing additional tests of the robustness of the relationship between the discounts and clientele ownership. Following these results, Chen, Kan and Miller point out that the covariation between closed-end fund discounts and size-based returns is no more than a trivial 4%[27]. Summing up, the key issues in this debate are the statistical and economic significance of the correlation between changes in the discounts on closed-end funds and excess returns on small stocks, as measured by incremental R-squared. Chopra, Lee, Shleifer and Thaler argue that,

---

[23]Lee, Shleifer, Thaler (1991) show that the average pairwise correlation of annual changes in discounts among domestic stock funds is 0.389 over the period 1965–1985.

[24]Levis and Thomas (1994) show that UK investment trust IPOs are subject to 'hot' issue periods, implying that they tend to occur when there is a marked narrowing in the discounts of seasoned trusts.

[25]The shift is most noticeable in the case of the Spain fund in September 1987 when many short sellers lost money because they were caught in a 'short squeeze' and were unable to obtain additional shares after being forced to close their positions. A 'short squeeze' occurs when shorted shares are called back from the short seller by the owner's broker.

[26]Small capitalization stocks are shown to earn returns different from large capitalization stocks.

[27]In their regression for the critical small-firm size group of decile 1, Chopra, Lee, Shleifer, and Thaler (1993) report a R-squared of 3.5% – smaller than the upper bound of 4% determined by Chen, Kan and Miller. However, they claim that the low R^2 is misleading because the dependent variable is effectively a change in the discount, which is purged of marker return, while the independent variable, decile 1, contains the market return. Correcting for this bias the R^2 rise to 6.9%.

with an R-squared of about 7 to 9%, the investor sentiment index explains small firm excess returns at least as well as 'fundamental' APT factors[28].

The investor sentiment and this contemporaneous correlation between closed-end fund discounts and small firm returns have been further investigated by Swaminathan (1996)[29]. He recognizes that any (mean-reverting) small investor sentiment should not only affect current stock prices but also forecast future stock returns[30]. The idea is that small investors' optimism pushes current stock prices above fundamentals, causing current returns to be high. Then as these temporary deviations are corrected, stock prices fall and revert to fundamentals. This in turn causes future returns to be low. The empirical tests produce reliable evidence that discounts forecast small firm returns better than they forecast large firm returns and that their forecasting power is independent of the movements tracked by other forecasting variables, such as the dividend yield, the default spread and the term spread[31].

Barber (1994) provides further evidence which supports the investor sentiment hypothesis. He documents several empirical facts which are consistent with the hypothesis that noise trading drives the time-series variation of the premium of Primes and Scores[32]. Barber (1994) shows that Primes and Scores are predominantly traded by individual investors, that the levels and changes of their premiums are correlated across trusts and, finally, that changes in the premium of Primes and Scores are correlated with the changes in the discount of closed-end funds as well as with small firm returns. However, during 1995-96 US small stocks did well while discounts remained high.

---

[28]Lee *et al.* show that the five 'fundamental' factors used by Chen, Roll and Ross together explain 12.5% of time series variation in small firm monthly excess returns. Adding the change in the value-weighted discount increases R^2 to 17.9%.

[29]Swaminathan (1994) develops a noisy rational expectations model of closed-end fund discounts with perfectly informed large investors and imperfectly informed small investors.

[30]This is strictly true only if the small investor sentiment is stationary and mean-reverting. However, it is hard to imagine a sentiment that is non-stationary and yet behaves, for instance, like a random walk.

[31]The default spread is a measure of the default risk premium in the economy and is defined as the difference between the yield on a portfolio of low grade corporate bonds and the yield on a portfolio of high grade corporate bonds. The term spread is a measure of the term risk premium in the economy and is defined as the difference between the yield on a portfolio of high grade bonds and the short term interest rates.

[32]Primes and Scores are derivative securities created by Americus Trust. Americus offered investors the opportunity to tender common stock of select DOW 30 companies in exchange for a Prime and a Score. A Prime entitles an investor to all cash distributions on the stock and a pre-set capita gains portion, fixed by a termination claim. A Score entitles an investor to the stock's capital appreciation above the specified termination claim. Despite the fact that a combined Prime and Score guarantees an investor the same cash flows as holding the underlying common stock, Prime and Scores trade, on average, at a 1% premium over the price of the underlying common stock.

Minio-Paluello

4.3   Discount and net redemption

Malkiel (1977) and Lee, Shleifer and Thaler (1991) provide additional evidence that lends further support to the view that changes in closed-end fund discounts reflect changes in individual investor sentiment. They show that discounts tend to increase with net redemption[33] from open-end funds, although the regression coefficient was not significant. This suggests that open-end fund investors, which are mainly individuals, are affected by the same investor sentiment. Furthermore, discounts do not seem to be highly correlated with measures of fundamental risk[34], which implies that this sentiment index is not a proxy for macroeconomics factors previously identified in the literature.

4.4   The marketing of closed-end funds

The fact that closed-end funds are characterized by a substantial price decline after the floatation cannot explain the behaviour of investors who buy in the first place. Weiss Hanley, Lee and Seguin (1995) attempt to explain this anomaly examining the aftermarket transactions for closed-end fund IPOs. They show that most closed-end fund trading in the first weeks is seller-initiated[35], that there is evidence of intense price stabilization and finally that a significantly higher proportion of the sells (buys) over the first 30 days are initiated by large (small) traders. These findings tend to suggest that closed-end fund IPOs are sold by professionals to less informed (irrational) investors. This 'marketing' hypothesis is consistent with US evidence of only small investors holding these shares in the long-run.

4.5   Investor sentiment in the international markets

Lee, Shleifer and Thaler (1991) speculate that the discount movements reflect individual investors' sentiment. However, using closed-end funds whose NAV are determined in the same market as the share prices (domestic funds), does not capture all the market-wide sentiment. Bodurtha, Kim and Lee (1992) investigate an extended form of the investor sentiment hypothesis using

---

[33]Net redemption is defined as the number of units redeemed in excess of the number of new units issued.

[34]Chen, Roll and Ross (1986) present a number of macroeconomics variables that affect stock returns in time-series regressions and expected returns in cross-section regressions. They interpret the variables to be risk factors. The variables include 'innovation' in: industrial production, risk premia on bonds, the term structure of interest rates, and expected inflation.

[35]Short-selling is not allowed during the first weeks of trading. Therefore, Weiss *et al.* relate this selling pressure with the presence of 'flippers' – investors who buy IPO shares during the pre-issue period and immediately resell them in the aftermarket.

closed-end country funds. They find that stock prices of country funds co-move with US market returns, but changes in their NAV do not. Bodurtha *et al.* also show that premiums on country funds tend to move together, but not with domestic closed-end funds premiums.

The investor sentiment hypothesis finds interesting support in the international market. Empirical evidence on the behaviour of country funds shows that discounts can be used to predict the future prices of the funds, but not of the underlying assets. This suggests that fund prices are driven by factors other than the assets' values. Moreover, this predictability seems linked to changes in world-wide and American stock returns, and not to changes in individual countries.

4.6 Limits to the investor sentiment hypothesis

The investor sentiment hypothesis provides an interesting explanation of the four-part discount puzzle. However, some papers do not seem to confirm it. Abraham, Elan and Marcus (1993) examine the sentiment hypothesis using the comparative performance of bond versus stock funds. They find that discounts on bond funds exhibit a systematic risk (the beta of the discount) almost as large as that on stock funds, despite the fact that bond funds hold assets whose values are far less subject to fluctuations of individual investors' sentiment. Furthermore, despite the similar level of systematic risk, bond funds on average do not trade at discounts.

Even more contradictory is Ammer's (1990) evidence. The limited rationality theory is grounded on the evidence that individual investors own the largest stake of US closed-end funds. However, despite the fact that British investment trusts go through periods of discount and premium similar in most respect to US funds[36], their clientele is, and has been over the last decade, almost entirely institutional (70 to 75% in 1990)[37]. Despite this inconsistency, Levis and Thomas (1995) show that UK closed-end fund IPOs are subject to 'hot' issue periods – IPOs tend to occur when there is a marked narrowing in the discounts of seasoned trusts – which is related to the implications of the De Long *et al.* (1990) noise trader model. UK closed-end fund IPOs disclose an aftermarket performance similar to that observed for industrial IPOs. When compared to US funds, the results show that the long-run underperformance is smaller. Larger institutional ownership in the UK is suggested as a possible explanation.

The noise trading hypothesis is met with considerable scepticism by financial economists and more research is undoubtedly required. However,

---

[36]The characteristics of UK fund discounts are very similar to those reported for American data, although discounts have been generally larger in the UK.

[37]The 1989–1990 Warburg Securities Investment Trust Manual reports that only 7 out of 102 funds have more than 50% of their shares registered to individuals.

no existing theories of asset pricing are able to explain the empirical results documented by Lee, Shleifer and Thaler (1991).

5 The Efficiency of the Closed-End Fund Market

Several US studies show that abnormal returns can be earned by following simple trading strategies based on the level of recent movements of the discount. However, there is still doubt that the existence and behaviour of discounts is evidence of persistent mispricing of assets resulting from market inefficiencies[38]. With respect to open-ending, the behaviour of closed-end funds is generally rational and the market for closed-end fund shares seems efficient. If these traits characterize closed-end funds in general, then the persistent discounts at which most funds sell must have a rational explanation in an efficient market.

Closed-end fund prices diverge consistently from NAV but there seems to be little opportunity for arbitrage. To some extent, it is possible to buy shares of a fund trading at a discount and sell short its portfolio. But the costs of only partial proceeds from a short sale and the risk of an increase in the discount result in a loss to an investor with a short horizon. A possible alternative to 'buy and hold' arbitrage for eliminating the discount is taking over the fund. However, this approach would tend to be resisted by fund managers and shareholders have an incentive not to tender unless the bid is at NAV (Grossman and Hart (1980)), which would leave no profit for the bidder.

5.1 Inefficiency

Several studies have tested various strategies for investing in closed-end funds using buy-and-hold rules, filter rules, rules based on 'open-ending' information, and rules exploiting discounts and expenses data. Often these rules appear to generate abnormal returns and, therefore, contradict the efficient market hypothesis. However, the contribution of mismeasured normal return benchmarks to such results is still unclear.

5.1.1 Abnormal returns

One of the most influential papers is Thompson (1978). He documents the empirical regularity that US closed-end funds trading at a discount (premium)

---

[38]In most cases tests of market inefficiency are tests of the joint hypothesis; market inefficiency (generally referring to semi-strong form of efficiency), and two-parameter equilibrium model of asset pricing.

accrue positive (negative) abnormal returns. Annual strategies based on this finding yield, over the period 1940 to 1975, abnormal risk-adjusted return of about 4%. These results suggest that high discounts tend to represent some sort of underpricing and that the market is inefficient as it does not recognize this[39]. Thompson is careful to emphasize that it is not possible to identify the extent to which his result reflect capital market inefficiency as opposed to deficiencies in the method of adjusting prices for risk[40] but confirmatory evidence from an earlier study by Zweig (1973) suggests a very real inefficiency[41]. Furthermore, Anderson (1986) reports evidence that abnormal returns could be earned by the use of filter rules involving the purchase of the shares of US closed-end funds of which the discount had widened and the sale of the shares of those companies of which the discounts had narrowed.

Evidence from UK investment trusts is both less comprehensive and less compelling. The investment performance study of Guy (1978) suggests that trusts do not outperform the market after suitable allowance for risk. However, Cheng, Copeland, O'Hanlon (1994) test for evidence that abnormal returns can be earned by holding investment trust shares in accordance with a simple discount based strategy. The results are not strong enough to infer that positive abnormal returns are available to round trip trading strategies, but very substantial returns would be available to a strategy of selling one's existing holding of low discount investment trusts and replacing it by high discount investment trusts[42]. The availability of such returns suggests the possibility of market inefficiency. The paper identifies positive and negative abnormal returns as being associated respectively with high and low discount investment trusts. Cheng *et al.* confirm Brickley, Manaster and Schallheim's (1991) finding that US discounts tend to narrow as the market rises and widen as it declines[43]. This empirical regularity might suggest the existence of some overreaction in the pricing of investment trust shares.

---

[39]Thompson tests relatively simple discount based trading rules ('premium', 'discount-equal weights', 'discount weighted') which he observes are unlikely to have used all of the information contained in the discounts, and finds that positive abnormal returns can be earned using theses rules.

[40]Thompson argues that the abnormal returns are likely to be due to inadequacies of the asset pricing model and not to market inefficiency since the data on closed-end fund discounts were widely available over the entire period and extensively discussed in the professional press.

[41]Zweig demonstrates the existence of sufficient forecasting properties in the investor's expectations.

[42]Returns are computed using monthly prices of investment trust shares, without adjusting for the bid-offer spread. However, the typical spread on UK investment trusts is lower than 2%.

[43]Brauer and Chang (1990) show that US closed-end funds display the typical size-related January effect, while their NAVs do not. The return of large fund share portfolios over the first four weeks of the year exceeds the average four-week rate of return over the rest of the year by 3.41%. Portfolios of small fund shares earn almost twice this differential, 6.67%.

5.1.2 Abnormal returns during the reorganization of the funds

A number of papers document the earning of positive abnormal returns by the holders of US closed-end funds which reorganize to allow shareholders to obtain the market value of the fund's assets. Brauer (1984) notes that a strategy of buying shares upon the announcement and holding them for three months would have been rewarded with abnormal returns. Brickley and Schallheim (1985) demonstrate this more rigorously and examine the possibility of exploiting the announcement of reorganizations, by investing on the last day of the month in which the announcement is made and holding until the fund is reorganized. The result of the strategy was a 15.3% average abnormal return although after adjustments for transaction costs and liquidity premiums during a liquidation, abnormal return are small. The existence of abnormal returns after the announcement of 'open-ending' is inconsistent with the joint hypothesis of market efficiency and that the market model is the correct return benchmark for funds undertaking reorganizations. It is possible for the market model not to capture the risks of whether or not the reorganization will take place, the costs of reorganization and the uncertainty about the true NAV. Consequently, the market prices these risks to yield higher expected returns than those given by the benchmark. The initial response to the announcement suggests that investors in closed-end funds are rational and awake to profit opportunities. Despite the evidence, neither study is able to determine whether the closed-end fund market is really inefficient or the market model is not the appropriate benchmark.

Brauer (1988) investigates further the returns earned during the restructuring of a closed-end fund, focusing on the valuation effects of the potential for 'open-ending'. The paper suggests a trading strategy based on the identification of likely candidates for open-ending which is based on both the size of the discount and the management expense ratio. Therefore, US closed-end funds' discounts contain information in the sense that they can be used in a model that predicts open-ending activity to construct portfolios that earn returns exceeding those predicted by the two-factor capital asset pricing model as well as those earned by a discount-only strategy investigated by Thompson (1978).

5.1.3 Predictability

Fraser and Power (1991) find significant autocorrelation in the excess returns[44] of UK investment trust fund shares, which tend to suggest the predictability

---

[44]Whiting (1984) finds evidence of the ability to trade profitably in UK investment trust shares using models based on discount levels. In forecasting and trading rules tests, the AR(1) model applied to the raw series gives the most consistent improvement upon the random walk model.

of these returns. Pontiff (1995) provides further evidence showing that US discounts have an economically strong ability to predict returns. However, this relationship remains puzzling and cannot be explained by factors that affect expected returns, such as multifactor risk exposure, bid-offer spreads, dividends and varying risk exposure. Pontiff attributes the correlation between fund discounts and future returns to discount-mean reversion, not to anticipated future performance. He finds that funds with 20% discount have expected 12-month returns that are 6% greater than non discounted funds.

5.1.4   Over-supply argument

An additional explanation for the inefficiencies in the UK investment trust market is the over-supply argument suggested by Arnaud (1983). He argues that the market is segmented and dominated by institutions that distort the prices[45]. The idea is that, consistent with the empirical evidence of the steady stream of sales by individual investors in particular over the period 1965 to 1985, institutions are prepared to buy the investment trust shares but can, to a large extent, influence the price in terms of the discount at which they are willing to buy.

5.1.5   Arbitrage

The level of inefficiency in the closed-end fund market can be measured by the presence of arbitrageurs. Weiss (1989) and Peavy (1990) document the gradual decline in the value of the shares over the first 100 trading days. If this gradual decline in value reflects market inefficiency, then investors possibly could take advantage of the mispricing by short selling the securities. Weiss Hanley and Seyhun (1994) investigates the profitability of arbitrage and provide evidence of closed-end fund pricing inefficiency during the initial public offering (IPO). They show that short sellers are interested in funds trading at a premium, but their profitability is limited to the first months of the IPO[46] – on average, short sellers earn significant abnormal return of 21% after 150 trading days.

---

[45]The ownership structure of the UK funds has changed considerably over the years. In 1964 individuals held almost 60% of the trusts but the stake was not larger than 25% in 1984. The increase in the proportion of institutional shareholders – 70 to 75% in 1990 – has, however, began to revert and individual investors are now more active, particularly since the introduction of the saving schemes in 1984.

[46]Short sellers are unable to take advantage immediately of the overpricing in closed-end fund IPOs since physical delivery of the securities does not occur until after the distribution is completed (at least 7 to 10 days later). Incentives to sell short investment trust IPOs are large, but Lee, Shleifer and Thaler (1991) report conversations with traders who say that they find it very difficult to execute short sales of closed-end fund IPOs.

Discounts persist because arbitrage[47] is costly and, therefore, not always profitable. Pontiff (1996) identifies factors that influence the arbitrage profitability and shows that cross-sectionally, the magnitude of discounts is most severe in closed-end funds that holds portfolios with high idiosyncratic risk, in funds with the lowest dividend yield[48], and in funds with the highest bid-offer spread. In time series, the average magnitude of discounts is shown to increase when interest rates increases – interest rates being an opportunity cost since arbitrageurs do not enjoy full access to short-sale proceeds.

5.2 Efficiency

The US evidence indicating the existence of profitable decision rules from investing in closed-end funds, together with a failure to explain the discount, has turned market inefficiency into the only possible explanation for the existence of the discount. Therefore, if the closed-end fund market is inefficient, then the share price is expected to respond slowly to new information. However, Brauer (1984) and Draper (1989) provide strong evidence supporting the efficient market hypothesis.

5.2.1 Price reaction to open-ending announcements

Brauer (1984) investigates the rationality and the informational efficiency of the market for closed-end fund shares in the US by examining 'open-ending' events that force the share price to its NAV. The paper reports that most of the positive abnormal returns associated with open-ending is exhausted by the end of the announcement month. This rapid market reaction suggests that the market for closed-end funds is generally efficient[49].

Draper (1989) investigates the UK investment trust industry and finds that share prices react rapidly to the announcement of takeovers, unitizations and liquidations. He shows that by the end of the announcement month all the information about the unitization has been incorporated in the price and no significant rise occurs thereafter[50]. The study shows that the adjustments to

---

[47]If investment trusts trade at a discount to NAV, an apparent arbitrage profit can be realized by shorting the fund's portfolio and holding the fund's shares. However, if the discount stays relatively constant over the investment horizon, the arbitrageurs make no profit. Furthermore, the exact portfolio composition is not known at every instant. In the UK, the Association of Investment Trust Companies (AITC) publishes monthly each fund's exposure, but not the detailed list of all shares held.

[48]Dividends are a benefit for the arbitrageur, since for funds with similar discounts, trading the fund with the higher dividend yield will result in larger expected returns.

[49]Brauer also reports that high discount companies and companies with low management expense ratios (expense ratios being a proxy for managerial resistance to open-ending) were more likely to open-end.

[50]No statistically significant increase in returns occur after details of the liquidation

the announcement of open-ending, as compared to Brauer's (1984) results for US closed-end funds, appear to be more concentrated and completed more rapidly.

Draper (1989) provides an additional test of market efficiency comparing investment trust market prices at the time of the announcement of open-ending (liquidating the fund's assets) with the value of the trust's asset at the actual open-ending, adjusting for transaction costs. On average, the difference between the announcement price and the NAV of the liquidated trust was very small. The difference was somewhat larger for the unitization announcement (transforming the investment trust into a unit trust), but attempts to derive a profitable decision rule were unsuccessful. These results show that investing in unitizing trusts from the day of the announcement to the end of the unitization yielded significant returns only if calculated using simply market prices. Draper demonstrates that considering, instead, asking prices adjusted for transaction costs this would reduce abnormal returns to levels not even approaching those found by Brickely and Schallheim (1985).

6 Conclusion

This literature review has attempted to show the breadth of empirical evidence on the closed-end fund puzzle, which has been presented over the last 20 years. Most emphasis has been placed on the behaviour of US funds but some research has centred on experience within the UK market. The research has presented evidence on the puzzle from a number of perspectives, suggesting a variety of plausible explanations. Several theories of the pricing of closed-end funds attempt to make sense of the discount within the framework provided by the efficient market hypothesis, but none can account for all parts of the puzzle – investors buy closed-end fund IPO shares despite evidence of a substantial price decline within the first few months, discounts vary cross-sectionally and fluctuate according to a mean-reverting pattern.

Agency costs (management performance and management fees) cannot account for the fluctuations in the discounts. The conjecture that NAVs are undervalued (as a result of letter stocks and tax liabilities related to unrealized capital gains) cannot explain the price rise at open-ending. The tax-timing option hypothesis attempts to explain the closed-end fund discount in terms of the loss of valuable tax-trading opportunities associated with idiosyncratic movements of the individual shares. The theory, however, is inconsistent with evidence that investors' trades are not motivated by tax incentives. Finally, the investor sentiment theory is extensively reviewed because of its ability to explain some parts of the puzzle.

---

or unitization become public. This suggests that investors were able to make accurate estimates of the value of the portfolio.

Several studies have shown that discount-based strategies can prove profitable. However, it is not clear that the existence and behaviour of discounts is evidence of persistent mispricing of assets resulting from market inefficiency. Closed-end fund prices react rapidly to the announcement of open-ending and there is no evidence of profitable arbitrage (except during the IPO months when the price decline is substantial).

References

Abraham, Abraham, Don Elan and Alan J. Marcus (1993) 'Does Sentiment Explain Closed-End Fund Discounts? Evidence from Bond Funds', *The Financial Review* **28** (4) 607–616.

AITC (1997), *Complete Guide to Investment Trusts*, The Association of Investment Trust Companies, London.

Ammer, John M. (1990) 'Expenses, yields, and excess returns: new evidence on closed-end fund discounts from the UK', *Group Discussion Paper Series* **108** (London School of Economics, London).

Anderson, Seth Copeland (1984), *Relationship between Value of an Investment Company's Shares and Value of the Underlying Net Assets*, PhD thesis (University of North Carolina at Chapel Hill).

Anderson, Seth C. (1986) 'Closed-End Funds versus Market Efficiency', *Journal of Portfolio Management* **13** 63–67.

Arkle, Jane (1997) 'Tax and Investments', Oracle Conference.

Arnaud, A.A. (1983), *Investment Trusts Explained*, Woodhead-Faulkner in co-operation with The Association of Investment Trust Companies.

Barber, Brad M. (1994) 'Noise trading and prime and score premiums', *Journal of Empirical Finance* **1** 251–278.

Barclay, Michael, Clifford Holderness and Jeffrey Pontiff (1993) 'Private benefits from block ownership and discounts on closed-end funds', *Journal of Financial Economics* **33** 263–291.

Bekaert, Geert and Michael Urias (1994) 'Diversification, Integration and Emerging Market Closed-End Funds', Working Paper, Stanford University.

Black, Fisher and Myron Scholes (1973) 'The Pricing of Options and Corporate Liabilities', *Journal of Political Economy* **81** 637–654.

Blume, M. and R.F. Stambaugh (1983) 'Biases in Computed Returns: An Application to the Size effect', *Journal of Financial Economics* **12** 387–404.

Bodie, Zvi, Alex Kane and Alan J. Marcus (1996), *Investments*, Third Edition, Irwin.

Bodurtha, J., E. Kim and C. Lee (1993) 'Closed-End Country Funds and US Market Sentiment', *Review of Financial Studies* **8** 879–918.

Bonser-Neal, Catherine, Greggory Brauer, Robert Neal and Simon Wheatley (1990) 'International Investment Restrictions and Closed-End Country Fund Prices', *Journal of Finance* **45** 523–547.

Boudreaux, Kenneth J. (1973) 'Discounts and Premiums on Closed-End Funds: a Study in Valuation', *Journal of Finance* **28** 515–522.

Brauer, Greggory A. (1984) 'Open-Ending Closed-End Funds', *Journal of Financial Economics* **13** 491–507.

Brauer, Greggory A. (1988) 'Closed-End Fund Shares' Abnormal Returns and the Information Content of Discounts and Premiums', *Journal of Finance* **43** 113–127.

Brauer, Greggory A. and Eric C. Chang (1990) 'Return Seasonality in Stocks and Their Underlying Assets: Tax-Loss Selling versus Information Explanations', *The Review of Financial Studies* **3** 255–280.

Brealey, Richard (1986) 'Comments to "Do we really know that financial markets are efficient?", by L.H. Summers. In *Recent Developments in Corporate Finance* edited by J. Edwards, J. Franks, C. Mayers, S. Schaefer, Cambridge University Press.

Brickley, James, Steven Manaster and James Schallheim (1991) 'The Tax-timing option and the discounts on Closed-End Investment Companies', *Journal of Business* **64** (3) 287–312.

Brickley, James A. and James S. Schallheim (1985) 'Lifting the Lid on Closed-End Investment Companies: a Case of Abnormal Returns', *Journal of Financial and Quantitative Analysis* **20** 107–117.

Brown, R.G (1962), *Smoothing, Forecasting and Prediction of Discrete Time Series*, Prentice-Hall.

Burch, T. Timothy and Kathleen Weiss Hanley (1996) 'When are Closed-End Funds Open? Rights Offers as a Response to Premiums', Working Paper, University of Maryland.

Burton, H. and D.C. Corner (1970) 'Capital gains tax liabilities and investment and unit trust', *Journal of Business and Finance* **2** 37–44.

Canina, Linda, Roni Michaely, Richard Thaler and Kent Womack (1998), 'Caveat Compounder: a Warning about Using the Daily CRSP Equal-Weighted Index to Compute Long-Run Excess Returns', *Journal of Finance* **53** 403–416.

Carhart, Mark M. (1996) 'On Persistence in Mutual Fund Performance', Working Paper, University of Southern California.

Chang, Eric, Cheol S. Eun and Kolodny Richard (1995) 'International Diversification through Closed-End Country Funds', *Journal of Banking & Finance* **19** 1237–1263.

Chay, Jong-Bom (1992), *The Pricing of Closed-End Funds: Discounts and Managerial Performance*, PhD thesis, State University of New York at Buffalo.

Chen, Nai-Fu, Raymond Kan and Merton H. Miller (1993) 'Are the Discounts on Closed-End Funds a Sentiment Index?', *Journal of Finance* **48** (2) 795–800.

Chen, Nai Fu, Richard Roll and Stephen Ross (1986) 'Economic forces and the stock market', *Journal of Business* **59** 383–403.

Cheng, A., L. Copeland and J. O'Hanlon (1994) 'Investment Trust Discounts and Abnormal Returns: UK Evidence', *Journal of Business Finance & Accounting* **21** (6) 813–831.

Chevalier, Judith and Glenn Ellison (1996) 'Are Some Mutual Fund Managers Better Than Others? Cross-sectional Patterns in Behavior and Performance', Working Paper, University of Chicago.

Chopra, Navin, Charles M.C. Lee, Andrei Shleifer and Richard H. Thaler (1993), 'Yes, Discounts on Closed-End Funds are a Sentiment Index', *Journal of Finance* **48** (2) 801–808.

Conrad, Jennifer and Gautam Kaul (1993) 'Long-Term Market Overreaction or Biases in Computed Returns?', *Journal of Finance* **48** 39–63.

Constantinides, George M. (1983) 'Capital Market Equilibrium with Personal Tax', *Econometrica* **51** 611–636.

Constantinides, George M. (1984) 'Optimal Stock Trading with Personal Taxes: Inplications for Prices and the Abnormal January Returns', *Journal of Financial Economics* **13** 65–90.

Corner, D.C. and J. Matatko (1982) 'Investment Trust Portfolio Performance. Measurement and Determinants', *The Investment Analyst* **61** 9–13.

Credit Lyonnais Laing (1997), *Investment Trust Yearbook*.

Cuthberston, Keith, Stephen Hall and Mark P. Taylor, *Applied Econometric Techniques*.

Datar, Vinay, Narayan Naik and Robert Radcliffe (1998) 'Cross-Section of Stock Returns Revisited: the Role of Liquidity and Size', *Journal of Financial Markets* **1** 203–219.

De Bondt, Werner F.M. and Richard Thaler (1985) 'Does the Stock Market Overreact?', *Journal of Finance* **40** 793–805.

De Long, J. Bradford and Andrei Shleifer (1992) 'Closed-end fund discounts', *Journal of Portfolio Management* **18** (2) 46–53.

De Long, J. Bradford, Andrei Shleifer, Lawrence H. Summers and Robert J. Waldmann (1990) 'Noise Trader Risk in Financial Markets', *Journal of Political Economy* **98** 703–738.

Deaves, Richard and Itzhak Krinsky (1994) 'A Possible Reconciliation of some of the Conflicting Findings on Closed-End Fund Discounts: a Note', *Journal of Business Finance & Accounting* **21** (7) 1047–1057.

Dickey, D.A. and W.A. Fuller (1979) 'Distribution of the Estimators for Autoregressive Time Series with a Unit Root', *Journal of the American Statistical Association* **74** 427–431.

Dickey, D.A and W.A Fuller (1981) 'Likelihood Ratio Statistics for Autoregressive Time Series with a Unit Root', *Econometrica* **49** (4) 1057–1072.

Dimson, Elroy and Paul Marsh (1983) 'The Stability of UK Measures and The Problem of Thin Trading', *Journal of Finance* **38** 753–783.

Dimson, Elroy and Paul Marsh (1986) 'Event Study Methodologies and the Size Effect. The Case of UK Press Recommendations', *Journal of Financial Economics* **17** 113–142.

Dimson, Elroy and Paul Marsh (1997) 'The Hoare Govett Smaller Companies Index 1955–1996', ABN-Amro, London.

Draper, Paul (1989), *The Investment Trust Industry in the UK: an Empirical Analysis*, Gower.

Elton, Edwin J., Martin J. Gruber and Christopher R. Blake (1996a) 'The Persistence of Risk-adjusted Mutual Fund Performance', *Journal of Business* **69** (2) 133–157.

Elton, Edwin J., Martin J. Gruber and Christopher R. Blake (1996b), 'Survivorship Bias and Mutual Fund Performance', *Review of Financial Studies* **9** (4), 1097–1120.

Elton, Edwin J., Martin J. Gruber, Sanjiv Das and Matt Hlavka (1993), 'Efficiency with Costly information: a Re-Interpretation of Evidence from Managed Portfolios', *Review of Financial Studies* **6** 1–21.

Emerson, Rebecca (1994) 'An Evaluation of UK Investment Trust as Contingent Claims: a Theoretical Investigation', Discussion Paper no. DP 23–94, London Business School, London.

Engle, R.F. and C.W.J. Granger (1987) 'Cointegration and Error Correction: Representation, Estimation and Testing', *Econometrica* **55** (2) 251–276.

Fama, Eugene F. and R. Kenneth French (1992) 'The Cross-Section of Expected Stock Returns', *Journal of Finance* **47** (2) 427–465.

Fama, Eugene F. and Kenneth R. French (1993) 'Common Risk Factors in the Returns on Bonds and Stocks', *Journal of Financial Economics* **33** 3–56.

Fama, Eugene F. and Kenneth R. French (1998) 'Value Versus Growth: the International Evidence', *Journal of Finance* (forthcoming).

Fama, Eugene F. and J. MacBeth (1973) 'Risk, Return and Equilibrium: Empirical tests', *Journal of Political Economy* **81** 607–636.

Finnerty, John D. and Dean Leistikow (1993) 'The Behavior of Equity and Debt Risk Premiums', *The Journal of Portfolio Management* **19**, Summer, 73–84.

Frankel, Jeffrey (1994), *The Internationalisation of Equity Markets*, University of Chicago Press.

Fraser, P and D.M. Power (1991) 'Predictability, Trends and Seasonalities: an Empirical Analysis of UK Investment Trust Portfolios, 1970–1990', University of Dundee.

Fredman, Albert and George Cole Scott (1991), *Investing in Closed-End Funds: Finding Value and Building Wealth*, New York Institute of Finance.

Fung, William and David A. Hsieh (1997) 'Empirical Characteristics of Dynamic Trading Strategies: the Case of Hedge Funds', *The Review of Financial Studies* **10** (2) 275–302.

Goetzmann, William N. and Roger G. Ibbotson (1994) 'Do Winners Repeat? Patterns in Mutual Fund Performance', *Journal of Portfolio Management* **20** 9–18.

Granger, C.W.J. (1981) 'Some Properties of Time Series Data and their Use in Econometric Model Specification', *Journal of Econometrics* **16** (1) 121–130.

Grinblatt, Mark and Sheridan Titman (1989) 'Mutual Fund Performance: an Analysis of Quarterly Portfolio Holdings', *Journal of Business* **62** (3) 393–416.

Grinblatt, Mark and Sheridan Titman (1992) 'The Persistence of Mutual Fund Performance', *Journal of Finance* **47** 1977–1984.

Grossman, Sanford J. and Oliver D. Hart (1980) 'Takeover bids, the Free-rider Problem, and the Theory of the Corporation', *Bell Journal of Economics and Management Science* **11**, Spring, 42–64.

Gruber, Martin J. (1996) 'Another Puzzle: The Growth in Actively Managed Mutual Funds', *Journal of Finance* **51** 783–810.

Guy, James R.F. (1978) 'The Performance of the British Investment Trust Industry', *Journal of Finance* **33** (2) 443–455.

Heisler, Jeffrey (1994) 'Recent Research in Behavioral Finance', *Financial Markets, Institutions and Instruments* **3** 78–105.

Hendricks, Darryll, Jayendu Patel and Richard Zeckhauser (1993) 'Hot Hands in Mutual Funds: Short-Run Persistence of Relative Performance, 1974–1988', *Journal of Finance* **48** (1) 93–130.

Hoare Govett ABN-AMRO (1997), *Le Patron Mange Ici*, Investment Trust Research.

Inland Revenue, (December 1997) 'The New Individual Savings Account', (Consultative Document).

Ibbotson, R.G. (1975) 'Price Performance of Common Stock New Issues', *Journal of Financial Economics* **2** 235–272.

Jegadeesh, Narasimhan (1990) 'Evidence of Predictable Behavior of Security Returns', *Journal of Finance* **45** 881–898.

Jegadeesh, Narasimhan and Sheridan Titman (1993) 'Returns to Buying Winners and Selling Losers: Inplications for Stock Market Efficiency', *Journal of Finance* **48** (1) 65–91.

Jensen, Michael (1978) 'Some Anomalous Evidence Regarding Market Efficiency', *Journal of Financial Economics* **6** (2/3) 95–101.

Kahn, Ronald N. and Andrew Rudd (1995) 'Does Historical Performance Predict Future Performance?', *Financial Analyst Journal* **51**, November–December, 43–52.

Keim, Donald B. (1983) 'Size-Related Anomalies and Stock Return Seasonality: Further Empirical Evidence', *Journal of Financial Economics* **12** 3–32.

Keim, Donald (1988) 'Stock Market Regularities: a Synthesis of the Evidence and Explanations'. In *Stock Market Anomalies*, Elroy Dimson (ed.), Cambridge University Press.

Kim, Chang-Soo (1994) 'Investor Tax-Trading Opportunities and Discounts on Closed-End Mutual Funds', *The Journal of Financial Research* **17**, Spring, 65–75.

Kothari, S.P. and Jerold B. Warner (1997) 'Measuring Long-Horizon Security Price Performance', *Journal of Financial Economics* **43** 301–339

Kumar, Raman and Gregory M. Noronha (1992) 'A re-examination of the relationship between closed-end fund discounts and expenses', *Journal of Financial Research* **15** (2) 139–147.

Lee, Charles M.C., Andrei Shleifer and Richard H. Thaler (1990) 'Anomalies: Closed-End Mutual Funds', *Journal of Economic Perspectives* **4** (4) 153–164.

Lee, Charles M.C., Andrei Shleifer and Richard H. Thaler (1991) 'Investor Sentiment and the Closed-End Fund Puzzle', *Journal of Finance* **46** 76–110.

Levis, Mario (1993) 'The Long-Run Performance of Initial Public Offerings: the UK Experience 1980–1988', *Financial Management*, Spring, 24–41.

Levis, Mario and Dylan C. Thomas (1995) 'Investment Trust IPOs: issuing behaviour and price performance. Evidence from the London Stock Exchange', *Journal of Banking and Finance* **19** 1437–1458.

Litzenberger, Robert H. and Howard B. Sosin (1977) 'The theory of recapitalizations and the evidence of dual purpose funds', *Journal of Finance* **32** (5) 1433–1455.

Malkiel, Burton and Paul Firstenberg (1978) 'A Winning Strategy for an Efficient Market', *Journal of Portfolio Management* **4** (4) 20–25.

Malkiel, Burton G. (1977) 'The Valuation of Closed-End Investment-Company Shares', *Journal of Finance* **32** 847–858.

Masey, Anthea (1988) *Investment Trusts* Financial Times Business Information.

Mendelson, M. (1977) 'Closed-End Fund Discount Revisited, Center for Study of Financial Institutions', University of Pennsylvania Law School.

Merton, R.C. (1973) 'Theory of rational option pricing', *Bell Journal of Economic and Management Science* **4** 141–183.

Miller, Edward M. (1977) 'Risk, Uncertainty, and Divergence of Opinion', *Journal of Finance* **3** (4) 1151–1168.

Minio-Paluello, Carolina (1996) 'The closed-end fund discounts as a predictor of managerial performance', Working Paper, London Business School, London.

Peavy, John W. (1990) 'Returns on Initial Public Offerings of Closed-End Funds', *Review of Financial Studies* **3** (4) 695–708.

Pontiff, Jeffrey (1995) 'Closed-End Fund Premia and Returns: Implications for financial market equilibrium', *Journal of Financial Economics* **37** 341–367.

Pontiff, Jeffrey (1996) 'Costly Arbitrage: Evidence from Closed-End Funds', *Quarterly Journal of Economics* **111**, November, 1135–1151.

Pontiff, Jeffrey (1997) 'Excess Volatility and Closed-End Funds', *American Economic Review* **87** (1) 155–169.

Pratt, Eugene J. (1966) 'Myths Associated with Closed-End Investment Company Discounts', *Financial Analysts Journal* **22** (4) 79–82.

Richards, R. Malcolm, Don R. Fraser and Groth John C. (1980) 'Winning Strategies for Closed-End Funds', *Journal of Portfolio Management* **6** (3) 50–55.

Ritter, Jay R. (1991) 'The Long-Run Performance of Initial Public Offerings', *Journal of Finance* **46** 3–27.

Rockinger, Michael (1995) 'Determinants of Capital Flows to Mutual Funds', Working Paper (HEC – School of Management).

Roenfeldt, Rodney and Donald L. Tuttle (1973) 'An Examination of the Discounts and Premiums of Closed-End Investment Companies', *Journal of Business Research* **1** 129–140.

Rozeff, Michael S. (1991) 'Closed-End Fund Discounts and Premiums.' In *Pacific-Basin Capital Markets Research*, Vol. II, S.G. Rhee and R.P. Chang (eds.), Elsevier, 503–522.

Risk Measurement Service, (January–March 1997), London Business School.

Schwert, G.W. (1989) 'Why does stock market volatility change over time?', *The Journal of Finance* **44** 1155–53.

Seltzer, David Fred (1990), *Closed-End Funds: Discounts, Premia and Performance*, PhD thesis, University of Arizona.

Seyhun, H. Nejat and Douglas J. Skinner (1994) 'How Do Taxes Affect Investors' Stock Market Realizations? Evidence from Tax-Return Panel Data', *Journal of Business* **67** (2) 231–262.

Sharpe, William F. (1966) 'Mutual Fund Performance', *Journal of Business* **39** 119–138.

Sharpe, William F. (1992) 'Asset allocation: Mangement Style and Performance Measurement', *Journal of Portfolio Management* **18** Winter, 7–19.

Sharpe, William F. (1996) 'The Styles and Performance of Large Seasoned US Mutual Funds, Working Paper, Stanford.

Sharpe, William F. and Howard B. Sosin (1974) 'Closed-end Investment Companies in the United States: Risk and Return', *Proceedings of European Finance Association*, 37–63.

Sias, Richard William (1992), *Closed-End Funds and Market Efficiency*, PhD thesis, The University of Texas at Austin.

Spiess, Katherine and John Affleck-Graves (1995) 'Underperformance in Long-Run Stock Returns Following Seasoned Equity Offerings', *Journal of Financial Economics* **38** 243–267.

Swaminathan, B. (1994) 'A Rational Model of Closed-End Fund Discounts', Working Paper, University of California, Los Angeles.

Swaminathan, B. (1996) 'Time-Varying Expected Small Firm Returns and Closed-End Fund Discounts', *Review of Financial Studies* **9** (3) 845–887.

Thompson, Rex (1978) 'The Information Content of Discounts and Premiums on Closed-End Fund Shares', *Journal of Financial Economics* **6** 151–186.

Walker, Matthew (1992), *An Agency Explanation of Closed-End Fund Discounts*, PhD thesis, Texas Tech University.

SG Warburg Securities (1988), *Investment Trusts Manual.*

SG Warburg Securities (1990), *Investment Trusts Manual.*

Weiss Hanley, Kathleen, Charles M.C. Lee and Paul J. Segui (1995) 'The Marketing of Closed-End Fund IPOs: Evidence from Transactions Data', Working Paper, University of Michigan.

Weiss Hanley, Kathleen and H. Nejat Seyhun (1994) 'The Profitability of Short Selling: Evidence From Closed-End Funds', Working Paper, University of Michigan.

Weiss, Kathleen (1989) 'The Post-Offering Price Performance of Closed-End Funds', *Financial Management* **18** (3) 57–67.

Whitting, A. (1984), *The Analysis of the Share Discounts of UK Investments Trust Companies: a Time Series Approach*, PhD thesis, University of Lancaster.

Woodbury, Denise (1992), *Compensation of Investment Advisors in Agency Framework*, PhD thesis, University of Utah.

Woodward, R.S. (1983) 'The Performance of UK Trusts as Internationally Diversified Portfolios over the Period 1968 to 1977', *Journal of Banking and Finance* **7** 417–426.

Wylie, Sam (1997), *Essays on the Decision of Investors and Fund Managers*, Doctoral Dissertation, London Business School.

Ziemba, William T. and Sandra L. Schwartz (1991), *Invest Japan*, Probus Publishing Company.

Zweig, Martin E. (1973) 'An Investor Expectations Stock Price Predictive Model Using Closed-End Fund Premiums', *Journal of Finance* **28** 67–78.

Stock Splits and Stock Returns for Nasdaq Stocks: The Effects of Investor Trading and Bid–Ask Spreads on Ex-Date Returns

Mark Grinblatt and Donald B. Keim

Abstract

Previous research has suggested that the abnormally large ex-date returns of stock splits and stock dividends may be caused by market microstructure effects. Specifically, if stocks have a tendency to trade at the ask price after the ex-date, measured returns may be upward-biased estimates of intrinsic returns, that is returns arising from revisions in the market's estimate of a firm's prospects. This paper explores returns around the ex-dates of stock splits and stock dividends, using a large sample of Nasdaq stocks. Because bid-ask effects may also be correlated with intrinsic returns, it is particularly difficult to separate out the components of the returns around the ex-date that are due to bid-ask effects and those that are due to more fundamental factors. By decomposing close-to-close returns into a pure market microstructure component and a pure intrinsic revaluation component, we are able to confirm that not all of the observed ex-date return is due to measured return biases driven by market microstructure.

1 Introduction

Stock splits and stock dividends should be purely cosmetic events. Unlike most cash dividend and capital structure changes, stock splits and stock dividends do not directly affect the after-tax cash flows of the firm, nor do they affect a shareholder's personal taxes or proportionate ownership of the firm. Also, in a market where transaction costs are set by competition, rather than by regulation, there are no apparent costs associated with these events.

This argument notwithstanding, empirical research over the last 25 years documents a large number of ways in which stock splits and stock dividends affect shareholder returns. Specifically, a strong positive announcement return effect[1] and an almost equally strong positive ex-date[2] return character-

---

[1]See Fama, Fisher, Jensen, and Roll (1969), Bar-Yosef and Brown (1977), Charest (1978), Foster and Vickrey (1978), Woolridge (1983a), and Grinblatt, Masulis, and Titman (1984). The study by GMT finds an average increase in stock prices of about 3.3% on the announcement of a split for a sample of stocks selected to avoid contamination by other contemporaneous firm-specific announcements.

[2]See Chottiner and Young (1971), Charest (1978), Foster and Vickrey (1978), Woolridge

izes these events. Other split and stock dividend effects have also been documented, including abnormal returns in the days around the announcement or ex-date, a change in the stock's risk,[3] an increase in the percentage bid-ask spread,[4] an increase in cash dividends, and an increase in reported earnings.[5]

By far the most puzzling of these effects is the large return on the ex-date, on the order of 2% according to Grinblatt, Masulis, and Titman (1984), henceforth GMT. Since the ex-date is known at the announcement date of the stock split or stock dividend, the return achievable by purchasing shares just prior to the ex-date appears anomalous. By contrast, announcement effects, being surprises, cannot be used by investors to earn economic profits.[6]

Beginning with GMT, researchers have hypothesized that the ex-date effect may be largely a bid-ask phenomenon. In other words, computation of stock returns using closing transaction prices may impart 'biases' into close-to-close returns because of the existence of systematic patterns in the observed relative frequencies of bid and ask transaction prices. Keim and Stambaugh (1984) examine such a bias as a possible explanation of the 'weekend effect', and Keim (1989) investigates the bias as partial explanation of abnormal stock returns at the beginning of January. Conrad and Conroy (1994) entertain the notion of a bid-ask bias explanation for the ex-date effect, building on the evidence in Conroy and Flood (1988). Conroy and Flood (1988) find that on the ex-date of a split, the number of reported transactions for a representative security increases unexpectedly by about 40%; at the same time, there is a slight decline in reported volume. These authors also report an increase in the number of shareholders following the ex-date of a split.

Conrad and Conroy (1994) argue that this evidence suggests that there are more buy orders (and hence observed prices at the ask) at the smaller, post-split lot sizes than there are sell orders (generating observed prices at the bid) in the larger lots derived from pre-split holdings. To demonstrate how this buying and selling behavior might translate into the observed return patterns, Conrad and Conroy give an example of a 2 for 1 split where the

(1983b), Eades, Hess, and Kim (1984), and Grinblatt, Masulis, and Titman (1984).

[3]Ohlsen and Penman (1985) document a permanent increase in the stock's return variance for splits, while Dravid (1987) identifies an opposite effect for stock dividends. Brennan and Copeland (1988a) identify a temporary increase in beta on the ex-date.

[4]See Copeland (1979).

[5]See McNichols and Dravid (1989) and Lakonishok and Lev (1987) for earnings effects, and Fama, Fisher, Jensen, and Roll (1969) and Grinblatt, Masulis, and Titman (1984) for dividend effects.

[6]Some research in the literature has suggested that there is real information content to the announcement of a split or stock dividend. Signaling models have been proposed by GMT, McNichols and Dravid (1990), and Brennan and Copeland (1988b) that account for this information content. These models are based on (exogenously assumed) differences in the cost of splitting for firm's of high and low quality. Irrespective of one's opinions of the merits of these models, one must admit that no similar explanation can account for the ex-date phenomenon.

closing price on the day before the ex-date is $50 and the bid and the ask prices on the ex-date are $24.75 and $25.25 respectively. If sellers were selling lots of 200 shares on the ex-date at the bid and twice as many buyers were buying lots of 100 shares on the ex-date at the ask, supply and demand would balance but we would observe twice as many post-split trades at the ask than at the bid. The net effect is an expected closing price on the ex-date of $25.085(24.75 × 0.33 + 25.25 × 0.67) which would result in a 0.34% return even though there is no change in supply and demand.

Conrad and Conroy examine a sample of 232 stock splits for NYSE and AMEX stocks during the period from October 1980 to October 1982 and find: (1) a significant increase in the number of trades on the ex-date; and (2) that on average the closing price on the ex-date is above the midpoint of the bid-ask spread, which they interpret as implying that buyers outnumber sellers for that day. They also look at Nasdaq stocks from 1983–1990. They conclude that it is the market maker's accommodation of the change in the nature of trades that increases the likelihood of observing a trade at the ask on the ex-date.

The bias described above makes it difficult to draw conclusions about 'true' average returns when orders to buy (at the ask price) and orders to sell (at the bid price) that are initiated at similar points in time are executed at substantially different prices.[7] This is possible in illiquid markets where securities trade with large bid–ask spreads. Clearly, some type of adjustment to measured returns is needed when asymmetries in the order flow result in a small buyer-initiated transaction being more likely to be observed than a large seller initiated transaction. Conrad and Conroy (1994) propose an order flow effect that measures the deviation of the bid-to-bid return from the close-to-close return (see Keim (1989)). In the context used here, however, the measure overstates the order flow effect because the bid-to-bid return is a downward-biased estimate of the intrinsic mean return when percentage bid-ask spreads are widening, as they are around stock split events. This paper corrects for this overstatement under the assumption that returns computed using the average of the bid and ask prices are reasonable proxies for 'true' returns, and decomposes this order flow effect into a spread widening component and a component that is due to the change in the tendency of the closing transaction price to occur at the bid or the ask.

Our insights into these issues are based on an examination of all stock splits and stock dividends for Nasdaq National Market System stocks for the years 1983 to 1994 for which we measure the impact of the bid-ask effect on returns around stock dividend and split ex-dates. We find that a portion of the ab-

---

[7]This would be like inferring that the price of automobiles went up after observing that the wholesale price at which the dealer purchased the automobile from the manufacturer was less than the retail price at which the automobile was sold to a customer the next day.

normal ex-date returns associated with stock dividends and splits is due to a disproportionate number of closing trade prices around the ex-date occurring at the ask and a widening of the percentage spread. While systematic trading patterns by investors, as discussed in Keim and Stambaugh (1984) and Keim (1989) appear to generate a portion of the abnormally positive returns at and around the ex-date, the effect is most pronounced for stocks that have low-price shares. For high-price stocks, these microstructure effects are less important. In the end, we find evidence of an upward revision in the value of splitting firms around the ex-date after controlling for microstructure effects. That is, return measures that are little influenced by market microstructure effects are significantly larger on ex-dates than on dates surrounding the ex-date.

The paper is organized as follows. Section 2 describes the data and methods employed in measuring the impact of systematic trading patterns. Sections 3 and 4 present the results, measuring the impact of the bias on computed transaction price returns around the ex-dates of stock splits and stock dividends. In section 5 we estimate a regression model designed to control for the microstructure effects. The paper concludes with a brief summary and suggestions for future research.

2 Data and Methods

We construct several series of returns in event time using the daily Nasdaq file from the Center for Research in Security Prices (CRSP). This file provides end-of-day bid and ask prices as well as closing transaction prices and daily returns (assuming reinvestment of cash dividends) from 1/1/83 to 12/31/94 for all Nasdaq stocks. The database lists the announcement days and the ex-dates for stock splits and stock dividends. For some of the analysis below, we distinguish between stock splits and stock dividends. Following GMT and generally accepted accounting principles, we distinguish a stock dividend from a stock split by the split factor (the percent increase in the number of shares). Split factors at or below 25% indicate a stock dividend, those above 25% a stock split.

Several screens are used to eliminate 'abnormal' stock splits or stock dividends. To qualify for inclusion in our sample, a stock had to satisfy the following criteria:

(1) To avoid thinly traded stocks, the stock had to be traded by at least 3 market makers on the ex-date.

(2) To avoid the unusual price patterns exhibited by IPO's in their first year,[8] the stock had to have been listed on the NMS for more than 250 trading days.

---

[8]See Ritter (1991) for documentation of these patterns.

(3) To eliminate small stock dividends that are periodic and therefore predictable, and to permit comparisons with the GMT sample, only stock dividends with split factors of at least 10% were included in the sample.

(4) To avoid confounding of the announcement and ex-date effects, the sample only contains events where the announcement date precedes the ex-date by at least three trading days.

In total, 1335 ex-date events passed all of the above screens.

We use several definitions of returns in our analysis. The close-to-close return is based on closing transaction prices on consecutive days. The returns provided on the CRSP files are close-to-close-returns and are, therefore, the return measure used in most empirical analyses. Close-to-close returns are affected by a combination of (1) changes in the bid-ask spread and (2) changes in the tendency of the closing price to occur at the bid or the ask. The L-ratio (LR) for an individual stock on a particular day, developed in Keim (1989), is the weight on the ask price that satisfies

Transaction Price = [LR × Ask Price] + [(1 - LR) × Bid Price]

Hence, if the closing transaction price on day t is at the bid price, the LR for day t is zero. The LR is one when the closing transaction price is at the ask price. Increases in the LR tend to generate higher close-to-close returns, *ceteris paribus*. Keim (1989) assumed constant percentage bid–ask spreads, an assumption that characterized the turn-of the-year period used in his analysis. If spreads are not constant, however, the within–spread movement captured by the LR will be confounded by a separate spread change effect. For example, if the LR is unchanged, then an LR above 0.5 tends to be associated with larger close-to-close returns if the percentage bid-ask spread widens. The percentage bid–ask spread is defined for day t as the difference between the day t closing ask price and the closing bid price, divided by half their sum (the day t bid-ask average or mid-spread price).

As alternatives to the close-to-close return, we compute three additional measures. The mid-spread return is computed using the average of the closing bid and ask prices on consecutive days. The ask return and the bid return are the returns based on consecutive-day closing ask and bid prices, respectively. If, on average, the bid-ask spread is widening between days, there is a tendency for the ask return to exceed the bid return.

3   The Ex-Date Effect: Findings for Nasdaq Stocks

Table 1 shows returns in event time for ex-date events, using the four above definitions of returns. For the return of an individual security to be included

Table 1. Average Daily Percentage Returns (Standard Errors), Percentage
Spreads, and *L*-Ratios on the Ex-Date and Surrounding Ten Trading Days
for Nasdaq Stock Splits and Stock Dividends for the Period 1983–1994
(Entire Sample of 1335 Events)

Event Day	Close-to-Close Return	Mid-Spread Return	Ask Return	Bid Return	Percent Spread	*L*-Ratio
−5	0.121 (0.072)	0.119 (0.059)	0.114 (0.059)	0.125 (0.060)	2.19	0.54
−4	0.022 (0.071)	0.076 (0.059)	0.086 (0.059)	0.066 (0.061)	2.21	0.51
−3	0.270 (0.075)	0.148 (0.066)	0.153 (0.066)	0.145 (0.068)	2.22	0.55
−2	0.371 (0.072)	0.308 (0.061)	0.301 (0.061)	0.318 (0.062)	2.20	0.57
−1	0.419 (0.073)	0.437 (0.062)	0.424 (0.062)	0.452 (0.063)	2.17	0.59
0	1.501 (0.111)	1.320 (0.097)	1.756 (0.098)	0.875 (0.100)	3.04	0.60
1	0.460 (0.099)	0.398 (0.082)	0.342 (0.083)	0.460 (0.085)	2.93	0.63
2	0.446 (0.094)	0.460 (0.081)	0.404 (0.080)	0.522 (0.085)	2.82	0.61
3	0.215 (0.086)	0.205 (0.072)	0.208 (0.073)	0.205 (0.075)	2.82	0.61
4	0.127 (0.084)	0.159 (0.071)	0.150 (0.072)	0.720 (0.073)	2.81	0.59
5	−0.000 (0.087)	−0.034 (0.069)	−0.021 (0.069)	−0.046 (0.072)	2.83	0.59

in the average return for a particular day, it has to have a closing transaction
price on both that day and the prior trading day. In addition to averaging
returns over events for the ten days surrounding the event and the event day
itself, this table reports each day's percentage bid–ask spread and *L*-ratio.

Consistent with previous research, Table 1 shows significantly larger returns on the ex-dates than on the surrounding days. For example, the average close-to-close return on the ex-date is 1.5%, which is more than three
times larger than the close-to-close returns on the surrounding days. The
LR, in the rightmost column, is virtually the same on both the day before
the ex-date and the ex-date. Because the average LR is virtually unchanged
between event days −1 and 0, any microstructure effect on returns is driven

primarily by the widening of the percentage bid-ask spread from 2.17 to 3.04 (in conjunction with the approximately 60% tendency of the closing transaction prices to be at the ask price on both of these two days). Consistent with this, the ask return is the largest of the four ex-date return measures, and at 1.76% is four to five times larger than the returns surrounding the ex-date. In contrast, the average bid return is 0.86%, about half the size of the ask return. The avergae mid-spread return, a return measure that is reasonably clean of microstructure-induced biases, is 18 basis points lower than the average close-to-close return on the ex-date – an indication of a non-trivial microstructure component in the close-to-close return. The fact that the mid-spread return on the ex-date is about three times the magnitude of the returns on the surrounding dates indicates that a substantial portion of the average ex-date effect is not microstructure-related. Nevertheless, the evidence in Table 1 points also to a sizeable microstructure component to the ex-date return: the significant difference between the average close-to-close and mid-spread returns $(T = 3.36)$ is a reflection of the widening spread on the ex-date.

While Table 1 indicates that closing transaction prices are, on average, close to the ask price in the ten days surrounding the event and on the event day itself, there appears to be a gradual increase in the tendency of the closing price to be at the ask price in the four days prior to the event. This partly explains the abnormally large close-to-close return on days -3 to -1.

If the closing transaction price is indicative of patterns throughout the day, then market makers tend to eliminate inventory in the days surrounding the ex-date, although this inventory decumulation is slightly more pronounced on and after the ex-date than in the 3 days prior to it. Although we do not report the data in this paper, a similar pattern for L-ratios occurs around the announcement date. It is not possible to determine whether this is due to investors desiring to acquire the stock as a result of the split announcement and ex-date or whether it is due to the desire of market makers to reduce inventory.

4 Partitioning the Sample by Split Factor and Share Price

To better characterize the ex-date returns, we partition the sample by stock split versus stock dividend.[9] We also partition the sample into four quartiles based on share price, with quartile breaks at $14.25, $19.00 and $24.875. Since the bid–ask spread is generally a larger fraction of the value of a low priced share, any bias created by the bid–ask spread can be expected to have

---

[9]The split factor is measured as the number of new shares created divided by the number of old shares. Hence, a 2-for-1 split has a split factor of 1. As suggested by GMT (1984) and Brennan and Copeland (1988), if the split factor was indeed being used as a signal by management to convey private information, one would expect to find an increase in anomalous returns as the split factor values increased.

the largest impact on the lowest price portfolio.

4.1   Stock Splits versus Stock Dividends

Panels A and B of Table 2 break the results of Table 1 into stock split and stock dividend subsamples. In contrast to the analysis in GMT, there is no significant difference in the ex-date returns between the stock split and stock dividend subsamples. One difference between the two samples is that the percentage bid–ask spreads are wider for the stock dividend subsample. This wider spread affects the difference in close-to-close returns between splits and stock dividends only if the changes in the LR around the ex-date are different for the two subsamples, but they aren't. As a consequence, for the remaining analysis, we do not disaggregate the sample by breaking it into stock splits and stock dividends.

4.2   Low versus High Share Price

Table 3 reports results for the sample partitioned by share price. To economize on space, we report results only for the two extreme quartiles. Panel A contains results for those stocks with share price in the lowest price quartile (less than $14.25) and Panel B contains those stocks with share price in the highest price quartile ($25 or more). Consider first the results for the low-price stocks in Panel A for which the ex-date effect is most pronounced. The average close-to-close return on the ex-date is 2.40% which is on the order of 5 to 8 times larger than the average returns on the surrounding days. That Keim (1989) found the bid–ask microstructure effect at the turn of the year to be most severe for low-price stocks suggests that the microstructure component may likewise be important for the low-price stocks in our split sample. Consistent with this, the percentage spreads in Panel A for the lowest price shares are more than twice as large as those in Panel B, for the highest-price shares, and also exhibit a significant increase of 1.37% from the day prior to the ex-date to the ex-date. Similar to our finding for the entire sample, the average ask return is twice the average bid return on the ex-date, a reflection of the widening percentage spread on that day. In addition, the LR increases from 0.57 to 0.65 on the ex-date, indicating that some of the average close-to-close ex-date return is a reflection of intra-spread price movement. The average difference between the close-to-close return and the mid-spread return on the ex-date is 49 basis points ($T = 3.06$), indicating a significant microstructure component in the average close-to-close returns on the ex-date. However, the fact that the average mid-spread return on the ex-date is approximately four times the average return on the surrounding days, indicates that much of the average ex-date effect is not microstructure related, even for these low-price stocks that are most susceptible to these effects.

Table 2. Average Daily Percentage Returns (Standard Errors), Percentage
Spreads, and *L*-Ratios on the Ex-Date and Surrounding Ten Trading Days
for Nasdaq Stock Splits and Stock Dividends for the period 1983–1994
(Separately for Stock Splits and Stock Dividends)

Panel A: Stock Splits (N = 1030)

Event Day	Close-to-Close Return	Mid-Spread Return	Ask Return	Bid Return	Percent Spread	*L*-Ratio
−5	0.044	0.064	0.060	0.069	2.00	0.53
	(0.080)	(0.067)	(0.068)	(0.069)		
−4	0.032	0.041	0.045	0.037	2.00	0.51
	(0.081)	(0.070)	(0.071)	(0.071)		
−3	0.280	0.138	0.153	0.124	2.03	0.55
	(0.086)	(0.078)	(0.078)	(0.079)		
−2	0.329	0.314	0.298	0.331	2.00	0.57
	(0.082)	(0.072)	(0.073)	(0.073)		
−1	0.415	0.463	0.442	0.487	1.95	0.59
	(0.084)	(0.074)	(0.074)	(0.075)		
0	1.502	1.313	1.792	0.823	2.91	0.60
	(0.128)	(0.113)	(0.115)	(0.116)		
1	0.479	0.409	0.367	0.456	2.83	0.62
	(0.119)	(0.101)	(0.101)	(0.104)		
2	0.484	0.499	0.440	0.564	2.71	0.60
	(0.108)	(0.096)	(0.096)	(0.101)		
3	0.133	0.118	0.128	0.111	2.73	0.62
	(0.098)	(0.085)	(0.085)	(0.088)		
4	0.076	0.132	0.128	0.140	2.72	0.59
	(0.099)	(0.085)	(0.085)	(0.087)		
5	0.042	−0.037	−0.027	−0.045	2.73	0.60
	(0.103)	(0.084)	(0.083)	(0.088)		

By contrast, the average close-to-close ex-date return for the highest-price shares in Panel B is 1.08% – much lower than for the low-price shares but, nevertheless, at least twice the magnitude of the average returns on the surrounding days. Interestingly, the average ex-date returns show much less variation across return definitions than for the low-price shares. This is no doubt an indication that microstructure effects related to changes in percentage spreads and systematic movements within the spread become much less important as the share price increases because the percentage spread represents a corre-

Table 2 (*cont.*). Average Daily Percentage Returns (Standard Errors), Percentage
Spreads, and *L*-Ratios on the Ex-Date and Surrounding Ten Trading Days
for Nasdaq Stock Splits and Stock Dividends for the period 1983–1994
(Separately for Stock Splits and Stock Dividends)

Panel B: Stock Dividends (N = 305)

Event Day	Close-to-Close Return	Mid-Spread Return	Ask Return	Bid Return	Percent Spread	*L*-Ratio
−5	0.383	0.303	0.295	0.312	2.84	0.58
	(0.158)	(0.120)	(0.120)	(0.127)		
−4	−0.009	0.193	0.225	0.163	2.90	0.49
	(0.146)	(0.107)	(0.104)	(0.116)		
−3	0.234	0.181	0.153	0.213	2.84	0.53
	(0.157)	(0.124)	(0.125)	(0.130)		
−2	0.516	0.291	0.310	0.272	2.88	0.59
	(0.149)	(0.109)	(0.111)	(0.113)		
−1	0.432	0.347	0.364	0.332	2.91	0.61
	(0.151)	(0.106)	(0.110)	(0.109)		
0	1.500	1.345	1.632	1.051	3.49	0.60
	(0.220)	(0.184)	(0.184)	(0.193)		
1	0.396	0.360	0.259	0.470	3.28	0.63
	(0.160)	(0.117)	(0.120)	(0.124)		
2	0.315	0.328	0.282	0.378	3.19	0.61
	(0.193)	(0.142)	(0.140)	(0.150)		
3	0.493	0.501	0.480	0.525	3.15	0.60
	(0.179)	(0.133)	(0.136)	(0.135)		
4	0.298	0.251	0.225	0.278	3.10	0.60
	(0.157)	(0.123)	(0.124)	(0.127)		
5	−0.146	−0.026	0.000	−0.050	3.15	0.58
	(0.153)	(0.103)	(0.105)	(0.108)		

spondingly smaller component of the return. Panel B contains evidence of
this: the differential between the average bid and ask returns is much smaller
than for low-price shares (Panel A), and the average mid-spread return is
insignificantly different from the average close-to-close return ($T = -0.57$).

Table 3. Average Daily Percentage Returns (Standard Errors), Percentage
Spreads, and L-Ratios on the Ex-Date and Surrounding Ten Trading Days
for Nasdaq Stock Splits and Stock Dividends for the Period 1983–1994
(Separately for Low-Price and High-Price Stocks)

Panel A: Share Price < $14.25 *(N = 324)*

Event Day	Close-to-Close Return	Mid-Spread Return	Ask Return	Bid Return	Percent Spread	L-Ratio
−5	−0.056	0.014	0.017	0.015	3.30	0.51
	(0.168)	(0.123)	(0.125)	(0.129)		
−4	−0.066	−0.117	−0.086	−0.146	3.36	0.53
	(0.185)	(0.151)	(0.153)	(0.156)		
−3	0.116	−0.008	−0.025	0.014	3.33	0.59
	(0.172)	(0.151)	(0.149)	(0.160)		
−2	0.270	0.243	0.269	0.218	3.38	0.59
	(0.181)	(0.143)	(0.145)	(0.148)		
−1	0.303	0.445	0.433	0.459	3.35	0.57
	(0.186)	(0.149)	(0.153)	(0.151)		
0	2.398	1.906	2.589	1.200	4.72	0.65
	(0.281)	(0.226)	(0.227)	(0.240)		
1	0.573	0.534	0.443	0.637	4.54	0.64
	(0.247)	(0.187)	(0.188)	(0.200)		
2	0.535	0.557	0.440	0.690	4.30	0.64
	(0.222)	(0.173)	(0.171)	(0.189)		
3	0.111	0.249	0.219	0.287	4.24	0.58
	(0.203)	(0.144)	(0.143)	(0.155)		
4	0.232	0.209	0.226	0.196	4.27	0.57
	(0.189)	(0.149)	(0.156)	(0.152)		
5	−0.002	−0.142	−0.127	−0.153	4.29	0.62
	(0.218)	(0.155)	(0.150)	(0.170)		

4.3   Conditioning on the Occurrence of an Ask or Bid Price at the Previous Day's Close

Table 4 reports the returns, percentage spreads, and L-ratios for the event
days in which the prior day's closing transaction price was at the ask price
(Panel A) or at the bid price (Panel B). This filtration downwardly biases
Panel A's close-to-close returns and upwardly biases Panel B's close-to-close

Table 3 (*cont.*). Average Daily Percentage Returns (Standard Errors), Percentage Spreads, and *L*-Ratios on the Ex-Date and Surrounding Ten Trading Days for Nasdaq Stock Splits and Stock Dividends for the Period 1983–1994 (Separately for Low-Price and High-Price Stocks)

Panel B: Share Price > $24.875 (N = 334)

Event Day	Close-to-Close Return	Mid-Spread Return	Ask Return	Bid Return	Percent Spread	*L*-Ratio
−5	0.228	0.248	0.222	0.276	1.39	0.52
	(0.123)	(0.109)	(0.107)	(0.113)		
−4	0.119	0.149	0.153	0.145	1.38	0.52
	(0.114)	(0.102)	(0.102)	(0.105)		
−3	0.208	0.212	0.235	0.189	1.43	0.47
	(0.147)	(0.144)	(0.145)	(0.144)		
−2	0.562	0.508	0.507	0.510	1.43	0.53
	(0.126)	(0.118)	(0.118)	(0.119)		
−1	0.593	0.521	0.509	0.533	1.40	0.59
	(0.127)	(0.116)	(0.116)	(0.117)		
0	1.080	1.120	1.385	0.851	1.93	0.55
	(0.202)	(0.193)	(0.194)	(0.195)		
1	0.270	0.165	0.143	0.189	1.88	0.59
	(0.169)	(0.151)	(0.153)	(0.151)		
2	0.478	0.434	0.404	0.467	1.82	0.57
	(0.159)	(0.148)	(0.148)	(0.151)		
3	0.186	0.095	0.087	0.105	1.81	0.64
	(0.152)	(0.136)	(0.135)	(0.140)		
4	−0.163	−0.051	−0.053	−0.047	1.80	0.56
	(0.140)	(0.124)	(0.124)	(0.127)		
5	0.134	0.066	0.072	0.062	1.81	0.59
	(0.145)	(0.126)	(0.127)	(0.128)		

returns. But the other three return measures remain interesting and unbiased.

The most noteworthy aspect of Table 4 is that the post ex-date returns, mid-spread returns, ask returns, and bid returns are much lower when the prior day's closing transaction price was a bid price than when it was an ask price. Indeed, post ex-date, the returns do not significantly differ from zero when the prior day's closing price is at the bid (Panel B). Also, except for

Table 4. Average Daily Percentage Returns (Standard Errors), Percentage
Spreads, and L-Ratios on the Ex-Date and Surrounding Ten Trading Days for
Nasdaq Stock Splits and Stock Dividends for the Period 1983–1994 (Number of
Events in Parentheses Underneath Event Day)

Panel A: Events Conditioned on Day t−1 Closing Price Being an Ask Price

Event Day	Close-to-Close Return	Mid-Spread Return	Ask Return	Bid Return	Percent Spread	L-Ratio
−4	−0.596	0.287	0.339	0.236	2.16	0.53
(579)	(0.102)	(0.084)	(0.090)	(0.091)		
−3	−0.508	0.303	0.343	0.265	2.19	0.58
(517)	(0.113)	(0.104)	(0.104)	(0.107)		
−2	−0.381	0.377	0.387	0.368	2.18	0.62
(550)	(0.102)	(0.095)	(0.095)	(0.097)		
−1	−0.229	0.676	0.648	0.707	2.18	0.62
(599)	(0.111)	(0.098)	(0.098)	(0.100)		
0	0.753	1.355	1.792	0.910	2.91	0.62
(624)	(0.147)	(0.137)	(0.138)	(0.142)		
1	−0.356	0.654	0.608	0.705	2.91	0.65
(648)	(0.134)	(0.177)	(0.116)	(0.122)		
2	−0.275	0.675	0.622	0.735	2.69	0.63
(690)	(0.126)	(0.133)	(0.112)	(0.119)		
3	−0.463	0.540	0.571	0.512	2.85	0.62
(658)	(0.113)	(0.098)	(0.099)	(0.101)		
4	−0.637	0.344	0.383	0.306	2.72	0.60
(643)	(0.119)	(0.107)	(0.107)	(0.109)		
5	−0.865	0.140	0.165	0.116	2.77	0.61
(630)	(0.119)	(0.101)	(0.100)	(0.107)		

the ex-date itself, the percentage spreads and L-ratios appear to be slightly
higher in Panel A than in Panel B. The latter finding implies that a closing
transaction at an ask price increases the next day's tendency to have a closing
transaction at the ask price.

One interpretation of this is that in the aggregate, investors are acquiring
some splitting stocks from market makers. If the observed L-ratio is indicative
of what is happening intraday, this disbursement of market maker inventory
to interested investors seems to be serially correlated. While transaction
prices are rising for these stocks in Panel A after the ex-date, market makers
do not seem to be adjusting their ask prices quickly enough to eliminate this

Table 4 (*cont.*). Average Daily Percentage Returns (Standard Errors), Percentage Spreads, and *L*-Ratios on the Ex-Date and Surrounding Ten Trading Days for Nasdaq Stock Splits and Stock Dividends for the Period 1983–1994 (Number of Events in Parentheses Underneath Event Day)

Panel B: Events Conditioned on Day t−1 Closing Price Being a Bid Price

Event Day	Close-to-Close Return	Mid-Spread Return	Ask Return	Bid Return	Percent Spread	*L*-Ratio
−4	0.829	−0.102	−0.07	−0.133	2.05	0.48
(451)	(0.118)	(0.101)	(0.102)	(0.104)		
−3	1.132	0.113	0.146	0.079	2.11	0.50
(487)	(0.127)	(0.116)	(0.115)	(0.119)		
−2	1.241	0.119	0.155	0.084	2.00	0.53
(435)	(0.119)	(0.095)	(0.098)	(0.096)		
−1	1.298	0.242	0.274	0.211	1.96	0.55
(409)	(0.124)	(0.104)	(0.107)	(0.105)		
0	2.264	0.970	1.433	0.496	2.99	0.55
(375)	(0.218)	(0.173)	(0.179)	(0.175)		
1	1.731	0.003	−0.019	0.028	2.75	0.60
(382)	(0.183)	(0.145)	(0.147)	(0.149)		
2	1.525	0.033	0.034	0.035	2.64	0.56
(360)	(0.171)	(0.150)	(0.150)	(0.154)		
3	1.336	−0.123	−0.095	−0.149	2.55	0.58
(383)	(0.167)	(0.142)	(0.144)	(0.145)		
4	1.327	−0.147	−0.127	−0.165	2.52	0.59
(357)	(0.157)	(0.126)	(0.127)	(0.131)		
5	1.347	−0.206	−0.121	−0.290	2.74	0.58
(396)	(0.169)	(0.132)	(0.131)	(0.137)		

depletion of inventory.

5   A Regression Model

In this section we estimate a regression model designed to control for the microstructure influences described above while estimating the effect of other variables on the intrinisic ex-date return. We construct the variable MICRO which is the component of the average return induced by market microstructure efects as reflected in changes in percentage spreads and changes in LR:

$$\text{MICRO} = (\text{LR} - 0.5) \times \Delta\text{SPD} + \Delta\text{LR} \times \text{SPD},$$

where SPRD is the percentage bid-ask spread of the stock at event day -1, LR is the L-ratio of the closing stock price at event day 0, and Δx is the change in any given variable x between day -1 and day 0. Note that udner the following assumptions, MICRO will be equivalent to the close-to-close return: (1) the mid-spread price represents the true intrinsic value of the firm; (2) the mid-spread price does not change between the day before the ex-date (event day -1) and the ex-date (event day 0); and (3) change in the bid-ask spread between event days -1 and 0 are uncorrelated with changes in the L-ratio over these same days.

We estimate the following regression model

$$R_i = a_0 + a_1 \text{MICRO}_i + a_2 \text{LR}_i + a_3\, 1/\text{P}_i + a_4 \text{SIZE}_i + a_5 \text{NDLR}_i + a_6 \text{SPLFAC}_i + e_i$$

where R_i is either the close-to-close or mid-spread return for event i, P_i is the average of the bid and ask prices at day -1 for event i, SIZE_i is the natural logarithm of the market capitalization of the stock at day -5 for event i, NDLR_i is the number of market makers in the stock associated with event i, and SPLFAC_i is the split factor for event i.

Table 5 reports OLS regression coefficients and t-statistics for several cross-sectional regressions. In Panel A, which reports two regression with the close-to-close ex-date return, the coefficient on MICRO is close to one. This is consistent with the constructed proxy for the market microstructure variable being simply an addition to the intrinsic return, where the intrinsic return is predicted by the variables LR, 1/P, SIZE, NDLR, SPLFAC. The coefficient on MICRO seems to be the same whether one controls for variables that explain the intrinsic return or not. As a robustness check, a measure of the intrinsic return, the mid-spread return, appears to have virtually the same coefficients on its predictors as does the close to close return, when controlling for the bid–ask bias proxied for by the market microstructure variable.[10]

6   Conclusion

This paper confirms that the occurrence of systematic trading patterns and market maker behavior around ex-dates impart biases to the computation of anomalous returns associated with stock splits and stock dividends. However, the combination of market makers widening their spreads and the greater tendency for stocks to close at the ask price on the ex-date does not completely explain the phenomenon. These combined microstructure effects primarily

---

[10]Splitting the microstructure variable into its two components also leads to similar, but not identical results. In regressions analogous to those in Panel A, the coefficients on the spread increase component are 0.83 in the univariate regression and 0.67 in the multivariate regression while the analogous coefficients for the L-ratio change component are 1.07 and 1.17 respectively.

Table 5. Estimated Coefficients (*T*-statistics) for a Regression Model of Ex-Date
Returns

$$R_i = a_0 + a_1 \text{MICRO}_i + a_2 \text{LR}_i + a_3 1/P_i + a_4 \text{SIZE}_i + a_5 \text{NDLR}_i + a_6 \text{SPLFAC}_i$$

Intercept	MICRO	LR	1/P	SIZE	NDLR	SPLFAC	Adj. R^2

Panel A: R = Close-to-Close Return

0.013	1.010	–	–	–	–	–	0.226
(13.76)*	(19.71)*						
0.007	1.030	0.005	0.176	−0.003	0.000	0.005	0.259
(0.99)	(18.34)*	(2.01)*	(4.28)*	(−2.70)*	(2.34)*	(2.50)*	

Panel B: R = Mid-Spread Return

0.003	–	0.005	0.183	−0.003	0.0003	0.007	0.043
(0.39)		(2.07)*	(4.50)*	(−2.60)*	(2.29)*	(3.65)*	

* = significant at the 5% level

impact low-price stocks for which the percentage bid-ask spreads is large;
but even for these stocks, the average mid-spread return, a return measure
that is relatively free of microstructure influences, is significantly larger on
the ex-date than on the surrounding days. For the larger-priced stocks in
our sample, the microstructure effects are negligible because the percentage
bid-ask spread is of second-order importance in the return calculation.

The results of this study were drawn from data for Nasdaq stocks, while
much of the previous evidence on patterns around stock split announcements
is based on data for NYSE and AMEX stocks. It would be useful to generalize
the results reported here to the NYSE and AMEX.

References

Bar-Yosef, S. and L. Brown (1977) 'A Re-Examination of Stock Splits Using Moving
Betas,' *Journal of Finance* **32** 1069–1080.

Blume, Marshall E. and Robert F. Stambaugh (1983) 'Biases in Computed Returns:
An Application to the Size Effect,' *Journal of Financial Economics* **12** 387–404

Brennan, Michael J. and Thomas E. Copeland (1988a) 'Beta Changes Around
Stock Splits: A Note,' *Journal of Finance* **43** 1009–1014.

Brennan, Michael J. and Thomas E. Copeland (1988b) 'Stock Splits, Stock Prices, and Transaction Costs,' *Journal of Financial Economics* **22** 83–101.

Charest, G. (1978) 'Split Information, Stock Returns, and Market Efficiency,' *Journal of Financial Economics* **6** 265–296.

Chottiner, S. and A. Young (1971) 'A Test of the AICPA Differentiation between Stock Dividends and Stock Splits,' *Journal of Accounting Research* **9** 367–374.

Conrad, Jennifer S. and Robert Conroy (1994) 'Market Microstructure and the Ex-Date Return,' *Journal of Finance* **49** 1507–1519.

Conroy, R. and M. Flood (1988) 'The Effects of Stock Splits on Marketability: Transaction Rates and Share Ownership'. Working paper, Darden School, University of Virginia.

Copeland, Thomas E. (1979) 'Liquidity Changes Following Stock Splits,' *Journal of Finance* **34** 115–141.

Dravid, Ajay R. (1987) 'A Note on the Behavior of Stock Returns around Ex-Dates of Stock Distributions,' *Journal of Finance* **42** 163–168

Eades, K., P. Hess and H. Kim (1984) 'On Interpreting Security Returns during the Ex-Dividend Period,' *Journal of Financial Economics* **13** 3–34.

Fama, E., L. Fisher, M. Jensen and R. Roll (1969) 'The Adjustment of Stock Prices to New Information, *International Economic Review* **10** 1–21.

Foster, T. and D. Vickrey (1978) 'The Information Content of Stock Dividend Announcements,' *The Accounting Review* **53** 360–370.

Grinblatt, Mark S., Ronald W. Masulis and Sheridan Titman (1984) 'The Valuation Effects of Stock Splits and Stock Dividends,' *Journal of Financial Economics* **13** 461–490

Keim, Donald B. (1989) 'Trading Patterns, Bid–Ask Spreads and Estimated Security Returns: The Case of Common Stocks at Calendar Turning Points,' *Journal of Financial Economics* **25** 75–98.

Keim, Donald B. and Robert F. Stambaugh (1984) 'A Further Investigation of the Weekend Effect in Stock Returns,' *Journal of Finance* **39** 819–835

Kryzanowski, Lawrence and Hao Zhang (1993) 'Market Behavior around Canadian Stock Split Ex-Dates,' *Journal of Empirical Finance* **1** 57–81.

Lakonishok, J. and B. Lev (1987) 'Stock Splits and Stock Dividends: Why, Who, and When,' *Journal of Finance* **42** 913-932.

McNichols, Maureen and Ajay R. Dravid (1990) 'Stock Dividends, Stock Split and Signalling,' *Journal of Finance* **45** 857–879.

Ohlson, J.A. and S. Penman (1985) 'Volatility Increase Subsequent to Stock Splits: An Empirical Aberration,' *Journal of Financial Economics* **14** 251–266

Patell, James M. and Mark A. Wolfson (1984) 'The Intraday Speed of Adjustment of Stock Prices to Earnings and Dividend Announcements,' *Journal of Financial Economics* **13** 223–252

Ritter, Jay (1991) 'The Long Run Performance of Initial Public Offerings,' *Journal of Finance* **46** 3–28.

Woolridge, R. (1983a) ' Ex-Date Stock Price Adjustment to Stock Dividends: A Note,' *Journal of Finance* **38** 247–255.

Woolridge, R. (1983b) 'Stock Dividends as Signals,' *Journal of Financial Research* **6** 1–12.

Part III
International Evidence

Canadian Security Market Anomalies

George Athanassakos and Stephen Foerster

Overview/Introduction

Despite similarities in the Canadian and US equity markets, there are also distinct differences emanating primarily from differences in the two countries' taxation and institutional frameworks. These differences may give rise to differential reaction of stock prices to calendar and price-related factors in Canada vs. the US. Therefore, in order to improve our knowledge across markets and provide useful insights to Canadian and international investors, this paper examines security market 'anomalies' in a Canadian context. The first section provides an overview of the Canadian environment. Section 2 reviews historical returns in contrast to US results. Section 3 examines calendar-based return patterns, while Section 4 examines price-related anomalies. Section 5 focuses on stock returns and economic factors, while Section 6 focuses on stock returns and political factors.

1 The Canadian Environment

1.1 The Economic Structure

The size of Canada's economy, as measured by Gross Domestic Product (GDP), is approximately $C 750 billion. It has a population of about 28 million and a labour force participation rate of around 65% (down from a peak of 69.2% in 1989).

The strong growth rate of the 1980s slowed substantially in recent years largely because of high interest rates and tax increases. There has also been a rise in the unemployment rate, which bottomed at 6.7% in 1989. The national average unemployment rate was 9.5% in 1995. However, the high level of the federal deficit (5% of GDP) and of net federal debt (72.8% of GDP)[1] that Canada amassed in the 1980s, made fiscal adjustment programs extremely difficult and very susceptible to derailment by adverse shocks, including higher than anticipated interest rates.

Agriculture accounts for about 2% of GDP and forestry, fishing and trapping for another 1%. Canadian agriculture has become increasingly capital intensive as agricultural employment dropped to about 4% of the work force.

---

[1] The gross public debt as a percentage of GDP is 94.6%.

Manufacturing accounts for about 20% of GDP and construction for another 6%. The latter two sectors employ about 25% of the work force. Most manufacturing production is located in British Columbia, Ontario and Quebec. In recent years, capacity utilization peaked in 1988 at 86.4%.

The US accounts for 75% of Canada's exports and 69% of imports. The main exports are motor vehicles and parts, metals and minerals and energy materials. The main destinations, other than US, are Mexico, the UK and Japan. The main imports are machinery and equipment, motor vehicles and parts and industrial materials. Canada is the world's second largest wheat exporter. Canada is also the world's largest exporter of market pulp and newsprint and the world's largest exporter of nickel, zinc, uranium, potash and asbestos and the second largest producer of molybdenum. Canada has substantial reserves in gas, oil and coal, mainly concentrated in western Canada, particularly in Alberta. Natural gas has become an increasingly important export to the US, with Canada supplying about 7% of the US market.

Ontario is the Canada's most populous province with the nation's finance and services capital Toronto, where the largest stock exchange, the Toronto Stock Exchange (TSE), is located. Other stock exchanges are located in Montreal, the Montreal Stock Exchange (MSE), in Calgary, the Alberta Stock Exchange (ASE), and in Vancouver, the Vancouver Stock Exchange (VSE).

1.2   The Political Structure

Canada is a federation of ten self-governing provinces and two territories. Queen Elizabeth II is the (figure head) head of state, represented locally by an appointed governor-general who (technically) appoints the prime minister and cabinet. As of 1996, the federal party in power is the Liberal party, elected in power in 1993, after an almost ten year domination of national politics by the Progressive Conservative (PC) party. As in the UK, the prime minister represents the leader of the party which wins the most elected seats in the House of Commons.

While Canada does not pose a political risk in the traditional sense, the emergence of secessionist tendencies in parts of Canada are quite disconcerting to investors. These tendencies have given rise to two new political parties which have regional support, the Reform party with a support base in Western Canada and the Bloc Québécois party with a support base in the Province of Quebec. These parties, which did not exist only a few years ago, now command opposition status in Ottawa. The Reform party is a right wing party, while the Bloc Québécois party has a mandate to separate Quebec from Canada. The emergence of these two parties has taken away support from the more traditional parties such as the PC and New Democratic Party (NDP), which have almost disappeared from national politics. Finally, the election of

Party Québéçois in the province of Quebec in 1994 has strengthened concerns about a possible break-up of Canada. The recent referendum vote in the province of Quebec in November 1995, that gave a razor thin victory to the 'No-Separation' camp, has only temporarily settled this matter.

1.3 Taxation of Investment Income

The tax treatment of ordinary cash and stock dividends is different in Canada than the US. First, as opposed to the relative constancy of the difference in the dividend and capital gains tax rates in the US between 1970 and 1986, the difference in tax rates between dividends and capital gains progressively declined in Canada for top-bracket investors. Second, the marginal tax rate per dollar of dividends received by Canadian taxed investors is lower than the marginal tax rate for ordinary income. Third, the Canadian tax system has never made a distinction between short-term and long-term capital gains. Finally, Canadian corporations pay no tax on dividends received from another publicly held Canadian corporation, whereas only part of their capital gains (the proportion varies depending on the tax regime) is taxed at the combined federal/provincial ordinary income tax rate. The Appendix outlines details pertaining to Canada's differing regimes related to dividend taxation, capital gains taxation and transaction costs.

2 Historical Returns in Canada

2.1 Stocks, Bonds, Bills and Inflation Since 1950

One of the most extensive studies of post-World War II Canadian securities is carried out in Hatch and White (1985, 1988). These studies follow the general analysis of US securities presented by Ibbotson and Sinquefield (1982) and subsequent updates.

Table 2.1 summarizes the major findings by Hatch and White (1988). They compare Canadian security returns to US returns presented in Ibbotson (1987), after adjusting US returns to reflect changes in the value of the Canadian dollar relative to the US dollar in order to reflect unhedged returns to Canadian investors in US markets.

Over the 37-year period, overall security returns in both Canada and the US are of similar magnitudes. Equity returns are slightly higher and long-term government bond returns are slightly lower in the US, while T-Bill returns and inflation are very similar in both countries. While the volatility of equity returns are virtually identical in both countries, the volatility of the other series are slightly higher in the US series. One interesting feature of both the Canadian and US series is the lower return and higher volatility of the long-term government bonds compared to treasury bills.

Table 2.1. Canadian and US (Total) Returns, 1950–1986

Canadian Returns	Geometric Mean	Arithmetic Mean	Standard Deviation
Treasury Bills	5.8%	5.9%	4.2%
Long–term Govt.	5.4%	5.8%	10.3%
Long–term Corp.	6.2%	6.6%	9.8%
Equities	11.5%	12.5%	17.9%
Inflation	4.7%	4.8%	3.8%

U.S. Returns	Geometric Mean	Arithmetic Mean	Standard Deviation
Treasury Bills	5.7%	5.9%	5.3%
Long-term Govt.	5.1%	5.7%	12.0%
Long-term Corp.	5.8%	6.4%	12.3%
Equities	12.8%	14.2%	17.9%
Inflation	4.8%	5.0%	5.0%

Foerster (1993) compares the return performance of individual Canadian stocks over a similar period. He compares two samples of stocks. One sample includes stocks listed on Canadian exchanges over the entire 1950 to 1987 period. The other sample includes stocks which are interlisted on US exchanges. He finds the performance of both groups of stocks are similar. The average annualized return of the non-interlisted group is 10.2%, while the return of the interlisted group is 10.0%.

2.2   Longer-term Studies

While the Toronto Stock Exchange/University of Western Ontario (TSE/UWO) database utilized in earlier studies provides extensive data on individual security prices (similar to the CRSP data for US securities), other data sources provide longer-term data relating to indices of prices and returns. Foerster (1994c) obtained data directly from the TSE consisting of monthly data of an index series (which eventually developed into the TSE 300 Index) from 1919 to 1993. This series does not include dividend information. However, the series does include important information covering the performance of stocks leading up to the 1929 stock market crash and the subsequent Great Depression. Average annual capital gains over the entire period are 6.9%. Since the dividend yield (post-1956) has been estimated to be around 3.5%, the total return can be estimated at close to 10.5%. This average annual return is about 2% lower than the estimate in the Hatch and White study for the shorter 1950 to 1986 period. The best single year was 1933, when prices rose 46.8%. The worst single year was 1931, when prices dropped 37.2%. The best five-year period was 1924 and 1928, when average

Table 2.2. Canadian (Total) Returns, 1924–1994

Overall Period	Geometric Mean	Arithmetic Mean	Standard Deviation
Treasury Bills	5.0%	5.1%	4.7%
Long–term Govt.	5.5%	5.8%	8.6%
Equities	10.1%	11.8%	19.1%
Inflation	3.2%	3.3%	4.5%

1924–1956	Geometric Mean	Arithmetic Mean	Standard Deviation
Treasury Bills	0.8%	0.7%	0.6%
Long–term Govt.	4.0%	4.2%	5.3%
Equities	11.0%	13.2%	21.7%
Inflation	1.4%	1.5%	4.9%

1957–1994	Geometric Mean	Arithmetic Mean	Standard Deviation
Treasury Bills	7.6%	7.7%	4.1%
Long–term Govt.	6.8%	7.3%	10.7%
Equities	9.4%	10.6%	16.5%
Inflation	4.8%	4.8%	3.4%

annual capital gains were 21.9%. The worst five-year period was 1928 and 1932, when average annual losses were 18.7%. The best twenty-year period was 1942 and 1961, when average annual capital gains were 10.3%. The worst twenty-year period was 1929 and 1948, when average annual losses were 1.8%. This study emphasizes the importance of sample period in studies which focus on security returns. An investor in Canadian stocks who bought near the market peak around 1929 would have needed to wait over twenty years before the initial investment was recaptured.

A recent study by Booth (1995) updates data from the *Report on Canadian Economic Statistics 1924-1993* by the Canadian Institute of Actuaries. While Booth (1995) recognizes that some of the early data are of poorer quality (particularly before 1956), there are interesting contrasts with these results versus the Hatch and White (1988) findings. Table 2.2 summarizes the main results, with pre-1956 and post 1956 results presented separately (the treasury bill data are from 1934).

In the overall period, as expected, the riskier securities provide the greatest returns. Unlike results in the Hatch and White (1988) study, the return on long-term government bonds exceeds the return on treasury bill investments. Treasury bills outperform inflation by 1.8% on an annual basis. The excess equity return over long-term government bonds is 6.0% based on arithmetic returns, and 4.6% based on geometric returns.

2.3   The Shrinking Equity Premium

Booth (1995) focuses his study on the shrinking equity premium (the difference between mean equity returns and mean returns on long-term government bonds). Equity returns are often viewed as the return on risk-free (government) bonds (long-term) plus a risk premium. Regulatory boards often use such a framework. Booth (1995) argues that Canadian equities are unlikely to return as large a premium as in the past. A closer look at table 2.2 reveals a dramatic difference between the equity premium in the first and second sample periods. The premium shrinks from 9.0% to 3.3% (based on arithmetic returns) due to lower equity returns (2.6% lower) and higher long-term government returns (3.1% higher). Inflation is 3.3% greater in the second period as well. Booth (1995) argues that a structural change occurred in the fixed-income market contributing to the shrinking equity premium.

Foerster (1995) compares equity returns with T-bill returns from 1950 to 1994. He divides the sample into nine five-year periods. Overall, average annual equity returns were 11.7% compared with T-Bill returns of 6.6%, resulting in an average difference of 5.1%. In each of the first four five-year subperiods, equities outperform T-Bills. In the first period, 1950 to 1954, equities (returns of 18.0%) outperformed T-Bills (returns of 1.1%) by 16.9% (which highlights the crucial differences among studies that start in 1950 versus, say, 1956). However, in the last five five-year subperiods, equities outperformed T-Bills only in two subperiods, 1975 to 1979 and 1985 to 1990. In the most recent subperiod, 1990 to 1994, equities *underperformed* T-Bills by 2.7%. Foerster contrasts real T-Bill returns in the 1950 to 1979 period – only 0.4% – with the real return, 5.2%, on T-Bills in the 1980 to 1994 period. A large part of the much higher real return appears attributable to both the fiscal problems (large federal debt and continuing deficits) as well as political problems (concerns, particularly by foreign investors, of Canada's political stability in light of the Quebec question).

3   Calendar-Based Return Patterns

3.1   The Day-of-the-Week Effect

Athanassakos and Robinson (1994) examine daily index return data for the TSE-300 price, TSE-300 total return and the TSE/UWO total return index for the period January 1975 to June 1989 and sub-periods. Table 3.1 reports day-of-the week tests using value-weighted indices to isolate any possible holiday-related effects. The table computes daily mean returns per day of the week by excluding days following holidays. Panel A shows significant negative Monday returns, insignificant positive Tuesday returns and significantly positive Wednesday, Thursday and Friday returns. The average return on Friday is greater than the average return on all other days of the week.[2]

Athanassakos and Robinson (1994) also carry out tests using an equally-weighted portfolio of all securities in the TSE/UWO data base. The results show certain similarities with the value-weighted index returns of Table 3.1. First, a significant negative return is still observed for Monday (except for the 1975-1977 period) and a significant positive return is observed for Wednesday, Thursday and Friday. The positive returns are greatest on Friday. An F-test rejects the equality of returns across days of the week. The results on Tuesday, however, are much different than Table 3.1. When an equally-weighted index is used for the calculation of daily returns per day of the week, a significant negative return for Tuesday is observed. Moreover, this negative Tuesday effect was small prior to 1981, but since then it has become larger (in absolute terms) than the negative Monday returns.

3.2   The January Effect

A. Monthly Data

Athanassakos (1995b) uses monthly TSE-300 total return data to test for seasonality in stock returns (i.e., the January effect) in Canada over the period January 1956 to December 1991, and sub-periods. Table 3.2 documents the existence of a strong January seasonal in the TSE-300 total return index over the whole sample period and, particularly, over the January 1956 to December 1977 sub-period[3]. Interestingly, over the period January 1978 to December 1989, there is no particularly strong January seasonal vis-á-vis the other months of the year. Athanassakos (1995b), however, demonstrates that the lack of seasonality, on average, in the latter sub-period may not be troubling if one is interested in identifying the causes of the seasonal behaviour in stock returns. What is more important is to examine whether there is a January seasonal when the market goes up and an opposite effect when the market goes down. Indeed, Table 3.2 shows that when the markets went up there is a strong seasonal in all periods tested. The opposite was the case

---

[2]Other results report mean returns per day of the week by excluding days following holidays and days around the turn of the year (i.e., from the last four days of December to the first four days of January). In Canada, the tax laws differ from those in the US, and, as a result, the year-end effect tends to be concentrated in the last four trading days of December (rather the last trading day in December) as well as the first four trading days of January (see Griffiths and White (1993)). These results reported differ only slightly from those reported in Table 3.1. Finally, other results report mean daily returns by day of the week including all days but by redefining 'day 1' as the first and 'day 5' as the last trading day of the week. This is an alternative to discarding returns for days following holidays. The results are only slightly different from those in Table 3.1. The 'day 1' effect is slightly larger than the Monday effect, while the 'day 5' effect is slightly larger than the Friday effect.

[3]These results are consistent with earlier studies of Canadian stock market seasonality by Berges *et al.* (1984) and Tinic *et al.* (1987, 1988).

Table 3.1. Average Index Returns, and t-Statistics, by Day-of-the-Week

Excludes Days Following Holidays[a]						
	Monday	Tuesday	Wednesday	Thursday	Friday	F-Stat.
TSE-300 Price	-0.1547^a	0.0170	0.1097^a	0.0837*	0.1370*	11.76^d
Index[b]	(–3.761)	(0.528)	(3.472)	(2.602)	(4.664)	
TSE-300 Total	-0.1239^a	0.0316	0.1200*	0.0938*	0.1510*	10.23^d
Return[b]	(–3.010)	(0.981)	(3.800)	(2.918)	(5.136)	
TSE/UWO	–0.0985*	0.0168	0.1146*	0.1029*	0.1512*	12.04^d
Value Weighted						
Total Return[c]	(–2.904)	(0.622)	(4.114)	(3.716)	(6.018)	

Notes:
[a] All returns are percentages, and t-statistics are in brackets.
[b] January 1977–June 1989.
[c] January 1975–June 1989.
[d] The equality of returns across days of the week is rejected at the one percent significance level.
* Return is significantly different from zero at the 1% significance level.

when the market went down. Something seems to transpire in January that causes returns to be very positive in good years and very negative in bad years. As shown in Table 3.2, the reason that there is no January seasonal in the 1978-1989 period is simply because there is an equal number of positive and negative Januarys. Consequently, one may still be able to benefit, if he/she is able to find a way to differentiate between positive and negative Januarys. To be able to differentiate between positive and negative Januarys, one ought to investigate the causes of stock return seasonality and the January effect. This issue is discussed next.

B. Quarterly Data (Portfolio Rebalancing Tests)

Athanassakos and Schnabel (1994) develop a theory of portfolio rebalancing which is based on the effects induced by the calendar year planning horizon of professional portfolio managers. A simple model of portfolio choice and leisure consumption is developed which shows that portfolio allocations early in the calendar year should be more heavily weighted towards the stock market and less heavily weighted towards cash and cash equivalents than later in the calendar year.

Their paper provides evidence on the impact systematic shifts in the portfolio holdings of institutional investors have had on the aggregate stock market in Canada over the period 1973:Q1 to 1992:Q4. As institutional data are unavailable on a monthly basis, the study examines the seasonal effect

Table 3.2. Average Monthly Returns for the TSE–300 Total Returns Index Over the Period 1956–1991 and Sub-Periods.

	Jan	Feb	Mar	Apr	May	June	July	Aug	Sept	Oct	Nov	Dec
Panel A: 1956:01 to 1991:12												
Overall Sample	.0247*	.0067	.0126	.0051	.0040	.0014	.0136	.0071	-.0110	-.0095	.0190*	.0268*
	(2.96)	(1.34)	(1.51)	(0.77)	(0.60)	(0.21)	(1.94)	(0.85)	(1.65)	(0.95)	(2.28)	(4.02)
	[36]	[36]	[36]	[36]	[36]	[36]	[36]	[36]	[36]	[36]	[36]	[36]
When TSE-300 Total Return was +ve	.0554*	.0286*	.0361*	.0311*	.0298*	.0271*	.0358*	.0395*	.0294*	.0292*	.0431*	.0361*
	(6.79)	(6.40)	(5.77)	(6.95)	(6.82)	(6.06)	(8.58)	(4.42)	(5.69)	(4.13)	(7.18)	(6.70)
	[24]	[20]	[23]	[20]	[21]	[20]	[23]	[20]	[15]	[18]	[25]	[31]
When TSE-300 Total Return was -ve	-.0365*	-.0221*	-.0307*	-.0289*	-.0338*	-.0307*	-.0256*	-.0333*	-.0392*	-.0483*	-.0356*	-.0304*
	(6.32)	(4.42)	(2.21)	(3.85)	(4.09)	(2.21)	(4.61)	(4.44)	(5.99)	(4.10)	(3.93)	(3.40)
	[12]	[16]	[13]	[16]	[15]	[16]	[13]	[16]	[21]	[18]	[11]	[5]
Panel B: 1956:01 to 1977:12												
Overall Sample	.0344*	.0018	.0134*	.0064	-.0080	-.0027	.0134*	-.0043	-.0084	-.0032	.0086	.0285*
	(3.20)	(0.28)	(2.10)	(0.75)	(0.94)	(0.32)	(2.10)	(0.50)	(0.98)	(0.38)	(0.81)	(3.34)
	[22]	[22]	[22]	[22]	[22]	[22]	[22]	[22]	[22]	[22]	[22]	[22]
When TSE-300 Total Return was +ve	.0535*	.0234*	.0295*	.0340*	.0230*	.0270*	.0321*	.0282*	.0348*	.0233*	.0378*	.0371*
	(5.51)	(8.11)	(5.52)	(6.13)	(7.63)	(4.48)	(6.00)	(4.68)	(5.22)	(2.80)	(7.07)	(5.40)
	[17]	[12]	[14]	[13]	[11]	[11]	[14]	[11]	[9]	[13]	[14]	[19]
When TSE-300 Total Return was -ve	-.0269*	-.0241*	-.0146*	-.0334*	-.0390*	-.0326*	-.0194*	-.0375	-.0383*	-.0414	-.0123*	-.0258*
	(6.02)	(3.81)	(4.13)	(3.34)	(4.31)	(3.60)	(2.74)	(3.11)	(4.60)	(6.21)	(2.99)	(2.23)
	[5]	[10]	[8]	[9]	[11]	[11]	[8]	[11]	[13]	[9]	[8]	[3]
Panel C: 1978:01 to 1989:12												
Overall Sample	.0168	.0127	.0174	.0123	.0174*	.0113	.0139	.0344	-.0094	-.0241	.0405*	.0229
	(0.97)	(1.10)	(0.86)	(1.42)	(2.01)	(0.98)	(0.96)	(1.92)	(0.65)	(1.04)	(3.51)	(1.95)
	[12]	[12]	[12]	[12]	[12]	[12]	[12]	[12]	[12]	[12]	[12]	[12]
When TSE-300 Total Return was +ve	.0735*	.0325*	.0488*	.0262*	.0359*	.0271*	.0491*	.0527*	.0213*	.0460	.0523*	.0350*
	(6.00)	(3.01)	(4.88)	(2.47)	(10.15)	(2.71)	(6.50)	(3.20)	(2.61)	(2.30)	(5.51)	(3.69)
	[6]	[8]	[9]	[8]	[8]	[9]	[7]	[9]	[6]	[4]	[10]	[10]
When TSE-300 Total Return was -ve	-.0398*	-.0268	-.0768	-.0154*	-.0195*	-.0362	-.0354*	-.0205	-.0401	-.0592*	-.0189	-.0372
	(3.25)	(2.68)	(1.66)	(3.08)	(3.90)	(1.25)	(7.92)	(3.55)	(1.95)	(2.39)	(2.67)	(2.63)
	[6]	[4]	[3]	[4]	[4]	[3]	[5]	[3]	[6]	[8]	[2]	[2]

Notes: * statistically significant, at least, at the 5% significance level. ( ) denotes t-statistics. [ ] number of observations.

on stock prices of portfolio rebalancing by institutional investors using quarterly data[4]. The institutional data are obtained from the CANSIM data base (Sector VII.3 for pension funds and Sector VIII.2 for mutual funds) and are deflated using the GDP deflator, also from CANSIM, to convert the series into real magnitudes and achieve comparability over time (1986=100). Moreover, comparability over time is enhanced by employing a dummy variable to control for the change in monetary policy in the 1980s vis-á-vis the 1970s. Canadian stock markets were adversely affected by the attempts of monetary authorities to control the growth in the money supply and the ensuing collapse of inflation and inflationary expectations in the 1980s.

Table 3.3 shows summary statistics of pension and mutual fund data (i.e., equity and cash and equivalents) for the period 1973:Q1 to 1992:Q4 not only for the total sample, but also for quarters when the market (the equally-weighted TSE 300 total return index) experienced 'down days' and quarters when the market went up. The results support the hypothesis that institutional investing affects first quarter returns. In an 'up' market, in particular, institutional investors are responsible for the strong performance of all indexes.

To further provide support for this hypothesis, Athanassakos and Schnabel (1994) run simple OLS regressions of the value-weighted and equally-weighted total returns of the TSE/UWO universe against measures of mutual fund and pension fund equity investing variables from the flow of funds data deflated by the GDP deflator. Each of the regressions tests for the joint effect of the time of the year and equity investing by institutional investors, differentiates between the first quarter and the rest of the year and controls for the change in monetary policy in 1979. All regression coefficients are expected to have positive signs and the slope coefficients of the first quarter influence of the mutual and pension fund variables to be larger than the corresponding coefficients for the rest of the year. All signs are found to be as expected and the regression coefficients for the first quarter exceed those for the rest of the year for both mutual fund and pension fund variables, although only the coefficient for the mutual fund variable in the first quarter is found to be statistically significant at traditional levels of significance. Hence, these results provide further support to their hypothesis, substantiating the findings of Table 3.3[5].

---

[4]The correlation coefficient between the first quarter and the January stock returns is 0.67, which is statistically significant at the 1% level. Hence, the same forces that drive the first quarter returns drive also the January seasonal.

[5]Athanassakos (1992) also uses quarterly data to test for seasonal stock market behaviour in Canada over the period 1957:Q1 to 1989:Q4. The time period for his portfolio rebalancing tests was limited by the unavailability of the institutional and/or individual investor data supplied by SEI Fund Evaluation Services of Toronto (pension fund data) and by the Investment Funds Institute of Canada (mutual fund data), which were not available over the whole 1957 to 1989 period. His findings are consistent with the findings of Athanassakos and Schnabel (1994).

Table 3.3. Summary Statistics and Characteristics of Mutual and Pension Fund Flow of Funds Data and Major Stock Indexes: 1973:Q1 to 1992:Q4

	Total Sample		Quarters with (+ve) Returns		Quarters with (−ve) Returns	
	First Quarter	Rest of Year	First Quarter	Rest of Year	First Quarter	Rest of Year
Panel A: Pension Funds (1986 $ Billion)						
Cash and Equivalents[a]	0.228	0.249	0.174	0.259	0.532	0.229
	[0.55]	[0.60]	[0.58]	[0.54]	[0.10]	[0.70]
	(1.85)***	(3.21)*	(1.23)	(2.95)*	(8.92)*	(1.50)
Equity[b]	0.555	0.581	0.654	0.544	0.537	0.651
	[0.42]	[0.56]	[0.45]	[0.58]	[0.20]	[0.54]
	(5.89)*	(7.97)*	(4.90)*	(5.87)*	(5.62)**	(5.46)*
Panel B: Mutual Funds (1986 $ Billion)						
Cash and Equivalents[a]	0.393	0.083	0.233	0.084	0.421	0.083
	[0.54]	[0.88]	[0.54]	[0.29]	[0.24]	[1.46]
	(3.48)*	(0.73)	(3.22)*	(1.80)***	(1.67)***	(0.26)
Equity[b]	0.140	0.098	0.173	0.077	−0.045	0.137
	[0.32]	[0.25]	[0.33]	[0.23]	[0.19]	[0.29]
	(1.91)***	(3.00)*	(2.10)**	(2.12)**	(−0.39)	(2.10)**
Panel C: Stock Index Returns						
TSE–300 Return	0.036	0.010	0.064	0.049	−0.117	−0.071
	[0.09]	[0.08]	[0.06]	[0.05]	[0.07]	[0.08]
	(1.81)***	(0.91)	(4.27)*	(5.68)*	(−2.83)**	(−4.09)*
TSE–300 Total Return	0.046	0.020	0.073	0.060	−0.108	−0.062
	[0.09]	[0.08]	[0.06]	[0.05]	[0.07]	[0.08]
	(2.29)**	(1.83)***	(4.92)*	(6.84)*	(−2.67)**	(−3.56)*
VW[c]–Index Return	0.044	0.019	0.059	0.055	−0.043	−0.055
	[0.08]	[0.08]	[0.07]	[0.06]	[0.03]	[0.08]
	(2.56)**	(1.71)***	(3.36)*	(5.74)*	(−2.41)***	(−2.99)*
EW[d]–Index Return	0.094	0.023	0.120	0.076	−0.067	−0.085
	[0.11]	[0.10]	[0.08]	[0.06]	[0.09]	[0.08]
	(3.86)*	(1.74)***	(6.29)*	(7.78)*	(−1.52)	(−4.59)*

Notes:
[a] Net change in real dollars invested in cash and equivalents.
[b] Net change in real dollars invested in equity.
[c] VW stands for the value weighted index of the TSE/Western data base.
[d] EW stands for the equally weighted index of the TSE/Western data base.
[ ] denotes standard deviation.
( ) denotes t–statistic.
* Statistically significant at the 1% level.
** Statistically significant at the 5% level.
*** Statistically significant at the 10% level.

C. Disentangling the Impact of Institutional Investing and Economic Factors on Stock Return Seasonality

The knowledge that institutional investors, on average, have an impact on stock market seasonality may be of little help to investors. What would be more helpful to know from an investment decision-making point of view is what determines institutional investing in any given January. While institutional investors do tend to commit more funds into the stock market in January, on average, what makes them demonstrate significant deviation around the mean?

Athanassakos (1995b) hypothesizes that institutional investing is affected by economic and risk variables. These variables are: the shape of the yield curve (i.e., the quarter to quarter change of the ratio of the ten-years-and-over to one-to-three-year government of Canada bond yields), **YIELD**; real corporate profits (i.e., the quarterly percentage change in real corporate profits before-tax), **REALPRO**; expected profits (i.e., the quarter to quarter change in the average forecast of the annual percentage change in nominal profits before-tax made by a group of economists surveyed by the Conference Board of Canada in year t for year $t + 1$), **EXPRO**; expected inflation (i.e., the quarter to quarter change in the average forecast of the annual percentage change in the GDP deflator made by the above-mentioned group of economists in year t for year $t + 1$), **EXPINF**; and a measure of default risk (i.e., the quarter to quarter change in the ratio of the ScotiaMcLeod long-term corporates to the ten-years-and-over government of Canada bond yields), **RISK**.

Athanassakos (1995a) constructs composite economic indicators consisting of the above mentioned economic and risk variables that could proxy for the indicators that institutional investors consider before investing in the stock market. The weight that each individual variable has in the indicators is based on each variable's coefficient from regressing total TSE-300 stock returns against these variables. Two indicators are composed. They are calculated as follows:

$$\text{INDICATOR1} = 0.5986 \times \text{YIELD} + 0.1477 \times \text{REALPRO}$$
$$+0.0017 \times \text{EXPRO} - 0.0124 \times \text{EXPINFL}$$
$$-0.4085 \times \text{RISK}$$
$$\text{INDICATOR2} = 0.5485 \times \text{YIELD} + 0.1994 \times \text{REALPRO}$$
$$+0.0014 \times \text{EXPRO}$$

Table 3.4 reports the regression coefficients of regressing **TOTALPF**, the quarter-to-quarter change in the natural logarithm of the percentage change of dollar invested in equities by Canadian pensions funds surveyed by SEI Funds Evaluation Services of Toronto, against the indicators for the first quarter and the rest of the year, as well as their corresponding lagged by a quarter values. It is the slope coefficients of lagged indicators, particularly for the first quarter, that are significant[6].

Having demonstrated the importance of the composite indicators, in their lagged form, in explaining institutional investing behaviour, and given the previous evidence of the effect of portfolio rebalancing on stock returns, it is only natural that we ask the following questions. First, how well would these indicators have anticipated the stock market performance over the sample period. Second, would investing based on the calls made by the indicators (an aggressive strategy) outperform an investment strategy that called for all Januarys to have positive returns (a naive strategy)? That is how does buying when the indicators are positive and investing in T-Bills when the indicators are negative fare against a strategy that invests in the stock market every January?

The in-sample performance of the indicators and the average January returns of the above stated strategies are examined. The indicators make a correct call eight out of the twelve Januarys between 1978 and 1989. Although the aggressive strategy (1.761%) beats the naive (1.68%), it is not by as much as one would have expected. This primarily because two of the four wrong calls were made in 1980 and 1982, when the markets rose and fell by 11.96% and −8.40%, respectively, large magnitudes by any count. However, one has to also consider the risk factors and the transaction costs of the two strategies. As the aggressive strategy trades much less frequently than the naive strategy, it will involve considerably less transaction costs. The aggressive strategy also involves lower risk, as it involves a portfolio of stocks and T-Bills over time. Hence, the risk-adjusted, after transaction costs difference in returns between the two strategies is much larger than it appears.

Finally, Table 3.5 reports an ex-ante test of the predictability of the composite indicators by comparing the lagged values of the indicators (LGIND1 and LGIND2) with TOTALPF and the TSE-300 performance in Januarys of 1976, 1977, 1990, 1991, 1992, 1993 and 1994. The indicators gave a wrong signal only in one out of the seven out-of-sample Januarys. The aggressive strategy beats the passive one by a wide margin (i.e., 2.76% vs. 1.40%)[7]. Again, in risk-adjusted, after transaction costs, the difference in favour of the aggressive strategy would be much larger that the reported one.

---

[6]These findings are consistent with the work of Hensel and Ziemba (1995b) on the 'January Barometer'. They tested the hypothesis that if the market rises in January, it will also rise in the rest of the year. If institutional investors are, on average, correct about the economy and invest in the stock market only when economic indicators are positive, we would expect tests to be unable to reject the above hypothesis. Indeed, Hensel and Ziemba (1995b) find support for their hypothesis in Canada.

[7]Due to the small sample size, the difference in means is not statistically significant at traditional levels of significance.

Table 3.4. Regression Results[a] with the Quarterly TOTALPF[b] as the Dependent
Variable for 1978:Q1 to 1989:Q4: The first Quarter vs. the Rest of the Year

Dependent Variable	TOTALPF			
Independent Variables / Equation	4.1	4.2	4.3	4.4
Intercept	0.0013	0.0014	0.0018	0.0023
	(0.15)	(0.16)	(0.21)	(0.27)
D_1	0.0109	0.0133	0.0112	0.0067
	(0.60)	(0.74)	(0.65)	(0.38)
D_1^{c*}INDICATOR1[d]	0.6198			
	(1.59)			
D_1INDICATOR2[e]		0.8293		
		(1.80)***		
D_5^{f*}INDICATOR1	−0.0170			
	(−0.06)			
D_5 *INDICATOR2		0.0531		
		(0.20)		
D_1 *LGIND1[g]			0.9491	
			(2.62)*	
D_1 *LGIND2				0.8999
				(2.48)*
D_5 *LGIND1			0.2545	
			(0.96)	
D_5 *LGIND2				0.3733
				(1.33)
R^2	.06	.08	.16	.16
Mean Square Error	0.0029	0.0029	0.0026	0.0026

Notes:

[a] t–statistics in parentheses.

[b] TOTALPF is the quarter-to-quarter change in the natural logarithm of the percentage of dollar amount invested in all equities (both Canadian and foreign) by Canadian Pension Funds (i.e., $\log_e$ (% equity$_t$) − $\log_e$ (% equity$_{t-1}$)).

[c] D_1 is a seasonal dummy variable which takes on the value of unity if first quarter, 0 otherwise.

[d] INDICATOR1 = $0.5986 \times$ YIELD + $0.1477 \times$ REALPRO +$0.0017 \times$ EXPPRO − $0.0124 \times$ EXPINFL − $0.4085 \times$ RISK.

[e] INDICATOR2 = $0.5485 \times$ YIELD + $0.1994 \times$ REALPRO + $0.0014 \times$ EXPPRO.

[f] D_5 is a qualitative variable which takes on the value of unity if 2nd, 3rd or 4th quarters, 0 otherwise.

[g] LGIND1 and LGIND2 are INDICATOR1 and INDICATOR2, respectively, lagged by one quarter.

* Statistically significant at the 1% level.

*** Statistically significant at the 10% level.

Table 3.5. An Ex-ante Test of the Predictive Ability of the Composite Economic
and Risk Indicators for the Januarys of 1976, 1977 and 1990–1994

Year	LGIND1[a]	LGIND2[a]	TOTALPF[b]	January TSE-300 Total Return
1976	0.0207	0.0227	N/A[d]	0.0961
1977	0.0150	0.0185	N/A[d]	–0.0070
1990	–0.0486	–0.0378	–0.0182	–0.0653
1991	0.0013	0.0024	0.0636	0.0067
1992	0.0340	0.0360	0.0272	0.0255
1993	–0.0247	–0.0176	0.0020[c]	–0.0124
1994	0.1512	0.1342	0.0130	0.0547
Strategy Results				
Naive[e]				0.0140
Aggressive[f]				0.0276

Notes:

[a] LGIND1 and LGIND2 are defined as in the text.

[b] TOTALPF is defined as in Table 3.4.

[c] The actual change in the natural logarithm of the percentage of dollar amount invested into Canadian (as opposed to both Canadian and foreign, i.e., TOTALPF) equities by pension funds was –0.0139 in the first quarter of 1993. This figure may be more reflective of the out-of-sample LGIND1 and LGIND2 forecasts reported in this Table. The indicators were developed in a period in which the percentage of foreign equity investment had reached its 10% allowable limit and did not fluctuate much. For example, for most of the years in the latter half of the 1980s, this percentage fluctuated between 9% and 10%. However, recent changes in the allowable exposure to foreign equities by pension funds has caused this variable to increase sharply within a short time period. For example, the percentage of exposure to foreign equities by pension funds jumped from 9.5% in December of 1991 to 14.1% in March of 1993. These large increases (vis-á-vis the historical changes) may have caused the TOTALPF figures reported in this Table to be biased upwards when compared with the out-of-sample forecasts of LGIND1 and LGIND2.

[d] N/A stands for not available.

[e] Buy the market portfolio at the end of December and sell at the end of January (Average January Return).

[f] Buy the market portfolio at the end of December and sell at the end of January only when the LGIND1 and LGIND2 indicators are positive and invest in T-bills when the indicators are negative (Average January Return).

4   Price-Based Anomalies

4.1   The Size-Effect

A. Daily Data

Athanassakos and Robinson (1994) formed twenty portfolios based on each security's market value at the beginning of a month and then calculate equally-weighted average daily returns for the securities in each portfolio over the next month. The averages of these monthly returns per day of the week were then calculated for each portfolio over the period January 1975 to June 1989, and are reported in Table 4.1. Large capitalization stocks have a larger negative return than small capitalization portfolios on Mondays. A similar result was found in the US by Keim and Stambaugh (1984, Table II). Regarding Tuesdays, Table 4.1 shows that small capitalization portfolios still had a negative return, whereas the average return of the large capitalization stocks becomes positive. Unlike Keim and Stambaugh's (1984) findings, Tuesdays' negative returns are also statistically significant. Overall, Athanassakos and Robinson (1994) find that the negative Tuesday returns dominate the negative Monday returns for the smaller capitalization portfolios, while the negative Monday returns dominate the Tuesday effect for all other portfolios. Finally, although in Canada we also observe significant differences in mean Friday returns across portfolios, with the exception of the smallest capitalization portfolio, there seems to be no relationship between the size of the mean returns and the capitalization of the portfolios. This result is also somewhat different from the US studies which show that the mean positive returns on Fridays decrease monotonically from the lowest to the highest capitalization portfolio.

Athanassakos and Robinson (1994) argue and provide evidence that the occurrence of a negative Tuesday effect for small stocks in Canada is a result of the high degree of thin trading that exists in Canadian stock markets. This thin trading causes information that is released over the weekend and impounded in the price of Canada's larger companies on Monday, to be reflected in the price of Canada's smaller stocks with a lag. Table 4.2 reports the average percentage of securities that did not trade by day of the week and size-sorted portfolios for the period January 1975 to June 1989. Indeed, almost two-thirds of the smallest capitalization stocks did not trade on Monday, as opposed to close to only one-tenth of the larger capitalization stocks. Along the same lines, Table 4.3 reports the average percentage of securities with negative returns on each day of the week for 20 size-sorted portfolios. This Table shows that for the majority of size-sorted portfolios, and overall, the proportion of securities with negative returns is significantly different across the days of the week and declines monotonically from Monday to Friday. The exception is the lowest capitalization portfolio which has a higher proportion of stocks with negative returns on Tuesday than on Monday. This

Table 4.1. Day-of-the-Week Effect for Market Value (Size-Sorted) Portfolios[a]

Portfolio[b]	Market Value of Equity[c]	Mon[f]	Tues[f]	Wed[f]	Thurs	Fri[f]	F-Stat
		Panel A: January 1975–June 1989					
2	$1.44	−0.0223	−0.2180[d]	0.0014	0.0607	0.2202[d]	13.05[g]
		(−0.492)	(−4.924)	(0.034)	(1.428)	(5.181)	
4	3.11	−0.1661[d]	−0.1940[d]	0.0492	0.0220	0.1160[d]	20.01[g]
		(−5.202)	(−6.037)	(1.629)	(0.755)	(4.013)	
6	5.86	−0.1006[d]	−0.1719[d]	−0.0061	0.0633[d]	0.1477[d]	25.15[g]
		(−3.804)	(−6.592)	(−0.246)	(2.664)	(5.916)	
8	9.70	−0.1558[d]	−0.1453[d]	0.0079	0.0651[d]	0.1862[d]	43.10[g]
		(−6.601)	(−6.547)	(0.373)	(3.076)	(8.665)	
10	16.09	−0.1339[d]	−0.1254[d]	0.0659[d]	0.0715[d]	0.1600[d]	46.37[g]
		(−6.711)	(−6.226)	(3.561)	(3.898)	(8.816)	
12	26.51	−0.1492[d]	−0.0520[d]	0.0506[d]	0.0641[d]	0.1909[d]	47.58[g]
		(−7.882)	(−2.808)	(2.880)	(3.494)	(10.335)	
14	44.66	−0.0995[d]	−0.0305[d]	0.0678[d]	0.0705[d]	0.1612[d]	34.41[g]
		(−6.479)	(−2.034)	(4.695)	(3.780)	(8.050)	
16	81.60	−0.1182[d]	−0.0244	0.1188[d]	0.1035[d]	0.1258[d]	66.93[g]
		(−8.226)	(−1.819)	(9.420)	(8.320)	(10.214)	
18	183.25	−0.1409[d]	0.0066	0.1026[d]	0.0866[d]	0.1545[d]	80.13[g]
		(−11.035)	(0.490)	(7.920)	(7.152)	(13.223)	
20	1,014.40	−0.1297[d]	0.0410[d]	0.1176[d]	0.0831[d]	0.1192[d]	88.60[g]
		(−10.830)	(4.110)	(10.060)	(7.768)	(13.156)	
All TSE/UWO Securities		−0.1180[d]	−0.0915[d]	0.0567[d]	0.0720[d]	0.1709[d]	484.53[g]
		(−20.538)	(−16.396)	(10.678)	(13.544)	(32.841)	

Notes:

[a] This table shows mean total returns for even numbered portfolios, expressed as a percentage, and *t*-statistics in brackets. (There were at least 26,000 observations for each portfolio/day of the week combination).

[b]. Portfolio 2 contains the lowest capitalization securities and portfolio 20 contains the highest capitalization securities.

[c] Market value of equity is measured by the average, across all years, of the median market value of the particular portfolio at the beginning of the year. All market values are expressed in millions of dollars.

[d] Return is significantly different form zero at a one percent significance level.

[e] Return is significantly different from zero at a five percent significance level.

[f] The equality of returns across all 20 portfolios is rejected at the 1% significance level.

[g] The equality of returns across days of the week is rejected at the 5% significance level.

Table 4.2. The Average Percentage of Securities that did not Trade[a] by
Day-of-the-Week and Size-Sorted Portfolios: January 1975–June 1989

Portfolio[b]	Market Value of Equity[c]	Mon[d]	Tues[d]	Wed[d]	Thurs[d]	Fri[d]	F-Stat
2	$1.44	64.45	63.22	64.06	64.31	64.73	2.40
4	3.11	55.44	53.97	54.71	55.28	55.77	2.68[f]
6	5.86	47.94	46.67	47.97	48.17	48.17	2.42
8	9.70	42.60	41.25	41.57	42.30	42.57	1.93
10	16.09	37.59	36.31	36.87	37.10	37.79	1.87
12	26.51	33.54	32.06	33.24	33.54	33.58	3.10[f]
14	44.66	27.82	26.57	27.23	27.45	27.95	2.27
16	81.60	20.91	19.73	19.91	20.07	20.36	2.31
18	183.25	13.26	12.38	12.66	13.09	13.11	2.24
20	1,014.40	13.37	13.15	13.29	13.39	13.57	1.07
All TSE/UWO Securities		37.34	36.15	36.76	37.07	37.39	8.49[e]

Notes:
[a] This table shows the average percentage of all TSE/UWO securities in even numbered portfolios that did not trade on each day of the week. The reported averages are determined by computing the equally-weighted portfolio averages for each trading day, then averaging across all trading days.
[b] Portfolio 2 contains the lowest capitalization securities and portfolio 20 contains the highest capitalization securities.
[c] Market value of equity is measured by the average, across all years, of the median market value of the particular portfolio at the beginning of the year. All market values are expressed in million of dollars.
[d] The equality of percentages across all 20 portfolios is rejected at the 1% significance level.
[e] The equality of percentages across days of the week is rejected at the 1% significance level.
[f] The equality of percentages across days of the week is rejected at the 5% significance level.

result, in combination with the evidence from Table 4.2, helps explain the much larger negative return on Tuesday than on Monday for the smallest capitalization stocks reported in Table 4.1.

B. Monthly Data

Foerster and Porter (1992) form five market value-based portfolios of stocks for the period 1975:1 to 1987:12. Their portfolio returns are calculated in three ways. The first method computes the arithmetic mean returns of the securities in each portfolio. The second method adjusts return calculations for the bouncing of transaction prices between a security's bid and ask prices. Finally, the third method computes the risk-adjusted returns of the portfolios based on CAPM, and adjusts stock returns for nonsynchronous trading problems using 'Dimson' betas.

Foerster and Porter (1992) report the average return on the smallest quin-

Table 4.3. The Average Percentage of Securities with Negative Total Returns[a] by Day-of-the-Week and Size-Sorted Portfolios: January 1975–June 1989

Portfolio[b]	Market Value of Equity[c]	Mon[d]	Tues[d]	Wed[d]	Thurs[d]	Fri[d]	F-Stat
2	$1.44	46.34	47.59	46.13	45.68	44.21	5.39[e]
4	3.11	46.14	45.89	44.44	44.37	43.01	7.25[e]
6	5.86	43.84	44.00	42.90	42.09	40.66	9.87[e]
8	9.70	43.24	42.53	40.93	40.38	39.25	14.94[e]
10	16.09	42.38	41.67	39.91	39.42	37.61	29.38[e]
12	26.51	41.02	39.88	38.12	38.00	36.19	20.23[e]
14	44.66	39.19	38.36	37.06	36.66	35.24	14.66[e]
16	81.60	39.20	37.34	35.62	36.05	34.79	16.70[e]
18	183.25	40.00	37.80	36.50	35.85	35.08	17.38[e]
20	1,014.40	43.46	40.07	39.29	39.10	38.19	9.59[e]
All TSE/UWO Securities		42.56	41.71	40.30	39.87	38.58	208.05[e]

Notes:
[a] This table shows the average percentage of all TSE/UWO securities in even numbered portfolios that had negative returns on each day of the week. The reported averages are determined by computing the equally-weighted portfolio averages for each trading day, then averaging across all trading days.
[b] Portfolio 2 contains the lowest capitalization securities and portfolio 20 contains the highest capitalization securities.
[c] Market value of equity is measured by the average, across all years, of the median market value of the particular portfolio at the beginning of the year. All market values are expressed in million of dollars.
[d] The equality of percentages across all 20 portfolios is rejected at the 1% significance level.
[e] The equality of percentages across days of the week is rejected at the 1% significance level.

tile portfolio (portfolio 1) is over 2.5 times larger than that of the adjacent portfolio and six times larger than that of portfolio 5. The size effect is more pronounced in January than the rest of the year.

They also report parallel results using the bid-ask adjustment technique. Average returns over the year are now half as large for portfolio 1, but virtually unchanged for portfolio 5. January returns for each portfolio decrease substantially less than the returns for the rest of the year. However, a significant decline in portfolio returns has taken place only for the smallest quintile. Despite that, a size-effect is still observed. A January effect is also observed for all size-sorted portfolios.

Finally, Foerster and Porter (1992) report the findings when returns are adjusted for risk based on CAPM and when betas have been adjusted for non synchronous trading problems. They show that the market risk-adjusted mean return difference between the smallest and largest portfolios is 0.359%, which is not substantially different from the unadjusted return difference of 0.369%. Combining the risk-adjustment with the bid-ask adjustment results

Table 4.4. Mean Returns of the Small (R_{small}) and Large (R_{large}) Stock
Portfolios when the Respective Portfolio Returns go Up or Down by Quarter of
the Year: 1962:Q1 to 1992:Q4

Quarter	R_{small}			R_{large}		
of the	When R_{small}	When R_{small}	Overall	When R_{large}	When R_{large}	Overall
Year	is +ve	is −ve		is +ve	is −ve	
Q1	.1838	−.0620	.1311	.0906	−.0495	0.0555
	(7.59)*	(2.00)***	(4.72)*	(5.43)*	(1.96)***	(3.32)*
Q2	.0972	−.0837	.0262	.0668	−.0885	.0117
	(4.89)*	(4.27)*	(1.18)	(5.04)*	(4.34)*	(.64)
Q3	.1258	−.1111	.0581	.0647	−.0728	.0205
	(4.38)*	(4.00)*	(1.94)***	(5.29)*	(−3.42)*	(1.26)
Q4	.1162	−.0634	.0456	.0466	−.0497	.0259
	(4.85)*	(2.72)**	(1.91)***	(5.37)*	(1.86)***	(2.26)**

Notes:
( ) denotes t–statistics.
* Statistically significant at the 1% level.
** Statistically significant at the 5% level.
*** Statistically significant at the 10% level.

in a 0.145% difference between the returns of the smallest and largest portfolios, which again is not significantly different from the 0.154% return difference based strictly on the bid-ask adjustment.

C. Quarterly Data

Athanassakos (1997) examines firm-size in relation to stock market seasonality over the period 1962:Q1 to 1992:Q4. He uses quarterly data as his purpose is to test for portfolio rebalancing, and individual (Sector I & II) and institutional data (Sector VII.3 for pension fund data, and Sector VIII.2 for mutual fund data) from the flow of fund matrixes of the CANSIM data base were available only on a quarterly basis. The questions he asked were: Is there a firm-size related seasonal in Canada? If yes, are individual investors responsible for the seasonal effect observed in small stocks and institutional investors for the seasonal effect observed in larger stocks?

Although five size-sorted portfolios are formed, Table 4.4 reports the mean quarterly returns only of the small and large stock portfolios. Table 4.4 divides the entire sample period into quarters when the small stock returns are positive or negative, and when the large stock returns are positive or negative. Both small and large stocks have their highest returns in the first quarter, regardless of whether their respective portfolios go up or down. Both portfolios' returns are statistically significant. This is evidence against the argument that individual investors, who sell small stocks for tax purposes in December and then buy them back in January, are the primary cause of the small-firm January effect. A firm-size effect is also evident. Small stocks experience an average return which is more than double the return of the larger stocks. We can conclude, therefore, that, in Canada, the January effect is more widespread than in the US. The difference may be explained by the fact that most Canadian stocks are small when compared with those in the US, and averaging the returns of large stocks in Canada may be equivalent to looking at the medium to small capitalization stocks in the US.

In attempting to associate the firm-size seasonal effect with the investing behaviour of individual and/or institutional investors, Athanassakos (1997) runs a single equation regression model that is equivalent to using a Zellner's Seemingly Unrelated Regression Specification on a set of four separate quarterly regressions, one per quarter (see Judge *et al.* (1985), 800–801). Table 4.5 reports the results from running that regression. Only the slope coefficients that are statistically significant are shown. The slope coefficient for the first quarter of the mutual fund variable (MUTF) is the only one that is positive and significant for the smallest quintile. None of the pension fund variables (PENF) are significant for the small stock portfolio. The most statistically significant variable for this portfolio is the individual investor variable (INDF). The slope coefficient for INDF in the third quarter is positive and significant at the 1% level, indicating that individual investors play an important role in explaining stock return variability in the third quarter. Individuals also play an important role in explaining stock price variability in the second quarter of the year. With regards with the largest stock portfolio, the slope coefficient for MUTF in the first quarter is positive and statistically significant. None of the other variables are significant. The results from this table can be interpreted as providing evidence that much of the firm-size seasonality in large stocks can be explained by institutional trading behaviour. There is no evidence that individual trading affects the variability of stock returns in the first quarter. However, there is strong evidence that individuals play an important role in the second and third quarter variability of smaller stocks in Canada. The institutional investing variables seem to affect more the large stock returns and the individual investing variable the small stock returns.

4.1.1 Price/Earnings Ratio

Every January for the period 1972 to 1988, Bourgeois and Lussier (1994) form six portfolios based on the P/E ratio on December 31 of the prior year. All stocks with a negative P/E ratio are allocated to one portfolio and all remaining stocks are allocated equally to the remaining five portfolios. The number of stocks vary from year to year depending on the number of stocks with negative P/E ratios for the year.

Table 4.5. Regression Results with the Quarterly Total Stock Returns (%) of the Smallest (R_{small}) and Largest (R_{large}) Size-Sorted Quintiles and their Difference ($R_{small}-R_{large}$) as the Dependent Variables for the Period 1962:Q1 to 1992:Q4: Quarterly Effect

Dependent Variable/ Independent Variables	R_{small}	R_{large}	$R_{small} - R_{large}$
Intercept	.165	—[a]	.134
	(4.19)*		(4.91)*
Q2	−.125	—	−.111
	(2.17)**		(2.77)*
Q3	—	—	—
Q4	—	—	−.078
			(1.94)**
MUTF1[b]	.00031	.022	—
	(1.80)***	(2.02)**	
MUTF2	—	—	—
MUTF3	−.00046	—	−.00033
	(2.15)**		(2.24)**
MUTF4	—	—	—
PENF1[c]	—	—	−.00013
			(1.84)***
PENF2	—	—	—
PENF3	—	—	—
PENF4	—	—	—
INDF1[d,e]	−.00005	—	—
	(1.67)***		
INDF2	.00004	—	—
	(1.69)***		
INDF3	.00011	—	.00009
	(3.15)*		(3.83)*
INDF4	—	—	—
DUM8092	—	—	—
R^2	.27	.14	.33
Durbin–Watson	1.62	1.68	2.00

Notes:
[a] — stands for not significant.
[b] Net quarterly change in real investment in equities by mutual funds (1986 $ billions).
[c] Net quarterly change in real investment in equities by pension funds (1986 $ billions).
[d] Net quarterly change in real investment in equities by individual investors (1986 $ billions).
[e] MUTF1, PENF1 and INDF1 capture the joint effect of the impact of mutual fund, pension fund and individual investor investing in the first quarter of the year on the dependent variables. An analogous interpretation is given to the other independent variables for the remaining quarters.
* Statistically significant at the 1% level.
** Statistically significant at the 5% level.
*** Statistically significant at the 10% level.

Bourgeois and Lussier (1994) attempt to find out whether the implementation of an investment strategy based on the P/E ratio and size could have resulted in a significantly higher average return than the TSE-300 stocks over the period April 1972 to March 1988. Table 4.6 reports the annual return of each of the six P/E ratio-based portfolios versus the SEI (Fund Evaluation Services of Toronto) Median portfolio. The lowest average returns over the period are recorded by the negative P/E ratio portfolio (portfolio 0) and the highest P/E ratio portfolio (portfolio 5). The lowest P/E ratio portfolio generate the highest mean return over the period studied and is superior to the SEI median and the TSE-300 portfolio. All risk-adjusted performance results confirm the above conclusion.

Bourgeois and Lussier (1994) also run OLS regressions of monthly excess portfolio returns against the excess market index returns. These results confirm their earlier results. The beta coefficients of the negative and high P/E ratio portfolios are larger than those of medium or low P/E portfolios. Moreover, the intercepts resulting from these regressions are negative for portfolios of negative and high P/E stocks and positive for all other portfolios. However, they are only statistically significant at the 5% level for portfolios 3 and 5. Hence, portfolios of low P/E stocks provide a statistically higher excess return than the TSE-300 index over the 16 year period studied.[8]

4.1.2 Other Price-Based Effects

Jog and Li (1995) test whether risk-adjusted portfolio returns are inversely related to a number of price related variables such as stock price level (P), market capitalization (MC), price to sales ratio (PSR), price to earnings ratio (PER), price to adjusted after-tax operating earnings (POER), price to cash flows ratio (PCR), price to book ratio (PBR) and price to debt ratio (PR). They expect the portfolios that are ranked lowest on any of the above price related variables, on average, to outperform the portfolios that are rated the highest on the respective price-related variables. The period of their study is January 1986 to December 1993. To account for any potential industry bias, they use 'industry standardized price related variables (ISPRV)' to construct portfolios. The ISPRV for a firm is defined to be the difference between its value for the firm and the corresponding cross-sectional industry average value, divided by the standard deviation of that variable across all sample firms within the industry. These industry adjusted variables are then used to create portfolio return series. The unadjusted variables are also used.

Table 4.7 reports the average returns, risks and the three risk-adjusted performance measures for portfolios constructed with each of the eight individual and two composite price related variables, the Pearson correlation coefficients

---

[8]This evidence is contrary to Bartholdy (1993) who could not find evidence that low P/E stocks have higher returns than high P/E ratio stocks.

Table 4.6. P/E Effect – Performance Results

	Annual Portfolio Returns (in percentage)							
Year Ending in March	#0	#1	#2	#3	#4	#5	SEI Median	TSE Index
1973	11.0	14.2	18.8	16.6	10.3	13.7	na	17.1
1974	−7.8	−1.9	−5.4	−7.3	−10.5	−9.0	−3.8	1.2
1975	−34.9	−24.22	−16.2	−12.4	−12.9	−20.0	−14.0	−14.4
1976	15.6	6.4	3.4	19.0	15.9	26.9	10.8	11.6
1977	0.6	0.7	7.1	8.2	6.4	2.4	−0.6	2.1
1978	15.1	15.2	24.2	16.4	24.7	33.3	8.4	9.3
1979	48.1	38.2	40.9	48.6	41.3	52.5	43.2	44.4
1980	10.3	50.4	39.6	14.2	16.4	14.7	23.5	28.1
1981	11.7	29.7	34.9	45.3	46.1	54.7	38.2	35.0
1982	−44.5	−50.1	−27.6	−27.2	−26.3	−19.3	−25.5	−28.8
1983	12.2	22.8	35.4	51.9	32.4	51.6	43.4	42.6
1984	6.3	21.4	25.6	17.2	26.9	14.4	13.8	14.7
1985	−9.4	−5.7	15.6	13.6	13.1	25.0	16.4	13.8
1986	5.0	7.4	23.7	33.9	26.8	36.7	25.6	20.6
1987	34.3	46.8	23.9	21.7	23.1	35.4	23.4	26.3
1988	−12.8	−18.9	−21.5	−20.6	−21.9	−18.5	−8.6	−8.7
Avg. yearly return	1.1	6.1	11.7	12.6	11.1	15.6	11.1[a]	
Without fees	3.0	7.4	13.2	14.1	13.0	17.2	–	11.6
Sharpe measure	−0.076	−0.025	0.050	0.075	0.060	0.119		
Treynor measure	−0.005	−0.002	0.003	0.004	0.003	0.007		
Jensen measure	−0.007	−0.003	0.001	0.002	0.001	0.004		
Median size	0.68	1.18	1.24	1.47	1.03	0.74		
$\tilde{R}_{pt} - RF_1 = \alpha_0 + \beta_p(\tilde{R}_{mt} - RF_t) + \mu_t.$								
Intercept (a_0) in %	−3.58 (1.45)	−0.068 (0.030)	0.253 (1.72)	0.280 (2.28)*	0.102 (0.85)	0.320 (1.95)*		
β	1.10 (23.84)*	1.11 (26.70)*	0.98 (35.59)*	0.81 (34.47)*	0.77 (34.21)*	0.81 (27.51)*		

Notes.
[a] The average return of 11.1% is for the period 1974 through 1988 only. During the same period portfolios #5 recorded an average performance of 15.7%.
* Statistically significant at the 5% significance level.

between the portfolios' returns and the returns on the TSE-300 index and the values of the portfolio means of the corresponding price-related variable. As shown in Table 4.7, the low ISPRV portfolios (Q1) tend to outperform the market on either raw or on a risk-adjusted return.

In contrast, the Q4 portfolios tend to underperform the market. In addition, most portfolios appear to be well diversified. The Pearson correlation coefficients between returns on these portfolios and returns on the TSE-300 index are around 0.83 and strategies based on price-related variables do not seem to trade off diversification benefit for return performance.

It appears that four price-related variables P, PER, PCR and POER ex-

Table 4.7. Industry Standardized Portfolio Risk-Return Characteristics and
Return Performance (1986:1–1993:12)

	Portfolio	Variable	Return	Standard Deviation	Beta	Sharpe Measure	Treynor Measure	Jensen Measure	PT–MKT Correlation
Panel A:									
Price Level	P1	$3.92	0.0163	0.0525	1.052	0.3102**	0.0155	0.0084***	0.83
	P2	$6.60	0.0129	0.0482	0.950	0.2684	0.0136	0.0051*	0.82
	P3	$13.03	0.0048	0.0446	0.970	0.1069	0.0049	−0.0030	0.90
	P4	$25.99	0.0034	0.0471	0.887	0.0813**	0.0038	−0.0044*	0.88
Panel B:									
Market	MC1	$56,201.50	0.0112	0.0466	0.885	0.2407	0.0127	0.0035	0.79
Capitalization	MC2	$85,828.80	0.0087	0.0478	0.905	0.1826	0.0096	0.0009	0.79
	MC3	$272,144.90	0.0097	0.0486	1.036	0.1999	0.0094	0.0019	0.88
	MC4	$1,772,610.00	0.0072	0.0460	1.015	0.1574	0.0071	−0.0006	0.91
Panel C:									
Price to Sales	PSR1	0.41	0.0101	0.0465	0.952	0.2171	0.0106	0.0023	0.85
Ratio	PSR2	0.81	0.0089	0.0455	0.971	0.1949	0.0091	0.0011	0.88
	PSR3	1.61	0.0104	0.0486	0.973	0.2132	0.0106	0.0026	0.83
	PSR4	3.63	0.0077	0.0480	0.956	0.1611	0.0081	−0.0001	0.83
Panel D:									
Price to	PER1	10.19	0.0188	0.0453	0.877	0.4146***	0.0214	0.0110***	0.80
Earnings	PER2	14.16	0.0078	0.0436	0.904	0.1791	0.0086	0.0000	0.86
Ratio	PER3	22.12	0.0090	0.0511	0.906	0.1767	0.0100	0.0013	0.73
	PER4	78.07	0.0056	0.0468	0.996	0.1190	0.0056	−0.0022	0.88
Panel E:									
Price to After	POER1	5.98	0.0126	0.0478	0.956	0.2637	0.0132	0.0048*	0.83
Tax Operating	POER2	10.46	0.0110	0.0447	0.961	0.2465	0.0115	0.0032	0.89
Earnings	POER3	15.82	0.0066	0.0435	0.922	0.1511	0.0071	−0.0012	0.88
Ratio	POER4	77.54	0.0078	0.0496	0.937	0.1567	0.0083	0.000	0.79
Panel F:									
Price to	PCR1	5.34	0.0130	0.0478	0.965	0.2728*	0.0135	0.0052*	0.84
Cashflows	PCR2	9.33	0.0104	0.0447	0.963	0.2331	0.0108	0.0026	0.89
Ratio	PCR3	13.43	0.0064	0.0459	0.963	0.1398	0.0067	−0.0014	0.87
	PCR4	72.77	0.0075	0.0462	0.880	0.1627	0.0085	−0.003	0.79
Panel G:									
Price to Book	PBR1	0.86	0.0104	0.0455	0.942	0.2282	0.0110	0.0052*	0.86
Ratio	PBR2	1.27	0.0067	0.0434	0.936	0.1550	0.0072	0.0026	0.89
	PBR3	1.86	0.0085	0.0460	0.944	0.1839	0.0090	−0.0007	0.85
	PBR4	4.01	0.0111	0.0518	1.021	0.2134	0.0108	0.0032	0.82
Panel H:									
Price to Debt	PDR1	0.55	0.0079	0.0496	1.023	0.1600	0.0078	0.0001	0.85
Ratio	PDR2	1.00	0.0087	0.0448	0.942	0.1948	0.0093	0.0009	0.87
	PDR3	2.00	0.0105	0.0526	1.015	0.1997	0.0104	0.0027	0.80
	PDR4	8.59	0.0093	0.0447	0.876	0.2077	0.0106	0.0015	0.81
Panel I:									
Equally	EWCV1	71.73	0.0165	0.0478	0.891	0.3457**	0.0185	0.0087***	0.77
Weighted	EWCV2	113.69	0.0092	0.0439	0.946	0.2096	0.0097	0.0014	0.89
Composite	EWCV3	146.23	0.0089	0.0446	0.939	0.1991	0.0095	0.0011	0.87
Variable	EWCV4	178.90	0.0065	0.0485	0.905	0.1345	0.0072	−0.0013	0.78
Panel J:									
Anomaly	AWCV1	71.24	0.0169	0.0471	0.852	0.3588**	0.0199	0.0092***	0.75
Weighted	AWCV2	113.67	0.0106	0.0449	0.968	0.2356	0.0109	0.0028	0.89
Composite	AWCV3	147.54	0.0093	0.0452	0.934	0.2056	0.0099	0.0015	0.86
Variable	AWCV4	177.81	0.0046	0.0487	0.924	0.0936	0.0049	−0.0032	0.79
Panel K:									
TSE300			0.0078	0.0415	1.000	0.1884	0.0078	0.0000	1.00

* Statistically significant at the 1% level.
** Statistically significant at the 5% level.
*** Statistically significant at the 10% level.

hibit the ability to generate 'abnormal' risk-adjusted returns compared with the TSE-300 index. It also appears that price-related variables tend to be inversely related to portfolio returns. These results provide evidence that price-related anomalies do exist in the Canadian stock market.

4.2 Price Momentum

Foerster *et al.* (1994) attempt to identify stocks that seem to 'ride the wave', i.e., that exhibit price momentum. Using monthly data for the period 1977 to 1992, they single-out from the TSE-100 stocks those stocks with the 'best outlook' and those stocks with the 'worst outlook' based on the weighted

Table 4.8. Summary Statistics, Net of a 2% Transactions Cost, for Portfolios of
10 'Best Outlook' and 'Worst Outlook' TSE-100 Stocks Over 60 Quarters
Between 1977 and 1992

	Best Outlook Portfolio	Best Outlook Less Transaction Costs	Worst Outlook Portfolio	TSE Total Return
Quarter Mean	10.3%	8.3%	2.7%	3.3%
Quarter Standard Deviation	13.7%	13.7%	13.2%	8.6%
Annual Mean	41.2%	33.2%	10.9%	13.1%
Annual Standard Deviation	27.4%	27.4%	26.4%	17.3
Portfolio β	1.09	1.09	1.20	1.00
Number of Quarters with Returns Greater than 0	50 of 60	45 of 60	35 of 60	45 of 60
Number of Quarters with Returns Greater than TSE	46 of 60	41 of 60	25 of 60	

average of the stocks' total returns for their preceding four quarters. The
most recent three month return is weighted twice as heavily as the return of
the other three quarters making up the twelve months. The 'best outlook'
portfolio consists of the stocks with the top ten weighted annual returns
over the last four quarters, while the 'worst outlook' portfolio consists of the
bottom ten stocks. At the beginning of each quarter, each of the ten stocks
in each portfolio are equally-weighted. If a particular stock is not sold at the
end of a quarter, it is rebalanced so that its dollar weighting equals the other
nine stocks in the portfolio.

Table 4.8 shows that the price momentum hypothesis is indeed successful. The average quarterly return for the 'best outlook' portfolio of stocks
was 41.2% annualized vs. 13.1% for the TSE and 10.9% for the 'worst outlook' portfolio.[9] Even after accounting for transaction costs, the 'best outlook' portfolio does statistically and economically better than the TSE or the
'worst outlook' portfolio. The results are equally dramatic after controlling
for risk using the Treynor index and Sharpe ratio for each portfolio. The 'best
outlook' portfolio continues to significantly outperform the TSE-300 and the
'worst outlook' portfolio.

[9]These results are overstated since Foerster *et al.* (1994) note a survivorship bias in
their sample by examining TSE-100 stocks as of the end of the sample period. Korkie and
Plas (1995) indicate a difference between 'best' and 'worst' portfolio returns of about 21%
after correcting for the survivorship bias.

Table 5.1. Business Cycles and Stock Returns

Expansion	Duration	TSE Peak	Cycle Peak	TSE Lead
Jan/47 – May/51	53 months	Apr/51	May/51	1 month
Jan/52 – May/53	17 months	Jul/52	May/53	10 months
Jul/54 – Jan/57	31 months	Aug/56	Jan/57	5 months
Feb/62 – May/74	160 months	Oct/73	May/74	7 months
Apr/75 – Jan/80	58 months	Feb/80	Jan/80	–1 month
Jul/80 – Jun/81	12 months	Nov/80	Jun/81	7 months
Dec/82 – Mar/90	88 months	Aug/89	Mar/90	7 months
Average	59.9 months		5.5 months	
Standard Deviation	47.6 months		3.5 months	
Recession	Duration	TSE Trough	Cycle Trough	TSE Lead
Jun/51 – Dec/51	7 months	June/51	Dec/51	6 months
Jun/53 – Jun/54	13 months	Sep/53	Jun/54	9 months
Feb/57 – Jan/58	12 months	Dec/57	Jan/58	1 month
Apr/60 – Jan/61	10 months	Jul/60	Jan/61	6 months
Jun/74 – Mar/75	10 months	Sep/74	Mar/75	6 months
Feb/80 – Jun/80	5 months	Mar/80	Jun/80	3 months
Jul/81 – Nov/82	17 months	Jun/82	Nov/82	5 months
Apr/90 – Dec/91	21 months	Nov/90	Dec/91	13 months
Average	11.9 months		6.1 months	
Standard Deviation	4.9 months		3.4 months	

5 Stock Returns and Economic Factors

5.1 Preliminary Findings

Business cycles describe expansion and contraction of economic activity as captured by such economic indicators as industrial production, service activity, corporate profits, and unemployment, which translate into changes in real GDP. In Canada, Statistics Canada documents economic activity and dates recessions (typically two consecutive quarterly declines in GDP growth) based on the depth or severity of economic decline, the duration, and the diffusion.

A history of business cycles in Canada since 1947 is outlined in Table 5.1 (see Foerster and Turnbull (1993)). The average business cycle expansion length is around five years, while the average recession is around one year. Canadian cycles are comparable in length to US cycles (see Seigel (1991)). There is much greater variability around the duration of expansions as opposed to contractions. Table 5.1 also indicates the peak (trough) in the TSE Index relative to the economic peak (trough). On average, the stock index tends to lead the business cycle by five to six months.

Over the 1956 to 1994 period, during expansions, the average annual capital gain on the stock index is 9.1% versus a 6.8% return on T-Bills. During recessions, the average annual capital gain on the stock index is –4.6% versus a 9.4% return on T-Bills. Consequently, any model able to predict business cycle turning points offers potential rewards versus a buy-and-hold equity strategy.[10]

5.2 Stock Returns and Economic Activity

While directly predicting business cycle turning points using subjective analysis can prove an elusive task, we can attempt to identify variables which are likely to be associated with business cycle turning points. We examine the relationship between a variety of economic variables and stock returns over the 1956 to 1993 period. The first variable is the **inflation rate** (see Fama and Schwert (1977) for a discussion of stock returns and inflation). The time period is divided into three categories of annual observations:

(1) 'inflation up' periods, defined as years in which year-over-year inflation (as of January 1) was at least 1% greater than inflation measured twelve months earlier;

(2) 'inflation down' periods, defined as years in which year-over-year inflation (as of January 1) was at least 1% less than inflation measured twelve months earlier; and

(3) 'inflation steady' periods, defined as years in which year-over-year inflation (as of January 1) was less than 1% different from inflation measured twelve months earlier.

Based on these definitions, 'inflation up' environment occurred 13% of the time. In only half of the years, the stock index rose. The average stock return (capital gains) was only 0.2%. The 'inflation down' environment occurred 13% of the time as well. However, the stock index rose in 70% of the years and the average gain was 12.7%. The 'inflation steady' environment occurred 46% of the time. The stock index rose most frequently — 78% of the years — although the average gain of 7.2% was lower than during the 'inflation down' environment. Thus changes in inflation rates appears to be related to future stock returns.

The second variable examined is the shape of the **yield curve** (see Campbell (1987)). This shape is captured by the difference between the long-term

---

[10]The data consist of monthly series of the TSE-300 Index of stock returns (capital gains only) and a variety of economic variables derived from CANSIM. The economic series include year-over-year inflation as measured by the Consumer Price Index, the 90-day Treasury-bill rate, the 90-day commercial paper rate, the long-term (over ten year) government of Canada bond yield, and the long-term corporate bond yield.

government yield and the 90-day T-Bill yield. A positive number represents an upper-sloping yield curve, while a negative number represents a downward-sloping or inverted yield curve. Over the sample period, the yield curve was upward-sloping during 82% of the months. Not surprisingly, this corresponds approximately with the percentage of time the Canadian economy is expanding since upward-sloping yield curves tend to be prevalent around the time of recoveries. In months with upward-sloping yield curves, the stock index increased, on average, by 7.6% (annualized), while during months with downward-sloping yield curves, the average gain was only 3.7%. There was also a tendency for stock returns to be even greater in months during which the yield curve tended to be even more positively sloped. For example, when long-term rates exceed short-term rates by at least 2% (which occurred 22% of the time), annualized returns were 10.8%.

The third variable examined attempts to capture bond **default risk** premiums (see Keim and Stambaugh (1986)). This risk premium is defined as the difference between the average yield on long-term corporate bonds and long-term government bonds. During the sample period, this spread is greater than 1% over 44% of the time. During these months, the average stock index return is -1.2% (annualized), while during months when the spread is less than 1% the average return is 12.7%.

Table 5.2 examines the average value of these three variables around business cycle turning points.[11] Near the peak of the business cycle, year-over-year inflation is increasing, the yield curve is inverted, and the risk premium spread is above 1%. Near the trough of the business cycle, year-over-year inflation is decreasing, the yield curve has become upward-sloping, and the risk premium is less than 1%.

5.3 A Tactical Asset Allocation Model

These three variables are combined in a regression equation to model expected stock returns. The regression equation is:

$$\text{TSE}_{t+1,t+6} = \alpha + \beta_1 \text{INFCHG}_t + \beta_2 \text{YIELD}_t + \beta_3 \text{RISK}_t + e_{t+1,t+6},$$

where $\text{TSE}_{t+1,t+6}$ is the price change on the TSE-300 Index over the next six months, α is an intercept term, INFCHG_t is the difference between the (year-over-year) inflation rate in a particular period and the rate one year earlier YIELD_t is the difference between the long-term government of Canada average bond yield and the 90-day Treasury-Bill yield, RISK_t is the difference between the 90-day rate on prime commercial paper and the 90-day Treasury-Bill yield, and $e_{t+1,t+6}$ is an error term. The regression results (tested over

---

[11]The risk premium variable is the difference between the 90-day T-Bill rate and the 90-day commercial paper rate.

Table 5.2. Economic Variables[a] Around Business Cycle Turning Points

Variable	Peak – 6 months	Peak	Trough – 6months	Trough
Inflation change	1.1%	2.0%	1.0%	−0.4%
Yield curve	−1.0%	−0.8%	0.4%	1.5%
Risk premium	1.0%	1.1%	0.9%	0.7%

Note:

[a] 'Inflation change' is the difference between the (year-over-year) inflation rate in a particular period and the rate one year earlier. 'Yield curve' is the difference between the long-term government of Canada average bond yield and the 90-day Treasury-bill yield. 'Risk premium' is the difference between the 90-day rate on prime commercial paper and the 90-day Treasury-bill yield.

the 1956 to 1992 period) are:

$$\text{TSE}_{t+1,t+6} = 4.33 - 0.52\text{INFCHG}_t + 1.06\text{YIELD}_t - 2.97\text{RISK}_t + e_{t+1,t+6},$$

$$(t = 5.1) \quad (t = 1.5) \qquad (t = 2.6) \qquad (t = 2.8).$$

Using these results and the results in Table 5.2, the expected six-month TSE Index return at typical business cycle peaks is 0.61% (or 1.2% annualized) and at troughs is 3.95% (or 8.1% annualized).

Using this model of expected returns, a simple tactical asset allocation model can be developed. If the expected six-month stock return exceeds the expected return from a T-Bill investment (based on the current T-Bill yield), then invest in the TSE Index; otherwise, invest in T-Bills. Over the sample period, the model results in switches between equities and T-Bills, on average, over one year apart. The strategy also produces returns 2.5% above a simple buy-and-hold equity strategy. Such a strategy offer potential gains, and decreases risk.

6   Stock Returns and Political Factors

6.1   The Relationship with Political Parties

The section examines the stock returns in Canada related to (1) the Canadian party in power, (2) the US party in power, (3) combinations of Canadian and US parties, and (4) majority versus minority governments.

Foerster (1994a) examines Canadian stock returns related to these factors. Since becoming a nation in 1867, Canada has experienced 35 federal elections. During the overall 1919 to 1992 period studied by Foerster, the Liberal party

Table 6.1. Average Annual Capital Gains in Canadian Stocks

	1919 to 1992	1919 to 1959	1960 to 1992
Overall	6.5%	6.1%	6.9%
Liberal	8.3%	9.8%	6.2%
Conservative	2.8%	−2.5%	8.1%
Democrat in US	−	−	10.4%
Republican in US	−	−	4.9%
Majority	6.2%	5.8%	6.7%
Minority	7.4%	7.2%	7.5%

was in power during fourteen terms or 68% of the time, while the Conservative party was in power during seven terms or 32% of the time. On a monthly basis, the Canadian stock market rose during 57% of all months, slightly more during Liberal terms (58%) versus Conservative terms (54%). However, overall capital gains by party in power is very sensitive to the sample period since the Conservative party governed during the Great Depression between 1930 and 1935. Thus while the stock market performed much better during Liberal terms between 1919 and 1959, the market performed better during Conservative terms between 1960 and 1992 (see Table 6.1).

During the 1960 to 1992 period, during Republican administrations, the US market rose by an average annual capital gain of 7.9% versus 6.5% during Democratic administrations. Paradoxically, Canadian markets performed better during Democratic administrations, gaining 10.4% annualized, versus 4.9% during Republican administrations. In order, the best performing Canadian markets occurred during Democrat/Conservative administrations, followed by Democrat/Liberal, Republican/Conservative, and Republican/Liberal administrations. These results suggest Canadian investors should welcome the election of Democratic administrations.[12]

Foerster (1994a) also examines returns during majority versus minority governments. Until the 1994 election, the third major federal party, the left-wing New Democratic Party (NDP), often played the role of 'spoiler' preventing either the Liberals or Conservatives with a majority of seats. Consequently, 24% of the time over the full sample period, minority governments were in power. These minority governments were often loosely based on coalitions, but were always in danger of loosing 'no confidence' votes which would force new elections. One might expect increased political uncertainty during such terms to have an adverse effect on stock performance, but such was not

---

[12]Hensel and Ziemba (1995a) document that US small cap investors should also welcome Democratic administrations.

the case. Not only was the market up during a particular month with a minority government (60% of the time versus 56% of the time during majority governments), but average returns were higher and standard deviations were lower.

6.2   Stock Market Reactions Around Canadian Elections

There are competing views concerning how stock markets are expected to react around elections. US evidence suggests markets react more favorably to the election of 'business-friendly' Republican presidents rather than 'labor-friendly' Democrat presidents. However, it is possible that investors simply react positively to any election results, perhaps due to the resolution of uncertainty regarding which party will govern. Foerster (1994a) examines the stock market return around nine elections between 1963 and 1992. Six of the elections were subsequently won by the Liberal party and three by the Conservative party. Canadian federal elections are generally called two months before the election is to occur (approximately 40 trading days). Foerster defines 'excess daily returns' as returns on the Toronto Stock Exchange (TSE) market in excess of US returns.

The cumulative excess return between the calling of an election and three trading days before the election is only 0.2%; between three days prior and three days subsequent to the election is 1.6%; and between four and thirty days subsequent to the election is 1.0%. The largest and most significant excess daily return occurs on the trading day prior to the release of election results (0.8%). The results are very similar whether the election is won by Liberals or Conservatives. These results support two (non-competing) hypotheses: (1) markets react positively to the resolution of uncertainty; and (2) markets interpret the policies of Liberals and Conservatives to be quite similar – certainly more similar than Democrats and Republicans. Further support for the first hypothesis comes from an examination of the ten changes of government between 1919 and 1992 (i.e., a different party forming the government). During such months, the average *monthly* gain is 2.7%; since 1960, such changes result in monthly gains of 3.9%.

6.3   Canadian Stock Market Performance Related to the US Election Cycle

Foerster (1994b) examines year-by-year capital gains on the TSE market between 1959 and 1994 (36 years). He defines Year 1 of the four-year US election cycle starting January first when the President begins his term. Defined in this manner, average annual returns during election Years 1 through 4 are

9.2%, −5.1%, 13.8% and 9.9% respectively. As in the US, the third year provides the highest returns. Interestingly, the TSE market is up in two-thirds of the Year 1s and Year 4s, in only one-third of Year 2s, and in all Year 3s.

Foerster and Schmitz (1995) examine the predictability of stock returns in most countries with well-established stock markets, including Canada, related to US cycles. In their study, year 1 of the four year US election cycle is defined as the twelve months commencing November 1 of a US election (e.g., 1992) and concluding October 31 of the next year. Year 2 begins on November 1 of the year *following* the most recent US election (e.g., 1993) and concludes on October 31 of the following year. Years 3 and 4 of the election cycle are defined in a similar manner.

The stock market capital gains returns cover the period from February 1957 to December 1994. As Tufte (1978) notes, this time period corresponds to the period in which US administrations have taken a more active role in managing the economy. The sample encompasses ten election cycles.

Over the period, the average annualized capital gain in Canada is 6.7% (including a dividend yield over the time period of approximately 3.5%, total returns were approximately 10.2%). The annualized capital gains returns for years 1 through 4 of the US election cycle were (in order) 6.8%, −3.8%, 12.6% and 13.0%. During the same period, the average US capital gain was 7.3% and the annualized capital gains returns for years 1 through 4 of the US election cycle were (in order) 4.3%, −2.8%, 21.6% and 8.8%. The most striking feature is the capital losses in both the US and Canada during year 2. In Canada, the average difference between year 2 returns and years 1, 3, and 4 is 14.5%, slightly higher than the 14.0% difference in the US. One interpretation of these results is that the US administration in power tends to influence any 'bad news' about the economy during the second year, then attempts to influence 'good news' such as easing of interest rates in years 3 and 4 leading up to the next election. Since the Canadian economy is tied closely to the US economy, a similar effect occurs in Canada.

The relative importance of the US election cycle is examined by means of regression analysis. Foerster and Schmitz (1995) expand on previous stock market predictability studies by including a dummy variable for the second year of the US Presidential election cycle. The year 2 dummy variable, denoted $ELECT2_t$, is set to one in each month during year 2 of the US election cycle and zero in all other years. They attempt to determine if this variable, after being added to common conditioning stock return information variables, captures any further explanatory power. They obtained support for the hypothesis that the second year of the US election cycle is an important economic effect and that it is not simply a proxy for other commonly used information variables.

One other information variable used as a predictor of local stock market returns is the lagged Canadian short-term interest rate, denoted SHT_{t-1}. The lagged short-term interest rate captures the level of expected inflation within the local economy. The second information variable is the lagged Canadian term structure of interest rates. The local term structure variable, denoted $TERM_{t-1}$, is defined as the difference between the local long-term government bond rate and the short-term interest rate. The term structure variable and stock returns are expected to be positively correlated. The third variable is a January dummy variable. Denoted JAN_t, the January dummy variable is set to one in January of each year and zero in all other months. The January effect in stock returns has been shown to be predominately positive.

The resulting (monthly) regression equation is:

$$R_t = 1.22 + 1.85JAN_t - 0.05SHT_{t-1} - 0.02TERM_{t-1} - 1.59ELECT2_t + e_t,$$

where R_t was the monthly capital gain on the Canadian (Toronto) stock market and e_t was a random error term. The January dummy was significant at the 10% confidence level and the election dummy was significant at the 1% level, while no other variable was statistically significant.

In another set of regressions, Foerster and Schmitz (1995) investigate whether the year 2 election dummy variable is simply a proxy for another US or 'global' factor. For example, it is possible that in the second year of the election cycle, as the result of an incumbent president influencing monetary or fiscal policy, some US financial variables such as the US short-term interest rate might be quite cyclical. Since the US possesses the largest and most influential financial markets in the world, an increase in, say, the US T-Bill rate may have a profound effect on international financial markets and, in particular, Canadian markets. They control for these factors by including US information variables such as the lagged short-term US interest rate, $USSHT_{t-1}$, the lagged US term structure, $USTERM_{t-1}$, and the lagged US dividend yield, $USDIV_{t-1}$, in addition to the previously described Canadian information variables. The resulting (monthly) regression equation was as follows:

$$
\begin{aligned}
R_t &= -2.47 + 1.79JAN_t + 0.16SHT_{t-1} + 0.11TERM_{t-1} - 0.67USSHT_{t-1} \\
&\quad -0.26USTERM_{t-1} + 1.79USDIV_{t-1} - 1.83ELECT2_t + e_t.
\end{aligned}
$$

Based on t-tests, the January dummy was significant at the 10% level (p-value of 0.056), both the US dividend variable and the election variable were significant at the 1% level (p-values of 0.000 and 0.001 respectively), and all other variables were not significant. Thus after controlling for these factors the $ELECT2_t$ variable is still significant in determining Canadian stock returns. In other words, the results support for the notion that the US election cycle is important in and of itself to Canadian stock returns, and is not simply capturing the residual correlations of other US financial variables.

Foerster and Schmitz (1995) also examine a simple tactical asset allocation strategy based on the observation that year 2 returns tend to be lower than returns in years 1, 3, and 4. The switching strategy tested is simply: *invest in Canadian equities in years 1, 3, and 4, and invest in Canadian T-Bills in year 2.* A buy-and-hold equity strategy (capital gains only) results in an annualized return of 7.3% compared with the election cycle switching strategy of 11.4% over the February 1961 to December 1994 time period.

Appendix: Canadian Taxation and Transaction Costs

A.1. Ordinary Cash Dividends

Since 1949, ordinary cash dividends have been given a special treatment in Canada in order to minimize the effects of double taxation. As a result of this special tax treatment of ordinary cash dividends, in combination with the various tax reforms of the Canadian tax system over time, the unfavourable taxation of dividends versus capital gains has steadily declined over time in Canada. Whereas the difference between the tax on dividends and that on capital gains was larger in Canada than the US before 1972, the situation has been subsequently reversed.

Specifically, the marginal tax rate on ordinary cash dividends over the different tax regimes in Canada[13] has been calculated as follows:

$$\text{Pre-1972} \quad T_d = t_{fp} - C_{fp}$$
$$\text{Post-1971} \quad T_d = (1 + G)(t_f - C)(1 + t_p)$$

where

T_d is the tax payable per dollar of dividend income,

G is the gross-up factor[14]

t_f is the marginal federal tax rate on ordinary income.

---

[13]Provincial taxes in Quebec are calculated differently in Quebec rather than a single percentage of the Federal tax rate.

[14]To encourage investments in equities and as part of the scheme to integrate corporate and shareholder income and eliminate double taxation of dividends, dividend income earned by Canadian individuals from taxable Canadian Corporations is afforded a special treatment. Investors include $(1 + G)\%$ of such dividends received in their computation of taxable income. The incremental $G\%$ was called Gross-up, a term invented by the Department of Finance. Federal tax was computed on this grossed-up amount. From the federal tax, the investors then subtract a dividend tax credit equal to $C\%$ of the dividend. Provincial tax is then calculated on the residual net federal tax payable.

C is the dividend tax credit.

t_p is the marginal provincial tax as a (constant) percentage of the federal tax payable.

t_{fp} is the combined federal/provincial marginal tax rate.

C_{fp} is the combined federal/provincial dividend tax credit.

A.2. Stock Dividends

The taxation of stock dividends in Canada has changed several times. On April 1, 1977, the Canadian government began to treat stock dividends as capital gains taxable only at the time of their disposition.[15] Thus, as in the US, stock dividends became tax neutral. On May 24, 1985, however, stock dividends became taxable like cash dividends in Canada.

Specifically, the marginal tax rate on stock dividends has been calculated as follows:

$$\text{Pre-March 31/1977:} \quad T_{sd} = (1 + G)(t_f - C)(1 + t_p)$$

$$\text{April 1/1977–May 23/1985:} \quad T_{sd} = \left(\frac{t_f}{2}\right)(1 + t_p)$$

$$\text{Post- May 23/1985:} \quad T_{sd} = (1 + G)(t_f - C)(1 + t_p)$$

where

T_{sd} is the tax payable per dollar of stock dividend.

B. The Taxation of Capital Gains

In Canada, realized capital gains have been taxed as follows:

$$\text{Pre-1972:} \qquad T_{cg} = 0$$

$$\text{Post-1971:} \quad T_{cg} = \left(\frac{t_f}{k}\right)(1 + t_p) \qquad \begin{array}{l}(k = 2 \quad \text{pre-1989}) \\ (k = 1.5 \quad 1989\text{--}1990) \\ (k = 1.333 \quad \text{post-1990}),\end{array}$$

where

-----

[15] A stock dividend paid to a Canadian resident shareholder by a public corporation was not included in income when received and the shares received under sub-section ITA 52(3) were deemed to have been acquired at a cost of nil. That meant that the whole dollar equivalent of the stock dividend was considered a capital gain and taxed as such in the year in which the stock was sold. If a shareholder already owned 'identical' shares to those distributed as the stock dividend (which is true for all stock dividends), then the cost of shares determined under the sub-section ITA 52(3) received as a stock dividend would be averaged under these rules with the Adjusted Cost Base of the other identical shares owned.

T_{cg} is the tax paid per dollar of capital gain.

In addition, on May 24, 1985, the Canadian government introduced an exemption on capital gains. The exemption was scheduled to rise from $20,000 in 1985 to $500,000 in 1990. In 1989, the Canadian government substantially reduced the relative attractiveness of capital gains versus dividends. The lifetime exemption was limited to $100,000 and the marginal tax rate on capital gains above the exemption for the top bracket investor was raised to approximate that of dividends. Finally, in 1995, the Canadian government discontinued altogether the life time exemption for capital gains. Capital gains are now taxed more heavily than dividends for the top bracket investor in Canada.

Transaction Costs

Pre-1983 Prior to 1983, commissions in Canada were fixed and Canadian exchanges had tough rules on what services a broker could provide to a client for commission business.

Post-1975 Despite the fixed commission regime prevailing in Canada, there were a number of developments which, following the introduction of negotiable commissions in the US (May 1975), resulted in reduced commissions in Canada in the post-1975 years vis-á-vis earlier years. They were initiated by Canadian exchanges and regulatory authorities in an attempt to become competitive with US exchanges, where many Canadian stocks were interlisted. First, investors could negotiate on transactions in excess of $500,000. This led to a substantial decline in commission rates for these transactions between 1976 and 1980. Second, exchange schedules provided for a tapered reduction in commission rates for orders of more than $25,000. Third, traders who sold a stock within 45 days from buying it could get a discount on selling commission of up to 50%.

Post-1982 On April 1, 1983 commissions became negotiable in Canada. The era of negotiated commissions is generally associated with (a) the introduction of discount brokerage firms into the Canadian securities brokerage field and (b) a new set of commission policies or guidelines for full service brokerage firms. Discount brokers have offered discounts of up to 85% off the full service brokerage firms commission.

D. Odd-Lot Costs

Unlike the US markets, there have generally been no odd-lot costs for stocks that trade on the TSE. This has been a consequence of the Computer Assisted

Trading System (CATS) which was installed in 1977. The system charges no premium or discount on odd-lot orders. Thus, on the TSE, investors do not face this additional transaction cost that US investors do.

E. Short-Sales

Two Canadian tax regulations pertaining to dividend treatment in the case of short sales are worth noting. First, the proceeds from short sales are generally treated as regular income and taxed at the marginal tax rate for ordinary income, which for taxed investors is higher than the rate applicable to either dividends or capital gains.[16] Second, in a long position, the dividends get preferential tax treatment, while in a short position, the dividend payout affects the adjusted cost base. The implication of these two points is that the borrower of shares in short sales can fully deduct, as carrying charge, any payment to the lender of the shares to cover dividend payments.[17]

F. Institutional Investors

Canadian Pension Funds and other Fund managers were not allowed by regulators to use the futures and options markets to hedge their risks prior to July 1, 1992.

References

Athanassakos, G., (1992), 'Portfolio Rebalancing and the January Effect in Canada', *Financial Analysts Journal*, November/December, 67–78.

Athanassakos, G., (1995a), 'The January Effect: Solving the Mystery', *Canadian Investment Review*, **8** (1) 23–29.

Athanassakos, G., (1995b) 'Is the January Effect Predictable: Evidence from Canada', Wilfrid Laurier University.

Athanassakos, G., (1996) 'Tax Induced Trading Volume Around Ex-Dividend Days Under Different Tax Regimes: The Canadian Experience 1970–1984', *Journal of Business Finance and Accounting* **23** 557–584.

Athanassakos, G., (1997), 'Firm Size, Stock Return Seasonality, Temporal Variability of Stock Returns and the Trading Pattern of Individual and Institutional Investors: the Canadian Experience', *Journal of Investing* **6** 75–86.

Athanassakos, G., and B. Smith, (1996), 'Odd-Lot Costs, Taxes and the Ex-Date Price Effects of Stock Dividends: Evidence from the Toronto Stock Exchange', *Journal of Business Finance and Accounting* **23** 989–1003.

---

[16]The pertinent rules are discussed in paragraph 18 of Revenue Canada interpretation Bulletin IT 479R, issued in February 1984.

[17]As of July 1, 1989, this is no longer true.

Athanassakos, G., and D. Fowler, (1993), 'New Evidence on the Behaviour of Canadian Stock Prices in the Days Surrounding the Ex-Dividend Day', *Quarterly Journal of Business and Economics* **32** 26–50.

Athanassakos, G., and M. Robinson, (1994), 'The Day-of-the-Week Anomaly: The Toronto Stock Exchange Experience', *Journal of Business Finance and Accounting* **21** 833–856.

Athanassakos, G., and J. Schnabel, (1994), 'Professional Portfolio Managers and the January Effect: Theory and Evidence', *Review of Financial Economics* **4**(1) 79–91.

Bartholdy, J., (1993), 'Testing for a Price-Earning Effect on the Toronto Stock Exchange', *Canadian Journal of Administrative Sciences* **10**(1) 60–67.

Berges, A., McConnell, J.J., and G.G. Schlarbaum, (1984), 'The Turn-of-the-Year in Canada', *The Journal of Finance* **39** 185–192.

Booth, L., (1995), 'On Shaky Ground', *Canadian Investment Review* **7** (1) 9–15.

Bourgeois, J., and J. Lussier, (1994), 'P/Es and Performance in the Canadian Market', *Canadian Investment Review*, Spring, 33–39.

Campbell, J., (1987), 'Stock Returns and the Term Structure, *Journal of Financial Economics*, **18** 373–399.

Chamberlain, T.W., C.S. Cheung and C.C.Y. Kwan, (1988), 'Day-of-the-Week Patterns in Stock Returns: The Canadian Evidence', *Canadian Journal of Administrative Sciences* **5** (4) 51–55.

Fama, E. and G. Schwert, (1977), 'Asset Returns and Inflation', *Journal of Financial Economics* **5** 115–146.

Foerster, S., (1993), 'The Daily and Monthly Return Behaviour of Canadian Stocks', *Canadian Capital Markets*, The Toronto Stock Exchange – University of Western Ontario.

Foerster, S., (1994a), 'Stock Market Performance and Elections: Made-In-Canada Effects?' *Canadian Investment Review* **7** (2) 39–42.

Foerster, S., (1994b), 'The US Presidential Cycle in Canadian Stocks', *Canadian MoneySaver*, December, 5–6.

Foerster, S., (1994c), 'The Performance of Canadian Stocks: Lessons from a 75-year Perspective', *Canadian MoneySaver*, July/August, 6–8.

Foerster, S., (1995), 'The Shrinking Equity Premium', *Canadian MoneySaver*, September, 3–5.

Foerster, S., and D. Porter, (1993), 'Calendar and Size-Based Anomalies in Canadian Stock Returns'. In *Canadian Capital Markets*, M. Robinson and B. Smith (eds.), 133–140.

Foerster, S., A. Prihar, and J. Schmidt, (1994/95), 'Back to the Future', *Canada Investment Review* **7** (4) 9–13.

Foerster, S. and A. Turnbull, (1993), 'The Key to Effective Tactical Asset Allocation', *Canadian Investment Review* **6** (1) 13–19.

Griffiths, M.D. and R.W. White, (1993), 'Tax Induced Trading and the Turn-of-the-Year Effect: An Intraday Study', *Journal of Finance* **48** 575–598.

Hatch, J. and R. White, (1985), 'Canadian Stocks, Bonds, Bills, and Inflation: 1950–1983', *The Financial Analysts Research Foundation*, Charlottesville, Virginia.

Hatch, J. and R. White, (1988), 'Canadian Stocks, Bonds, Bills, and Inflation: 1950–1987', *The Research Foundation of the Institute of Chartered Financial Analysts*, Charlottesville, Virginia.

Hensel, C. and W.T. Ziemba, (1995a), 'United States Investment Returns during Democratic and Republican Administrations, 1928–1993', *Financial Analysts Journal* **51**, (2) 61–69.

Hensel, C. and W.T. Ziemba, (1995b), 'The January Barometer: European, North American, Pacific and Worldwide Results', *Finanzmarket and Portfolio Management* **9** (2) 187–196.

Ibbotson, R. and R. Sinquefield, (1982), 'Stocks, Bonds, Bills, and Inflation: The Past and the Future', *The Financial Analysts Research Foundation*, Charlottesville, Virginia.

Ibbotson, R., (1987), *Stocks, Bonds, Bills, and Inflation: 1986 Yearbook*. R.G. Ibbotson Associates. Chicago, Illinois.

Jaffe, J., D.G. Keim, and R. Westerfield, (1989), 'Earnings Yields, Market Values, and Stock Return', *Journal of Finance* **44** 135–149.

Jog, V., and B. Li, (1995), 'Price Related Anomalies on the Toronto Stock Exchange', *ASAC 1995 Conference Proceedings, Finance Division*, **16**(1) 47–59.

Judge, G., W. Griffiths, C. Hill, H. Lutkephol and T. Lee, (1985), *The Theory and Practice of Econometrics*. Wiley.

Keim, D. and R. Stambaugh, (1986), 'Predicting Returns in the Stock and Bond Markets', *Journal of Financial Economics* **17** 357–390.

Korkie, B. and J. Plas, (1995), 'Back to Reality: Another Look at Share Price Momentum Strategies', University of Alberta Working Paper.

Nordhaus, W., (1975), 'The Political Business Cycle', *The Review of Economic Studies* **42** 169–190.

Seigel, J., (1991), 'Does It Pay Stock Investors to Forecast the Business Cycle?', *The Journal of Portfolio Management* **18** (1) 27–34.

Tinic, S., G. Barone-Adesi and R. West, (1987), 'Seasonality in Canadian Stock Prices: A Test of the "Tax Loss-Selling Hypothesis" ', *Journal of Financial and Quantitative Analysis* **22** (1) 51–64.

Tinic, S. and G. Barone-Adesi, (1988), 'Stock Return Seasonality and the Tests of Asset Pricing Models: Canadian Evidence'. In *Stock Market Anomalies*, E. Dimson, (ed.), Cambridge University Press, 129–146.

Tufte, E., (1978), *Political Control and the Economy*. Princeton University Press.

Seasonal Anomalies in the Italian Stock Market, 1973–1993*

Elio Canestrelli and William T. Ziemba

Abstract

This paper investigates the existence of a number of seasonal anomalies in the Italian stock market from 1973 to 1993. The results show that effects that have been found in the US, Japan and other markets such as the weekend, turn-of-the-month, monthly, holiday and the January barometer were present in Italy during this period.

1   Introduction to the Italian Stock Market

The major stock market in Italy is in Milan. Its capitalization amounts to about 1.5% of the world equity index or 5% of the New York Stock Exchange. The COMIT index, which does not include dividends and is computed by the Banca Commerciale Italiana, measures the performance of the Milan exchange. The index had a base of 100 on December 31, 1972. Its values from January 1974 to December 1993 are shown in Figure 1. In general equity performance from 1986 to 1993 was poor. A major part of the reason for this has been high and steady bond yields which provided a very competitive asset class with a high mean return and low standard deviation during this period, see e.g. Banca d'Italia (1983–1997). After the period of this study, ending in 1993, bond yields and other interest rates have fallen dramatically and equity returns have been much higher; the 30-year bond fell to 5.69% in March 1998 versus 8.71% in March 1997 and 10.93% in March 1996.

The COMIT index is calculated daily based on closing prices and utilizes all securities traded on the Milan stock exchange. All dividends are generally paid on the first (zero) day of the monthly account so that day is considered separately or omitted from price comparisons. This is discussed below. For securities that have had no changes in the amount and composition of their nominal capital since the base date, the index is a Laspeyres type or *starting year method* and is defined by

$$I_{\text{COMIT}} = \frac{\text{Today total capitalization}}{\text{Initial (basic capitalization)}} \times 100 = \frac{\sum_{i=1}^{n} p_{1i} q_{01}}{\sum_{i=1}^{n} p_{0i} q_{0i}} \times 100$$

where

---

*Thanks are due to Stefano Frezza for research assistance.

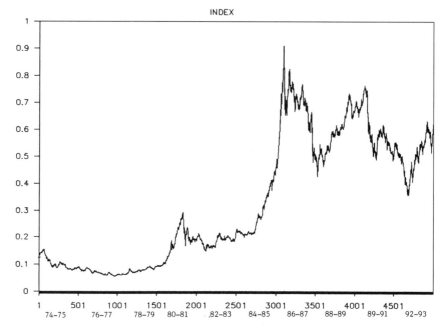

Figure 1. The COMIT Index, January 1, 1974 to December 31, 1993

n = number of securities,

p_{0i} = base price of the ith security,

p_{1i} = current price of the ith security,

q_{0i} = number of shares in the nominal base capital of i,

$p_{0i}q_{0i}$ = base capitalization of the ith security, and

$p_{1i}q_{0i}$ = current capitalization of the ith security.

For shares of companies that have had changes in the amount or composition of the nominal capital since the base time, the index uses a Paache type or *given year method* and is defined by

$$I = \frac{\sum_{i=1}^{n} p_{1i}q_{1i}}{\sum_{i=1}^{n} p_{0i}q_{1i}} \times 100$$

where

q_{1i} = number of shares of i after a change in capital.

This formula is modified so that values which are homogeneous in time can be compared. Paache's formula becomes

$$\frac{p_{1i}q_{1i}}{p_{0i}q_{0i} + p_{0i}q_{ai}}$$

where

$p_{0i}q_{ai}$ is the additional investment in security i computed at the mean price level at the base time, and

q_{0i} is the number of new shares issued by i so $q_{1i} = q_{0ii} + q_{ai}$.

Hence, the initial value $p_{0i}q_{0i}$ becomes

$$\frac{p_{0i}}{p_{1i}} \times p_{si}q_{ai}$$

where $p_{si}q_{ai}$ is the increase in capital, namely the subscription price p_{si} times q_{ai}, the number of new shares issued. Thus with capital increases, the initial capitalization in the COMIT index becomes

$$p_{0i}q_{1i} = p_{0i}q_{0i} + \frac{p_{0i}}{p_{1i}} \times p_{si}q_{ai}.$$

The COMIT index is then

$$I_{\text{COMIT}} = \frac{\sum_{i=1}^{n} p_{1i}q_{1i}}{\sum_{i=1}^{n} \left(p_{0i}q_{1i} + \frac{p_{0i}}{p_{1i}} \times p_{si}q_{ai} \right)} \times 100.$$

The number of separate companies on the exchange varied from about 150 to 200 with a slightly higher number of individual equity stocks in the index. As of the end of 1993 there were 324 equities worth 232,005 billion lira to give an index value of 619.48. The index year end peaked in 1986 at 722.78; see Table 1.

The data used in this study consist of daily returns on the COMIT index from January 3, 1973 to December 31, 1993. This period had of 7668 days of which 5258 were trading days with the balance Saturdays, Sundays, holidays and 33 days where trading was suspended. The highest daily return was 8.03% and the lowest −10.02%. Sample statistics on the data appear in Table 2. Returns are defined as the log price relatives namely $100 \log[p_{ti}/p_{t-1,i}]$.

Figure 2 provides the daily return distribution with the best Gaussian fit superimposed. Like most other equity markets, the Gaussian underestimates the center of the distribution and both extreme event tails.

Table 1. Milan stock exchange market capitalization (in billions of lira),
1975–1993

Year	Number of Companies	Number of Stocks	Market Capitalization	Yearly Trading Volume	Ave. Daily Trading Volume	COMIT Index (31/12/72=100)
1975	153	171	7426	1118	5	82.79
1976	155	173	6994	1092	4	74.89
1977	145	168	5371	754	3	55.68
1978	143	165	8147	1620	7	68.83
1979	139	164	10339	2875	12	82.45
1980	134	174	23541	7343	29	170.20
198 1	132	178	28749	12334	49	192.43
1982	138	190	27299	3770	15	164.06
1983	139	201	34698	5880	23	191.48
1984	143	213	49793	7143	28	230.31
1985	147	214	98195	26315	104	457.04
1986	184	284	190472	66571	265	722.78
1987	205	316	140721	41566	164	487.99
1988	211	317	176827	41264	163	589.72
1989	220	328	215244	53238	210	687.58
1990	223	332	168135	50596	204	514.70
1991	225	337	177359	31010	126	507.59
1992	228	343	169533	33075	146	446.33
1993	216	324	232005	103908	406	619.48

Settlement procedures

The settlement procedure in Italy during the period of study was an interesting one. Understanding it enables us to understand the existence of the turn-of-the-month effect in Italy although it is very different from those in the US and elsewhere as discussed in other papers in this volume. For a typical period, say February–March, we have the *operating cycle of March*. It starts from the '0th' day of the cycle and goes up to the trading day prior to the '0th' day of the following cycle. The '0th' day is the first day of the monthly account; it is on the 18th calendar day (approximately, with minor changes around this date). All trades on or after the '0th' trading day of cycle of March (approximately 18th calendar February) up to the trading day prior to the '0th' day of cycle of April (aproximately 18th calendar March) are settled on the last trading day of calendar March. Hence trading in the calendar second half of February and the calendar first half of March are all settled on the last trading day of calendar March. Thus the trading month is roughly calendar 18–17 versus the calendar 1–31. Not surprisingly there are higher

Table 2. Data and sample statistics on the COMIT index,
January 3, 1973 to December 31, 1993

Trading Days		Sample Statistics, %	
Mondays	1039	Sample size	5258
Tuesdays	1052	Mean	0.0328
Wednesdays	1059	Median	0.02
Thursdays	1030	Mode	−0.09
Fridays	1077	Variance	1.7467
Saturdays & Sundays	2184	Standard dev	1.3216
Holidays	194	Minimum	−10.02
Suspended days	33	Maximum	8.03
		Skewness	−0.4809
Total days	7668	Kurtosis	5.7195

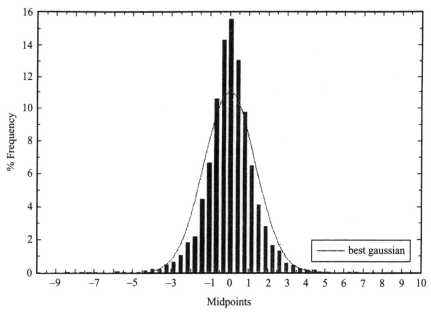

Figure 2. Daily return distribution of the COMIT,
1974–1993 with Best Gausian fit

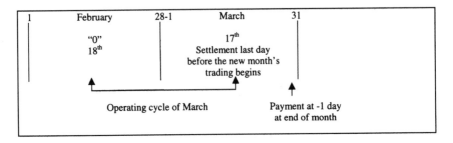

Italian Settlement Procedure. The 18th of calendar February is the '0th' (zero) day of first day of trading (operating) cycle of March. The 17th of calendar March is the last day of the same operating cycle (so-called cycle of March). One could buy and sell during this cycle and settle both at the end (last trading day of calendar) of March (usually 31st of March, or if it is a holiday, 30th of March).

returns in the second half and lower returns in the first half of the month as Italy's turn-of-the-month effect. One could buy then sell the same stock and settle both at the end of the month. The data for this study ended in December 1993 just prior to settlement rules changes made in January 1994 that were meant to eventually yield a more standard 5-day settlement akin to the US 3-day system. This 5-day electronic market system was fully phased in on February 16, 1996.

The seasonality patterns studied in this paper consist of

- the weekend effect where raw returns and returns adjusted for auto correlation are used to investigate the existence of the Monday effect and determine if the weekly patterns were consistent over time,

- the January barometer and the January effect,

- the turn-of the-month effect,

- monthly effects, and

- the holiday effect.

This research follows that in the US, Japan and elsewhere reported and referenced in other papers in this volume including that in Ziemba and Schwartz (1991) and the earlier study in Italy by Barone (1990) that covers the period January 2, 1975 to August 22, 1989.

2   Day of the week effects

During the period of this study there was a strong day of the week effect. Mondays and Tuesdays had negative mean returns and Wednesdays to Fridays

Table 3. The Day of the week effect

Time Period	Monday	Tuesday	Wednesday	Thursday	Friday
1973–1993	−0.1955	−0.1729	0.0263	0.0662	0.1297
St. Dev.	1.4686	1.2399	1.2359	1.2359	1.1296
Observations	971	1023	1007	992	1013
t-Stat	−4.5587	−4.4550	0.6759	1.6863	3.6545
1973–1979	−0.2864	−0.2817	−0.0668	0.0547	0.1465
St. Dev.	1.3281	1.1669	1.2127	1.0567	1.0383
Observations	316	340	323	329	338
t-Stat	−4.0634	−4.4518	−0.9897	0.9391	2.5941
1980–1989	−0.1298	−0.1351	0.1204	0.0477	0.1493
St. Dev.	1.5747	1.3370	1.2905	1.3845	1.2319
Observations	467	487	490	479	485
t-Stat	−1.7813	−2.2294	2.0653	0.7547	2.6683
1990–1993	−0.2059	−0.0782	−0.0563	0.1346	0.0499
St. Dev.	1.4187	1.0959	1.1158	1.1212	1.008
Observations	188	196	194	184	190
t-Stat	−1.9900	−0.9986	−0.7027	1.6287	0.6822

had positive mean returns. Mondays had the lowest mean returns and the mean return of each succeeding day was monotonically higher; see Figure 3 and Table 3. The negative Monday and Tuesday and positive Friday mean returns were statistically significant. The mean returns are similar across the sub-periods 1973-79, 1980-89 and 1990–93; see Figure 4.

A regression model that separates out the independent effects of the various days of the week is shown in Table 4a. The regression is significant for the entire period and each of the sub-periods (except the last). This is tested via the hypothesis

$$H_0 : a_2 = a_3 = a_4 = a_5 = 0.$$

For the model

$$R_t = a_t + \sum_{i=2}^{5} a_i d_i + u_t \qquad , d_i = \{0.1\},$$

a_1 = Monday, and $a_2, \ldots, a_5$ are the differences from Monday, for Tuesday to Friday. In this analysis the first trading day of the month was excluded to avoid settlement biases. The equity market was closer to efficient after 1990

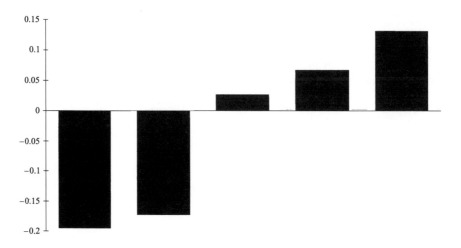

Figure 3. The day of the week effect

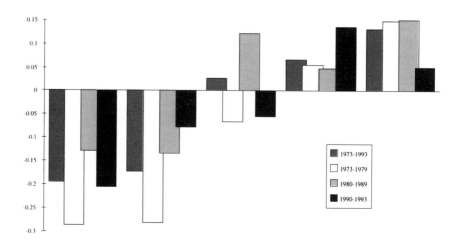

Figure 4. The day of the week effect: comparison of subperiods

according to the 2.34 F-statistic which is not significant at the 5% level ($p = 0.0534$) as it was in previous periods.

There is autocorrelation in the equity returns. Hence it is useful to consider an autoregressive model of order one which adjusts for the one day autocorrelation ρ which is about 0.2. In January $\bar{\rho} = 0.218$ ($t = 4.53$) and non-January $\bar{\rho} == 0.219$ ($t = 15.16$) hence the hypothesis that $\rho = 0$ is rejected.

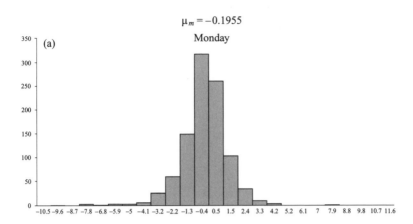

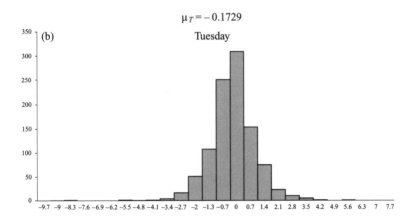

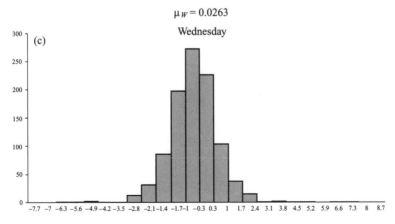

Figure 5a–c. Daily return distributions of the COMIT index, 1973–1993

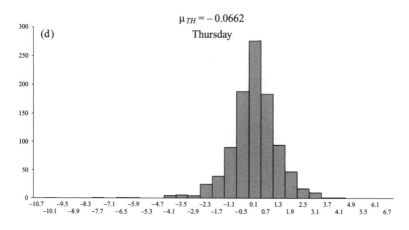

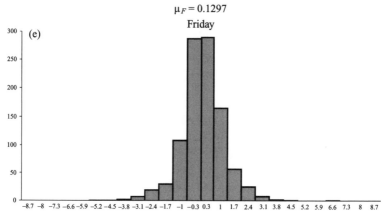

Figure 5d–e. Daily return distributions of the COMIT index, 1973–1993

The model is

$$R_t = a_1 + a_2 d_2 + a_3 d_3 + a_4 d_4 + a_5 d_5 + u_t,$$

where $u_t = \rho u_{t-1} + \varepsilon_t$.

This model improves the estimation. For example, the Durbin–Watson statistic is then close to 2. Table 4b shows the results. We can now reject the null hypothesis that all coefficients are equal to zero in all periods.

The Italian equity market had many market closures; see Table 2. Was the weekend effect (high Friday, low Monday and Tuesday) really a closed market effect? The results confirm that there was a separate closed market-weekend effect distinct from the day of the week effect. This is demonstrated by computing the returns after market closures; that is Monday's mean returns

Table 4a. Regression model of day of the week effects

Period	$a_l(t)$	$a_2(t)$	$a_3(t)$	$a_4(t)$	$a_5(t)$	D.F.	F	Confidence Level	D.W.
1973–1993	−0.1955 (−4.81)	0.0226 (0.40)	0.2218 (3.90)	0.2617 (4.58)	0.3252 (5.72)	(4,5001)	13.38	0.0000	1,559
1973–1979	−0.2864 (−4.38)	0.0046 (0.05)	0.2196 (2.39)	0.3411 (3.72)	0.4329 (4.76)	(4,1641)	9.35	0.0000	1,539
1980–1989	−0.1298 (−2.05)	0.0053 (0.06)	0.2502 (2.83)	0.1776 (2.00)	0.2791 (3.15)	(4,2403)	4.74	0.0008	1,563
1990–1993	−0.2059 (−2.44)	0.1277 (1.08)	0.1496 (1.26)	0.3405 (2.83)	0.2558 (2.14)	(4,947)	2.34	0.0534	1,576

Table 4b. Autoregressive model of order one for day of the week effects

Period	$a_l(t)$	$a_2(t)$	$a_3(t)$	$a_4(t)$	$a_5(t)$	D.F.	F	Confidence Level	D.W.
1973–1993	−0.1898 (−4.70)	0.0182 (0.36)	0.2083 (3.82)	0.2587 (4.72)	0.3182 (6.33)	(45,001)	13.81	0.0000	1.945
1973–1979	−0.2775 (−4.27)	−0.0016 (-0.02)	0.1901 (2.16)	0.3442 (3.94)	0.4220 (5.27)	(41,641)	9.74	0.0000	1.938
1980–1989	−0.1234 (-1.96)	−0.0073 (-0.09)	0.2433 (2.87)	0.1651 (1.94)	0.2690 (3.43)	(42,403)	5.16	0.0004	1.937
1990–1993	−0.2056 (−2.44)	0.119 (1.13)	0.1456 (1.27)	0.3498 (3.03)	0.2597 (2.45)	(4,947)	2.71	0.0291	1.974
January 1973–1993	−0.0523 (−0.43)	0.082 (0.55)	0.2621 (1.62)	0.3568 (2.21)	0.4292 (2.84)	(4,413)	2.49	0.0428	1.97
Non January 1973–1993	−0.202 (−4.74)	0.012 (0.23)	0.2045 (3.55)	0.2484 (4.29)	0.3072 (5.78)	−44,583	1 1.76	0.0000	1.942

Table 5. Mean returns after market closures by day of the week in percent

Day	Mean Return	Standard Deviation
Monday	−0.1193	1.0388
Tuesday	0.0309	0.9929
Wednesday	0.0417	0.8141
Thursday	0.2115	1.2234
Friday	0.4519	1.1871

when Friday is closed, Tuesday's mean returns when Monday is closed, etc. The results indicate that Monday then has a mean return of −0.1193% and Friday's return 0.4519% on average; hence a weekend effect did exist in Italy during this period. Table 5 details the results.

3   The January barometer

The concept of the January barometer is due to Hirsch (1986) who updates his results annually for the U.S. stock market in his almanacs, see e.g. Hirsch (1998). Hirsch defines the January barometer to be: 'as January goes so goes the whole year with January part of the year.' Hensel and Ziemba (1995a,b) investigated this for US and for international data, respectively. They considered two hypotheses

(a) if January rises does the rest of the year rise? and

(b) if it falls does the rest of the year fall?

Hensel and Ziemba found that for the US and some international markets:

(a) seems to be a statistically significant signal but

(b) is noise.

They also found that January's return provides a signal regarding the rest of the year's return magnitude, e.g. positive rest of the years have lower returns following negative Januarys, etc. Tables 6 and 7 show the Italian results for 1973 to 1993.

Table 6. January barometer results, 1973–93

Year	January	Rest of Year	Jan Signal	Rest of Year	Did Barometer Work?
1973	-2.7%	15.7%	DOWN	UP	NO
1974	8.9%	-46.4%	UP	DOWN	NO
1975	13.6%	-18.3%	UP	DOWN	NO
1976	5.4%	-16.3%	UP	DOWN	NO
1977	-4.6%	-23.9%	DOWN	DOWN	YES
1978	6.3%	15.0%	UP	UP	YES
1979	5.5%	13.2%	UP	UP	YES
1980	8.9%	63.9%	UP	UP	YES
1981	19.6%	-8.4%	UP	DOWN	NO
1982	-1.8%	-13.7%	DOWN	DOWN	YES
1983	11.9%	3.5%	UP	UP	YES
1984	17.6%	0.6%	UP	UP	YES
1985	14.1%	55.2%	UP	UP	YES
1986	4.3%	41.5%	UP	UP	YES
1987	-2.5%	-37.3%	DOWN	DOWN	YES
1988	-3.8%	23.4%	DOWN	UP	NO
1989	1.0%	13.7%	UP	UP	YES
1990	-0.6%	-27.9%	DOWN	DOWN	YES
1991	-3.7%	2.3%	DOWN	UP	NO
1992	6.5%	-19.3%	UP	DOWN	NO
1993	7.2%	24.1%	UP	UP	YES

Hypothesis (a) had nine successes in the 14 positive January years with nine successes in the last 11 years. Hypothesis (b) had 4 successes in the 7 years with negative Januarys. So the results, while not as good as in other countries, more or less indicate that (a) works and (b) is noise.

If January was positive and the rest of the year was positive then the returns were much higher on average, 25.6% versus 13.8%. If January returns were negative and the rest of the year's returns were negative then they were slightly more negative on average, -25.7% versus -21.7%. Again the signal (a) seems to be valuable and (b) is noise. The small sample precludes reliable statistical analysis, although the results are similar to what Hensel

Table 7. January barometer success rates

	All Januarys	Positive Januarys	Negative Januarys
1973–1993	61.9%	64.3%	57.1%
1973–1979	42.9%	40.0%	50.0%
1980–1993	71.4.%	77.8%	60.0%

	January very positive Rest of the year positive	January negative Rest of the year positive
1973–1993	25.6%	13.8%
1973–1979	14.1%	15.7%
1980–1993	28.9%	12.8%

	January very negative Rest of the year negative	January positive Rest of the year negative
1973–1993	–25.7%	–21.7%
1973–1979	–23.9%	–27.0%
1980–1993	–26.3%	–13.8%

and Ziemba found. The January effect in the Italian market was also studied by Caparelli *et al.* (1992).

4 Monthly effect

There was a significant monthly effect in Italian stock prices from 1973 to 1993. January and February returns were much higher than in other months. These months had significantly positive returns while the other ten months had returns not statistically different from zero. Table 8, which includes the first day of the monthly account, details this. The first trading day of the new monthly account which is calendar day 15–19 was significantly positive as shown in Table 9 which gives the monthly returns, which are now lower, without this first day when dividends are paid. The differences across months are further investigated in Tables 10 and 11.

The model
$$R_t = a_1 + a_2 d_2 + \cdots + a_{12} d_{12} + u_t,$$
where $d_i = \{0, 1\}$ for $i = 2, \ldots, 12$ is used. The hypothesis

$$H_0 : a_2 = a_3 = \cdots = a_{12} = 0,$$

tests if February (2), ..., December (12) have different mean returns than January. The results in Table 10 indicate that over the whole sample pe-

Table 8. Mean monthly returns in various sub-periods, t-statistics in ( ) for $H_0 : R = 0$.

	Observations	1973–1993	1973–1979	1980–1989	1990–1993
January	439	0.2528	0.2121	0.3332	0.1852
		(4.61)	(2.44)	(3.93)	(1.40)
February	423	0.2051	0.1786	0.2332	0.1814
		(3.65)	(2.01)	(2.58)	(1.65)
March	456	0.0672	0.0109	0.1771	−0.1109
		(1.24)	(0.13)	(2.07)	(−1.01)
April	418	0.0135	−0.1382	0.0741	0.1206
		(0.25)	(−1.47)	(0.90)	(1.17)
May	443	−0.0006	0.0007	−0.0717	0.1742
		(−0.01)	(0.01)	(−0.71)	(1.83)
June	434	−0.0827	−0.0152	−0.0998	−0.1479
		(−1.03)	(−0.12)	(−0.73)	(−1.53)
July	462	−0.0094	−0.1864	0.1686	−0.1409
		(−0.33)	(−1.66)	(1.47)	(−1.24)
August	443	0.1166	0.1308	0.1997	−0.1159
		(1.77)	(1.27)	(2.20)	(−0.62)
September	448	−0.1168	−0.0718	−0.0629	−0.3314
		(−1.91)	(−0.67)	(−0.73)	(−2.16)
October	462	−0.0207	−0.1431	−0.0002	0.1506
		(−0.31)	(−1.36)	(0.00)	(1.10)
November	426	−0.0388	−0.0489	0.0178	−0.1470
		(−0.58)	(−0.49)	(0.16)	(−1.10)
December	404	0.0058	−0.1400	0.0213	0.1764
		(0.10)	(−1.27)	(0.26)	(1.44)

Table 9. Mean monthly returns in various sub-periods when the first trading day of each month is excluded, *t*-statistics in ( ) for $H_0 : R = 0$.

	Observations	1973–1993	1973–1979	1980–1989	1990–1993
January	418	0.1775 (3.39)	0.1362 (1.64)	0.2279 (2.88)	0.1240 (0.99)
February	402	0.1371 (2.47)	0.0962 (1.11)	0.1543 (1.75)	0.1663 (1.45)
March	435	−0.0172 (−0.34)	−0.0706 (−0.87)	0.0644 (0.82)	−0.1300 (−1.17)
April	397 −0.0523	−0.2259 (−0.99)	0.0237 (−2.48)	0.0532 (0.29)	(0.53)
May	422	−0.0435 (-0.72)	−0.0342 (-0.38)	−0.1366 (−1.35)	0.1726 (1.75)
June	413	−0.1534 (−1.89)	−0.0984 (−0.78)	−0.1641 (−1.18)	−0.2151 (−2.33)
July	441	−0.0406 (−0.57)	−0.2484 (−2.19)	0.1436 (1.22)	−0.1342 (−1.15)
August	422	0.0811 (1.31)	0.0681 (0.68)	0.1229 (1.39)	−0.0014 (−0.01)
September	427	−0.1509 (−2.44)	−0.0883 (−0.82)	−0. 1052 (−1.22)	−0.3786 (−2.51)
October	441	−0.0947 (−1.44)	−0.2213 (−2.13)	−0.0809 (−0.79)	0.1021 (0.73)
November	405	−0.1123 (−1.73)	−0.1044 (−1.08)	−0.0816 (−0.76)	−0.1819 (−1.34)
December	383	−0.0708 (−1.22)	−0.2165 (−1.94)	−0.0687 (−0.88)	0.1217 (0.99)

riod January returns were statistically higher than the other months except February and August.

Table 12 investigates this using the autoregressive model:

$$R_t = a_1 + a_2 d_2 + \cdots + a_{12} d_{12} + u_t,$$

where $u_t = \rho u_{t-1} + \varepsilon_t$. The estimates of ρ are

	ρ	t
1973–1979	0.22	9.13
1980–1989	0.21	10.54
1990–1993	0.19	5.95

The F-test is for $H_0 : a_2 = a_3 = \cdots = a_{12} = 0$.

January had the highest returns of any month. In fact, each day of the week in January had higher returns than in the rest of the year. The January effect dominated the weekend effect and the mean returns were monotonically higher as the week progressed in January and in the rest of the year. This is detailed in Table 12 for the standard and autoregressive models

For the autoregressive model of order 1

$$R_t = a_1 + a_2 d_2 + \cdots + a_5 d_5 + u_t,$$

where $u_t = \rho u_{t-1} + \varepsilon_t$, we have

January $\quad \hat{\rho} = 0.218 \quad (t = 4.53)$

Rest of Year $\quad \hat{\rho} = 0.219 \quad (t = 15.16)$.

Hence $H_0 : \rho = 0$ is rejected and the Durbin–Watson statistics indicate a better model. The estimates are

	a_1	a_2	a_3	a_4	a_5	Degrees of Freedom	F	Confidence Level	DW
January 1973–93	−0.0523 (−0.43)	0.0820 (0.55)	0.2621 (1.62)	0.3568 (2.21)	0.4292 (2.84)	(4,413)	2.49	0.0428	1.970
Rest of Year 1973–93	−0.2020 (−4.74)	0.0120 (0.23)	0.2045 (3.55)	0.2484 (4.29)	0.3072 (5.78)	(4,4583)	11.76	0.0000	1.942

Table 10. The monthly effect when the first day of the monthly account is
excluded, *t*-statistics in ( ) for $H_0 : R = 0$.

	1973–1993	1973–1979	1980–1989	1990–1993
January	0.1775	0.1362	0.2279	0.1240
	(2.86)	(1.37)	(2.35)	(0.96)
February	−0.0404	−0.0529	−0.0653	0.0423
	(−0.46)	(−0.37)	(−0.47)	(0.23)
March	−0.1947	−0.2067	−0.1634	−0.2540
	(−2.24)	(−1.48)	(−1.21)	(−1.41)
April	−0.2298	−0.3621	−0.2042	−0.0708
	(−2.58)	(−2.53)	(−1.47)	(−0.38)
May	−0.2210	−0.1704	−0.3645	0.0486
	(−2.52)	(−1.21)	(−2.66)	(0.27)
June	−0.3309	−0.2346	−0.3920	−0.3391
	(−3.76)	(−1.64)	(−2.87)	(−1.86)
July	−0.2181	−0.3846	−0.0843	−0.2582
	(−2.52)	(−2.77)	(−0.62)	(−1.43)
August	−0.0966	−0.0681	−0.1056	−0.1254
	(−1.10)	(−0.48)	(−0.77)	(−0.69)
September	−0.3284	−0.2245	−0.3331	−0.5026
	(−3.76)	(−1.60)	(−2.45)	(−2.75)
October	−0.2723	−0.3575	−0.3088	−0.0219
	(−3.14)	(−2.59)	(−2.28)	(−0.12)
November	−0.2901	−0.2406	−0.2894	−0.3915
	(−3.28)	(−1.70)	(−2.10)	(−2.09)
December	−0.2483	−0.3527	−0.2815	0.0039
	(−2.77)	(−2.44)	(−2.0l)	(0.02)
D.F	11,4994	11,1634	11,2396	11,940
F	2.97	1.70	1.83	2.06
Confidence Level	0.0006	0.0678	0.0446	0.0205
DW	1.571	1.559	1.579	1.618
Mean of R_t	−0.029	−0.085	0.012	−0.032

Table 11. The monthly effect when the first day of the month is excluded using the autoregressive model, t-statistics in ( ) for $H_0 : R = 0$.

	1973–1993	1973–1979	1980–1989	1990–1993
a_1	0.1843	0.1393	0.2312	0.1446
	(2.43)	(1.14)	(1.96)	(0.94)
a_2	−0.0579	−0.0696	−0.0737	0.0048
	(−0.46)	(−0.40)	(−0.44)	(0.02)
a_3	−0.1999	−0.1899	−0.1793	−0.2681
	(−1.88)	(−1.10)	(−1.09)	(−1.24)
a_4	−0.2393	−0.3789	−0.1945	−0.1120
	(−2.20)	(−2.16)	(−1.15)	(−0.51)
a_5	−0.2264	−0.1744	−0.3660	0.0315
	(−2.11)	(−1.01)	(−2.20)	(0.14)
a_6	−0.3366	−0.2343	−0.4032	−0.3421
	(−3.13)	(−1.33)	(−2.43)	(−1.57)
a_7	−0.2196	−0.3817	−0.0756	−0.2898
	(−2.07)	(−2.24)	(−0.46)	(−1.35)
a_8	−0.1035	−0.0720	−0.1158	−0.1306
	(−0.97)	(−0.42)	(−0.70)	(−0.60)
a_9	−0.3331	−0.2374	−0.3236	−0.5285
	(−3.12)	(−1.38)	(−1.95)	(−2.42)
a_{10}	−0.2804	−0.3644	−0.3148	−0.0375
	(−2.65)	(−2.14)	(−1.91)	(−0.17)
a_{11}	−0.2970	−0.2323	−0.2949	−0.4246
	(−2.75)	(−1.33)	(−1.76)	(−1.89)
a_{12}	−0.2593	−0.3574	−0.2900	−0.0205
	(−2.38)	(−2.03)	(−1.72)	(−0.09)
DF	11,4994	11,1634	11,2396	11,940
F	1.96	1.13	1.24	1.48
C.L.	0.0282	0.3293	0.2571	0.1347
DW	1.942	1.933	1.936	1.971

Table 12. Day of the week effect in January and the rest of the year for the statistical model $R_t = a_1 + a_2 d_2 + a_3 d_3 + a_4 d_4 + a_5 d_5 + u_t$.

	a_1	a_2	a_3	a_4	a_5	Degrees of Freedom	F	Confidence Level	DW
January 1973–93	−0.0630 (−0.52)	0.1151 (0.68)	0.2666 (1.59)	0.3566 (2.13)	0.4403 (2.60)	(4,413)	2.26	0.0615	1.561
Rest of Year 1973–93	−0.2067 (−4.83)	0.0134 (0.22)	0.2163 (3.60)	0.2510 (4.16)	0.3143 (5.24)	(4,4583)	11.53	0.0000	1.563

Mean Percentage Returns

	Ordinary Regressions		Autoregressive Model	
	January	Rest	January	Rest
Monday	$a_1 = -0.0630$	−0.2067	−0.0523	−0.2020
Tuesday	$a_1 + a_2 = 0.0521$	−0.1933	0.0297	−0.1900
Wednesday	$a_1 + a_3 = 0.2036$	0.0096	0.2098	0.0025
Thursday	$a_1 + a_4 = 0.2936$	0.0443	0.3045	0.0464
Friday	$a_1 + a_5 = 0.3773$	0.1076	0.3769	0.1052

5   Turn-of-the-month effect

During the period of study there was a strong turn-of-the-month effect in Italy. The papers by Hensel, Sick and Ziemba (1999) and Comolli and Ziemba (1999) in this volume document this effect in the US and Japan, respectively. The effect refers to the observed fact that returns at the beginning of the month are higher than in the rest of the month because of cash flows, wage payments, dividend payments, institutional practices such as stock purchases, etc. The dates differ by country depending on the timing of these cash flows. In Italy they lead to high returns at the beginning of the trading month (calendar days 15–21) and lower returns in the latter part of the trading month (calendar days 22–29 and 1–14). Table 13 shows the mean returns

Table 16. Data on the 100 highest and 100 lowest return days.

	Number of Days Per Year	% of Total Days	% of largest and smallest returns		
First day of monthly account	12	4.7	33	3.8788	1.1364
–9 to –5	60	23.7	6	3.9417	0.8783
Rest of month	181	71.6	61	3.7551	1.1498
Rest of Year	253	100	100	3.8070	1.1240
First day of monthly account	12	4.7	3	–4.7060	2.2812
–9 to –5	60	23.7	34	–4.1644	1.6200
Rest of month	181	71.6	63	–4.5000	1.6808
Entire Year	253	100	100	–4.3970	1.6669

and t-values for the hypothesis that the daily mean return around the turn-of-the-month is different from the over all daily mean return, 0.0442. In the table '0' refers to the first trading day, '+1' to the second, etc. Table 14 shows these results by month. Table 15 gives the percentage of positive returns by trading day around the turn-of-the-month by month and decade. The 0th trading day had the highest returns and probability of positive returns. As in the US (see Hensel, Sick and Ziemba, 1999), the strong turn-of-the-month is a combination of slightly higher probability of positive returns coupled with a slight shifting of the mean of the return distributions.

Table 16 displays the 100 highest and 100 lowest daily returns. The periods day '0', trading days –9 to –5 and the rest-of-the-month amount to 4.7%, 23.7% and 71.6% of the trading days. Of the 100 largest gains, 33% occurred on the 0th day, 6% during –9 to –5 and 61% during ROM. The 100 lowest returns occurred with frequencies 3%, 34% and 63% in these three respective periods. These results are consistent with what Hensel, Sick and Ziemba (1999) found for the S&P500. That is, more of the highest returns and less of the lowest returns occurred during the turn-of-the-month, however, these rare events have a random, in time character based on the event that caused the large price change.

6   Holiday effects

There was a strong holiday effect in Italy during the sample period similar to that in other markets during this era; see Cervera and Keim (1999). Figure 6

Table 13. Mean returns and standard deviation by trading day around the turn-of-the-month.

	-9	-8	-7	-6	-5	-4	-3	-2	-1	0	1	2	3	4	5	6	7	8	9	-9 to +9	-9 to -5
January	0.0819 (1.13)	-0.0248 (0.77)	0.1295 (1.37)	0.0457 (0.62)	-0.0852 (1.09)	0.3110 (0.89)	0.3833 (1.23)	0.5610* (1.22)	0.8338** (1.07)	1.9986** (1.60)	0.0071 (0.69)	-0.1971 (0.94)	0.0614 (1.39)	0.3310 (1.05)	-0.3067 (1.49)	0.0971 (0.87)	0.1276 (0.99)	0.0605 (0.68)	0.2590 (0.96)	0.2434** (1.17)	0.0294 (1.01)
February	0.1895 (1.53)	0.3567* (0.82)	0.0590 (1.30)	0.1962 (0.93)	-0.2219 (0.98)	0.0890 (0.71)	0.2548 (0.95)	0.4838** (0.2590)	0.4838* (1.09)	1.5008** (1.22)	0.5433 (1.52)	0.4876** (0.88)	0.1029 (1.28)	0.4714* (0.79)	0.4519 (1.29)	-0.2857 (1.01)	0.1557 (1.13)	0.2729 (0.80)	0.1086 (1.26)	0.2737** (1.14)	0.1219 (1.14)
March	-0.2038 (1.21)	0.1081 (1.11)	0.0614 (0.96)	-0.1476 (1.06)	-0.1767 (0.91)	-0.0038 (0.92)	0.0090 (1.00)	0.1395 (1.05)	-0.3425 (1.33)	1.8148** (1.62)	0.1424* (1.01)	0.2450 (1.03)	0.1057 (1.01)	-0.2648 (1.36)	0.0490 (0.85)	-0.3000 (0.96)	0.0657 (0.87)	0.1395 (1.40)	-0.2356 (0.87)	0.0635 (1.16)	-0.0717 (1.04)
April	-0.2110 (0.85)	-0.2519 (1.35)	-0.1414 (0.99)	-0.3281 (1.00)	-0.4924 (1.59)	0.1738 (0.86)	0.1795 (0.94)	0.2055 (0.94)	-0.0648 (1.10)	1.2576** (1.35)	0.5714 (1.27)	0.1700 (1.02)	0.0657 (1.10)	0.1448 (0.75)	-0.2076 (0.84)	-0.0704 (0.96)	0.1443 (0.84)	-0.1548 (0.58)	-0.2410 (0.73)	0.0396 (1.03)	-0.2850 (-2.88)
May	-0.1000 (1.09)	-0.1157 (0.38)	-0.1419 (0.66)	-0.3305 (0.94)	-0.2457 (0.84)	-0.0805 (0.78)	0.1752 (0.90)	0.2062 (1.20)	0.1205 (1.28)	0.8614** (1.39)	-0.0305 (1.28)	-0.4110 (1.44)	0.0505 (1.39)	0.3424 (1.32)	-0.0814 (1.56)	-0.0119 (1.21)	-0.5462 (1.26)	-0.4271 (2.38)	0.6357 (1.73)	-0.0059 (1.31)	-0.1868 (0.84)
June	-0.4986 (1.03)	-0.4376 (0.91)	-0.3871 (1.11)	-0.2625 (2.21)	0.2514 (1.20)	0.0235 (1.60)	-0.4457 (1.82)	-0.0014 (1.43)	-0.8457 (2.81)	1.3086** (1.61)	0.6443 (1.67)	-0.6038 (2.09)	-0.5071 (1.95)	-0.0162 (1.06)	0.4871 (1.94)	0.1833 (0.88)	0.6671 (2.19)	-0.1495 (1.29)	0.0267 (0.94)	-0.0999 (1.69)	-0.2670 (1.38)
July	-0.2357 (1.07)	-0.4219 (1.28)	-0.5643 (1.72)	-0.2957 (1.52)	-0.2333 (1.91)	0.1352 (1.22)	-0.0448 (1.94)	-0.0419 (1.06)	0.0267 (0.82)	0.6471* (1.42)	0.4800 (1.23)	-0.3295 (2.26)	-0.3505 (1.79)	-0.1690 (0.83)	0.4705 (2.25)	0.0433 (1.76)	0.1819 (2.04)	0.2110 (0.81)	0.4776 (1.25)	-0.0028 (1.56)	-0.3502 (1.51)
August	0.1667 (1.43)	-0.2245 (1.52)	-0.1671 (1.25)	-0.0743 (0.98)	-0.0880 (1.55)	-0.0995 (1.10)	0.2414 (1.19)	0.1538 (0.98)	0.4643 (1.50)	0.8290 (2.77)	-0.0257 (1.69)	0.0681 (1.25)	0.1048 (1.25)	0.1257 (1.12)	-0.1367 (1.09)	0.0148 (1.75)	0.4433* (1.00)	0.5095* (0.79)	-0.0038 (1.00)	0.1211 (1.39)	-0.0775 (1.34)
September	0.2148 (1.20)	0.4881 (1.41)	-0.2755 (1.06)	-0.3076 (1.38)	-0.1033 (1.18)	-0.5276 (1.19)	0.0071 (1.09)	0.0943 (1.31)	-0.1667 (1.54)	0.5890 (1.75)	-0.1181 (1.68)	0.0162 (1.43)	0.1367 (1.21)	-0.2867 (1.62)	0.0524 (1.33)	-0.4033 (1.05)	-0.0176 (0.98)	0.6005* (1.25)	0.2055 (0.85)	-0.0410 (1.32)	-0.1928 (1.25)
October	-0.1629 (1.26)	-0.1105 (1.09)	-0.1619 (1.31)	-0.0143 (1.17)	-0.3476 (1.31)	-0.1819 (1.35)	-0.1643 (0.90)	-0.3614 (1.03)	-0.3086 (2.00)	1.5343** (1.82)	-0.2014 (1.77)	-0.1467 (2.23)	0.1757 (1.87)	0.4671 (1.32)	-0.1462 (0.96)	-0.2319 (1.05)	-0.4429 (1.44)	0.1862 (1.13)	0.1629 (1.03)	-0.0240 (1.45)	-0.1594 (1.21)
November	-0.5300 (1.18)	-0.8200 (2.09)	-0.2619 (0.89)	-0.5267 (1.11)	-0.7586 (1.48)	0.3735 (1.97)	0.0338 (1.50)	0.0105 (0.91)	-0.1433 (1.34)	1.3138** (1.87)	0.2971 (1.35)	0.1533 (1.13)	-0.4452 (1.20)	-0.2729 (1.17)	0.2190 (1.22)	0.2786 (1.20)	0.4648 (1.26)	0.5239* (0.97)	0.2924 (1.07)	-0.0599 (1.42)	-0.5792 (1.40)
December	-0.2619 (1.39)	-0.1400 (1.09)	-0.6719 (1.34)	-0.6035 (1.35)	-0.4129 (1.35)	-0.1895 (0.95)	-0.2550 (1.06)	-0.3148 (0.90)	-0.4250 (0.85)	1.4043* (0.94)	0.6657* (1.34)	-0.0429 (1.08)	0.2195 (0.76)	0.5781* (0.99)	0.1853 (1.10)	-0.0852 (0.94)	0.3243 (0.88)	0.6514** (0.51)	0.1595 (1.05)	0.0241 (1.16)	0.4181 (1.30)

Table 14. Mean returns and standard deviation by trading day around the turn-of-the-month by month.

	-9	-8	-7	-6	-5	-4	-3	-2	-1	0	1	2	3	4	5	6	7	8	9	-9 to +9	-9 to -5
January	0.0819	-0.0248	0.1295	0.0457	-0.0852	0.3110	0.3822	0.5610*	0.8338**	1.9986**	0.0071	-01971	0.0614	0.3310	-0.3967	0.0971	0.1276	0.0605	0.2990	0.2434**	0.0294
	(1.13)	(0.77)	(1.370)	(0.62)	(1.09)	(0.89)	(1.23)	(1.22)	(1.07)	(1.60)	(0.69)	(0.94)	(1.39)	(1.05)	(1.49)	(0.87)	(0.99)	(0.68)	(0.96)	(1.17)	(1.01)
February	0.1895	0.3867*	0.0590	0.1962	-0.2219	0.0890	0.2548	0.2590	0.4838*	1.5068**	0.5433	0.4876*	0.1029	0.4714*	0.4519	-0.2857	-0.1557	0.2729	0.1086	0.2737**	0.1219
	(1.53)	(0.82)	(1.30)	(0.93)	(0.98)	(0.71)	(0.95)	(1.09)	(0.98)	(1.22)	(1.52)	(0.88)	(1.28)	(0.79)	(1.29)	(1.01)	(1.13)	(0.80)	(1.26)	(1.14)	(1.14)
March	-0.2038	0.1081	0.0614	-0.1476	-0.1767	-0.0038	0.0090	0.1395	-0.3429	1.8148**	0.1424*	0.2490	0.1057	-0.2648	0.0490	-0.3000	0.0657	0.1395	-0.2386	0.0635	-0.0717
	(1.21)	(1.11)	(0.96)	(1.06)	(0.91)	(0.92)	(1.00)	(1.05)	(1.33)	(1.62)	(1.01)	(1.03)	(1.01)	(1.36)	(0.85)	(0.96)	(0.87)	(1.40)	(0.87)	(1.16)	(1.04)
April	-0.2110	-0.2519	-0.1414	-0.3281	-04924	0.1738	0.1795	0.2095	-0.0648	1.2576**	0.5714	0.1700	0.0657	0.1448	-0.2076	-0.0704	0.1443	-0.1548	-0.2410	0.0396	-0.2850
	(0.85)	(1.35)	(0.99)	(1.00)	(1.59)	(0.86)	(0.94)	(0.94)	(1.10)	(0.35)	(1.27)	(1.02)	(1.10)	(0.75)	(0.84)	(0.96)	(0.84)	(0.58)	(0.73)	(1.08)	(-2.88)
May	-0.1000	-0.1157	-0.1419	-0.3305	-0.2457	-0.0805	0.1752	0.2062	0.1205	0.8614**	-0.0305	-0.4110	0.0505	0.3424	-0.0814	-0.0119	-0.5462	-0.4271	0.6357	-0.0069	-0.1868
	(1.09)	(0.38)	(0.66)	(0.94)	(0.84)	(0.778)	(0.90)	(1.20)	(1.28)	(1.39)	(1.28)	(1.44)	(1.39)	(1.32)	(1.56)	(1.21)	(1.26)	(2.38)	(1.73)	(1.31)	(0.84)
June	-0.4986	-0.4376	-0.3871	-0.2629	0.2514	0.0238	-0.4457	-0.0014	-0.8457	1.3086**	0.6443	-0.6038	-0.5071	-0.0162	0.4871	0.1833	-0.6671	-0.1495	0.0267	-0.0999	-0.2670
	(1.03)	(0.91)	(1.11)	(2.21)	(1.20)	(1.60)	(1.82)	(1.43)	(2.81)	(1.61)	(1.67)	(2.09)	(1.95)	(1.06)	(1.94)	(0.88)	(2.19)	(1.29)	(0.94)	(1.69)	(1.38)
July	-0.2357	-0.4219	-05643	-0.2957	-0.2333	0.1352	-0.0448	-0.0419	0.0267	0.6471**	0.4800	-0.3295	-0.3905	-0.1690	0.4705	0.0433	0.1819	0.2110	0.4776	-0.0028	-0.3502
	(1.07)	(1.28)	(1.72)	(1.52)	(1.91)	(1.22)	(1.90)	(1.06)	(0.82)	(1.42)	(1.23)	(2.26)	(1.79)	(0.83)	(2.25)	(1.76)	(2.04)	(0.81)	(1.25)	(1.56)	(1.51)
August	0.1667	-0.2248	-0.1671	-0.0743	-0.0880	-0.0995	0.2414	0.1538	0.4643	0.8290	-0.0257	0.0681	0.1048	0.1257	-0.1367	0.0148	0.4433*	0.5095*	-0.0038	0.1211	-0.0775
	(1.43)	(1.52)	(1.25)	(0.98)	(1.55)	(1.00)	(1.19)	(0.98)	(1.50)	(2.77)	(1.69)	(1.26)	(1.26)	(1.12)	(1.09)	(1.75)	(1.00)	(0.79)	(1.00)	(1.39)	(1.34)
September	0.2148	-0.4881	-0.2795	-0.3076	-0.1033	-0.5276	0.0071	0.0943	-0.1667	0.5890	-0.1181	0.0162	0.1367	-0.2867	0.0524	-0.4033	-0.0176	0.6005*	0.2095	-0.0410	-0.1928
	(1.20)	(1.41)	(1.06)	(1.38)	(1.18)	(1.19)	(1.09)	(1.31)	(1.54)	(1.75)	(1.68)	(1.43)	(1.21)	(1.62)	(1.33)	(1.05)	(0.98)	(1.25)	(.085)	(1.32)	(1.25)
October	-0.1629	-0.1105*	-0.1619	-0.0143	-0.3476	-0.1819	-0.1643	-0.3614	-0.3086	1.5343**	-0.2014	-0.1467	0.1757	0.4671	-0.1462	-0.2319	-0.4429	0.1862	0.1629	-0.0240	-0.1594
	(1.260)	(1.09)	(1.31)	(1.17)	(1.31)	(1.35)	(0.90)	(1.03)	(2.00)	(1.82)	(0.77)	(2.230)	(1.87)	(1.32)	(0.96)	(1.05)	(1.44)	(1.13)	(1.03)	(1.45)	(1.21)
November	-0.5300	-08200	-0.2619	-05257	-0.7586	-0.3778	0.0338	0.0105	-0.1433	1.3138**	0.2971	0.1533	-0.4452	-0.2729	0.2190	0.2686	0.4648	0.5238*	-0.2924	-0.0599	-0.5792
	(1.18)	(2.09)	(0.89)	(1.11)	(1.48)	(1.97)	(1.50)	(0.91)	(1.34)	(1.87)	(1.35)	(1.13)	(1.20)	(1.17)	(1.22)	(1.20)	(1.26)	(0.97)	(1.07)	(1.42)	(1.40)
December	-0.2619	-0.1400	-0.6719	-06038	-0.4129	-0.1895	-0.2590	-0.3148	-0.4290	1.4043**	0.6657	-0.0429	0.2195	0.5781	0.1833	-0.0852	0.3243	0.6524**	-0.1595	0.0241	-0.4181
	(1.39)	(1.09)	(1.34)	(1.35)	(1.35)	(0.95)	(1.06)	(0.90)	(0.85)	(0.94)	(1.34)	(1.08)	(0.76)	(0.99)	(1.10)	(0.94)	(0.88)	(0.51)	(1.05)	(1.16)	(1.30)

Numbers in ( ) mean standard deviation (no *t*-test)
* indicates significance at 5% level
** indicates significance at 1% level

Table 15. Percent of positive returns by trading day around the turn-of-the-month for the various months of the year and by decade.

Month	-9	-8	-7	-6	-5	-4	-3	-2	-1	0	+1	+2	+3	+4	+5	+6	+7	+8	+9
January	57%	52%	67%	52%	52%	76%	62%	52%	67%	90%	57%	43%	48%	52%	29%	62%	38%	43%	62%
February	52%	57%	62%	57%	43%	48%	62%	62%	62%	95%	67%	76%	43%	67%	67%	33%	48%	62%	52%
March	48%	57%	38%	43%	33%	57%	52%	48%	43%	81%	67%	57%	38%	43%	52%	33%	48%	71%	38%
April	43%	57%	33%	38%	38%	67%	57%	57%	57%	90%	67%	57%	62%	57%	48%	52%	48%	48%	33%
May	48%	33%	48%	43%	33%	43%	67%	57%	48%	76%	62%	38%	52%	62%	57%	57%	33%	48%	67%
June	29%	24%	43%	57%	62%	52%	33%	48%	38%	86%	76%	38%	48%	48%	62%	48%	33%	43%	62%
July	48%	43%	52%	48%	48%	57%	38%	43%	57%	71%	71%	48%	43%	43%	48%	38%	62%	57%	62%
August	57%	52%	52%	48%	29%	48%	57%	38%	57%	71%	57%	48%	52%	43%	48%	38%	33%	62%	62%
September	57%	43%	43%	52%	33%	33%	48%	57%	52%	71%	48%	43%	52%	43%	43%	19%	33%	67%	67%
October	62%	52%	33%	48%	29%	43%	38%	38%	43%	86%	48%	67%	52%	52%	52%	43%	38%	38%	52%
November	33%	29%	48%	38%	29%	52%	43%	57%	48%	71%	52%	67.%	38%	57%	71%	67%	62%	71%	48%
December	52%	52%	24%	29%	43%	38%	38%	38%	38%	95%	62%	48%	67%	81%	67%	43%	67%	90%	24%

Period	-9	-8	-7	-6	-5	-4	-3	-2	-1	0	+1	+2	+3	+4	+5	+6	+7	+8	+9
1973–1993	48%	46%	45%	46%	39%	50%	50%	50%	51%	82%	62%	52%	50%	55%	54%	45%	48%	58%	53%
1973–1979	45%	46%	42%	45%	43%	45%	43%	45%	49%	81%	57%	52%	55%	56%	48%	38%	50%	61%	45%
1980–1989	47%	47%	47%	46%	37%	53%	57%	53%	54%	85%	64%	53%	48%	61%	61%	53%	48%	59%	58%
1990–1993	58%	46%	48%	48%	37%	50%	44%	48%	46%	77%	63%	48%	44%	38%	46%	38%	44%	52%	54%

shows the results. The returns on pre-holidays are much higher than on non–pre-holidays (see Figure 6a) and the return on each day of the week is higher if it falls on a pre-holday (Figure 6b). The pre- and post-holiday returns for individual holidays are shown in Figure 6c; in most but not all cases, the pre-holiday return was higher than the post-holiday return.

7 Final remarks

The standard seasonal anomalies that have been present in the US, Japanese and other markets in most sample periods were present in Italy during the sample period of the study 1973–1993. For a discussion of the efficient market hypothesis in the Italian market see Caprio (1990). Similar to Ziemba and Schwartz (1991), who ranked the anomalies in order to form seasonality calendars the following regression was run:

$$R_t = a_1 D_{\text{Monday}} + a_2 D_{\text{Tuesday}} + a_3 D_{\text{Friday}} + a_4 D_{\text{1stDay}} + a_5 D_{\text{Easter}} +$$
$$a_6 D_{\text{Christmas}} + a_7 D_{\text{1stNov}} + a_8 D_{\text{Jan}} + a_9 D_{Day30} + a_{10} D_{Day31} + u_t.$$

	Mon.	Tues.	Fri.	1st Day	Easter	Xmas	1st Nov.	Jan.	Day 30 Cal.	Day 31 Cal.
Mean	−0.2221	−0.2082	0.0705	1.2667	0.6832	0.3048	0.6672	0.1419	0.3150	0.2386
St Dev	0.0442	0.0431	0.0439	0.0809	0.2945	0.29801	0.2820	0.0653	0.1003	0.1355
t-stat	−5.03**	−4.83**	1.61	15.66**	2.32*	1.09	2.37*	2.17*	3.14**	1.76

Cost = 0.00; R^2 = 0.0582; R^2_{adj} =0.0566; D.W. = 1.58; F = 32.44**
Referring to $H_0 : a_1 = a_2 = a_3 = \cdots = a_{10} = 0$.
* indicates 5% confidence level
** indicates 1% confidence level

The results yield the following ranking of the anomalies by importance of their effect on daily stock returns:

1 First day of the monthly account

2 Monday

3 Tuesday

4 Day 30 (calendar, not trading)

5 1st November

6 Easter

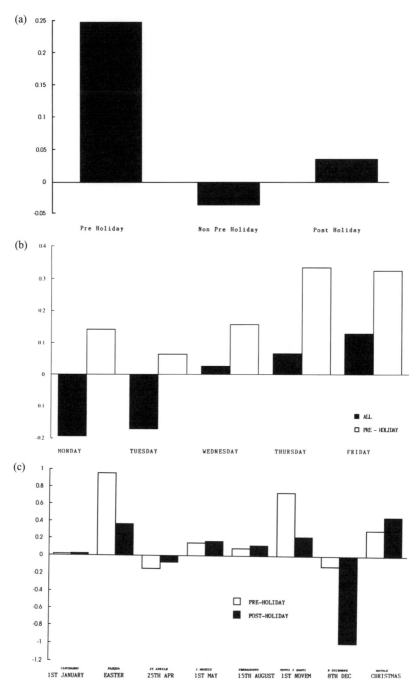

Figure 6 The holiday effect in Italy: (a) Pre-holidays versus non–pre-holidays and post-holidays; (b) The effect of pre-holidays on day of the week returns; (c) Pre-versus post-holiday returns for individual holidays.

7 January

8 Day 31 (calendar, not trading)

9 Friday

10 Christmas

References

Banca d'Italia (Bank of Italy) (1983–1997) Bollettino Statistico (Statistical Bulletin), 1–28, Roma.

Barone, E. (1990) 'The Italian stock market: efficiency and calendar anomalies. Centre for Research in Finance', IMI Group, February.

Caparelli F., d. De Simone and U. Calcagnini (1992) 'Mercato efficiente ed effetto gennaio', *Il Risparmio* 1 33–74.

Caprio, L. (1990) 'Gli studi sull'efficienza informativa dei mercati dei capitali: il dibattito sull'evidenza empirica e I modelli teorici', *Finanza Imprese e Mercati* 2 145–179.

Cervera, A. and D.B. Keim (1999) 'The international evidence on the holiday effect', this volume, 512–531.

Comolli, L.R. and W.T. Ziemba (1999) 'Japanese security market regularities, 1990–1994', this volume, 458–491.

Hensel, C.R. and W.T. Ziemba (1995a) 'The January barometer', *Journal of Investing* 4(2) 67–70.

Hensel, C.R. and W.T. Ziemba (1995b) 'The January barometer: Swiss, European and global results', *Finanzmarket and Portfolio Management* 9(2) 187–196.

Hensel, C.R., G.A. Sick and W.T. Ziemba (1999) 'A long term examination of the turn-of-the-month effect in the S&P500', this volume, 218–246.

Hirsch, Y. (1986) *Don't sell stocks on Monday.* Facts on File Publications, New York

Hirsch, Y. (1998) *The Stock Traders Almanac.* Facts on File Publications, New York.

Mondani, A. (1990) 'Metodologia dell'indice di Borsa Comit', *Il Risparmio* 6 1349–1375.

Ziemba, W.T. and S.L. Schwartz (1991) *Invest Japan.* Probus Publishers.

Efficiency and Anomalies in the Turkish Stock Market*

Gülnur Muradoğlu

1 Introduction

A small security market, an unfamiliar environment, but an excellent opportunity for international investors! The Istanbul Stock Exchange (ISE) can provide international investors with diversification benefits as well as profit opportunities due to a number of inefficiencies currently within the system. Investors who want to exploit profit opportunities raised by inefficiencies observed in emerging markets like the ISE, should consider the rapidly changing characteristics of those markets. Therefore, investment strategies based on overall results, that ignore the potential for rapid change can yield misleading inferences.

The Turkish stock market is described in terms of its historical background, its structures, the market participants, and its performance over time, focusing on anomalies and regularities which have arisen due to various context-specific factors. Economic and institutional rationales are also discussed. A main premise of this paper is that, in an emerging market like Turkey, one has to be cautious as institutions, macroeconomic factors, technology and information can change rapidly.

Financial markets in Turkey were highly inefficient and strictly regulated until 1980. Attempts at the liberalization of the country in general, and financial markets in particular, started at the beginning of 1980s with the introduction of a liberalization package supported by the World Bank and the IMF. The establishment of the legal framework and regulatory agencies for the stock market were completed in 1982, but it was four more years before the Istanbul Securities Exchange (ISE), the only stock exchange in Turkey, became operational in 1986.

During the first two years of the ISE operations, individual investors were permitted to enter and place their orders on the trade floor. Trade floor activities were limited to licensed brokers and a manual system was established in November 1987. Also, until 1987, there was no auditing requirement regarding the financial statements of companies listed on the ISE, and employees of the ISE could hold stock portfolios without notification.

*The author would like to acknowledge the constructive comments and criticisms of William Ziemba and Suzan Kamburoglu on earlier drafts of the manuscript, and the assistance of Amir Aghaty regarding the preparation of the figures.

By the end of 1989, interim financial statements were required from listed companies and, in 1990, the principles for going public were established. It was also in this year that legislation against insider trading was passed for the first time. Since then, serious steps have been taken to implement internationally accepted mechanisms for tracking and monitoring trading patterns as well as identifying and penalising attempts at insider trading (Tezcanli, 1996).

In August 1989, with the target of further liberalization of the system, non-residents were given permission to invest in Turkish securities and, concomitantly, Turkish investors were also permitted to invest in foreign securities through authorized financial intermediaries. In October 1990, the settlement date was decreased from a period of three days to one. Currently, settlement is specified as necessary within a maximum time period of two days. At the end of 1993 computer-aided procedures had also been established. Since November 1994, all listed stocks have been traded using these computer-assisted systems.

Trading takes place throughout the week in two sessions ; the morning session (10:00 am to 12:00 pm) and the afternoon session (2:00 pm to 4:00 pm local time – two hours ahead of GMT). Currently, the ISE is a fully-automated, order-matching market, and investors provide liquidity by submitting limit orders. There are no market makers or specialists assigned to stocks, and traders do not need to hold inventories in order to make the market (Onder and Guner, 1997). There is a limit on the maximum change in prices for each session which is plus or minus 10% of the average price of the previous session. Information on the companies listed on the ISE is provided via their interim reports announced every three months and four major indices – ISEcomposite–100, ISEcomposite–30 financial and industrial – are reported on a daily basis. The futures market will be operational by the first quarter of 1998 on the ISEcomposite–30 index.

The ISE's growth pattern can be characterized by the improvements in the institutional framework and in the regulations of trading activities; see Tables 1 and 2. The ISE is one of the 8 largest stock exchanges in Europe in volume. International investors hold 50% of total market capitalization but are responsible for approximately 7% of daily transactions.

In terms of trading volume and market capitalization, 1990 and 1993 were the critical years in characterizing the different phases of development in the Turkish stock market due to the permission given to international investors and the introduction of computer aided trading mechanisms respectively. The ISE composite index, which represents more than 80% of market capitalization, increased 6 times in USD terms during the ten year period since 1986. This increase, however, has continued with annual increases of up to 350% followed by corrections amounting to 70%. During the past ten years, dividend

Table 1. Development of the Turkish Stock Exchange. Source: ISE, Annual
Fact Book, 1996.

Year	Trading Volume	Av. Daily Volume	Market Capitalization	No. of Intermediary Members	No. of Listed Companies
1986	13	0.05	938	47	80
1987	118	0.47	3,125	73	82 (50)
1988	115	0.45	1,128	80	79 (50)
1989	773	3.03	6,756	94	76 (50)
1990	5,854	21.0	18,737	116	110
1991	8,502	33.7	15,564	165	134
1992	8,567	33.4	9,922	172	145
1993	21,771	86.5	37,824	176	160
1994	23,203	91.8	21,785	175	176
1995	52,537	208.6	20,782	165	205
1996	37,737	152.8	30,797	162	228

* All money figures are in USD

Table 2. Turkish Stock Exchange Main Indicators. Source: ISE Annual Fact
Book, 1996

Year	Composite Index (TL) 01/86=100	Composite Index ($) 01/86=100	Annual Return	Average Dividend Yield % (December)	Average Price/Earnings Ratio (December)
1986	170.90	131	31%	9.15	5.07
1987	673.00	384	196%	2.82	15.86
1988	373.90	119	−69%	10.48	4.97
1989	2,217.70	560	357%	3.44	15.74
1990	3,255.70	642	15%	2.62	23.97
1991	4,369.20	501	−20%	3.95	15.88
1992	4,004.18	272	−47%	6.43	11.39
1993	20,682.90	833	198%	1.65	25.75
1994	27,257.14	413	−52%	2.78	24.83
1995	40,024.57	382	-8%	3.56	9.23
1996	90,588.80	534	40%	2.87	12.15

yields have remained low and relatively stable but P/E ratios have increased
from 5 to 25.

The majority of Turkish corporations listed on the ISE are either organized
into family-owned combinations called 'Holdings' or formerly state-owned

enterprises, now operating as corporations. Therefore, between 70–80% of shares are cross-held and not traded. These groups act as market makers for their shares and, since they are also influential in the economy as a whole, market behavior is affected by macroeconomic policies, including money regime and privatization.

2 Investor Profile

The average stockholder in Turkey is middle-aged, male and in his early fifties with a university degree. He earns more than $20,000 a year and saves half of it, retaining approximately $50.000 in a variety of assets (Muradoglu, 1989). The average annual income is slightly more than $2.500 in Turkey, so this investor profile represents the upper classes. He has a diversified stock portfolio and intends to invest in the future following a similar pattern (Muradoglu, 1989).

A survey conducted by the ISE in 1994 showed that this investor profile had not changed much during the past 6 years (see Tables 3 and 4). However, the composition of current asset portfolios have changed over this period as new instruments have been introduced and the share of stocks have decreased from 50% to 25%. Stock ownership is still very much a target for upper class investors. Considering the limited size of potential individual investors, the impetus for the market to grow in depth is expected to be realized through the participation of more institutional investors. Insurance companies have become increasingly active at the ISE during recent years due to the fact that this market has enjoyed a rapid expansion as consumers have become increasingly dissatisfied with the provision of the state-based social security system.

Other financial institutions prefer T-bills for various economic reasons, predominantly due to their acceptance regarding the reserve requirements of the central bank and their viability as deposits at state bids, as well as their high real yields (between 30% and 45% during 1995). The Turkish stock market is still thin, in a critical phase of development, hence there are opportunities for foreign investors to have fundamental roles as market makers.

3 Market Microstructures

Onder and Guner (1997), investigating the bid-ask spreads at the ISE, report that the average spread is 1.84%, which is comparable to the bid-ask spread at the NYSE and smaller than that for NASDAQ securities. Considering that brokerage fees are around 0.04%, the bid-ask spread as a large part of transactions costs. The spread in the afternoon session is significantly higher than that in the morning session, and proportional spread also decreases as

Table 3. Demographic Characteristics of Individual Investors. Source:
Muradoglu (1992) and ISE (1995)

AGE DISTRIBUTION OF INDIVIDUAL INVESTORS

(a)

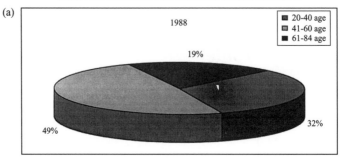

(b)

EDUCATION LEVEL OF INDIVIDUAL INVESTOR

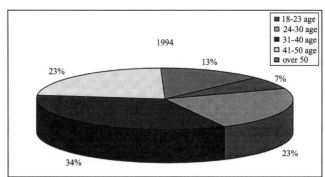

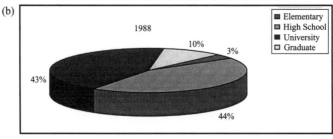

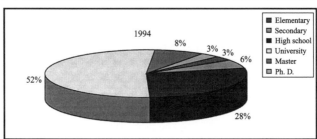

Table 3. *Cont.*

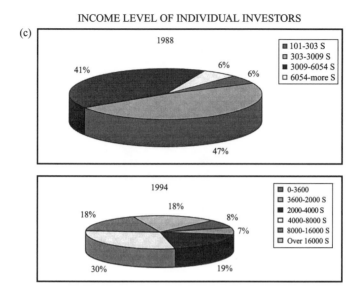

both volume of trading and size of the firm increases. Trading is information related.

Guner and Onder (1997) examine the volatility during trading and non-trading hours at the ISE. The authors show that the open-to-close per hour volatility is 13 times higher than the close-to-open per hour volatility assuming that the volatility during the break is the same as that during the trading hours. When it is assumed that the volatility during the break is the same as that during the non-trading hours this number increases to 20. Also, the variance ratio of open-to-open returns is greater than that for close-to-close returns for all size and volume quintiles. This indicates higher volatility for open-to-open returns as well. After testing for the effect of noise, the authors conclude that the higher volatility during trading hours is also caused by information related trades.

4   Stock Returns and Macroeconomic Variables

In the emerging markets of developing countries, macroeconomic variables constitute an important set of information. In developing countries like Turkey, capital accumulation and economic activity are initiated by govern-

Table 4. Preferences of Individual Investors. Source: Muradoglu (1992) and
ISE (1995)

PREFERENCE FOR FUTURE ASSET OWNERSHIP

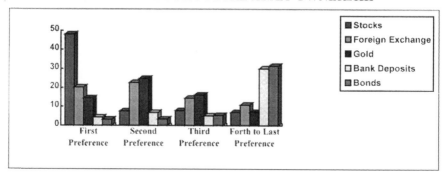

PERCENT DISTRIBUTION OF FINANTIAL HOLDINGS

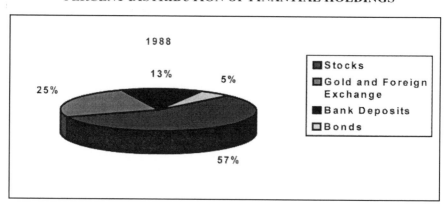

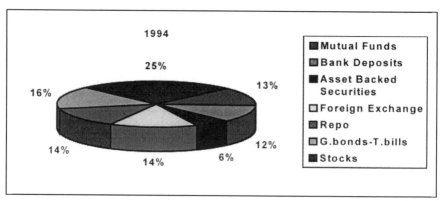

Table 5. Volatility during Trading and Non-Trading Hours (variance ratios of log-returns, and returns respectively assuming volatility of returns during the day break is the same as the volatility during trading hours)

	Variance Ratio (OO, CC)	Variance Ratio (OC, CO)	Variance Ratio (OO, CC)	Variance Ratio (OC, CO)
Average	1.40	13.67	1.39	13.68
Stand. Dev.	0.20	5.26	0.24	5.47
Minimum	0.98	2.22	0.97	1.52
Maximum	2.83	30.06	3.52	30.27

Note: OO: Open-to-Open; CC: Close-to-Close; OC: Open-to-Close; CO: Close-to-Open. Source: Guner and Onder (1997)

ment policies. In Turkey, more than 60% of industrial production is directly controlled by the state, and private companies are organized as Holdings through inter-company share holdings which makes them sensitive to government policies. In Turkey, especially during the initial phases of its market development, information on company-specific performance was generally limited and not timely. Therefore, similar to the stock markets of other controlled economies, the ISE is also expected to accommodate fiscal and monetary changes as important sets of information.

Muradoglu and Onkal (1992) developed equations to estimate monetary and fiscal actions, with their expected and unexpected components. A stock return model was estimated for the period January 1988 to December 1991 using monetary and fiscal policy as explanatory variables. The results indicate the presence of a significant lagged relation between fiscal policy (two months) and monetary policy (one month) and stock returns.

Muradoglu and Metin (1996) using cointegration analysis and a data set covering the period between January 1988 and December 1993, tested the long-run relationship between stock prices and inflation together with short-run dynamics. Inflation proxies for other monetary variables in the short run, but in the long run the negative relation between stock prices and inflation persists. Muradoglu, Metin and Argac (1999), investigating the long-run relationship between stock prices and monetary variables at different phases in the development of the ISE, conclude that inefficiencies are observed mainly during the later stages of the ISE development, due to the changing characteristics of the market structures and participants. Metin and Muradoglu (1999) also show that macroeconomic variables can be used to forecast stock prices by using Vector Autoregressive, Error Correction and Stochastic Seasonal Models.

Table 6. Monetary and Fiscal Policy and Stock Returns

A. Regression results for the monetary policy model. Dependent Variable: M_t

Variable	Estimated Coefficient	t-statistic	p-value
β_{10}	0.0407	5.14	0.000
M_{t-1}	−0.3276	−4.12	0.000
M_{t-2}	0.0635	0.78	0.436
I_{t-1}	−0.0118	−0.08	0.938
I_{t-2}	0.1702	1.15	0.254
D_{t-1}	−0.0050	−0.28	0.783
Y_{t-1}	0.0708	2.16	0.036
Y_{t-2}	0.0659	1.86	0.068
Y_{t-3}	0.0274	9.13	0.000

$R^2 = 70.0\%$, $R^{2(\text{adj})} = 65.5\%$, SER $= 0.02255$, F(8,54) $= 15,72$
with p-value $= 0.000$

B. Regression results for the fiscal policy model. Dependent Variable D_t

Variable	Estimated Coefficient	t-statistic	p-value
β_{20}	−0.2072	−0.4	0.681
D_{-1}	−0.0713	−0.47	0.639
D_{t-2}	−0.3878	−2.8	0008
D_{t-3}	0.0043	0.03	0.978
U_{t-1}	2.4000	0.62	0.540
Y_{t-1}	−0.1063	−0.40	0.689
Y_{t-2}	−0.8166	−3.10	0.004
IBR_{t-1}	1.8871	2.54	0.016
IBR_{t-2}	−0.7313	−1.01	.322
IBR_{t-3}	−1.0776	−1.48	0.149
M_{-1}	0.7306	1.25	.221

$R^2 = 49.0\%$, $R^{2(\text{adj})} = 33.1\%$, SER $= 0.1263$, F(10,32) $= 3.08$
with p-value $= 0.008$

The crisis periods in the Turkish stock market are also shown to be sensitive to government policy actions and macroeconomic variables. Ozer and Yamak (1992), for example, have analyzed the behavior of stock returns during the Gulf Crisis. Defining the January to July 1990 as the before-crisis period and the August 1990 to March 1991 period as the crisis period, they report that during the crisis period volatility increased and returns declined. Also, as risk-return relationships were distorted, it was observed that the effect of the

Table 6. *cont.*

C. Regression results for stock returns. Dependent Variable: Stock returns

i. Unexpected variables as independent variables

Variable	Estimated Coefficient	t-statistic	p-value
β_{30}	0.0863	1.96	0.057
UM_t	−1.7360	−0.91	0.368
UD_t	0.0461	0.11	0.910

$R^2 = 2.1\%$, $R^{2(\text{adj})} = 0.0\%$, SER $= 0.2882$, $F(2,40) = 0.43$
with p-value $= 0.651$

ii. Lagged expected variables as independent variables

Variable	Estimated Coefficient	t-statistic	p-value
β_{40}	−0.1773	−1.62	0.115
AM_t	0.5010	0.48	0.633
AM_{t-1}	3.6730	3.18	0.003
AM_{t-2}	1.9720	1.81	0.079
AD_t	0.2118	0.67	0.509
AD_{t-1}	−0.2378	−0.81	0.424
AD_{t-2}	−0.5484	−1.73	0.093

$R^2 = 37.1\%$, $R^{2(\text{adj})} = 26.0\%$, SER $= 0.1899$, $F(6,34) = 3.34$
with p-value $= .011$
M_t = growth rate in money stock;
I_t = inflation rate;
D_t = budget deficit;
Y_t = growth rate of GNP;
U_t = unemployment rate;
IBR_t = interest rates;
U stands for unexpected and A for anticipated components of the variables.
Source: Muradoglu and Onkal (1992)

crisis on stock returns were different for different industries.

Muradoglu, Berument and Metin (1997) investigated the macroeconomic determinants of risk and the risk-return relationship using a GARCH(1,1)–M specification. During the 1994 economic crisis, risk was determined by the government policy action of tight monetary policy which was associated with higher volatility in the stock market. However, before the crisis, the depreciation of the exchange rate was an indicator of political and economic instability and thus increased volatility, while, during the recovery period, after the

Table 7. Stock Returns and Monetary Variables

1. Test of Cointegration Between Stock Prices and Macroeconomic variables
a. ADF tests

Independent variables	ADF tests t-statistics
LP	−2.84541
LA	−3.59338*
LE	−2.39152
LR	−3.48798*
LM1	−3.31262*
LM2	−2.92940*
LA,LR	−2.16497
LA,LR,LP	−3.95284*
LA,LR,LP,LE	−3.69078
LP,LA,LE,LR,LM1	−4.23357**
LP,LA,LE,LR,LM2	−3.10000

Note: (1) ADF test statistics are based on regressions with twelve lags. (2) The critical values for the ADF test statistics are obtained from Engle and Granger (1987), Table 2. (3) ADF tests are based on the residuals of the following static equations presented at Table 2.b [also see equation (3) in the text].

b. Static equations [Dependent Variable = LISE]

Eq. No.	Constant	Trend	LA	LE	LR	LM1	LP	R^2	DW	F
Eq1~	7.46	0.80*	−0.10	−2.08**	−0.27	3.13**	−1.89*	0.95	0.513	93.77
Eq2	33.43**	0.16**	0.15	−2.88**	−1.55**	1.00	−	0.94	0.324	75.27
Eq3	28.98**	0.16**	0.0	−2.78**	−1.07*	−	−0.45	0.94	0.285	78.90
Eq4	26.76**	0.15**	0.08	−2.95**	−1.06*	−	−	0.94	0.289	84.91
Eq5	26.21**	0.17**	0.04	−3.79**	−	−	−	0.94	0.263	85.64
Eq6	8.07**	0.07**	−0.50**	−	−	−	−	0.91	0.197	60.21
Eq7	27.59**	0.17**	−	−3.66**	−	−	−	0.94	0.263	93.23
Eq	14.40**	0.06**	−	−	−2.48**	−	−	0.93	0.324	77.23
Eq9~	−22.30**	−0.08**	−	−	−	3.39**	−	0.90	0.257	56.71
Eq10	31.81**	0.21**	−	−	−	−	−3.85**	0.90	0.163	58.29

~Eleven deterministic dummies for seasonality are added to Equations 1 through 10, and (~) denotes equations with significant deterministic seasonality.
* significant at 5%
** significant at 1%

crisis, government induced risk persisted but expansionary monetary policy rather than contractionary, was positively associated with risk in the stock market. Besides risk, macroeconomic variables that explain the behavior of stock returns also were observed to change before, during and after the crisis.

Another opportunity for international investors is to follow the macroeconomic variables that lead stock returns to attempt to market time.

Table 7. cont.

2. The Short Run Dynamic Model (t values reported in parenthesis)

$$
\begin{aligned}
\text{dLISE}_t = \quad &-.00109 &+\ &0.05303\ \text{dLISE}_{t-6} &-\ &0.57274\ \text{dLR}_t \\
&(-.03) & &(.49) & &(-1.97) \\
&+0.45894\text{dLR}_{t-2} &-\ &0.65238\text{dLR}_{t-3} &-\ &0.76963\text{dLR}_{t-5} \\
&(-1.47) & &(-2.12) & &(-2.38) \\
&-0.48818\text{dLR}_{t-6} &-\ &0.47392\text{dLR}_{t-7} &-\ &0.68376\text{dLR}_{t-9} \\
&(-1.56) & &(-1.50) & &(-2.19) \\
&+0.74715\text{dLm1}_{t-1} &+\ &0.47176\text{dLm1}_{t-2} &+\ &0.34118\text{dLm1}_{t-7} \\
&(2.51) & &(1.71) & &(1.44) \\
&-1.83687\text{dLE}_{t-7} &+\ &2.45310\text{dLE}_{t-8} \\
&(2.18) & &(-1.51)
\end{aligned}
$$

$R^2 = 0.3825385$; $d = 0.1750084$; $F(13,69) = 3.29$; $D - h = 2.078$;
$FPE = 0.035794$

ISE= Istanbul Securities Exchange Composite Index;
A= budget deficit;
R= interest rates; E= exchange rates;
P= inflation;
M1 and M2 = money supply;
L stands for the natural logarithm of the variables, and d stands for first derivative
Source: Muradoglu and Metin (1996)

5 Distributional Properties of Return Series and Naive Strategies

Yuce (1996), examining price changes between 1988 and 1992, rejects normality and strict white-noise. Turkish stocks are heavily leptocurtic, and linear dependence as well as dependence in the squared series is rejected. Similar results are reported by Muradoglu and Unal (1994) who studied the distributional properties of return series by examining the independence, randomness and normality of the price series for each stock traded for more than 95% of the days during January 1988 to December 1991. Stocks traded on the Turkish stock exchange indicated strong deviations from independence; empirical distributions of successive returns were leptocurtic and non-normal; and deviated from random walk. Price forming information is not disseminated rapidly in the thinly traded Turkish Stock exchange and it seems possible to earn excess returns by following past price sequences. Considering the time period the data set covers, the distributional properties of stock returns can be explained by the lack of sophisticated communication both in terms of

Table 8. Stock Returns and Volatilities during the Gulf Crisis

1. Daily returns of ISE composite index before and
during the Gulf crisis

Period	Number of Cases	Mean Return	Standard Deviation
Entire Period	296	0.0023	0.0328
Before crisis	14	0.0051	0.0288
After crisis	155	−0.0003	0.0359

2. Inter-period variations of Individual stocks
(Numbers in parenthesis are the standard deviations of
respective distributions)

	Before Crisis	During Crisis	t-value
Mean Return	0.0656 (0.03)	−0.020 (0.03)	16.85
Standard Deviation	0.0614 (0.03)	0.0399 (0.02)	4.98
Coefficient of Variation	0.9359	−1.995	

Source: Ozer and Yamak (1992)

technology and in terms of content. Altug (1994), using neural networks and
a number of monetary variables as inputs in forecasting stock prices, showed
that prices can be forecast with smaller mean errors than moving averages
and APT models.

   Tuntas (1992), constructing efficient portfolios as suggested by Markowitz's
(1959) mean-variance approach with daily returns during 1990-1991, shows
that daily returns up to 0.8% are attainable. Kurdoglu (1994), forming portfolios using historical and adjusted Betas, demonstrates that the two type of
portfolios are made of essentially the same stocks and both beat the market. Kalkan (1994) constructed an equally weighted market portfolio of gold,
USD,DM, T–bills, CDs, commercial paper and ISE index and compared its
return with the returns of the 18 professionally managed mutual funds in
Turkey. During the January 1988 to April 1992 period, the equally weighted

Table 9. Modelling Risk and Return Jointly using Macroeconomic Variables
(GARCH–Model Estimates for Stock Returns)

1. Dependent Variable: Stock Returns

	Constant	Own-lag1	Own-lag-2	Interest Rates	Money Supply	Deprecia-tion of $	Conditional Variance
Before Crisis	−0.0429	0.2611**	−0.0821**			−0.0168	0.1429
Before Crisis	0.4568	0.2593	−0.0841**	−0.0098**			0.1615
Before Crisis	−0.0635	0.2617	−0.0834		−0.0635		0.15574
1994– Crisis	0.8448	0.4274**	−0.1234			0.1514	−0.3688
1994– Crisis	0.7427	0.3768**	−0.1106	−0.0005			−0.2722
1994– Crisis	1.5803	0.4496**	−0.1460		−1468		−0.5185
After Crisis	0.3049	0.0043	0.0497				0.2164
After Crisis	1.5587	−0.0505	−0.0316	−0.0154**		0.2434	0.1388
After Crisis	0.0656	−0.0125	0.0388		−0.2708*		0.2614

2. Dependent Variable: Conditional Variance

	Constant	Own-lag1	Error Squared	Interest Rates	Money Supply	Depreciation of $
Before Crisis	0.4681	0.6725	0.2815		0.7122**	
Before Crisis	0.3769	0.6662	0.2861	0.0040		
Before Crisis	0.6515	0.6669	0.2812		−0.0942	
1994–Crisis	0.6485	0.8763**	0.0626			0.6663
1994–Crisis	0.6059	0.8298	0.1172	0.0014		
1994–Crisis	0.6470	1.0277	−0.0508		−0.8794**	
After Crisis	0.8258	0.6833**	0.1967			−0.6577
After Crisis	−0.1839	0.8868**	0.1025	0.0039		
After Crisis	0.0417	0.9458**	0.0303		0.3790**	

Notes: (1) * denotes 5% significance level and ** denotes 1% significance
level; (2) Results reported here in (i) and (ii) are estimated jointly, by using
daily dummies as well, which are not reported here.
Source: Muradoglu, Berument and Metin (1997)

Table 10. Distributional Properties of Stock Returns

1. Autocorrelation coefficient and Kolmogorov–Smirrov Statistic

Stock No.	1 Day Lag	2 Day Lag	3 Day Lag	4 Day Lag	5 Day Lag	10 Day Lag	15 Day Lag	20 Days Lag	On
1	0.0813	0.041	0.06	0.0226	0.0207	0.0296	0.006	00.0183	0.276
2	0.0344	−0.013	−0.063	−0.019	0.0365	0.0624	−0.0331	0.0308	0.147
3	0.0474	−0.072	−0.023	0.0169	0.0293	0.0651	0.013	10.0165	0.237
4	0.1070	−0.068	−0.041	0.0094	−0.005	0.0035	0.0026	0.0123	0.248
5	0.1489	0.0756	0.0619	0.0374	0.0313	0.0742	0.0123	0.0183	0.311
6	0.1707	−0.016	−0.045	−0.005	0.0889	0.0315	0.0114	0.0216	0.503
7	0.1301	0.0448	0.0078	0.0561	0.0466	−0.033	−0.015	0.0431	0.107
8	0.1045	−0.012	−0.031	0.0221	−0.014	0.0343	0.0026	0.0699	0.148
9	−0.012	−0.061	−0.027	0.0239	0.0308	0.0413	0.1095	−0.006	0.245
10	0.1446	0.0021	0.0297	0.0357	0.0162	0.0599	0.0276	0.0126	0.269
11	0.1248	−0.013	0.0401	0.0087	−0.013	0.0313	0.0398	0.0126	0.231
12	0.0994	−0.018	−0.039	−0.013	0.0195	0.0152	0.0712	0.0481	0.189
13	0.1335	−0.058	−0.058	0.027	0.0349	0.0461	0.0253	0.0983	0.262
14	0.0977	0.0221	0.0559	0.0004	0.0141	0.0261	0.0284	0.0107	0.169
15	0.1158	−0.003	−0.026	0.0025	−0.026	0.0361	0.0665	0.0051	0.209
16	0.0892	0.0157	0.0266	0.0356	0.0146	0.0201	0.0485	−0.010	−0.302
17	0.1321	−0.003	−0.099	−0.003	0.0226	0.0509	0.0242	0.0223	0.312
18	0.0964	0.0143	0.0133	0.0193	−0.017	0.0769	0.0464	0.0013	0.114
19	0.1739	0.0005	−0.004	0.036	0.0227	0.0623	0.0072	0.0667	0.332
20	0.0538	−0.069	−0.059	0.0467	0.0102	0.0321	0.0334	0.0343	0.325

(*): statistically significant at $p = 5\%$ level

2. Results of Runs Tests

Stock No.	Total No. of runs (R)	Expected No. of Runs (Re)	Standard Error (SE)	Standard Variable(Z)
1	472	492.9	15.62	−1.31
2	434	477.1	15.07	−2.83
3	435	481.7	15.32	−3.05
4	445	489.2	15.45	−2.82
5	393	478.4	21.35	−3.97
6	392	490.4	24.6	−3.98
7	418	486.2	17.05	−4.00
8	409	467.5	17.99	−3.24
9	460	465.1	14.53	−0.31
10	450	488.9	15.42	−2.48
11	462	484.4	15.34	−1.42
12	464	489.2	15.45	−1.59
13	423	484.1	18.77	−3.22
14	360	455.6	23.91	−3.97
15	426	487.6	18.94	−3.22
16	427	479.3	15.37	−3.36
17	431	484.4	15.25	−3.46
18	439	484.8	15.3	−2.98
19	416	479.3	15.81	−3.96
20	443	462.6	14.53	−1.31

(*): Statistically Significant at $p = 5\%$ level
Source : Muradoglu and Unal (1994)

Table 10. *Cont.*

3. Results of Distribution Tests

Stock No.	Coefficient of Skewness (1)	Standard Error of Skewness (2)	I(1)/(2)I	Coefficient of Kurtosis (3)	Standard Error of Kurtosis (4)	I(3)/(4)I
1	−0.003(*LS*)	0.077	0.039	0.516	0.154	3.35
2	0.247(*RS*)	0.077	3.208	21.642	0.154	140.6
3	0.114(*RS*)	0.077	1.481	0.418	0.154	2714
4	0.143(*RS*)	0.077	1.857	3.527	0.154	22.90
5	0.951(*RS*)	0.077	51.31	54.37	0.154	3530
6	0.332(*RS*)	0.077	4.312	2.038	0.154	13.23
7	0.112(*RS*)	0.077	1.455	0.059	0.154	0.383
8	−0.041(*LS*)	0.077	0.532	−0.034	0.154	0.221
9	0.081(*RS*)	0.077	1.052	1.709	0.154	11.10
10	0.047(*RS*)	0.077	0.610	0.170	0.154	1.104
11	0.107(*RS*)	0.077	1.389	0.587	0.154	3812
12	0.126(*RS*)	0.077	1.636	1.168	0.154	7.584
13	0.119(*RS*)	0.077	1.545	0.382	0.154	2.481
14	0.061(*RS*)	0.077	0.792	0.058	0.154	0.377
15	−0.013(*LS*)	0.077	0.169	0.040	0.154	0.259
16	*0396(*LS*)	0.077	5.143	4.216	0.154	27.38
17	0.033(*RS*)	0.077	0.429	0.481	0.154	3.123
18	0.178(*RS*)	0.077	2.312	0.832	0.154	5.403
19	0.051(*RS*)	0.077	0.662	0.040	0.154	0.259
20	0.264(*RS*)	0.077	3.429	0.357	0.154	2.318

(RS) = Right Skew; (LS) = Left Skew; (*) = Statictically significant at p=
5% level
Source : Muradoglu and Unal (1994)

portfolio had higher returns than all of the mutual funds.

Table 11. Mean–Variance Efficient Portfolios. Source: Tuntas (1992)
PERFORMANCE OF PORTFOLIOS

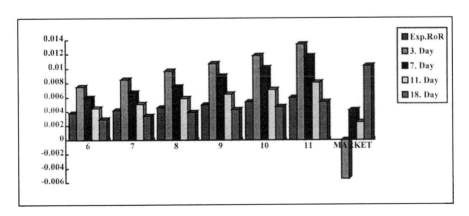

Table 12. Weekly Returns to Portfolios Constructed by using CAPM*.
Source: Kurdoglu (1994)

Period	Market Portfolio	Portfolios by using Adjusted Betas	Portfolios by using Historical Betas
1988 –1	−5.980	−2.106	−2.106
1988 –2	7.500	8.780	7.857
1989 –1	3.879	5.878	5.716
1989 –2	1.724	2.487	2.901
1990 –1	−2.280	1.190	1.204
1990 –2	−0.531	1.356	1.463
1991 –1	−0.132	−3.505	−4.897
Average	0.600	2.010	1.730

*Weekly returns reported here are in excess of the risk free rate and divided by the related standard deviations .
** For every year –1 denotes the first six months and –2 denotes the last 6 months of the year.

Yuce (1996), examining risk reduction via portfolio formation, reports that effective diversification is possible by forming portfolios of 15–20 stocks. Investors can diversify approximately 80% of the variance by constructing a portfolio with 10 stocks. Using a principal component analysis, shows that more stocks are needed for effective diversification since the commencement of foreign investor trading at the ISE in 1989. However the cumulative variance explained by fifteen stocks was 91% before 1989, it declined to 61% after the commencement of foreign investor trading at the ISE.

Around the world, it is difficult to beat market averages. A third possible opportunity for investors to beat the market averages in Turkey is by following price sequences, using naive investment strategies and by constructing mean variance efficient portfolios.

6   Anomalies

Research on anomalies in the Turkish market is recent and limited because the market is young, and analysis of trading rules has only recently become popular. However, some of the anomalies observed in the US, European and East Asian Markets have also been observed in Turkey.

Muradoglu and Oktay (1993), investigating the calendar anomalies at the ISE, had found that similar to Japan (Ziemba and Schwartz 1991), France (Solnik and Basquet 1990) and Italy (Barone 1990) the weekend effect in

Table 13. Mutual Funds and Equally Weighted Market Portfolio. Source: Kalkan (1994)

1. Average portfolio structures of mutual funds in Turkey (as of December 31, 1989)

Government Bonds	10.33%
Treasury Bills	14.16%
Foreign Currency Indexed Bonds	4.76%
Corporate Bonds	36.01%
Commercial Paper	23.46%
Common Stock	3.83%
Others	7.45%

2. Average returns for mutual funds and market portfolio

Mutual Funds		Equally Weighted Market Portfolio	
Fund Number	Average Return	Asset	Average Return
1	4.04%	ISE Index	7.89%
2	3.83%	Gold	2.84%
3	3.73%	US Dollar	3.29%
4	4.23%	German Mark	3.49%
5	3.75%	T-bills	4.29%
6	4.06%	Certificate of Deposits	4.21%
7	3.61%	Commercial Paper	4.58%
8	3.63%	Corporate Bonds	4.58%
9	3.89%		
10	6.54%	Equally Weighted Portfolio	4.40%
11	3.90%		
12	3.72%		
13	3.74%		
14	3.84%		
15	3.69%		
16	3.77%		
17	4.03%		
18	2.91%		

Turkey is an extended one from Friday to Tuesday. Metin, Muradoglu and Yazici (1997) show that, after 1994, and after the settlement date was increased to two days, the extended week-end effect shifted to being from Thursday to Monday. In both cases the contribution of the significant Friday effect

Table 14. Winner–Loser Effect. Source: Cetiner, 1993

Differences in cumulative average excess returns between winner (W) and loser (L) portfolios

Portfolio formation period	No. of weeks after portfolio formation	ACAER(L) – ACAER(W)	t values
47 Weeks Period	14 weeks	0.214	5.33
	15 weeks	0.258	3.10
	16 weeks	0.248	3.88
	17 weeks	0.248	5.75
48 Weeks Period	11 weeks	0.236	2.29
	12 weeks	0.302	5.29
	13 weeks	0.259	6.51
	14 weeks	0.196	3.78
	15 weeks	0.168	6.09
49 Weeks Period	8 weeks	0.271	3.66
	9 weeks	0.379	10.48
	10 weeks	0.337	7.40
	12 weeks	0.234	7.49
50 Weeks Period	5 weeks	0.230	3.22
	6 weeks	0.324	8.25
	7 weeks	0.338	4.99
	8 weeks	0.358	3.58
	9 weeks	0.284	2.64

is more than 0.5% per day.

As with other markets, the winner–loser effect (De Bondt and Thaler 1987) is also observed in the Turkish stock market. Cetiner (1993), using daily data for all the stocks listed on the ISE during the January 1988 to December 1992 period, formed winner–loser portfolios based on market adjusted excess returns. The performance of those portfolios was tested for periods of 1 to 52 weeks after portfolio formation. For loser portfolios, excess returns up to 16% for six weeks and up to 105% for 32 weeks were observed. Overreaction is asymmetric; i.e., excess returns to loser portfolios are higher than those to winner portfolios which oscillate around zero. Sayin (1993), using the same data set and risk adjusted excess returns, shows that this effect is emphasized if the first month of test period is January; in this case the difference between winner and loser portfolios reach up to 22%.

Civelekoglu (1993) tests the effect of size and Earnings to Price (E/P) ratio during the period January 1988 to December 1992. For the sample stocks neither the portfolios based on size (measured in terms of market value) nor the portfolios based on E/P ratio are different in terms of market risk (Beta). Although no size effect is observed, a significant E/P effect is reported; stocks with lower P/E ratios outperform those with higher P/E ratios with monthly returns up to 8% and 2.5% respectively. Aydogan and Guney (1997), analyzing the ten year period 1986–1995, show that both the average P/E ratios and the dividend yields have predictive power for returns during the subsequent 3–12 months after portfolio formation. The P/E generated portfolios strictly dominated 5 out of 7 naive portfolios of stocks and bills in terms of having mean-variance efficiency. Stock returns in general are very high after periods of very low P/E ratios and very high dividend yields. They conclude that in that regard, besides being the base of successful asset allocation strategy, P/E ratio and dividend yield can also be good indicators for market timing.

Aydogan (1992) examined the abnormal returns to initial public offerings (IPO) during the 1990–1991 period. During this period, average first day abnormal returns reached 11% with maximum underpricing of 230% and a minimum of 8%. The average underpricing of state-owned, privatized companies was 38% and for private issues, 2%. However, these figures are not significant and the long-term performance of IPOs are not significantly different from the index. Market timing is important in going public; the initial average underpricing of 17% in bull markets is reversed and reaches zero around day 120 and an initial underpricing of 3% in the bear market is reversed and reaches 22% around day 180.

On average, January is the strongest month and Friday is the strongest day.

7   Stock Dividends and Rights Offerings

A distinct characteristic of the Turkish stock market is the frequency and volume of stock dividends and rights offerings. Stock dividends are declared from retained earnings or a re-evaluation fund (an equity account created as a result of a two or three digit inflation adjustment of fixed assets in operation since 1983). Since corporations are limited to issuing debt of up to 600% of paid-in capital, under the high inflation rates experienced in Turkey most corporations convert the re-evaluation fund and retained earnings to paid-in capital by declaring stock dividends so that they can maintain consistent debt to paid-in-capital ratios. Therefore, accounting treatment of their transaction is similar to a stock dividend, yet the split factor can be quite high. Many corporations also issue new shares at par value (TL 1,000) through rights offering, usually paid simultaneously with declared cash dividends, making them identical to a stock split. In most cases, rights offerings are accompanied by stock dividends.

Aydogan and Muradoglu (1998) investigated excess returns due to the

Table 15. Price reaction to Stock Dividends and Rights Offerings: Announcement of Board Decision. Source: Aydogan and Muradoglu (1998)

| | | Average cumulative abnormal returns (ACARs) Numbers in parentheses represent t statistics | | | |
	Period	ACAR (-10)	ACAR (0)	ACAR (10)	ACAR (30)
	1988–93	2.69%	5.09%	4.94%	3.55%
		(1.49)	(2.39)	(2.05)	(1.21)
Board	1988–90	3.60%	8.36%	10.75%	12.04%
Meeting		(1.14)	(2.26)	(2.75)	(2.35)
	1991–93	2.33%	3.57%	2.16%	-0.64%
		(1.07)	(1.37)	(0.74)	(-0.18)

board announcement of rights issues and stock dividends in the ISE during the January 1988 to December 1993 period. Until 1990, positive price reactions to board decision disclosures were observed up to 13% and up to an 18 day event window. During this period, the magnitude of price reactions to the announcement of stock dividends and rights offerings were positively correlated with split percentages. As the market matures, and investor mix changes, price reactions to board decisions on rights issues and stock dividends fade away. They also show that the changing mix of investors from institutional to individual at the later stages of the development of the market shifted the event date from announcement to the implementation of stock dividends and rights offerings (Muradoglu and Aydogan, 1999). As the ISE has matured, the quality and quantity of interim reports released by listed companies have improved considerably, and firm–specific information has become less costly and more accessible. Results indicate that, after 1993, it was possible to earn abnormal profits of up to 18% throughout a 30 day period, by following the implementation of rights offerings and stock dividends at ISE.

8   Integration with Other Exchanges

Financial integration with other more developed markets was pronounced as one of the main purposes of the financial liberalization process that Turkey started during the early 1980s. Regulations concerning non-residents' investments in Turkey and Turkish investments abroad are aimed at increasing interdependencies between the Turkish stock exchange and European and American Exchanges. The integration of the Turkish stock market with its more developed counterparts is an important factor in the process of financial liberalization, but the lack of such integration will create an excellent opportunity for international diversification.

Table 16. Price Reaction to Stock Dividends and Rights Offerings: Implementation at ISE. Source: Muradoglu and Aydogan (1999)

		Average Cumulative Abnormal Returns (ACARs) Numbers in parentheses represent t statistics			
	Period	ACAR (-10)	ACAR (0)	ACAR (10)	ACAR (30)
Implement- ation at ISE	1988–94	1.29% (1.49)	8.87% (5.43)	9.09% (5.08)	9.97% (5.02)
	1988–89	0.04% (0.02)	5.09% (2.13)	4.79% (1.67)	3.72% (1.10)
	1990–92	0.75% (0.64)	7.26% (2.65)	6.37% (2.16)	4.81% (1.53)
	1993–94	2.23% (1.47)	11.72% (4.91)	13.60% (5.11)	17.99% (5.79)

Akdogan (1993) tested the empirical validity of a European Union benchmark portfolio for pricing capital assets in Turkey. Considering the monthly returns during the period January 1987 to February 1992 and using a standard Fama–McBeth procedure, it was revealed that the Turkish stock market was not integrated with that of EU. Systematic risk measured in terms of a value-weighted EU portfolio was unpriced. A more recent study by Bokesoy (1994), on the other hand, shows that price movements in the NYSE and the ISE are interdependent. This study considers the daily returns for the period April 1992 to December 1993 using co-integration analysis. The ISE was included in the foreign investment markets by the NYSE in November 1993.

Although the Turkish stock market is open to foreign investors, the fact that it was segmented from its more developed counterparts for a long time can be explained by the erection of indirect barriers concerning the difficulty of obtaining information. For foreign investors, difficulties associated with obtaining information on particular stocks, the reluctance of brokerage firms in dealing with them, language difficulties and difficulties in accounting disclosure requirements might constitute barriers to market entry and subsequent segmentation. Possible gains from international diversification by investing in Turkish stocks needs further attention. For the first time in its history, the ISE responded to a global shock simultaneously with a 20% decline in two days, during the crisis following the Hong Kong Stock Exchange crash in November 1997. International investors should note that Turkey is in the process of financial integration and opportunities for windfall profits are expected to decline in the future.

Table 17. Integration with the European Union, 1987–1992. Source
Akdogan (1993)

1. Summary statistics: Cross-correlation coefficient of share prices
The correlation coefficients reported here are computed using the monthly
country share price data published by the International Monetary Fund in its
International Financial Statistics. For Turkey, the data source is the Capital
Market Board. The period on which cross correlations are computed is based
is from January 1987 to February 1992.

	UK	Germany	France	NL	Belgium	Denmark	Italy	Spain
Germany	0.901							
France	0.922	0.932						
NL	0.909	0.975	0.941					
Belgium	0.805	•0.917	0.915	0.953				
Denmark	0.913	0.792	0.783	0.814	0.691			
Italy	0.829	0.912	0.923	0.910	0.912	0.668		
Spain	0.214	0.506	0.453	0.554	0.696	0.116	0.647	
Turkey	0.319	0.230	0.175	0.240	0.320	0.295	0.425	0.195

**2. Summary statistics for the country excess returns: Variances
and autocorrelation coefficients**
The returns are calculated as changes in share prices as published in the
International Financial Statistics. The country excess returns are computed
in excess of the 3-month Treasury Bill rates in the UK as reported by the
Morgan Bank of New York. The sample period is from January 1987 to
February 1992.

Market		p1	p2	p3	p4	p12	p24
UK	0.00515	0.06905	0.00049	0.08533	003515	−0.0048	0.0480
Germany	0.00116	0.23259	0.11095	0.17065	0.04274	0.01118	0.1828
France	0.00676	−0.1489	−0.1238	−0.1455	−0.0297	0.00105	0.0658
NL	0.00146	0.30564	0.12431	0.07309	−0.0267	0.19676	0.1655
Belgium	0.00249	−0.0268	0.00113	0.05637	−0.0103	0.17475	0.1116
Denmark	0.00292	0.17016	0.19093	0.21274	0.17152	−0.0733	−0.012
Italy	0.00492	0.29117	−0.0092	0.07086	0.09128	0.14930	0.0915
Spain	0.00351	0.14728	0.08344	0.09676	0.09313	0.01590	0.0695
Turkey	0.01098	0.17688	0.04791	0.03426	0.02047	0.09734	0.0568

9   Concluding Remarks

The Turkish stock market has fallen by 10% several times. For an international investor, the major risk in the Turkish stock exchange is the lack of
information about Turkey in general. Consequently, market timing is difficult. Crucial factors such as inflation rates, yield curve and need for public

borrowing must be considered for successful timing. Political stability is also very important for a state-initiated economy like Turkey's. For example, unpredictably, the ISE index has risen sharply twice as a result of military actions due to uncertainties being diminished; when US troops started military action during the Gulf crisis and when Turkish troops started military action against the PKK, both in Iraqi territories.

Using an approach that considers fundamentals, under-valuation or anomalies, investors may outperform the market averages. A focus on state-owned companies in the process of privatization, and large holding companies' stocks may also yield gains in the long-term. Monetary variables are important for successful investments. Forming portfolios based on CAPM or mean-variance models seem to outperform the benchmark ISE index. Buy-hold-switch strategies can be used for stocks with low returns, or low price–earnings ratios and calendar anomalies should be considered in this process. The Turkish stock market, as it is less integrated with other major stock exchanges, also provides interesting diversification opportunities for international investors.

The greatest difficulty in investing in emerging markets is the inherent state of continuous change in which they operate, rather than the regularities that are observed in its market structures, market participants, availability and accessibility of information, quality of information and related market behavior. The Istanbul Stock Exchange is no exception. Interpretations in one period, may lead to incorrect evaluations in the preceding one as the basic characteristics of the stock market change. The variables that explain stock prices can change over time. Diligent investors, who want to exploit a variety of profit opportunities at the ISE are therefore advised to consider the rapidly changing characteristics of the market, rather than searching for regularities which might be very few.

References

Akdogan, H. (1993), 'Testing the Stock Market Integration Between a Developing Country and its Core: The case of Turkey and the European Community', Discussion Paper, Department of Economics, Bilkent University, Ankara.

Altug, S. (1994), *Price Prediction in IMKB Using Neural Networks*, Unpublished MBA Thesis, Department of Management, Bilkent University, Ankara.

Aydogan, K. (1992), 'Initial Public Offerings in an Emerging Market: The Turkish Case', Proceedings of the 3rd. Izmir Iktisat Kongresi, 4-7 June, Izmir.

Aydogan, K. and Guney, A. (1997), 'P/E Ratio and Dividend Yield as Forecasting Tools in ISE', *ISE Review* 1/1, 83–96.

Aydogan, K. and Muradoglu, G (1998), 'Do Markets Learn from Experience? Price Reaction to Stock Dividends in Turkish Market', *Applied Financial Economics* **8** 41–49.

Barone, E. (1990), 'The Italian Stock Market: Efficiency and Calendar Anomalies', *Journal of Banking and Finance* 14 483–509.

Bókesoy, A. (1994), 'Interdependency Between Istanbul Stock Exchange and New York Stock Exchange', Unpublished MBA Thesis, Department of Management, Bilkent University, Ankara.

Cetiner, Y. (1993), 'Test of Overreaction in Istanbul Stock Exchange', Unpublished MBA Thesis, Department of Management, Bilkent University, Ankara.

Civelekoglu, H. (1993), 'An Investigation of Anomalies at Istanbul Securities Exchange: Size and E/P Effects', Unpublished MBA Thesis, Department of Management, Bilkent University, Ankara.

Debondt, W. and Thaler, R. (1987), 'Further Evidence on Investor Over-reaction and Stock Market Seasonality', *The Journal of Finance* **42** 793–805.

Guner, N. and Onder, Z. (1997), 'Stock Market Returns and Volatility during Trading and Non-trading Hours: Evidence from ISE', paper presented at the European Financial Management Association Meeting June 1997, June, Istanbul.

ISE (1995), *Survey Prepared to Guide Investor Education*, Unpublished Manuscript, Istanbul Stock Exchange, Istanbul.

ISE (1996), *Annual Fact Book 1996*, ISE Publications, Istanbul.

Kalkan, H, (1994), 'Performance of Mutual Funds: An Empirical Investigation', Unpublished MBA Thesis, Department of Management Bilkent University, Ankara.

Kurdoglu, G.C. (1994), 'Performance Measurement of Portfolios Constructed by the Single Index Model with Historical and Adjusted Betas', Unpublished MBA Thesis, Department of management, Bilkent University, Ankara.

Markowitz, H.M. (1959), *Portfolio Selection: Efficient Diversification of Investments*, Wiley.

Metin, K. and Muradoglu, G., (1999), 'Forecasting Stock Prices by Using Alternative Time Series Models', *METU Studies in Development*, forthcoming.

Metin, K. Muradoglu, G., and Yazici, B. (1997), 'Anomalies in Istanbul Stock Exchange: Day of the Week Effect', *ISE Journal* **1** 25–27.

Muradoglu, G. (1989), 'Factors Influencing Stock Demand in Turkey', Unpublished PhD Thesis, Institute of Social Sciences, Bogazici University, Istanbul.

Muradoglu, G. (1992), 'Factors Influencing Demand for Stocks by Individual Investors', Proceedings of 3rd Izmir Iktisat Kongresi, Vol. 1.

Muradoglu, G. and Aydogan, K. (1999), 'Do Price Reactions to Stock Dividends and Rights Offerings Change as ISE Matures?', *Bogazici Journal*, forthcoming.

Muradoglu, G., Berument, M. and Metin, K., (1997), 'An Empirical Investigation of Stock Returns and Determinants of Risk in an Emerging Market: GARCH–M Modelling at ISE', paper presented at the 20th Meeting of the EURO Working Group on Financial Modelling, April 1997, Dubrovnik, Croatia.

Muradoglu, G. and Metin, K. (1996), 'Efficiency of the Turkish Stock Exchange with respect to Monetary Variables: A Co-integration Analysis', *European Journal of Operational Research* **90** 566–576.

Muradoglu, G., Metin, K., and Argac, R., (1999), 'Are There Trends towards Efficiency for Emerging Markets? Co-integration between Stock Prices and Monetary Variables at ISE', *Applied Financial Economics*, forthcoming.

Muradoglu, G. and Oktay, T. (1993), 'Türk Hisse Senedi Piyasasinda Zayif Etkinlik Takvim Anomalileri' (Weak Form Efficiency in the Turkish Stock Exchange: Calendar Anomalies), *Hacettepe Iktisadi ve Idari Bilimler Fakïltesi Dergisi* **11** 51–62.

Muradoglu, G. and Onkal, D. (1992), 'Türk Hisse Senedi Piyasasinda Yari-Güclü Etkinlik' (Semi-Strong Form Efficiency in the Turkish Stock Market), *METU Studies in Development* **19** (2) 197–107.

Muradoglu, G. and Unal, (1994). 'Weak Form Efficiency in the Thinly Traded Turkish Stock Exchange', *The Middle East Business and Economic Review* **6** (2) 37–44.

Onder, Z. and Guner, N. (1997), 'The Bid–Ask Spread and its Determinants for Securities Traded on the ISE', paper presented at the European Financial Management Association Meeting June 1997, Istanbul.

Ozer, B. and Yamak, S. (1992), 'Effect of Gulf Crisis on Risk–Return Relationship and Volatility of stocks in the Istanbul Stock Exchange', *METU Studies in Development* **19** (2) 209–224.

Sayin, G. (1993), 'An Investigation of Anomalies at the Istanbul Securities Exchange: Winner-Loser Effect', Unpublished MBA Thesis, Department of Management, Bilkent University, Ankara.

Solnik, B. and Basquet, L. (1990), 'Day of the Week Effect on the Paris Bourse', *Journal of Banking and Finance* **14** 461–468.

Tezcanli, M.V. (1996), *Insider Trading and Market Manipulation*, ISE Publications, Istanbul.

Tuntas, M.C. (1992), 'Portfolio Selection Methods: An Application to the Istanbul Securities Exchange', Unpublished MBA Thesis, Department of Management, Bilkent University, Ankara.

Yuce, A. (1996), *An Examination of an Emerging Stock Exchange: the Case of the Turkish Stock Market*, ISE Publications, Istanbul.

Ziemba, W.T. and Schwartz, S.L. (1991), *Invest Japan*, Probus Publishing Company, Chicago.

Security Market Anomalies in Finland*

Teppo Martikainen

Abstract

This paper provides a review and new evidence on the seasonal and cross-sectional anomalies in the Finnish stock and derivatives markets. The main goal of the paper is to review whether the most pronounced regularities observed in major financial markets exist in one of the smallest European stock markets, and to investigate what are the main reasons that make the Finnish anomalies to differ from anomalies found in major markets. It appears that most of the anomalies observed in major markets appear also in Finland. However, regularities based on short return intervals, such as the time-of-the day and day-of-the-week are not as evident as in many larger markets. This may be because of the infrequent trading and consequently problematic return measurement in short return intervals. Also several specific institutional factors, such as peculiar Finnish accounting practices cause Finnish anomalies to be different from those observed in large markets.

Key words stock markets, derivatives markets, anomalies, thin trading, Finland

1 Introduction

Why should international investors invest in small security markets? One obvious reason is that small markets can provide international investors diversification benefits. Several studies suggest that these benefits may be of large magnitude. Hietala (1989), for instance, reports zero-betas for Finnish stocks with respect to a worldwide market portfolio using data from the 1980s. Although more recent studies, such as Booth, Martikainen and Puttonen (1993) and Bos et al. (1995), find that the co-movements between Finnish and global markets may have increased in recent years, the diversification benefits are

---

The author was visiting Louisiana State University when this research was undertaken. He acknowledges the financial support from the Väinö Tanner and Gustaf Svanljung Foundations as well as The Academy of Finland to support the visit. In addition, he acknowledges Helsinki Stock Exchange for providing some of the data needed in this paper and Bill Ziemba for helpful comments.

still apparent. It also seems that foreign investors have recognized these benefits in Scandinavian markets, because the foreign investors' in trading in small Scandinavian markets has considerably increased in recent years.[1]

In addition to diversification opportunities, small markets are interesting to international investors because of the profit opportunities these markets may offer to professional traders. It is a common belief that small markets behave more inefficiently than the major stock markets. This is, for instance, because of considerable information asymmetries between informed and uninformed traders, often less restrictive trading rules and less developed institutions for investment analysis in small markets. All these factors may lead to mispricings and undervaluation of assets in these markets.

This paper provides a review and new evidence on the calendar and cross-sectional anomalies in the Finnish stock and derivatives markets. The main goal of the paper is to review whether the most pronounced regularities observed in major financial markets exist in one of the smallest European markets, and to investigate what are the main reasons that make the Finnish anomalies to differ from the anomalies found in major markets. The Finnish market has recently become an interesting investment opportunity for foreign investors, because Finland joined the European Union with no foreign ownership restrictions for stocks or derivatives in 1995. The rest of the paper is organized as follows. Section 2 provides a general description of the Finnish markets. Trading activity, mechanisms and legislation are illustrated. The third section discusses the seasonal regularities in Finland. The cross-sectional anomalies, especially earnings related anomalies are presented in the fourth section. Finally, concluding remarks are provided in the fifth section.

2 Institutional Description

2.1 Finnish Stock and Stock Index Derivatives Markets

Liberalization of capital movements and the securitization of national markets have been the major trends in the development of financial markets in the 1980s and 1990s. Advanced computer technology and improved worldwide processing of news have made the international financial transactions easier and cheaper than before. The current ongoing integration in Europe has further accelerated the development in European countries. Deregulation of

---

[1]Up to 1993 two broad classes of shares existed in the Finnish stock market: unrestricted and restricted shares. The restricted stocks could be purchased only by domestic investors, while the unrestricted shares could be bought by both domestic and overseas investors. Booth, Chowdhury and Martikainen (1994) report empirical evidence supporting the hypothesis foreign investors were willing to pay a premium of the unrestricted stocks because of the diversification benefits they offered.

capital movements and the innovation of new financial products have led to sharper competition between both financial institutions and different national market places.[2]

2.1.1   The Finnish Stock Market

The Helsinki Stock Exchange (HSE) is the only stock exchange in Finland. The HSE is a small market comprised, for the most part, of infrequently traded stocks. At the end of 1992, the market capitalization of stocks listed on the HSE was $12 billion. At the same time, the capitalizations of the NYSE, Tokyo and London stock exchanges were $3,798 billion, $2,397 billion and $928 billion, respectively (International Federation of Stock Exchanges, 1992). Ranked by the market value of listed firms, HSE was the second smallest of the 16 European stock exchanges. Although the market value of listed stock increased considerably in Finland up to the level of $41 billion (FIM 192 billion) by the end of 1995, the market is still small.[3] In 1995 the trading volume of stocks in the HSE was $18 billion (FIM 83 billion). The number of listed firms at the end of the year was 73.[4] The ownership of Finnish firms is highly concentrated in institutions: at the end of 1995, households owned only 18% of the market value of the HSE. At the same time foreign investors owned 24% of the market value of listed firms in the HSE.

After a very strong bullish period, Finnish stock prices peaked in April 1989. After that a dramatic price drop followed not only in stock, but in all asset prices as demonstrated by Booth, Glascock *et al.* (1994). Finnish market index dropped more than 60% from its peak value by autumn 1992. On September 8th, 1992 the Finnish markka was put into float and fell about 30% in five months. During the same months stock prices started to rise again, largely because the international competitiveness of Finnish firms improved significantly. At the same time interest rates declined and the Finnish markka devaluated. By April 1995 stock prices doubled compared to the pre-floating period.

2.1.2   The Finnish Derivatives Market

In Finland the derivative contracts on shares are traded only on the HSE.

Trading in new options and futures contracts starts every second month. Index options are European and stock options are American. The index

---

[2]Booth, Glascock *et al.* (1994) provide a description on the liberalization of Finnish financial markets and Kallunki and Martikainen (1996) analyze its effects on stock prices.

[3]The market value of the largest firm, Nokia Corporation, represented 26.8% of the market value of the HSE in December 1995.

[4]This figure refers to the main list of the HSE. The number of firms in brokers' and OTC-lists are 33 and 18, respectively.

Table 1. The Nontrading Probability of the Stocks Included in the FOX index in 1988–1993. (Martikainen, Perttunen and Puttonen, 1995b).

Period	Min	Q1	Med	Q3	Max	Average	Weighted Average
8805–8807	0.000	0.000	0.000	0.000	0.047	0.005	0.004
8808–8901	0.000	0.000	0.008	0.023	0.208	0.029	0.014
8902–8907	0.000	0.008	0.008	0.024	0.669	0.063	0.022
8908–9001	0.000	0.008	0.031	0.063	0.227	0.048	0.021
9002–9007	0.000	0.032	0.081	0.194	0.573	0.134	0.081
9008–9101	0.000	0.016	0.024	0.143	0.437	0.086	0.040
9102–9107	0.000	0.000	0.024	0.105	0.371	0.066	0.023
9108–9201	0.000	0.016	0.056	0.128	0.312	0.093	0.046
9202–9207	0.000	0.008	0.032	0.073	0.298	0.054	0.024
9208–9301	0.000	0.000	0.019	0.038	0.264	0.036	0.020
9302–9307	0.000	0.000	0.000	0.000	0.064	0.006	0.003
9308–9401	0.000	0.000	0.000	0.000	0.043	0.002	0.000

The nontrading probability is the proportion of days in which there was no trade for a specific stock during the period. Min, Q1, Med, Q2 and Max refer to the minimum, lower quartile, median, upper quartile and maximum of the nontrading probabilities of the 25 stocks included in the FOX index in the period. Average and Weighted average are the averages of the non trading probabilities across stocks. In the weighted average the weights of the FOX index are used.

contracts are based on the value-weighted FOX-index, which is calculated on the basis of top 25 shares in terms of median trading during the previous half a calendar year. The FOX shares' proportion of the total trading volume of the HSE has been approximately 70%. However, the thin trading in the HSE can be seen in the nontrading probabilities of the shares in the FOX index. Table 1 shows that, for instance, in 1989 and 1990 the FOX index included stocks which were very infrequently traded.

2.2 Trading Mechanisms for Stocks and Derivatives

2.2.1 Stocks

Trading in the Finnish stock market is regulated by the Security Market Act, which was established in 1989 to improve investor protection. The Act applies to public issuance, markets and exchange in securities. The main areas of the legislation consist of disclosure requirements, provisions in insider trading and public bids requirements. It also states that the shareholder is obliged to purchase the shares of other owners if his stake exceeds two-thirds level and that securities intermediaries have to follow conventional business practices. For stock exchange activities, a license is required.

Until February, 1989 trading in the HSE was organized through a 'calling out' system in which stocks were auctioned issue by issue in the same order every day. The number of shares traded during the 'calling out' was restricted to be one block that was separately determined by the Exchange. After the conclusion of the price fixing auction, free trading took place at prices within the closing bid-ask spread.[5] An entirely new trading and information system HETI (Helsinki Stock Exchange Automated Trading and Information Systems) was implemented gradually so that all trades were performed within the system by April, 1990. The HETI system comprises automated trading in shares and other listed securities and the electronic transmission of trading data on these instruments. The system resembles with some minor differences the CATS system used in many markets, including Toronto, Paris, Brussels and Barcelona. The HETI system is an entirely open market-by-order type of limit order book in which every order is displayed individually to all dual-capacity dealers (brokers/dealers) of the exchange. The orders are put into the book in price and book priority.

Trading in the HSE starts at 9:30 a.m. with an opening call in order to fix opening prices for the day (pre-trading). The unmatched orders from the pre-trading form the basis for the continuous (free) trading session, which starts at 10:30 a.m. and closes at 5:30 p.m. The downstairs trades in the continuous market are either round-lot or odd-lot trades. The odd-lot trades are trades with a smaller number of shares than the usual minimum amount required for trading on the stock exchange, and are matched at regular intervals during the continuous trading. The price of the odd-lot trades is the most recent price in round-lot trading. The number of shares needed for a round-lot varies between stocks and is updated by the Exchange regularly. To illustrate, on March 1, 1993, the trade size varied between 50 and 5,000 shares, being dependent on the market value per share. The medium minimum size for a round lot was 500 shares and about FIM 22,000. Before that the trade sizes for round-lots were revised on August 16, 1991 and January 1, 1990.

The after-hours (batch) trading starts at 5:35 p.m. and ends at 6:00 p.m.

---

[5]Berglund and Liljeblom (1989) describe the 'calling out' system and its implications to stock prices in more detail. One of their main conclusions is that the trading system caused high autocorrelation in Finnish stock returns. A high level of autocorrelation has, however, been observed also during the new trading system as well as in Finnish stock index futures markets (see e.g., Martikainen and Puttonen, 1994). This high level of autocorrelation in Finnish stock returns has been exploited by several Finnish researchers to predict stock indices (see e.g., Virtanen and Yli–Olli, 1987, and Booth *et al.*, 1992, and Booth, Martikainen, Sarkar *et al.*, 1994).

In addition, after-hours trading continues for 20 minutes from 9:00 a.m. to 9:25 a.m. in the next morning. In this session, the price limits set in the continuous trade must be followed so that the trades have to be carried out inside the maximum price range bounded by the closing bid-ask spread and the high-low transactions prices during the trading day. The quotation and trading of securities is conducted under the HETI system in the office of the authorized brokerage firm and in the Stock Exchange Hall (pre-trading and continuous trading) and in the office of the authorized brokerage firm (after-hours or pre-trading).

Two major ways exists for trading in the HSE, i.e. the upstairs and downstairs market. The upstairs market (prearranged trades) in the HSE refers to the market for buyers and sellers who negotiate off the exchange floor. The downstairs market refers to the exchange floor where trades are anonymously executed (batch and continuous). In the upstairs trades large investors typically contact an upstairs brokerage firm who then attempts to locate counterparts to the trade. The upstairs brokerage firm usually tries to find the counterparts among its own customers, but it may also contact other brokerage firms so that the trade can be executed. The two markets are not independent of each other, however. Brokers/dealers on the upstairs market may use the price information provided from the downstairs market to set the price schedule, and vice versa.

Booth, Lin *et al.* (1995) investigate the trading volume of different types of trades in the HSE in 1993–1994. They report that most of the trading takes place in the continuous trading session and the proportion of pre-trading is small. In the continuous trading session the upstairs trades represent almost half of the total number of stocks traded in the HSE, while the respective figure for large-block transactions on the NYSE found by Cheng and Madhavan (1995) was about 11%. Therefore, it appears that the upstairs market plays a significant role in the HSE. Similar to the US market, the upstairs trades are larger than downstairs trades, the average trading volumes in continuous trading being 8,901 and 2,710 shares, respectively. The median upstairs and downstairs trades are 2,000 and 1,000 stocks, respectively.

Table 2 presents the trading activity of different brokers/dealers on upstairs and downstairs markets separately for buys and sells for the continuous trading session in HSE in 1993–1994. The amounts of individual brokers' buys and sells do not considerably differ from each other. However, it appears that the activities of different brokers/dealers vary significantly in the upstairs market. The two most active brokerage firms in the upstairs market were Kansallis–Meklarit (KM) and SYP–Pankkiirilike (SYP). The former was owned by Kansallis–Osake–Pankki (KOP) and the latter one by Unitas (UNI), which were the two largest commercial banks in Finland, but have recently merged their activities, and are nowadays called Merita.

Table 2. Trading Activity of Brokers/Dealers (see Booth, Lin *et al.*, 1995).

Dealer/ Broker	Directly Negotiated Trades				Anonymously Transacted Trades			
	Buys		Sells		Buys		Sells	
	(n)	(%)	(n)	(%)	(n)	(%)	(n)	(%)
ABB Aros	2167	4.7	2129	4.6	10178	6.5	11102	7.1
AG Pankkiiriliike	1142	2.5	1152	2.5	4672	3.0	5373	3.5
Aktia	545	1.2	559	1.2	3613	2.3	3568	2.3
Alfred Berg	4222	9.1	4186	9.0	16779	10.8	17918	11.5
Arctos	1757	3.8	1731	3.7	8084	5.2	7298	4.7
BBL Securities	0	0.0	0	0.0	33	0.0	6	0.0
Carnegie–Suomi	2931	6.3	2893	6.2	11836	7.6	10016	6.5
Evli	1343	2.9	1391	3.0	10240	6.6	6169	4.0
FIM–Pankkiiriliike	972	2.1	972	2.1	3934	2.5	4477	2.9
Hiisi	64	0.1	68	0.1	1760	1.1	1810	0.0
Interbank	827	1.8	826	1.8	5125	3.3	4246	2.7
Kansallis–Meklarit	13960	30.0	13970	30.0	22515	14.5	23308	15.0
Opstock	2653	5.7	2693	5.8	13195	8.5	13754	8.8
Postipankki	1601	3.4	1621	3.5	6634	4.3	6633	4.3
Protos	977	2.1	978	2.1	3379	2.2	1890	1.2
Erik Selin	1951	4.2	1953	4.2	8206	5.3	8052	5.2
Sofi	232	0.5	212	0.5	1469	0.9	1691	1.1
Suomen FSB	7	0.0	7	0.0	122	0.1	90	0.1
SP–Meklarit	1150	2.5	1160	2.5	4747	3.1	6726	4.3
Sv. Handelsbanken	9	0.0	9	0.0	55	0.0	12	0.1
SYP–Pankkiiriliike	7465	16.0	7463	16.0	14815	9.5	16573	10.7
United Bankers	537	1.2	539	1.2	4134	2.7	4668	3.0
Total	46512	100.0	46512	100.0	155525	100.0	155525	100.0

Buys and sells are identified directly from the intraday tape published by the HSE. BBL Securities gave up brokering activities in 1993, and Suomen FSB and Sv. Handelsbanken started in 1994.

2.2.2   Finnish Derivatives

The trading system for derivatives is close to the one used in the neighboring Stockholm Options Market. Customers give their buy or sell orders to brokers. Each customer is obliged to open a clearing account with the Exchange through the broker. The Exchange does not know the name of the customer, only the account number is revealed to the exchange by the broker. Most of the trading takes place by telephone. The quotes are recorded manually and entered into a computer. Other market participants receive the quotes by terminal. Market makers are connected with the exchange over an open telephone line. The minimum size of telephone order is 20 contracts. For smaller orders, brokers transmit quotes to the exchanges via computer terminals. (see Jokivuolle and Koskinen, 1991, for more details). There were 17 brokers operating in index options and futures markets in 1994. The market share of the four largest brokers was over 60%. The state-owned Leonia Ltd represented a market share of 74% among the market makers. The remaining share 26% was shared by three brokerage firms.

Trading and clearing take place within the same organization. They are governed by the trading rules applying to each product traded and cleared at the Exchange, and by laws and other regulations pertaining to such trading. Rules are complemented by instructions of the president of the Exchange including trading and clearing practices. The Exchange requires that the sellers and buyers of futures deposit the required margin in a custodian bank approved by the Exchange. The acceptable kinds of collateral and other collateral conditions are provided by the president of the Exchange.

The Exchange requires that the sellers and buyers of futures deposit the required margin in a custodian bank approved by the Exchange. Acceptable collateral and collateral conditions are provided by the Exchange. The Exchange knows whether the broker is making the trade on his own or on the customer's account. Regarding the trades made by the customers through brokerage firms, the Exchange calculates the margin requirements every evening and informs brokerage firms the next morning about the margin their customers have to place. By 11.00 a.m. the margin has to be placed in the custodian bank by the brokerage firm or the custodian itself. At that time, the brokers have to inform the Exchange of each customer's position and where the collateral is placed. The brokers may require customers to post a temporary margin before the official margin requirement is announced by the Exchange at the end of the day. This is because in the case of a customer default, the broker is responsible for his customers' positions to the Exchange. For this intraday margining, the Exchange has provided a computer program to the brokers calculating the margin requirements throughout the day. In the case that the broker trades for his own account, the Exchange requires that the broker posts a margin at the time of the trade. Intraday margin calls are therefore possible for both brokers and customers. For brokers, the intraday margin calls are made by the Exchange. For customers, an intraday margin call may be made by a broker, but not by the Exchange, because the Exchange checks the customers' positions only at the end of each day. Thus, the frequency of actual margin calls received by a customer is dependent on how frequently the broker actually updates its own margin requirements.[6]

---

[6]Booth, Broussard *et al.* (1997) provide a more detail description of Finnish margin requirements.

3   Seasonal Anomalies in Finland

3.1   Some Issues of Market Microstructure

3.1.1   Intraday U-Shape in Trading Volume

Several studies suggest that the prices and volumes in the US stock market follow an intraday U-shape pattern. For instance, Jain and Joh (1988) report that the first-hour trading volume is at least 50% higher on the NYSE than the average hourly volume during the day. Hedvall (1994) finds that the intraday trading pattern is more flat than in the US. At the open, the trading volume during the two first half-hours in the HSE is only 10% higher than the average half-hourly volume in 1990–1991. At the close, the upstairs trades do not display any peak prior to the close of continuous trading, while the peak is clearly observable in the downstairs market.

3.1.2   Clustering of Stock Prices

Harris (1991) showed that equity prices in the US market tend to cluster on round fractions. Booth, Kallunki *et al.* (1996) investigate the effect in the HSE. The Rules and Regulations of the Helsinki Stock Exchange (HSE, 1992) specify that the minimum price variation is 10 markkas for stocks at or above FIM 1000, one markka for stocks under FIM 1,000 and at or above FIM 100, 10 pennies for stocks under FIM 100 and at or above FIM 10, and one penni for stocks under FIM 10. Although the regulations on minimum price variations in Finland are different from those in NYSE and AMEX, the clustering of stock prices is obvious. The phenomenon appears to be stronger in Finland than in the US. For instance, in the most popular price class where stocks are priced between FIM 10.00 and FIM 99.90, prices cluster round markkas and 50 pennies. To illustrate, on trades between 60.00–60.90, 70% of the trades are made at 60.00 and 18% at 60.50. Booth, Kallunki *et al.* (1996) find that infrequent trading is an important determinant of price clustering the HSE. Moreover, they report that the level of price clustering is positively correlated with standard deviation of returns and market price per share. These results are similar to the US; see Harris (1991). The results also indicate that there exists more clustering in trades directly negotiated between two parties than in trades anonymously transacted on the floor.

The high level of observed discreteness in Finnish stock prices has important implications for empirical research using data measured at short return intervals (such as transaction data) in small markets. The discreteness increases the variance of observed returns and induces autocorrelation to return series. It appears that the effect may be especially strong in small markets. The methods correcting the discreteness in measuring variance and serial covariance may thus be of great importance in these markets when transaction data is used (see e.g., Gottlieb and Kalay, 1985).

3.1.3   Price Effects of Trades in Upstairs and Downstairs Markets

Booth, Lin *et al.* (1995) provide evidence on the information content and price effects of trades in the Finnish upstairs and downstairs markets in 1993–1994. Supporting the hypothesis that direct negotiations between traders and brokers/dealers in the upstairs market may reduce market makers' adverse information costs, the empirical results from the Helsinki Stock Exchange in Finland indicate that the permanent price effects are lower in the upstairs compared to the downstairs markets. They also find that the permanent price effects are different if the bank's broker/dealer is associated with the trade of its own bank's stocks or not. Moreover, this price effect is lower if the buying and selling are arranged through the same brokerage firm in the upstairs market. An important finding by Booth, Lin *et al.* (1995) is that the permanent price effect is larger for large than for small trades. This suggests that the Finnish market may not be able to accommodate large trades without price pressure.

3.1.4   Comparison of Stock Index Futures and Cash Markets

Considerable empirical evidence suggests that the futures and options prices lead the underlying cash index prices in all major markets. This appears to be the case also in Finland. There are several plausible reasons for this lead–lag effect. These include lower transaction costs on derivatives and less restrictive rules for short selling. Moreover, it has been suggested that the infrequent trading of the component stocks in the cash index might explain the effect. While the lead–lag effect in the US market seems to be only a few minutes, Puttonen (1993) reports that the respective in Finland is as long as 1–2 days. As demonstrated by Martikainen, Perttunen and Puttonen (1995a) the effect is visible for both frequently and infrequently traded stocks. Moreover, Martikainen and Puttonen (1995) suggest that the call and put volumes in the Finnish stock index options market can be used to predict cash prices.

The long lead–lag effect has interesting implications for investors. In particular, futures prices can be used to predict cash prices. Martikainen and Puttonen (1994a) use this observation to predict the levels and volatilities of Finnish index prices. Moreover, Booth, Martikainen and Puttonen (1993) and Martikainen and Puttonen (1994b) show that international information is first transmitted to Finnish futures markets.

3.2   Day-of-the-Week Effects

The day-of-the-week (DOW) effect seems to be different in different countries. While significantly low returns on Mondays are found in many studies in the US and UK markets, for instance, certain European markets, such as Spain, Sweden, Belgium, Italy, France and Ireland seem to exhibit significantly low returns on Tuesdays (see Martikainen and Puttonen, 1996, for a review).

Martikainen and Puttonen (1996) investigate the DOW effect in the Finnish stock and stock index derivatives markets during 1988–1990. Figure 1 shows the returns in the three markets. In the cash market, significant negative Tuesdays ad Wednesdays are observed. Meanwhile, negative Mondays are strong in index derivatives markets. The authors suggest that this evidence is consistent with the hypothesis by Miller (1988) and empirically supported by Ziemba (1993) in the Japanese market. Miller (1988) and Ziemba (1993) suggest that the increase in self-initiated sell orders during the weekend cause negative Mondays in large markets. Martikainen and Puttonen (1995) report that the increased selling pressure may be reflected in thin cash markets with a lag because of infrequent trading. This is also supported by the observation that if the cash market returns are adjusted for the lagged derivatives returns, the Tuesday and Wednesday effects disappear in cash markets.

3.3   Turn-of-the-Month Effects

Martikainen, Perttunen and Ziemba (1994) find that the turn-of-the-month (TOM) effect reported in several studies and reviewed in Ziemba (1994) exists for most of 24 different stock markets and 12 regions investigated. The effect is clearly observable in large markets, such as the US and UK, but less observable in smaller markets including Finland. They state that this may be because of three main reasons: (i) the short sample period 1988–1990 used, (ii) the small number of stocks and infrequent trading in the indices investigated, and (iii) measuring the TOM only between the last trading day of the month (-1) and the four first trading days of the month $(+1-+4)$.

Martikainen, Perttunen and Puttonen (1995b) investigate the TOM effect in more detail for the Finnish stock, stock index options and stock index futures markets using data from the beginning of the derivatives market, May 2nd, 1988, to October 14th, 1993. While the results on individual days suggest that most significant return for cash markets is found on day -1, the futures and put-call parity implied returns suggest the effect to be most significant one day earlier. For all markets, the effect is found to be significant: the average daily returns between days -5 and $+5$ are higher than in the rest of the month (see Table 3). A similar effect is also found in the implied volatilities, as well as in the call-put ratios in the options market. This suggests that investors may have more positive expectations in the TOM than in the rest

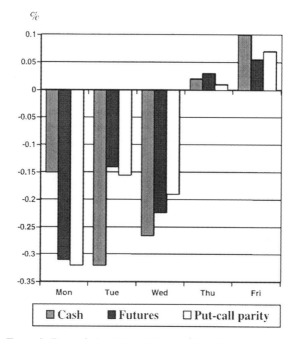

Figure 1. Finnish Day-of-the-Week Effect. (Martikainen and Puttonen, 1996).

of the month. Booth, Kallunki and Martikainen (1996b) suggest that this is mainly because of the trading behavior of individual investors. They also observe that the TOM effect is strongest for most frequently traded stocks.

3.4 Holiday Effects

Although several studies have found that stock returns are significantly different around holidays than in other periods of the year in different markets, there exists no published evidence on Finnish holiday effects. Therefore, I investigated the behavior of the value-weighted market index developed in Berglund, Wahlroos and Grandell (1984) around the most important public holidays in Finland in 1970–1990. The results in Figure 2 suggest that the holiday effects are not very strong in Finland. For the five most important holidays, mean adjusted-returns for the market index are calculated for five individual trading days before the holiday. Mean-adjusted returns before Christmas are significantly different from zero, but this may also be because of the turn-of-the-year (TOY) effect.

Table 3. The Finnish Turn-of-The-Month Effect [−5, +5]. May,
1988–October, 1993. (Martikainen, Perttunen and Puttonen, 1995b).

	Cash	Futures	Options	Volatility	Call–put
Mean	0.103	0.118	0.118	0.003	0.011
(t)	(2.389)	(2.095)	(2.089)	(3.081)	(2.175)
(p)	(0.017)	(0.037)	(0.037)	(0.002)	(0.030)

Cash is the average mean-adjusted daily stock market return based on
the FOX index.
Futures is the average mean-adjusted daily return of the nearby futures
contract.
Options is the average mean–adjusted daily put–call parity implied
return based on the options contracts on the FOX index.
Volatility is the average mean–adjusted implicit volatility of the options
contacts on the FOX index.
Call–put is the average mean–adjusted relative proportion of
call–option contacts of all option contracts on the FOX index.

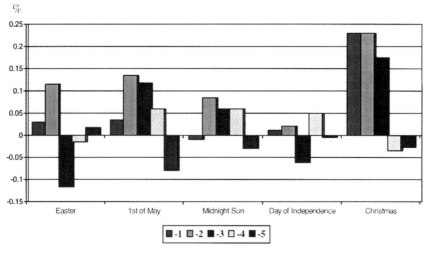

Figure 2. Finnish Holiday Effects in 1970–1990.

3.5   Monthly Effects

The monthly effect indicating high returns in January in Finland was observed
already by Berglund (1986) who noted that the Finnish results on monthly
seasonalities correspond closely to those obtained for other exchanges around

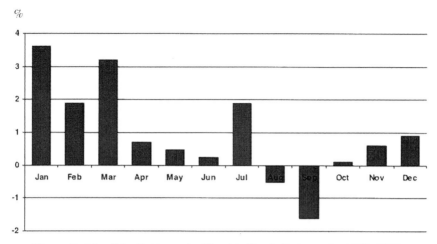

Figure 3. Monthly Patterns in Finnish Stock Returns in 1971–1992.

the world. The main difference in Finland, however, appears to be that the TOY effects and monthly seasonalities are similar for both small and large firms (see Berglund, 1986 and Kauppi and Martikainen, 1994). Figure 3 illustrates the average monthly returns based on my calculations on the value-weighted index by Berglund, Wahlroos and Grandell (1984) to the end of 1990 and on the value-weighted HEX-index calculated by HSE for 1991–1992. The returns are highest in January and lowest in August–September.

4 Cross-Sectional Anomalies in Finland

4.1 Earnings Anomalies

4.1.1 Finnish Earnings Measurements

Since the 1974 Accounting Act, Finnish firms have been under legal obligation to publish a yearly financial statement in accordance with a fixed scheme.[7] Moreover, the Business Tax Act and the Companies Act set additional requirements for limited companies in Finland. Although the legislation stipulates a fixed scheme for the reported income figures, accounting conventions and principles significantly affect the income figures. Therefore, firms usually follow the Business Tax Act, which is more restrictive, but still gives ample discretion for the determination of annual taxable net income.

The survey by the International Accounting Standard Committee (IASC, 1988) indicates that the Finnish accounting practice has had the lowest con-

---

[7]See also Salmi (1994) for a description of Saario's theory in English, and Räty (1992) for a discussion on the recent trends in harmonizing the Finnish accounting legislation.

formity with International Accounting Standards among the 54 countries investigated. The main differences between Finnish and IAS accounting rules are as follows (see Kasanen, Kinnunen and Niskanen, 1992). First, the inclusion of overhead in the inventory values is not allowed. Second, while in Finland the geometrically degressive declining balance method of depreciation for fixed long-term assets has been used, the IAS recommends the straight-line depreciation. Third, contrary to the IAS recommendations, the Finnish regulations do not allow financial leases to be capitalized on the lessee's balance sheet. Fourth, the cost method has to be applied when valuing the owned affiliates in Finnish balance sheets (in the IAS the equity method of accounting is used). Fifth, contrary to the IAS practices, unfunded pension obligations are not treated as balance sheet debt. Sixth, the Finnish system has allowed firms to create many kinds of untaxed reserves. Seventh, in Finland only 'the completed contract method' of accounting is normally allowed. Moreover, in Finland the reporting of foreseeable losses is not permitted during the project. As a consequence of these differences, IAS profits tend to be larger than Finnish profits.[8]

As listed by Kinnunen, Kasanen and Niskanen (1996), there are several instruments that can be used to manage reported earnings numbers in Finland. The most important instrument is the level of depreciation. While the Accounting Act does not regulate the amount of annual depreciation, nor does it specify any particular depreciation method, the Business Tax Act contains detailed instructions on the maximum depreciation rates for machinery and plants, which are 30% and 10%, respectively. Therefore, firms can in principle select any depreciation level between zero and the maximum depreciation rates. Second, firms are allowed to create untaxed reserves, such as an inventory reserve, a bad debt reserve, an operational reserve, a warranty reserve and an investment reserve. Third, firms do not have to record their pension liabilities contributed to pension foundations on accrual basis, and can thereby affect reported earnings figures. Fourth, firms are allowed either to expense their exchange losses in the year the exchange rate has changed, or to capitalize them to the balance sheet until the corresponding amount has been paid. Firms can also expense or capitalize the costs relating to their research and development activities. Finally, firms are allowed to deduct income taxes from earnings in income statements or from retained profits in the balance sheet. They can also add certain tax-free revenues, such as dividends, directly to the shareholders' equity without presenting them in the income statement.

Since Finnish accounting rules have provided the firms with exceptionally large opportunities to intentionally smooth income, it is necessary to adjust

---

[8]See Kasanen, Kinnunen and Niskanen (1992) for an empirical comparison of IAS adjusted and reported earnings in Finland.

the reported earnings figures for these actions to define whether the firm has a smooth income stream or not. For this purpose, the recommendations provided by the Finnish Committee for Corporate Analysis (COC) (see Finnish Committee for Corporate Analysis, 1990) are commonly used. The COC consists of representatives of major lending institutions, authorities, universities, etc. The main purpose of the COC adjustments is to improve the comparability of Finnish firms by standardizing accounting figures. In this process, the earnings are de-smoothed by taking into account the firms' most important ways to intentionally smooth income. The most important adjustment recommended by the COC concerns depreciation. The depreciation is adjusted to meet with the maximum depreciation allowed by the Business Tax Act. The other adjustments concern annualizing accounting periods, changes in actuarial deficiencies of pension funds, and extraordinary income.

The COC adjustments have been commonly used for research purposes when investigating the relationship between accounting figures and the stock market. They are also very commonly used by practitioners working with financial statement analysis. Martikainen, Ankelo and Ruuhela (1990) report correlations that are far below unity between reported and COC earnings. Moreover, as reported by Kinnunen, Kasanen and Niskanen (1996), reported earnings of Finnish firms are typically close to zero and have low variability over time. This is because taxation is based on the reported earnings figures and tax rate has been of higher magnitude than in many other western countries. As a result, Finnish firms have incentives to systematically reduce reported earnings figures to avoid taxes. Therefore, as shown by Martikainen, Ankelo and Ruuhela (1990), the reported earnings as such have little information content to investors.

4.1.2 The Post-Announcement Drift of Earnings

Booth, Kallunki and Martikainen (1996a) provide evidence on the information content of earnings releases in Finland in 1990–1993. First they classify firms to smoothers and non-smoothers. A firm is classified as a income smoother if its coefficient of variation of annual change in COC adjusted income is less than its coefficient of variation of an annual change in sales, using 12 years of data. It appears that smoothing appears to be common in Finland. Of the 100 observations, income smoothing is reported in 40 cases. Consistent with the earlier US evidence, the results suggest that the post-announcement period unexpected return of firms with positive earnings surprises is higher than the return of firms reporting negative earnings surprises. It appears that most of the return difference during the post-announcement period is due to the market reaction to the earnings surprises of the firms that do not have smooth income series. The results summarized in Table 4 suggest that there is delay on how returns caused by accounting earnings releases appear in the

Table 4. Unexpected Returns Around Earnings Announcements: Firms with Non-Smooth Income Streams. (Booth, Kallunki and Martikainen, 1995a).

	UNIFORM RETURNS					LUMPED RETURNS				
Window	unexp. return	P t-stat.	(prob.)	BMP t-stat.	(prob.)	unexp. return	P t-stat.	(prob.)	BMP t-stat.	(prob.)
Positive earnings surprises										
[0,10]	0.032	2.738	(0.011)	2.520	(0.018)	0.035	2.216	(0.035)	2.295	(0.029)
[1,10]	0.020	1.886	(0.070)	2.131	(0.042)	0.027	1.865	(0.073)	2.103	(0.045)
[0,5]	0.028	3.216	(0.003)	2.725	(0.011)	0.029	2.761	(0.010)	2.818	(0.009)
[1,5]	0.016	2.129	(0.042)	2.059	(0.049)	0.021	2.374	(0.025)	2.414	(0.023)
Negative earnings surprises										
[0,10]	−0.023	−3.399	(0.002)	−3.066	(0.143)	−0.016	−2.489	(0.018)	−2.117	(0.042)
[1,10]	−0.019	−2.819	(0.008)	−2.806	(0.093)	−0.018	−2.410	(0.022)	−2.374	(0.024)
[0,5]	−0.024	−4.441	(0.000)	−3.326	(0.002)	−0.013	−3.021	(0.005)	−2.122	(0.042)
[1,5]	−0.020	−3.809	(0.001)	−3.739	(0.002)	−0.014	−3.025	(0.005)	−2.636	(0.013)

P t-stat. is based on the test by Patell (1976). BMP t-stat. is based on the test by Boehmer, Musumeci and Poulsen (1991). When no trades have occurred, uniform as well as lumped daily returns are used. The uniform returns are calculated by assigning returns equally over the days in the multiperiod return intervals. The lumped returns procedure assigns all of the multiperiod return to the day actually trades and sets the rate of return over the intervening no-tarde days to zero. Booth, Kallunki and Martikainen (1995) discover that 33% of the firms included in the sample are not traded on the event date.

stock market. Hence, at least a portion of the post-announcement drift in Finland can be explained by information processing costs. This hypothesis is supported by Booth, Kallunki and Martikainen (1999), who report that the reaction of small investors is delayed around earnings announcements in Finland.

4.1.3 The Earnings-Price Anomaly

Several studies (see e.g., Basu, 1983, and Jaffe, Keim and Westerfield, 1989) suggest that there exists a distinct positive relation between E/P-ratios and average returns in excess those predicted by the CAPM. Kauppi and Martikainen (1994) investigate the existence of the E/P-anomaly in the Finnish stock market in 1975–1990. All continuously listed stocks are included in the analysis. Three portfolios are created based on the E/P-ratios and them are rebalanced annually in the beginning of April. Only the portfolio which includes firms with high E/P-ratios has both positive market- and sample-adjusted returns. The cumulative return difference between the extreme portfolios during the 15–year period is approximately 60%. Even after taking into transaction costs, the return difference between portfolio is clearly above 50%. The difference cannot be explained by the difference in systematic risk between the extreme portfolios. The market model betas for the portfolios of high and low E/P ratio firms are 0.81 and 0.77, respectively. A similar effect is reported for the Cash Flow/Price-ratio.

Martikainen and Gunasekaran (1994) investigate the cross-sectional determinants of the E/P-ratios in Finland. They observe that the accounting-based instrumental risk variables including financial and operating leverage, growth and accounting betas can explain a substantial part of the differences in E/P-ratios between firms. They suggest that the observed E/P-anomaly in Finland may at least partly be because of the serious risk estimation problems using market-based data. Moreover, consistent with US results, Booth, Martikainen, Perttunen and Yli–Olli (1994) report that the E/P-anomaly may be closely related to the size effect.

4.2 Dividends

Finnish firms have traditionally tried to pay out constant dividends. Martikainen, Rothovius and Yli–Olli (1993) investigate the individual and incremental significance of dividend releases in Finland in 1974–1987. They find out that a simple trading strategy in which stocks that have increased dividends are bought and stocks that have decreased dividends are sold yields abnormal returns. They also report that dividends may be incrementally important with respect to earning and cash flows when making portfolio strategies in Finland.

4.3 The Size Effect

4.3.1 Problems of Measuring Risk for Thinly-Traded Small Stocks

The market model remains an important specification of the return-generating process in stock markets when investigating stock returns around an economic event. Recent empirical studies have extensively investigated the problems of misestimation of market model parameters, especially the beta coefficient measuring the asset sensitivity to the stock market. There are many reasons why beta estimates are dependent on the return interval. Infrequent trading may cause downwards bias in betas estimated on short return intervals. This is, because of thin trading, the 'true' covariance between the returns of the asset and the market portfolio is underestimated. Since the beta coefficient embodies this covariance in its numerator, downwards bias is obvious (see Hawawini, 1983). Moreover, if the degree of serial correlation is not the same across stocks, betas may be sensitive to the return interval because for buy-and-hold returns the covariance of an asset's return with the market and the variance of the market return do not increase proportionately (see Handa, Kothari and Wasley, 1989).

Martikainen and Perttunen (1995) provide new evidence on how the selection of return interval affects the beta estimates in the Finnish market. While Berglund, Liljeblom and Löflund (1989) discover that little gain is obtained in the Finnish stock market if different procedures to correct the beta estimates based on daily data are employed, Martikainen and Perttunen (1995) use return intervals up to six months when measuring betas. All 24 continuously listed stocks for the period 1970–1990 are analyzed. The stocks are classified to three equal-weighted portfolios based on the average market value of equity during the period. The average market value is measured by using the consumer price index deflated annual equity figures. Returns on individual stocks are measured by logarithmic price index differences adjusted for cash dividends, stock dividends and right issues assuming that all proceeds from a given stock are reinvested in the same stock with no transaction costs. When no trade has occurred, the true price has been proxied by the bid quotation. A value-weighted market index is used when measuring market returns. Return intervals of one day, one week, two weeks, one month, two months, three months, four months, five months and six months are investigated. Weekly returns are based on Wednesday–to–Wednesday returns. Monthly returns are measured as logarithmic changes in month-end price indices.

The betas of the portfolios are estimated by using the following market model:

$$R_{it} = \alpha_i + \beta_i R_{mt} + \varepsilon_{it}, \tag{1}$$

where R_{it} is the return of portfolio i in return interval t, R_{mt} is the market return for the same period, α_i is the intercept term, β_i is the estimated beta for portfolio i and ε_{it} is the error term. The beta estimates for the Finnish portfolios are graphed in Figure 4. The results show clear return interval effects on betas, especially for the portfolio of small firms. While the beta of the portfolio of small stocks for the daily return interval is about 0.56, it increases to 0.84 when the return interval of two months is used. For longer return intervals similar remarkable changes do not continue.

In general, the results cast serious doubt on the use of betas based on short return intervals in Finland. Given the results in Figure 4, it is not surprising that Berglund, Liljeblom and Löflund (1989) do not advance very far by using different corrections for betas based on daily data in Finland.[9] Even betas based on monthly return intervals seem to be clearly underestimated for small stocks. An interesting observation is that the correlation between firm size

---

[9]See also Luoma, Martikainen and Perttunen (1993a), (1993b) and (1996) for the use of daily data in beta estimation in Finland.

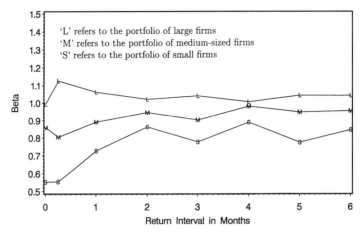

Figure 4. Betas for market value-ranked portfolios in Finland. (Martikainen and Perttunen, 1995).

and betas is positive. This opposite to the negative correlation reported in the US (see e.g., Handa, Kothari and Wasley, 1989). It appears that the positive correlation is also evident for long return intervals, and is not entirely due to underestimation of the betas of small stocks in short return intervals.[10] Kallunki (1996) reports that the misestimation of betas may explain at least some of the estimated price adjustment delays in the Finnish stock market.

4.3.2 The Lead–Lag Effect Between Large and Small Firms

Lo and MacKinlay (1990) report a positive correlation between the returns of lagged large market value portfolios and this period's small market value portfolio returns. A similar result in the Finnish market is observed by Martikainen, Perttunen and Puttonen (1995c). They report that yesterday's returns of large stocks can predict the returns of small stocks in Finland. However, they also find out that the prices of small and large stocks are cointegrated. Cointegration indicates that the prices of large and small stocks have a common trend in the long run. The authors suggest that this observed cointegration may means that the co-movements of stocks using short return intervals may be underestimated in Finland.

---

[10]See also Martikainen *et al.* (1994) on the impact of return interval on Finnish APT systematic risk components.

4.3.3   Investing in Small Stocks in the Long Run

In Finland the size effect, i.e. the smaller the market value of equity, the larger
the expected rate of return on a stock, other things being equal was first found
by Berglund (1986). Kauppi and Martikainen (1994) test a simple trading
strategy based on this anomaly using Finnish data for 1970–1990. Three
portfolios of the continuously listed stocks are created so that the portfolios
are rebalanced in the beginning of each January so that the portfolios are
created based on the market value of the last trading day of the previous
year. The results suggest that the portfolio of largest firms has performed
much worse than the market in general, having a cumulative market adjusted-
return of −75%. On the other hand, the portfolio of smallest firms gained a
market-adjusted return of 120% during the period. Differences in systematic
risk cannot explain the effect since the portfolio of small stocks had a smaller
beta (0.516) than the portfolio of large firms (0.984). The differences are
clearly over and above transaction costs.

5   Concluding Remarks

There are significant seasonal and cross-sectional regularities in the Finnish
stock and derivatives returns. The most prominent existing empirical evidence can be summarized as follows.

Seasonal anomalies

Intraday Pattern:	More flat than in the US
Day-of-the-Week Effect:	Negative Tuesdays
Turn-of-the-Month-Effect:	Significantly positive returns
Holiday Effects:	Weak
Monthly Effects:	Positive Januaries

Cross-sectional anomalies

Earnings Announcements:	Post-announcement drift between firms with unexpectedly positive and negative returns
E/P Anomaly:	Large E/P ratio, large abnormal returns
Cash Flow/Price Anomaly	Large cash flows, large abnormal returns
Dividend Anomaly:	Positive change in dividends, positive abnormal returns
Size Effect:	Small firms, large abnormal returns

There are certain points that are characteristic to the Finnish markets.
The first one is the very thin and infrequent trading. As a consequence of
thin trading, stock prices are more clustered on round fractions than in large
markets. It also seems that the permanent price effects of individual trades are

significantly larger for large than for small trades suggesting that the Finnish market cannot accommodate large trades without price pressure. The impact of thin trading is clearly evident when daily returns are used. Market indices consisted of even most frequently traded stocks Finnish stocks are biased because of infrequent trading. Also risk estimation problems are serious in Finland, especially when short return intervals and firms with small market value are investigated.

The importance of stock index derivatives in the transmission of information is evident in Finland. Studies investigating the lead–lag relations between stock and stock index derivatives markets suggest that the returns of stock index derivatives lead the underlying cash price movements with a lag of one or two days. The lead–lag effect is not observable only because of thin trading, since the effect is evident also for the most frequently traded stocks.

Finally, the diversification benefits of using Finnish markets seem to be evident for international investors. The correlations between the returns of Finnish stocks and US and UK stock returns are low. The Finnish stock index futures markets seem to react to global information more rapidly than the Finnish cash markets, but the co-movements between Finnish derivatives are foreign stock markets remain low, economically speaking.

References

Basu, S. (1977) 'Investment performance of Common Stocks in Relation to Their Price-Earnings Ratios: A Test of the Efficient Market Hypothesis', *Journal of Finance* **32** 663–682.

Berglund, T. (1986) *Anomalies in Stock Returns in a Thin Security Market*, Doctoral Dissertation at the Swedish School of Economics and Business Administration, Helsinki, Finland.

Berglund, T., and E. Liljeblom (1988) 'Market Serial Correlation on a Small Security Market: A Note', *Journal of Finance* **43** 1265–1274.

Berglund, T, E. Liljeblom and A. Löflund (1989) 'Estimating Betas on Daily Data for a Small Stock Market', *Journal of Banking and Finance* **13** 41–64.

Berglund, T., B. Wahlroos and L. Grandell (1984) 'The KOP and UNITAS indexes for the Helsinki Stock Exchange in the Light of a New Value-Weighted Index', *Finnish Journal of Business Economics* **32** 30–41.

Boehmer, E., J. Musumeci and A.B. Poulsen (1991) 'Event–Study Methodology under Conditions of Event–Induced Variance', *Journal of Financial Economics* **30** 253–272.

Booth, G.G., J.P. Broussard, T. Martikainen and V. Puttonen (1997) 'Prudent Margin Levels in The Finnish Stock Index Options Market', *Management Science* **43** 1177–1188.

Booth, G.G., M. Chowdhury and T. Martikainen (1994) 'The Effect of Foreign Ownership Restrictions on Stock Price Dynamics', *Review of World Economics* **130** 730–746.

Booth, G.G. J. Glascock, T. Martikainen and T. Rothovius (1994) 'The Financing of Residential Real Estate in Finland. An Overview', *Journal of Housing Research* **5** 205–227.

Booth, G.G., J. Hatem, I. Virtanen and P. Yli–Olli (1992) 'Stochastic Modeling of Security Returns: Evidence from the Helsinki Stock Exchange', *European Journal of Operational Research* **56** 98–106.

Booth, G.G., J.-P. Kallunki and T. Martikainen (1996a) Income Smoothing and the Post-Announcement Drift of Earnings. Finnish Evidence', *Journal of Business Finance and Accounting* **23** 1197–1211.

Booth, G.G., J.-P. Kallunki and T. Martikainen (1996b) The Finnish Turn-of-the-Month Effect and Buy-Sell Patterns of Individual and Institutional Traders'. Paper Presented at the 25th Annual Financial Management Association Meeting, New Orleans, Louisiana, October 1996.

Booth, G.G., J.-P. Kallunki and T. Martikainen (1999) 'Earnings News and the Behavior of Different Types of Traders in the Finnish Market', *Applied Economics Letters*, forthcoming.

Booth, G.G., J.-P. Kallunki, J.-C. Lin, T. Martikainen and J. Perttunen (1996) 'Stock Price Clustering: Empirical Evidence'. Paper Presented at the 8th Annual Nothern Finance Association Meeting, Quebec, Canada, September 1996.

Booth, G.G., J.-C. Lin, T. Martikainen and J. Perttunen (1995) 'The Information Content and Price Effects of Trades in Finnish Upstairs and Downstairs Markets'. Paper Presented at the 22nd Annual European Finance Association Meeting, Milan, Italy, August 1995.

Booth, G.G., T. Martikainen, J. Perttunen and P. Yli–Olli (1994) 'The Functional Form of Earnings and Stock Prices. International Evidence and Implications to the E/P-Anomaly', *Journal of Business Finance and Accounting* **21** 395–408.

Booth, G.G., T. Martikainen and V. Puttonen (1993) 'The International Lead-Lag Effect Between Market Returns: Comparison of Stock Index Futures and Cash Markets', *Journal of International Financial Markets, Institutions and Money* **3** 59–71.

Booth, G.G., T. Martikainen, S. Sarkar, I. Virtanen and P. Yli–Olli (1992) ' Nonlinear Dependence of Finnish Stock Returns', *European Journal of Operational Research* **74** 273–283.

Bos, T., T. Fetherston, T. Martikainen and J. Perttunen (1995) 'The International Co-Movements of Finnish Stocks', *European Journal of Finance* **1** 95–111.

Cheng, M. and A. Madhavan (1995) 'In Search of Liquidity: Block Trades in the Upstairs and Downstairs Markets, NYSE'. Working Paper.

Finnish Committee for Corporate Analysis (1990) *Financial Statement Analysis*, Painokaari, (in Finnish).

Gottlieb, G. and A. Kalay (1985) Implications of the Discreteness of Observed Stock Prices', *Journal of Finance* **40** 135–153.

Handa, P., S.P. Kothari and C.E. Wasley (1989) 'The Relation Between Return Interval and Betas: Implications for the Size Effects', *Journal of Financial Economics* **23** 79–100.

Harris, L. (1991) 'Stock Price Clustering and Discreteness', *Review of Financial Studies* **4** 389–415.

Hawawini, G. (1983) 'Why Beta Shifts as The Return Interval Changes', *Financial Analysts Journal* 73–77.

Hedvall, K. (1994) *Essays on the Market Microstructure of the Helsinki Stock Exchange.* PhD Dissertation at the Swedish School of Economics and Business Administration, Helsinki, Finland.

Hietala, P. T. (1989) 'Asset Pricing in Partially Segmented Markets: Evidence from the Finnish Market', *Journal of Finance* **44** 697–718.

International Accounting Standard Committee (1988) 'Survey of the Use and Application of International Accounting Standards'. IASC, London.

Jain, P.-C. and G.-H. Joh (1988) 'The Dependence between Hourly Price Changes and Trading Volume', *Journal of Financial and Quantitative Analysis* **23** 269–283.

Jokivuolle, E. and Y. Koskinen (1991) 'Financial Options and Futures Markets, *Bank of Finland Bulletin* 23–29.

Kallunki, J.-P. (1996) *Earnings-Related Anomalies in a Thin Security Market: An Accounting-Based Risk Estimation Approach.* PhD Dissertation at the University of Vaasa, Finland.

Kallunki, J.-P. and T. Martikainen (1996) 'Financial Market Liberalization and the Relationship between Stock Returns and Financial Leverage in Finland', *Applied Economics Letters* **4** 19–21.

Kasanen, E., J. Kinnunen and J. Niskanen (1992) 'The Prediction of International Accounting Standards Profits from Financial Statements of Finnish Firms', *Advances in International Accounting* **5** 47–73.

Kauppi, M. and T. Martikainen (1994) 'Some Simple Trading Strategies on the Helsinki Stock Exchange'. Research Paper **179**, University of Vaasa, Finland.

Kinnunen, J., E. Kasanen and J. Niskanen (1996) 'Earnings Management and the Economy Sector Hypothesis: Empirical Evidence on a Converse Relationship in the Finnish Case', *Journal of Business Finance and Accounting*, (Forthcoming).

Lo, A. and A.C. MacKinlay (1990) 'When Are Contrarian Profits Due to Stock Market Overreaction?', *Review of Financial Studies* **3** 175–205.

Luoma, M., T. Martikainen and J. Perttunen (1993a) Thin Trading and Estimation of Systematic Risk: An Application of an Error-Correction Model', *Annals of Operations Research* **45** 297–305.

Luoma, M., T. Martikainen and J. Perttunen (1993b) 'Adjusting for the Intervalling Effect Bias in Beta Coefficients on a Thin Security Market: Application of a Lag Distribution Model', *International Journal of Systems Science* **24** 2391–2398.

Luoma, M., T. Martikainen and J. Perttunen, (1996) 'A Pseudo Criterion for Security Betas in the Finnish Stock Market', *Applied Economics* **28** 65–69.

Martikainen, T., T. Ankelo and R. Ruuhela (1990) 'Stock Returns and Corporate Earnings Adjusted for Alternative Depreciation Methods', *Finnish Journal of Business Economics* **39** 109–127.

Martikainen, T. and A. Gunasekaran (1994) 'Modelling the Cross-Sectional Variation of E/P-Ratios: Implications for the E/P-Anomaly', *International Journal of Systems Science* **25** 1899–1909.

Martikainen, T. and J. Perttunen (1995) 'Long Return Intervals and Beta Estimates, Scandinavian Evidence', *International Journal of Finance* **7** 1195–1205.

Martikainen, T., J. Perttunen and V. Puttonen (1995a) On the Dynamics of Stock Index Futures and Individual Stock Returns', *Journal of Business Finance and Accounting* **22** 87–100.

Martikainen, T., J. Perttunen and V. Puttonen (1995b) 'Finnish Turn-of-the-Month Effects: Returns, Volume and Implied Volatility', *Journal of Futures Markets* **15** 605–615.

Martikainen, T., J. Perttunen and V. Puttonen (1995c) The Lead–Lag Effect Between Large and Small Firms. Evidence from Finland', *Journal of Business Finance and Accounting* **22** 449–454.

Martikainen, T., J. Perttunen, P.Yli–Olli and A. Gunasekaran (1994) 'The Impact of Return Interval on Common Factors in Stock Returns. Evidence from a Thin Security Market', *Journal of Banking and Finance* **18** 659–672.

Martikainen, T., J. Perttunen and W.T. Ziemba (1994) 'The Turn-of-the-Month Effect in the World's Stock Markets', *Financial Markets and Portfolio Management* **8** 41–49.

Martikainen, T., and V. Puttonen (1994a) 'A Note on the Predictability of Finnish Stock Market Returns. Evidence form Stock Index Futures Markets', *European Journal of Operational Research* **73** 27–32.

Martikainen, T., and V. Puttonen (1994b) 'International Price Discovery in Finnish Stock Index Futures and Cash Markets', *Journal of Banking and Finance* **18** 809–822.

Martikainen, T., and V. Puttonen (1995) 'Option Volume and Market Timing Effectiveness'. Research Symposium Prodeedings, Chicago Board of Trade, 1995, 103–121.

Martikainen, T., and V. Puttonen (1996) 'Finnish Day-of-the-Week Effects', *Journal of Business Finance and Accounting* **23** 1019–1032.

Martikainen, T., T. Rothovius and P. Yli–Olli (1993) 'On the Individual and Incremental Information Content of Accrual Earnings, Cash Flows and Cash Dividends in the Finnish Stock Market', *European Journal of Operational Research* **68** 318–333.

Martikainen, T., P. Yli–Olli and A. Gunasekaran (1991) 'The Finnish Stock Market: An Investigation for Empirical Research and Future Research Directions', *Finnish Journal of Business Economics* **40** 253–284.

Miller, E.M. (1988) 'Why a Weekend Effect?', *Journal of Portfolio Management* **14** 43–48.

Patell, J. (1976) 'Corporate Forecasts of Earnings Per Share and Stock Price Behavior: Empirical Tests', *Journal of Accounting Research*, 246–276.

Puttonen, V. (1993) 'Short Sale Restrictions and The Temporal Relationship Between Stock Index Futures and Cash Markets', *Journal of Futures Markets* **13** 645–664.

Räty, P. (1992) 'Reforming Finnish Accounting Legislation', *European Accounting Review* **1** 413–420.

Securities Market Legislation, *Bulletin of the Bank of Finland*, 51–56.

Virtanen, I. and P. Yli–Olli (1987) 'Forecasting Stock Market Prices in a Thin Security Market', *Omega* **15** 145–155.

Ziemba, W.T. (1993) 'Comment on *Why a Weekend Effect?*', *Journal of Portfolio Management* **19** 93–99.

Ziemba, W.T. (1994) 'World Wide Security Market Regularities', *European Journal of Operational Research* **74** 198–229.

Characteristics-Based Premia in Emerging Markets: Sector-Neutrality, Cycles and Cross-Market Correlation

Sandeep A. Patel

Abstract

This paper provides evidence on the realized premia to low price-to-book, low price-to-earnings, and small size portfolios in excess of the local stock market index in 22 emerging markets. The emerging markets evidence on excess returns based on predetermined characteristics augments similar evidence from developed international markets. For emerging markets, sector-neutrality helps reduce risk, and hence increases the significance of excess returns. There is significant time variation in excess returns to sector-neutral portfolios of stocks with low price-to-book, low price-to-earnings and small size. There is also considerable cross-country variation in the magnitudes of excess returns. Nonetheless, abnormal portfolio performance related to sector-neutral low price-to-book, low price-to-earnings and small size tends to be correlated across emerging markets.

1 Introduction

Following the mean-variance portfolio optimization given in Markowitz (1959), Sharpe (1964), Lintner (1965), Mossin (1966) and Treynor (1961) formalized the relationship between the expected stock returns and systematic risk, beta, as the Capital Asset Pricing Model, CAPM. Early empirical tests of the CAPM, as in Black, Jensen, and Scholes (1972), Blume and Friend (1973), and Fama and MacBeth (1973) provided evidence in support of a positive relationship between beta of a stock and returns realized on the stock. For the last two decades, however, evidence has accumulated that cross-sectional differences in stock returns can be better explained by a number of predetermined and publicly available characteristics such as size, January, leverage, past returns, dividend yield, earnings-to-price ratios, and book-to-market ratios.

The debate over the interpretation of the cross-sectional explanatory power of these predetermined variables is rich. The debated issues include: (i) whether this evidence implies semi-strong and weak-form inefficiency of markets as defined by Fama (1970); (ii) the independence and power of tests since some research relies on the results and data from previous studies (Lo and

MacKinlay 1990); (iii) the link between empirical factors and economic factors, e.g. the relation of size with labor share (Jagannathan *et al.* 1998), or of book-to-market with financial distress (Dichev 1998); (iv) whether a firm's book-to-market ratio and size are proxies for the firm's loading on priced risk factors (Fama and French 1996) or whether investors can realize book-to-market and size premia without taking exposure to common factors (Daniel and Titman 1997).

Considerable research has recently addressed the data-snooping bias by examining cross-sectional returns in international markets. Hawawini and Keim (1999) summarize the earlier research on the performance of characteristics-based portfolios in developed international markets. Examination of the relation between characteristics and returns in international markets has recently been extended to emerging markets (Claessens, Dasgupta, and Glen 1995, Fama and French 1998, Patel 1993, 1996, 1998, and Rouwenhorst 1998). These studies generally find evidence of characteristics-based premia in many of the markets examined. There are differences across studies, though, related primarily to methodological differences. As an example, Claessens *et al.* use regressions whereas other researchers form portfolios to study the relationship between excess returns and characteristics such as size, earning-to-price ratio, price-to-book ratio and dividend yield. Claessens *et al.* conclude that larger companies outperform smaller companies, whereas portfolio-based studies conclude that smaller companies outperform larger companies. Such differences in conclusions between the portfolio and regression based examinations is possibly due to influence of outliers or non-normality of returns in emerging markets.[1] The conclusions from portfolio-based examinations of excess returns to characteristics are broadly similar, but many questions remain. For example, there is variation across studies in the premia associated with size, price-to-book ratio and momentum. However, much of the variation appears to be related to research design and sample period. Specifically, Fama and French (1998) and Rouwenhorst (1998) examine returns to long-short portfolios, whereas Patel (1993, 1996, 1998) examines returns to value-weighted long-only portfolios in relation to a benchmark. Patel (1993) does not find excess returns to momentum for the January 1988 to December 1993 time period, whereas Rouwenhorst (1998) finds significant excess returns to momentum for the January 1982 to April 1997 time period, suggesting time variation in the momentum effect in emerging markets. The average excess returns and associated t-values of low price-to-book, low price-to-earnings and small size portfolios for the January 1988 to December 1995 time period (Fama and French 1998) and Patel 1996) are higher than the

---

[1] In a related study, Bekaert and Harvey (1995) reject the null hypothesis of unconditional normality for the time series of the index returns at 5% level of significance in fifteen of the twenty markets they study.

average excess returns and t-values for the January 1988 to March 1997 time period (Patel 1998), suggesting time variation in the size, price-to-book and price-to-earnings premia in emerging markets.

Second, there is the issue of correlation of the premia across markets. In developed markets, the characteristics-based premia display interesting variation through time, but time variation of the premia is not synchronous across markets. For example, Hawawini and Keim (1999) show that correlations among characteristics-based premia in developed markets are close to zero, and question risk-based explanations of characteristics-based premia. The issue of cross-market correlation of characteristics-based premia has not been addressed in emerging markets.

Third, there is considerable discussion in academic and practitioner research on the role of economic sectors in determining stock returns, as reviewed in Heckman, Narayanan, and Patel (1999). Thus, it is important to examine whether the excess returns earned by characteristics-based portfolios result from the conscious design of portfolios tilted toward specific characteristics, or from inadvertent concentration in economic sectors associated with those characteristics. To this end, this paper compares excess returns to sector-neutral and non-sector-neutral characteristics-based portfolios.

We provide here additional evidence on the performance of characteristics-based portfolios in emerging markets from January 1988 to November 1998. The findings extend earlier research in three ways. First, they emphasize the role of sector-neutrality on the magnitude and riskiness of excess returns of characteristics-based portfolios. Second, the analysis addresses time-series variation in excess returns. Third, the analysis examines cross-market correlation in excess returns and provides some preliminary evidence for emerging markets on the question of risk-based explanations of excess returns (Hawawini and Keim 1999).

The rest of the paper is organized as follows. The second section describes the data available from IFC and reports summary statistics for simple value-weighted characteristics-based portfolio returns. The role of sector-neutrality in reducing the riskiness and hence increasing the statistical significance of excess returns is discussed in the third section. The fourth section reports the time-variation in the excess returns and the fifth section examines co-movement in excess returns for emerging markets. Conclusions are presented in the sixth section.

2 Data and Preliminary Portfolio Results

The cross-sectional data for equities in emerging markets is available from a number of sources and leading providers of the data are: International Finance Corpiration (IFC), Morgan Stanley Capital International (MSCI),

Barings and Salomon Smith-Barney.[2] This study examines cross-sectional returns of the companies in International Finance Corporation's (IFC) Global index available from the IFC. Claessens, Dasgupta and Glen (1995), Fama and French (1998), Patel (1993, 1996, 1998) and Rouwenhorst (1998) also use IFC as their sample. Table 1 provides some details of this sample; IFC (1996) provides much detailed descriptions.

Table 1 shows the twenty-two countries covered in this sample, the first monthly observation, and number of monthly observations. The table shows that thirteen of the twenty-two markets have 119 months of history. Of the remaining markets, Jordan and Turkey have 106 and 82 months of data respectively, Indonesia has 95 months, Nigeria and the Philippines each have 94 months, Sri Lanka and Zimbabwe each have 58 months, and China and Peru have 54 and 46 months of data respectively.

Columns 3 to 8 in Table 1 show the market capitalization in US dollars, number of securities constituting the index and index (IFC Global) capitalization in US dollars as of December 1987 and December 1994. These columns show that the emerging markets analyzed in this paper account for approximately 90% of the market and index at both points in time. Table 1 shows the phenomenal growth of emerging markets. The total market capitalization grew from $184 billion at the end of 1987 to $1.6 trillion at the end of 1994. Over the same period, the number of securities constituting the index rose from 423 to 1,266 and the index capitalization rose from $64 billion to $959 billion.

The emerging markets sample provides an important extension to the data for developed markets. The construction of the IFC emerging market data ensures that the sample does not suffer from survivorship bias. The companies included in the index are determined once a year without a hindsight bias. The difficulty in timely compiling of the data, however, necessitates some data restatements. Data restatements are also made when discrepancies are discovered.

Despite the high quality of IFC data, replicating published index returns from the individual securities data requires great care as discussed in Patel (1998) and Barry, Peavey, and Rodriquez (1997). Patel (1998) points out the problems caused by indeterminate timing of corporate events and data transmission errors, both of which can result in internally inconsistent data (e.g. shares outstanding at the beginning of a month and all types of share issuance for the month do not sum to the shares outstanding at the end of month.)

---

[2]Peskin (1997) provides the country and security selection criteria for the three leading indexes: BEMI, IFCI, and MSCI Free. Peskin describes, for the three indexes, the adjustments made to market capitalization for cross-holdings, local government ownership, and foreign ownership restriction. Peskin also details the industry weightings, volatilities, and BARA risk exposures of each index. Peskin suggests that IFCI and MSCI are broadly defined, and BEMI and IFCI have provided high level of transparency.

Table 1. Summary descriptions of emerging markets data available from IFC and emerging markets analyzed in this study. The market capitalization, index capitalization, and number of securities for each of the IFC markets are given at two points in time, Dec. 1987 and Dec. 1994. The markets analyzed in this study are represented in bold in column 1, and the first date of available data is given in column 2. Each market studied has data to Nov. 1997, and column 3 gives the number of monthly observations for each emerging market analyzed in this study.

Country	First Observation	Number of Observations	Market Capitalization (million dollars)		Number of Securities		Index Capitalization (million dollars)	
			Dec. 87	Dec. 94	Dec. 87	Dec. 94	Dec. 87	Dec. 94
Argentina	8712	119	1,519	36,863	24	30	1,031	18,751
Brazil	8712	119	16,899	189,281	30	83	4,397	111,905
Chile	8801	118	5,341	68,194	25	40	2,833	45,006
China	9305	54		3,520		117		19,307
Colombia	8712	119	1,255	15,445	22	25	1,188	11,436
Greece	8712	119	4,463	14,920	10	36	2,232	8,021
Hungary				1,604		13		746
India	8712	119	17,057	127,515	40	119	5,816	65,357
Indonesia	8912	95		47,240		47		22,198
Jordan	8901	106	2,643	4,593	10	42	1,159	2,786
Korea	8712	119	32,905	191,778	23	162	8,024	125,065
Malaysia	8712	119	18,531	199,275	40	104	9,980	125,883
Mexico	8712	119	1,370	130,245	26	80	3,085	83,255
Nigeria	9001	94	973	2,711	15	28	502	1,964
Pakistan	8712	119	1,960	12,263	51	71	654	7,667
Peru	9401	46		8,177		35		5,271
Philippines	9001	94	2,947	55,518	18	41	1,862	30,221
Poland				3,057		12		1,478
Portugal	8712	119	8,857	16,248	12	32	3,675	11,185
South Africa				225,718		63		137,866
Sri Lanka	9301	58		2,883		32		1,710
Taiwan, China	8712	119	48,633	247,325	30	92	12,205	160,212
Thailand	8712	119	5,485	131,478	10	61	2,678	79,660
Turkey	9101	82	3,220	21,605	14	40	1,388	15,243
Venezuela			2,277	4,110	12	17	1,403	3,352
Zimbabwe	9301	58	718	1,828	11	24	324	1,341
Total Capitalization of markets in sample			168,060	1,566,194			55,956	953,444
Total Capitalization of markets not in sample (excluding South Africa)			15,993	8,771			8,480	5,576

I continue to adopt the procedure suggested in Patel (1998) to correct for internal inconsistency of data. This procedure replicates the index returns in the following steps: (i) calculate returns making two alternative assumptions on the timing of dividend payment relative to other corporate events; (ii) verify internal consistency of shares outstanding, and repeat the first step; (iii) calculate returns using capital adjustment rates supplied by IFC; (iv) summarize aggregate returns by country, month using all five calculation methods; (v) choose one calculation method per country and month by comparison with the index levels published by IFC, prioritizing in the order of the return calculations described above and penalizing the use of capital adjustment rate.

The portfolio construction method adopted in this paper is similar to the method used in Reinganum (1981) and Jaffe, Keim and Westerfield (1989).[3] I divide the period with available data into investment horizons of three months; this rebalancing frequency reduces the possibility of upward bias in portfolio return calculations stemming from bid-ask bounce as identified in Blume and Stambaugh (1983). At the beginning of each quarter, I sort all securities within a country on the variable of interest (price-to-book, price-to-earnings or size). Then I form a portfolio, for each market, having the lower half values of the variable of interest within the market.[4]

Table 2 provides value-weighted annualized excess returns in US dollars, over the local index returns in US dollars, calculated from individual company returns and their market values, for portfolios of companies in the lower-half of price-to-book ratio, price-to-earnings ratio and size within each market. Table 2 also shows value weighted average annualized returns of the twenty-two markets. Table 2 shows that low price-to-book stocks have outperformed the local index in 13 of 22 markets, and low price-to-earnings stocks have outperformed in 16 of 22 markets.[5] Individual t-values are significant in only a few markets. Excess returns for the aggregate portfolios of low price-to-book and low price-to-earnings are large at 2.6% and 3.0% respectively, however t-values are significant for low price-to-earnings stocks only. The evidence on size is less convincing, with small stocks outperforming in nine of 22 countries. Excess returns for the aggregate portfolio of smaller companies is positive, but small and statistically insignificant at 1.0%.

---

[3]The portfolio approach does not require explicit distributional assumptions on cross-sectional returns, as the cross-sectional regression approach does. The distributional properties of emerging markets are little known. Bekaert and Harvey (1995) study and reject the index return normality, as noted in Footnote 1.

[4]Excess returns and variance of excess returns of portfolios of lower third of price-to-book, price-to-earnings, and size are higher than excess returns to portfolios of lower half of price-to-book, price-to-earnings, and size, making the significance measured by t-values similar.

[5]The systematic risk of low price-to-book, low price-to-earnings and small size portfolios as measured by beta do not explain the magnitudes of excess returns for the twenty-two emerging markets and for the aggregate of markets.

Table 2. Average annual excess returns for value-weighted portfolios based on low price to book, low price to earnings, and small market cap. At the beginning of each quarter, companies in a country are sorted on price-to-book, price-to-earnings or size. Returns to portfolios with lower half of values are compared to the index returns, calculated from individual company returns and their market values at the beginning of the month. The reported returns are annualized dollar returns. The market capitalization weighted averages are given separately for 11 larger markets with data beginning January/February 1988 and for 22 emerging markets, including data for 11 smaller emerging markets. t-values of excess monthly dollar returns are also shown, with three levels of significance indicated as: 1% by ***; 5% by **; and 10% by *.

Country	First Observation	Number of Observations	Price to Book Ratio Average Excess Return	t-value	Price to Earnings Ratio Average Excess Return	t-value	Market Cap Average Excess Return	t-value
Argentina	8712	119	1.6	0.2	6.0	1.0	2.4	0.2
Brazil	8712	119	19.1	2.6***	20.1	3.2***	2.7	0.3
Chile	8801	118	4.9	1.9*	0.3	0.1	1.4	0.4
Greece	8712	119	6.3	1.4*	7.6	1.6*	-5.4	-0.9
India	8712	119	-1.6	-0.3	1.4	0.4	-5.5	-1.2
Korea	8712	119	5.7	1.7**	4.3	1.8**	-3.5	-0.7
Malaysia	8712	119	3.0	1.1	2.1	1.0	2.0	0.4
Mexico	8712	119	-0.7	-0.2	6.5	2.3**	-3.6	-0.6
Portugal	8712	119	-4.6	-1.2	0.3	0.1	-6.8	-1.4*
Taiwan	8712	119	-2.9	-0.5	-3.4	-0.7	1.4	0.2
Thailand	8712	119	-3.6	-0.7	3.8	0.9	-8.4	-1.5*
Market cap. weighted average of 11 markets	8712	119	2.3	1.0	2.9	1.6*	0.3	0.1
China	9305	54	3.0	0.3	8.6	1.0	5.9	0.9
Colombia	8712	119	7.2	1.6*	0.3	0.1	-9.6	-1.7**
Indonesia	8912	95	0.9	0.1	-1.1	-0.1	5.5	0.5
Jordan	8901	106	-2.2	-0.6	1.9	0.8	-7.4	-1.3*
Nigeria	9001	94	3.7	0.7	11.5	3.9***	10.3	1.8**
Pakistan	8701	119	-2.2	-0.7	-3.1	-1.0	-5.5	-1.1
Peru	9401	46	1.5	0.1	0.2	0.0	-3.0	-0.2
Philippines	9001	94	-4.3	-0.7	-0.5	-0.1	-3.7	-0.4
Sri Lanka	9301	58	-2.5	-0.6	-1.7	-0.4	-3.8	-0.7
Turkey	9101	82	22.7	2.0**	10.4	1.3*	10.2	1.0
Zimbabwe	9301	58	-8.8	-0.8	-6.6	-0.6	-15.2	-1.3*
Market cap. weighted average of 22 markets	8712	119	2.6	1.2	3.0	1.8**	1.0	0.3

3 Sector-neutral characteristics-based portfolios

It may be argued that in some instances, forming portfolios of a subset of stocks based on characteristics like price-to-book or price-to-earnings unwittingly translates into forming portfolios by sectors. For example, December 1997 low price-to-book portfolio for Brazil has 20% higher exposure to utilities than the Brazil index constructed from individual company market-value weights. The evidence from international markets is that deviations in sector/industry exposures from the index are important in outperforming the index, and such deviations increase variance of excess returns (for example, Heckman, Narayanan and Patel (1999) and Griffin and Karolyi (1995)). Hence, I also form characteristics-based portfolios within each market with sector exposures equal to the sector exposures of the local index constructed from individual company market-value weights.

I consider relative values within sectors with at least four securities and form a portfolio having the lower half of the sector-relative values of price-to-book, price-to-earnings and size. To account for the possibility that by considering sectors with at least four securities a significant proportion of the market might be left out, I include the remaining securities in the characteristics-based portfolios. Next, sector weights in the characteristics-based portfolio are adjusted to equate the sector weights in the index within each market.

Table 3 provides value-weighted annualized excess returns in US dollars, over the local index returns in US dollars, for sector-neutral portfolios of companies in the lower-half of price-to-book ratio, price-to-earnings ratio and size within each market. Table 3 also shows value weighted average annualized returns of the twenty-two markets. Table 3 shows that low price-to-book stocks have outperformed the local index in 15 of 22 markets, and low price-to-earnings stocks have outperformed in 16 of 22 markets. Individual t-values are significant in a few markets. Excess returns for the aggregate sector-neutral portfolios of low price-to-book and low price-to-earnings at 2.5% and 3.5% respectively are comparable to the excess returns for the aggregate non-sector-neutral portfolios of low price-to-book and low price-to-earnings. The statistical significance of sector-neutral portfolios, however, is higher; t-values are significant for low price-to-book portfolios at 5% and for low price-to-earnings portfolios at 1%. Sector-neutral portfolios of small stocks outperform in 12 of 22 countries. Annualized excess returns for the aggregate portfolio of smaller companies are positive at 2.8%, and statistically significant at 10%. The results presented here suggest that value-weighted sector-neutral portfolios of low price-to-book, low price-to-earnings, and small capitalization stocks outperform the value-weighted index. The magnitudes of results presented here, however, are lower than the magnitudes in earlier studies, including Patel (1996) and Patel (1998). A possible explanation for this difference in results

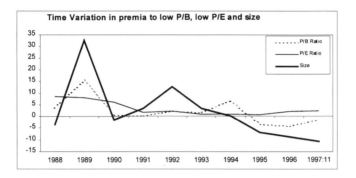

Figure 1. Time variation in excess returns to price-to-book ratio,
price-to-earnings ratio and size Sector-neutral portfolios are formed each
quarter as described in Table 3. This figure presents value-weighted annual
excess returns for eleven large markets with data beginning Jan./Feb. 1988
compared to the index returns calculated from company returns and their
market value weights at the beginning of the month.

is that excess returns to characteristics-based strategies in emerging markets
are time varying.

4   Time variation in excess returns for sector-neutral portfolios

Figure 1 shows the annualized excess returns to the low price-to-book, low
price-to-earnings and small capitalization for each calendar year. The time
variation in excess returns is considerable. For example, small companies
outperformed the index by 32.6% in 1989, whereas they haven't outperformed
the index since 1994. In 1997, small companies under-performed the index by
10.9%. Excess returns earned by sector-neutral portfolios composed of low
price-to-book and low price-to-earnings have also varied over time. The range
of excess returns for low price-to-book portfolios is 15.2% in 1989 to –4.3%
in 1996, and the range of excess returns for low price-to-earnings portfolios is
8.4% in 1988 to 0.6% in 1995. It is interesting to note that for all strategies,
higher returns are in the earlier time periods and the lower returns are in the
later time periods. This observed time variation in excess returns to low price-
to-book, low price-to-earnings, and small capitalization raises an interesting
question: are the annual excess returns influenced by single/few markets, or
are characteristics-based excess returns correlated across emerging markets?

Table 3. Average annual excess returns for sector-neutral portfolios based on low price to book, low price to earnings, and small market cap. At the beginning of each quarter, companies in a country and sector are sorted on price-to-book, price-to-earnings or size. Portfolios are formed to equate the final portfolio sector weights to the index sector weights within each country. All companies in a sector with less than four companies are included in low price-to-book, low price-to-earnings, and small size portfolios. Returns to portfolios with lower half of values are compared to the index returns, described in Table 2. The reported returns are annualized dollar returns. The market capitalization weighted averages are given separately for 11 larger markets with data beginning January/February 1988 and for 22 emerging markets, including data for 11 smaller emerging markets. t-values of excess monthly dollar returns are also shown, with significance levels: 1% by ***; 5% by **; and 10% by *.

Country	First Observation	Number of Observations	Price to Book Ratio		Price to Earnings Ratio		Market Cap	
			Average Excess Return	t-value	Average Excess Return	t-value	Average Excess Return	t-value
Argentina	8712	119	4.2	1.1	5.8	1.3	2.6	0.5
Brazil	8712	119	12.3	1.7**	10.3	2.3**	7.6	0.9
Chile	8801	118	3.2	1.2	2.1	1.0	3.1	0.9
Greece	8712	119	0.4	0.1	1.7	0.5	-6.0	-1.4*
India	8712	119	-0.6	-0.1	2.1	0.6	-3.7	-0.9
Korea	8712	119	2.5	1.1	2.8	1.8**	-2.5	-0.8
Malaysia	8712	119	4.1	1.8**	2.2	1.7**	1.7	0.4
Mexico	8712	119	2.9	1.0	3.4	1.5*	0.8	0.2
Portugal	8712	119	0.2	0.1	1.0	0.5	-3.6	-1.1
Taiwan	8712	119	0.7	0.2	4.1	1.8**	3.7	0.8
Thailand	8712	119	-6.0	-1.5*	4.6	1.7**	-3.2	-0.7
Market cap. weighted ave. of 11 markets	8712	119	2.2	1.6*	3.4	3.3***	1.7	0.8
China	9305	54	3.1	0.4	5.7	0.6	3.8	0.6
Colombia	8712	119	1.2	0.4	-5.2	-1.2	-8.6	-1.9**
Indonesia	8912	95	2.7	0.4	2.6	0.4	25.0	1.0
Jordan	8901	106	-1.9	-0.5	-1.4	-1.1	-14.3	-3.1***
Nigeria	9001	94	-0.7	-0.2	8.1	3.2***	-5.1	-1.5*
Pakistan	8701	119	-3.6	-1.2	-1.5	-0.7	0.0	0.0
Peru	9401	46	-2.9	-0.8	-1.8	-0.6	1.8	0.3
Philippines	9001	94	-2.6	-0.9	-0.9	-0.3	-1.3	-0.3
Sri Lanka	9301	58	2.1	0.5	1.1	0.3	0.6	0.1
Turkey	9101	82	17.8	1.9**	9.9	1.6*	17.6	2.0**
Zimbabwe	9301	58	6.9	0.8	-2.7	-0.3	-8.0	-0.7
Market cap. weighted ave. of 22 markets	8712	119	2.5	1.9**	3.5	3.6***	2.8	1.4*

5 Cross-market correlation of the sector-neutral portfolio returns

Tables 4, 5, and 6 present, for ten calendar years, excess returns to sector-neutral low price-to-book, low price-to-earnings, and small capitalization portfolios respectively for the eleven larger markets with data since December 1987. Tables 4, 5, and 6 also show the cross-sectional range of excess returns each year, the number of countries out-/under-performing the index each year and annualized excess return, arithmetic and geometric, for the entire time period for each country.

Table 4 shows that the cross-country range of excess returns to sector-neutral low price-to-book stocks is considerable each year, and there doesn't appear to be a time trend in the cross-country range of excess returns. The number of countries out-/under-performing, however, gives a different picture. Before 1993, the sector-neutral low price-to-book portfolios out-/under-perform in eight or more markets only once in six years; since 1993, eight or more markets have simultaneously out-/under-performed in three of four years. Table 5 shows the annual excess returns earned by sector-neutral low price-to-earnings stocks. The range of excess returns earned by low price-to-earnings is considerable, from 17.7% in 1994 to 101.8% in 1990, without a significant time trend. The low price-to-earnings stocks, however, out-/under-perform simultaneously in the eleven markets; the probability of observing the number of countries out-/under-performing is greater than 20% in only one of ten years, and that probability is less than 10% in six of ten years. The bunching of out-/under-performance of low price-to-earnings stocks appears to be greater before 1993 than since 1993.

Table 6 shows the cross-country variation in annual returns to sector-neutral portfolios of small capitalization stocks. The range of out-/under-performance varied from 79.7% in 1989 to 27.1% in 1991. The range of out-/under-performance is wide in the beginning as well as the end of the time period studied. The convergence in the number of markets out-/under-performing, however, has increased recently. In 1995, 1996, and 1997, sector-neutral small size has under-performed in nine, nine, and ten countries respectively. The evidence in this section suggests that the out-/under-performance of sector-neutral low price-to-book and small size strategies, though different in magnitude, has recently occurred simultaneously in emerging markets. The evidence also suggests that sector-neutral low price-to-earnings strategies have out-/under-performed simultaneously in emerging markets over the entire time period; however, the magnitudes of excess returns differ across emerging markets.

Table 4. Annual excess returns for sector-neutral portfolios formed on price-to-book ratio Sector-neutral portfolios are formed each quarter as described in Table 3. This table presents annual excess returns compared to the index returns calculated from company returns and their market value weights at the beginning of the month. Also reported are the within-year value-weighted mean excess returns across the eleven markets reported here, and the within-year range of excess returns. The bottom two rows show the number of countries outperforming, and the probability of observing that number if the country excess returns were independent, and the underlying binomial probability of out-/underperformance was 0.5. The two rightmost columns present annualized geometric and arithmetic mean excess returns for 119 months from Jan. 1988 to Nov. 1997.

Country	1988	1989	1990	1991	1992	1993	1994	1995	1996	1997:11	Annualized Since 1988 Geometric Mean	Annualized Since 1988 Arithmetic Mean
Argentina	30.9	-0.7	9.2	10.9	-9.5	2.4	4.6	2.5	1.5	-12.7	3.4	4.2
Brazil	10.5	13.3	51.3	-0.3	8.6	25.0	5.9	-12.3	-3.1	0.0	8.8	12.3
Chile	11.0	-0.1	-11.0	4.3	-0.2	3.3	21.7	12.7	-7.9	-0.8	2.9	3.2
Greece	-2.1	5.1	-2.8	-7.2	-5.2	1.5	9.8	-1.7	-16.0	6.7	-1.5	0.4
India	11.4	-2.3	4.5	-13.7	12.1	0.4	-1.4	-9.4	-11.8	-13.2	-2.7	-0.6
Korea	2.4	4.0	-0.9	3.6	5.3	-0.2	10.2	-4.8	7.2	-6.8	1.9	2.5
Malaysia	1.0	20.0	8.6	10.1	-11.2	2.4	1.7	-4.1	3.4	6.9	3.5	4.1
Mexico	21.1	1.7	8.2	-8.0	4.1	-10.2	4.0	-3.2	0.3	11.9	2.7	2.9
Portugal	7.3	-2.1	6.6	-6.2	10.1	-5.5	1.2	-6.0	-2.3	-2.7	-0.2	0.2
Taiwan	-4.7	27.8	-7.3	-3.1	-3.1	4.9	9.9	3.0	-13.7	-7.6	0.1	0.7
Thailand	-1.9	-13.9	0.4	0.9	5.6	-8.4	5.3	-5.4	-17.2	-27.7	-6.9	-6.0
Market cap. weighted ave. of 11 markets	3.7	15.2	0.6	0.2	2.2	1.6	6.5	-3.6	-4.3	-1.5	1.9	2.2
Range	35.6	41.7	62.3	24.6	23.3	35.2	23.1	25.0	24.4	39.6		
No. of countries (out of 11) outperforming	8	6	7	5	6	7	10	3	4	3	7	9
Probability of observing no. of countries out-/underperforming	0.073	0.387	0.194	0.387	0.387	0.194	0.003	0.073	0.194	0.073	0.194	0.019

Table 5. Annual excess returns for sector-neutral portfolios formed on price-to-earnings ratio. Sector-neutral portfolios are formed each quarter as described in Table 3. This table presents annual excess returns compared to the index returns calculated from company returns and their market value weights at the beginning of the month. Also reported are the within-year value-weighted mean excess returns across the eleven markets reported here, and the within-year range of excess returns. The bottom two rows show the number of countries outperforming, and the probability of observing that number if the country excess returns were independent, and the underlying binomial probability of out-/underperformance was 0.5. The two rightmost columns present annualized geometric and arithmetic mean excess returns for 119 months from Jan. 1988 to Nov. 1997.

Country	1988	1989	1990	1991	1992	1993	1994	1995	1996	1997:11	Annualized Since 1988 Geometric Mean	Annualized Since 1988 Arithmetic Mean
Argentina	9.9	4.2	3.3	16.9	0.1	1.0	-3.4	10.1	-2.7	-5.8	3.2	5.8
Brazil	8.5	6.6	97.4	26.2	1.1	-17.1	-8.2	-4.7	8.4	-4.3	8.2	10.3
Chile	2.7	7.9	-2.8	7.5	-4.8	4.6	1.4	2.6	0.7	-2.1	1.7	2.1
Greece	39.6	-2.9	8.5	-13.4	5.2	4.5	4.8	-2.3	-11.7	-0.9	2.3	1.7
India	8.3	27.8	6.1	-8.0	6.6	7.6	1.2	-5.3	-11.7	-15.7	1.1	2.1
Korea	-1.5	5.0	-4.4	3.6	1.1	4.4	9.5	1.4	1.8	10.4	3.1	2.8
Malaysia	5.8	-1.6	2.8	5.9	-5.1	0.3	4.0	2.3	9.7	-2.0	2.1	2.2
Mexico	3.0	11.1	8.9	2.3	-2.6	3.4	-3.2	-12.6	3.7	10.4	2.3	3.4
Portugal	6.2	-5.1	1.0	-2.1	23.6	1.7	-5.6	-2.5	0.3	-5.4	0.8	1.0
Taiwan	7.9	10.4	5.1	-4.7	5.3	0.4	1.3	3.9	-3.4	14.8	4.1	4.1
Thailand	1.1	6.4	7.8	10.0	14.7	0.9	1.7	8.7	0.3	-8.9	4.1	4.6
Market cap. weighted ave. of 11 markets	8.4	8.0	6.0	1.8	2.3	0.8	0.9	0.6	1.9	2.3	3.3	3.4
Range	41.1	32.9	101.8	39.6	28.7	24.7	17.7	22.7	21.4	30.5		
No. of countries (out of eleven) outperforming	10	8	9	7	8	10	7	6	7	3	11	11
Probability of observing no. of countries out-/underperforming	0.003	0.073	0.019	0.194	0.073	0.003	0.194	0.387	0.194	0.073	0.000	0.000

Table 6. Annual excess returns for sector-neutral portfolios formed on market cap. Sector-neutral portfolios are formed each quarter as described in Table 3. This table presents annual excess returns compared to the index returns calculated from company returns and their market value weights at the beginning of the month. Also reported are the within-year value-weighted mean excess returns across the eleven markets reported here, and the within-year range of excess returns. The bottom two rows show the number of countries outperforming, and the probability of observing that number if the country excess returns were independent, and the underlying binomial probability of out-/underperformance was 0.5. The two rightmost columns present annualized geometric and arithmetic mean excess returns for 119 months from Jan.1988 to Nov.1997.

Country	1988	1989	1990	1991	1992	1993	1994	1995	1996	1997:11	Annualized Since 1988 Geometric Mean	Annualized Since 1988 Arithmetic Mean
Argentina	-3.9	-12.2	17.0	11.2	13.7	-1.2	5.1	1.5	-2.8	-27.3	-7.0	2.6
Brazil	9.5	23.1	7.6	-7.1	25.4	34.7	5.6	-20.7	-8.0	-14.4	4.1	7.6
Chile	5.0	0.1	-2.4	3.4	8.5	1.0	-4.2	16.5	-2.8	-2.6	2.1	3.1
Greece	10.0	-7.7	7.5	-13.0	-4.4	-5.0	-6.8	-0.5	-15.7	-11.4	-5.1	-6.0
India	12.4	6.3	-12.0	5.8	5.5	8.7	2.3	-13.0	-24.5	-24.2	-4.3	-3.7
Korea	-4.2	2.2	-0.5	-7.0	10.1	-2.1	-0.4	-13.0	12.5	-23.4	-3.1	-2.5
Malaysia	-14.4	19.5	10.8	3.9	-0.9	9.1	-2.4	-5.4	-2.6	-16.1	-3.8	1.7
Mexico	9.1	9.1	-4.5	-6.4	17.4	-6.7	7.9	-7.7	-0.7	2.5	1.8	0.8
Portugal	3.8	-9.4	3.9	1.1	-0.2	-14.1	-9.6	-1.5	1.6	-7.6	-3.5	-3.6
Taiwan	-8.7	67.5	-3.5	14.1	5.0	-4.6	0.2	-1.7	-16.4	-8.9	2.4	3.7
Thailand	-5.7	-3.3	5.2	5.3	24.2	3.1	-14.4	-9.2	-32.8	-21.4	-6.2	-3.2
Market cap. weighted ave. of 11 markets	-3.5	32.6	-1.6	3.2	12.5	3.2	-0.1	-7.1	-8.9	-10.9	1.3	1.7
Range	26.8	79.7	29.0	27.1	29.8	48.8	22.3	37.2	45.3	29.8		
No. of countries (out of eleven) outperforming	6	7	6	7	8	5	5	2	2	1	4	6
Probability of observing no. of countries out-/underperforming	0.387	0.194	0.387	0.194	0.073	0.387	0.387	0.019	0.019	0.003	0.194	0.387

6 Conclusion

This paper provides evidence on the out-/under-performance of low price-to-book, low price-to-earnings, and small size portfolios over local index in 22 emerging markets: 11 larger ones with data from January/February 1988 to November 1997 and 11 smaller ones covering shorter periods. The evidence shows that portfolios formed on predetermined characteristics outperform the total market index in emerging markets. These results hold for the excess returns of both non-sector-neutral and sector-neutral low price-to-book, low price-to-earnings, and small size portfolios. The variation in excess returns is lower for sector-neutral portfolios, and hence the significance of excess returns is higher for sector-neutral portfolios. As in developed markets, the characteristics-based premia display considerable time variation in emerging markets. Portfolios of smaller stocks have under-performed since 1994, and portfolios of low price-to-book stocks have under-performed since 1995. Portfolios formed on low price-to-earnings have outperformed the index in each year since 1988. I also examine whether return premia are correlated across emerging markets. The range of excess returns across eleven larger emerging markets is wide every year and the range does not have a time trend; however, out-/under-performance of portfolios formed on sector-neutral low price-to-book, low price-to-earnings and small size tends to occur together in emerging markets.

Acknowledgements
I thank Donald Keim for helpful comments on this draft and Alejandro Baez-Sacasa, Douglas Dooley, Michael Granito, Thomas Madsen, Satyen Mehta, Shiv Mehta, Marc Roston, and Jian Yao for comments on earlier research.

References

Barry, Christopher B, John W. Peavy III, and Mauricio Rodriquez (1997) 'A Convenient Way to Invest in the Emerging Markets', *Emerging Markets Quarterly* **1** (1) 41–48.

Bekaert, Geert, and Campbell R. Harvey (1997) 'Emerging Equity Market Volatility', *Journal of Financial Economics* **43** 29–77.

Black, Fisher, Michael C. Jensen, and Myron Scholes (1972) 'The Capital Asset Pricing Model: Some Empirical Tests'. In *Studies in the Theory of Capital Markets*, Michael C. Jensen (ed.), Praeger, New York.

Blume, Marshall E., and Irwin Friend (1973) 'A New Look at the Capital Asset Pricing Model', *Journal of Finance* **28** 19–33.

Blume, Marshall E., and Robert F. Stambaugh (1983) 'Biases in Computed Returns: An Application to the Size Effect', *Journal of Financial Economics* **12** (3) 387–404.

Claessens, Stijn, Susmita Dasgupta and Jack Glen (1995) 'The Cross-Section of Stock Returns: Evidence from the Emerging Markets', Working Paper, The World Bank.

Daniel, Kent, and Sheridan Titman (1997) 'Evidence on the Characteristics of Cross Sectional Variation in Stock Returns', *Journal of Finance* **LII** (1) 1–33.

Dichev, Ilia D. (1998) 'Is the Risk of Bankruptcy a Systematic Risk?', *Journal of Finance* **LIII** (3) 1131–1147.

Fama, Eugene F. (1970) 'Efficient Capital Markets: A Review of Theory and Empirical Work', *Journal of Finance* **XXV** (2) 383–417.

Fama, Eugene F. and J. MacBeth (1973) 'Risk, Return and Equilibrium: Empirical Tests', *Journal of Political Economy* **71** 607–636.

Fama, Eugene F. and Kenneth R. French (1996) 'Multifactor Explanations of Asset Pricing Anomalies', *Journal of Finance* **LI** (1) 55–84.

Fama, Eugene F. and Kenneth R. French (1998) 'Value versus Growth: The International Evidence', *Journal of Finance* **LIII** (6) 1975–1999.

Griffin, John M. and G. Andrew Karolyi (1995) 'Another Look at the Role of the Industrial Structure of Markets for International Diversification Strategies', Ohio State University Working Paper.

Hawawini, Gabriel, and Donald B. Keim (1999) 'The Cross Section of Common Stock Returns: A Review of the Evidence and Some New Findings', this volume, 3–43.

Heckman, Leila, Singanallur R. Narayanan and Sandeep A. Patel (1999) 'Country and Industry Importance in European Returns', Working Paper, Salomon Smith-Barney and J. P. Morgan.

International Finance Corporation, (1996) 'The IFC Indexes: Methodology, Definitions, and Practices', June.

Jaffe, Jeffrey, Donald B. Keim, and Randolph Westerfield (1989) 'Earnings Yields, Market Values and Stock Returns', *Journal of Finance* **45) 135–148.**

Jagannathan, Ravi, Keiichi Kubota, and Hitoshi Takehara (1998) 'Relationship Between Labor-Income Risk and Average Return: Empirical Evidence From the Japanese Stock Market', Working Paper, University of Minnesota.

Lintner, John (1965) 'The Valuation of Risky Assets and the Selection of Risky Investment in Stock Portfolios and Capital Budgets', *Review of Economic Statistics* **14** 13–37.

Lo, Andrew and Craig MacKinlay (1990) 'Data-Snooping Biases in Tests of Financial Asset Pricing Models', *Review of Financial Studies* **3** 431–468.

Markowitz, Harry (1959) *Portfolio Selection: Efficient Diversification of Investments*, Wiley, New York.

Mossin, J. (1966) 'Equilibrium in a Capital Asset Market', *Econometrica* **34** 768–783.

Patel, Sandeep A. (1993) 'Predictability of Equity Returns in Emerging Markets', working paper, J.P. Morgan Investment Management (September).

Patel, Sandeep A. (1996) 'Performance and Risk of Value based portfolios in emerging markets: 1988–1995', working paper, J.P. Morgan Investment Management, (February).

Patel, Sandeep A., (1998) 'Cross-Sectional Variation in Emerging Markets Equity Returns: January 1988–March 1997', *Emerging Markets Quarterly* **2** (1) 57–70.

Peskin, K. Stuart (1997) 'Emerging Markets Benchmarks: Understanding an Evolving Asset Class', *Emerging Markets Quarterly* **1** (1) 27–32.

Reinganum, Marc R. (1981) 'A Misspecification of Capital Asset Pricing: Empirical Anomalies Based on Earnings Yields and Market Values', *Journal of Financial Economics* **9** 19–46.

Rouwenhorst, Geert K. (1998) 'Local Return Factors and Turnover in Emerging Stock Markets', working paper, Yale University.

Treynor, Jack (1961) 'Toward a Theory of Market Value of Risky Assets', unpublished manuscript.

Anomalies in Asian Emerging Stock Markets

Seng-Kee Koh and Kie Ann Wong

1 Introduction

We survey anomalies of the static capital asset pricing model (CAPM) for Asian emerging stock markets. Since its discovery by Sharpe (1964), Lintner (1965), Mossin (1966) and Treynor (1961), the CAPM has been widely embraced by academics and practitioners for its simplicity and applicability to a range of financial issues. Its usefulness is demonstrated by applications in areas such as capital budgeting, valuation and capital structure decisions. The CAPM states that the expected return on security i ($E(R_i)$) equals the sum of risk free rate (R_f) and a risk premium which is the product of the beta risk (β_j) of the security and the market risk premium,

$$E(R_i) = R_f + \beta_i(E[R_m] - R_f) \tag{1.1}$$

where $E(R_m)$ is the expected return on the market portfolio. The beta risk of a security represents the relative covariance risk of the security to an aggregate market risk factor.

The CAPM was extensively tested and empirically supported in early years by Black, Jensen, and Scholes (1972), Blume and Friend (1973) and Fama and MacBeth (1973). They showed that mean returns had a significant positive relation to estimated portfolio betas. A recent survey on the CAPM is Ferson (1995).

Since the 1980s, researchers have documented a number of deviations from the CAPM. These deviations are generally termed as 'anomalies' of the CAPM primarily because, beside betas, other financial variables such as the size of firm, ratio of book to market values, earnings yield and firm's prior returns explain some variation in the cross-sectional returns of stocks. Furthermore, there is strong evidence of seasonalities in daily and monthly returns not just in US stock market but also in numerous other stock markets.

Many of these anomalies have been documented for the US and European stock markets. The question that comes immediately to one's mind is whether these anomalies are peculiar to the US and European markets or they are universal and pervasive in most stock markets throughout the world. Moreover, are the anomalies of the CAPM a by-product of the unique institutional structures of stock markets? To help us answer these questions, we survey the literature on anomalies in the Asian emerging stock markets. Such a survey provides us several benefits. Firstly, if the well-documented anomalies found in the US and European stock markets are also present in Asian emerging stock markets, then there is evidence to imply that either the CAPM is not the appropriate equilibrium model or that stock markets are generally inefficient. The joint hypotheses problem does not allow us to specifically identify the direct cause of rejection, be it misspecification of the equilibrium model or markets are inefficient or both. However, given that stock markets studied by researchers are of different degrees of maturity, liquidity and development, the presence of similar anomalies does lead one to wonder whether the CAPM is a good working model for practical financial applications. There may be an urgent need to develop another parsimonious asset pricing model which could better explain the cross-sectional variation in securities returns.

Secondly, as the institutional structures of stock markets in Asia are different from those in Western countries, the presence of anomalies in Asian markets allows us to verify whether some of the explanations previously provided by researchers for anomalies in the US and European markets are credible.

The paper is organized as follows. In Section 2, we provide an overview of seven Asian emerging stock markets. We present empirical evidence for the month-of-the-year effect, turn-of-the-month effect, day-of-the-week effect, holiday effect, size effect and earnings-yield effect in Sections 3–8 respectively. Section 9 concludes the paper.

2   Asian Emerging Stock Markets

The Asian stock markets (with the exception of Japan) are considered to be emerging markets because of their short history, small number of listed companies and low market capitalization. Compared to the United States and Japan which had 2907 and 1792 listed companies, most of the seven Asian emerging stock markets in Table 1 had fewer than 800 companies listed on their exchanges as of December, 1996. Singapore had the smallest number of listed companies with 289 while India the largest with 1439.

The market capitalization of the seven Asian stock markets ranged from a low of US$46 billion for Thailand to a high of US$556 billion for Hong Kong. These seven stock markets were small in size compared to the market capitalizations of the Tokyo Stock Exchange and the New York Stock Exchange of US$2822 billion and US$7300 billion respectively. The aggregate market capitalization of the seven stock markets in Table 1 was only 23% and 58% of the New York and Tokyo stock exchanges, respectively.

The Taiwan stock exchange had the highest annual trading volume of US$135.8 billion while the Thailand stock exchange had the lowest of US$2.1

billion. In spite of their lower capitalization, the annual trading volumes of Hong Kong and Taiwan exchanges exceeded that of Japan. Table 1 also showed that most of the seven emerging markets had relatively low turnover to market capitalization ratios (below 0.10) except for Taiwan (0.39) and Hong Kong (0.15). The relatively higher volume of trading in Taiwan and Hong Kong was due to the relatively lower transaction costs of trading in these two countries (see Table 3).

While the Asian stock markets were generally lowly capitalized, their average P/E ratios (reciprocal of earnings yield) were quite varied. As of the end of 1996, the P/E ratios ranged from a low of 8.4 for Thailand to a high of 38.4 for Taiwan. The majority of the countries' P/E ratios were in the teens. The average dividend yields in the emerging markets ranged from 1% to 4.8%.

The trading of stocks in the emerging Asian stock markets is moving towards fully automated trading systems. An automated trading system has numerous benefits such as improved market efficiency, increased operational efficiency, easier monitoring of trading by stock exchanges and increased trading volume through orderly matching and clearing of trading orders.

Trading of shares on the Hong Kong Stock Exchange is done through the Automatic Order Matching and Execution System (AMS) which was set up in 1993. This trading system allows orders to be matched automatically. In contrast, the National Stock Exchange of India still uses the traditional auction system for trading of shares. In Malaysia, trading is done semi-automatically through the System on Computerized Order Routing and Execution (SCORE). Trading orders of clients are input at the stockbroking houses' premises and then routed to the Kuala Lumpur Stock Exchange's matching room where matching of orders are executed by officials of the exchange. The Stock Exchange of Singapore uses the Central Limit Order Book (CLOB) to match trading orders. Trading orders of clients are input directly into a computer system by dealers and remisiers of stockbroking houses. The matching and confirmation of orders are fully automated and transactions are scripless which means that no physical shares are exchanged. Instead, trading of shares are recorded as book entries with Central Depository (Pte) Limited. Stock trading in South Korea is executed through the Stock Market Automated Trading System (SMATS). Trading orders of clients are input directly into SMATS by dealers of brokerage houses and sent electronically to be matched automatically or semi-automatically by officials of the exchange. In Taiwan, trading is executed through the Computer Assisted Trading system (CATS). The matching and execution of trading orders are fully computerised. Trading in Thailand is done through the Automated System in Stock Exchange of Thailand (ASSET). The trading orders from clients are classified as either automated order-matching deals or put-through deals. The former

Table 1. Summary characteristics of Asian stock markets as at December, 1996

Country	Number of listed firms	Market capitalization (US$ billion)	Total turnover (US$ billion)	Ratio of turnover to market cap.	P/E ratio	Dividend yield (%)	Trading methods	Trading hours	Trading hours per day	Trading days	Delivery period
Hong Kong	625	556	81.7	0.15	17.2	2.5	AMS	10.00–12.30; 14.30–15.55	4	Mon.–Fri.	T+2
India[1]	1439	151	10.9	0.07	N.A.	N.A.	Auction	10.00–15.30	5.5	Mon.–Fri.	T+7[2]
Malaysia	439	175	7.8	0.04	15.8	2.3	SCORE	9.30–12.30; 14.30–17.00.	5.5	Mon.–Fri.	T+7
Singapore	289	228	7.8	0.03	17.0	1.6	CLOB	9.30–12.30; 14.00–17.00.	6	Mon.–Fri.	T+7
South Korea	774	143	14.1	0.10	21.2	1.5	SMATS	9.30–11.30; 13.00–15.00.	5	Mon.–Sat.	T+2
Taiwan	396	347	135.8	0.39	38.4	2.7	CATS	9.00–12.00; 9.00–11.00	3	Mon.–Fri. Sat.	T+1
Thailand	455	46	2.1	0.05	8.4	4.8	ASSET	10.30–12.30; 14.30–16.30.	4	Mon.–Fri.	T+3

Source: Fact books and official web pages of the various stock exchanges.
1. The information for India pertains to those of National South East India Stock Exchange.
2. T+7 settlement is for shares in group A and B1 while T+14 settlement is for shares in group B2.

Table 2. Opening and closing hours of Asian stock exchanges

US Eastern Standard Time	Opening and Closing hours of stock exchanges
7.00 p.m.	Tokyo stock exchange opens
7.30 p.m.	South Korea stock exchange opens
8.00 p.m.	Taiwan stock exchange opens
8.30 p.m.	Kuala Lumpur Stock Exchange opens Singapore stock exchanges opens
9.00 p.m.	Hong Kong stock exchange opens
10.30 p.m.	Thailand Stock Exchange opens
11.00 p.m.	Taiwan Stock Exchange closes
1.00 a.m.	Tokyo Stock Exchange closes South Korea Stock Exchange closes
2.55 a.m.	Hong Kong Stock Exchange closes
4.00 a.m.	London Stock Exchange opens Kuala Lumpur Stock Exchange closes Singapore Stock Exchange closes
4.30 a.m.	Thailand Stock Exchange closes
9.30 a.m.	New York Stock Exchange opens
11.00 a.m.	London Stock Exchange closes
4.00 p.m.	New York Stock Exchange closes

Source: Fact books and official web pages of the various stock exchanges.

type of order is matched automatically by the computerised trading system while the latter type of order is executed by internally matching two orders from the same broker or by negotiation between different brokers outside the trading system. The order is considered executed only after approval from the exchange is obtained.

The delivery period for trading of shares varies widely across the seven exchanges, ranging from one day delivery for Taiwan stock exchange to 7 days for the India, Malaysia and Singapore stock exchanges.

Most of the Asian stock exchanges trade from Monday to Friday with the exception of the South Korean and Taiwanese exchanges which also trade on Saturday. The daily trading hours of the seven exchanges range from 3 hours in Taiwan to 6 hours in Singapore. The opening and closing hours of the various exchanges with respect to those of Tokyo Stock Exchange and New York Stock Exchange are given in Table 2.

The transaction costs and the tax regimes applicable to trading in the seven Asian stock markets are presented in Table 3. The lowest transaction cost of 0.1425% for a one-way trade is levied for share transactions in Taiwan stock exchange. The other stock exchanges that impose relatively low transaction costs are Hong Kong, South Korea and Thailand. As for tax regimes, there is no capital gains tax in Hong Kong, Malaysia, Singapore, Taiwan and Thailand while India and South Korea impose capital gains tax of 20% and 10 or 25%, respectively. There are no withholding taxes for dividends in Hong Kong, Malaysia and Singapore. India, South Korea, Taiwan and Thailand impose withholding taxes of 10%, 15%, 15–35% and 10%, respectively.

3   Month-of-the-year effect

As early as 1976, Rozeff and Kinney (1976) found that the average stock return in January was higher than in other months for the US stock market. This phenomenon has been labeled the month-of-the-year effect. Subsequent research by Keim (1983) showed that the month-of-the-year effect and the size effect were related. He provided empirical evidence to show that the size effect was concentrated in January. In particular, approximately 50% of the return differential between small and large firms was in January. Furthermore, 50% of this January effect occurred in the first 5 trading days of January (+1 to +5 of January). Refer to Table 4 of Keim (1983).

Other researchers such as Gultekin and Gultekin (1983) demonstrated that the month-of-the-year effect was not confined to the US market but occurred in 12 of the 13 countries they examined.

To test the null hypothesis of equal expected returns for each month of the year, researchers estimate the regression:

$$R_t = a_l + \sum_{k=2}^{12} a_k D_{kt} + e_t \qquad (3.1)$$

where R_t is the average monthly excess return for a country's stock index and D_{kt} represents the dummy variable for the kth month. D_{kt} equals 1 if the return is in the kth month and 0 otherwise. The intercept a_1 measures the average return for January while the rest of the coefficients $a_2, a_3, \ldots, a_{12}$ measure the average returns for February, March and so on to December in excess of January's return.

Researchers have offered a number of explanations for the month-of-the-year effect, the most prominent being the tax loss hypothesis. This hypothesis postulates that investors sell stocks which have declined in value in December to obtain short-term capital loss which is then used to offset capital gains or taxable income. Due to the selling pressure on stocks, the price is depressed

Table 3. Transaction costs and tax regimes of Asian stock markets as at
December, 1996

Country	Commision rate	Taxes
Hong Kong	Minimum commision of higher of 0.25% or HK$50;	No capital gains tax;
	Transaction levy of 0.013%;	No withholding tax for dividends.
	Stamp duty of 0.15%; Transfer deed duty of HK$5 payable by first seller	
India	Maximum commission of 2.5%; Minimum commission of 25 rupees	Dividends taxed at 10%; Long-term capital gains taxed at 20%.
Malaysia	1/2 sen per share for share price less than 50 sen	No capital gains tax;
	1 sen per share for share price between 50 sens and RM1.00	No witholding tax for dividends.
	1% of transaction on the first RM500,000 0.75% of transaction on the next RM500,000 to RM2,000,000 0.5% of transaction on amounts exceeding RM2,000,00	
Singapore	1% of transaction on the first S$250,000 0.9% of transaction on the next S$250,000	No capital gains tax; No witholding tax for dividends.
	0.8% of transaction on the next S$250,000 0.7% of transaction on the next S$250,000 0.5% of transaction on the next S$500,000 Minimum of 0.3% for transaction exceeding S$1.5 million	
South Korea	Brokerage fee of between 0.3% and 0.6%	Floor-based transaction tax of 0.30%; Special tax of 0.15%; OTC-based transaction tax of 0.5%; Dividend and interest income withholding tax of 15%; Capital gains tax of 10% or 25%.
Taiwan	0.1425% of transaction value;	Securities transaction tax of 0.3%;
	Minimum fee of TWD20.	Cash dividends witholding tax of 15% to 35%; No capital gains tax.
Thailand	0.5% of transaction value; Minimum fee of THB50.	No capital gains tax; 10% witholding tax for dividends.

Source: Fact books and official web pages of the various stock exchanges.

at the end of the year. The prices of these stocks rebound to equilibrium levels when the excessive selling is over thereby giving rise to the January effect.

The tax loss hypothesis had received some empirical support from Reinganum (1983), Roll (1983) and Constantinides (1984). Schultz (1983) found that prior to 1917 before the current US tax codes were implemented, there was no January effect. Although the January effect was documented for Canada, Japan and the United Kingdom by Berges *et al.* (1984), Kato and Schallheim (1985) and Jagadeesh (1991) respectively, they found that the anomaly cannot be attributed to the tax loss explanation. For updated results, see also the paper by Comolli and Ziemba in this volume.

For the tax loss hypothesis to be validated, these countries must have a tax year-end that precedes the month of abnormally high return and capital gains must be taxed.

The empirical evidence for the month-of-the-year effect in Asian stock markets is presented in Table 4. Based on the F-test for equality of returns for all the months of the year, there was evidence of seasonality of monthly returns only in Malaysia and Singapore. There was no evidence to support the month-of-the-year effect in Hong Kong, India, Philippines, South Korea, Taiwan and Thailand. In fact, for countries such as India, Malaysia, South Korea and Thailand, January returns were lower than those in some other months of the year. However, for Hong Kong, Singapore and Taiwan, the January return was significantly larger than those of the remaining eleven months of the year. For the Philippines, January had the highest mean return but it was not significantly different from zero.

Ho (1990) studied the month-of-the-year effect in 12 countries, 8 of which were Asian countries (Hong Kong, Japan, Malaysia, Philippines, Singapore, South Korea, Taiwan and Thailand). Unlike other studies which analyzed the mean returns across the 12 months of a year, Ho's study focused on the difference between the January mean return and the mean returns of the other months of the year. His results indicated that the January return was significantly larger than those of the other months in Hong Kong, Japan, Malaysia, Philippines, Singapore, South Korea and Taiwan stock markets.

Although Singapore and Taiwan both have tax year-ends in December, they do not impose taxes on capital gains. Hence, the abnormally high returns in January in both countries cannot be attributed to the tax loss hypothesis. Furthermore, as the tax year-end in Hong Kong is March, any seasonality which is consistent with the tax loss hypothesis should surface in the month of April. The April return for Hong Kong was only 1.12% which was smaller than several other months, particularly that in January. Given that there is capital gains tax in India and South Korea, the tax loss hypothesis implies that we should observe higher abnormal returns in the month following the

Table 4. Mean monthly percentage returns of the seven Asian stock markets for the various months of the year

Country	Jan.	Feb.	March	April	May	June	July	Aug.	Sept.	Oct.	Nov.	Dec.	F-test for equal returns
Hong Kong[1]	8.96**	0.89	-3.92	1.12	3.62	0.93	0.47	-1.99	-4.77	4.69	-1.99	2.86	1.76
India[2]	1.67	6.30**	1.69	2.84	0.14	3.29	4.41	1.75	4.49	-1.65	-1.07	3.83	0.98
Malaysia[2]	3.17*	-0.62	2.24	3.46*	0.40	-1.98	-1.11	-2.69	1.88	-3.58**	2.69	3.97**	3.36**
Philippines[4]	7.32	0.22	0.81	-0.83	-0.64	3.28	1.84	-2.42	-0.42	-0.57	5.92**	3.55	0.93
Singapore[2]	6.85**	1.07	0.20	0.46	3.06**	0.40	0.06	-1.90	-0.62	0.10	-0.80	2.43*	4.42**
South Korea[3]	0.42	2.10	3.72	0.73	1.65	2.54	2.91	-0.13	-0.01	-0.67	3.02	3.34	1.00
Taiwan[3]	6.26*	3.41	2.40	4.97	3.49	1.24	0.88	3.83	4.01	-3.70	0.47	1.63	1.24
Thailand[2]	2.94	2.59	3.27*	3.58*	1.49	2.49	3.23*	2.14	1.71	0.60	-0.90	3.27*	0.75

Figures marked with ** (*) were means which were significantly different from zero at the 1% (5%) level based on a 2-tail test. The last column of the table showed the F-statistic for a test of equal means across all twelve months of the year. Those marked with ** (*) indicated a rejection of the null hypothesis of equal means at the 1% (5%) level.

1. The results for Hong Kong were based on Table 6 of Aggrawal and Tandon (1994). The index and time period used were Hang Seng Index from February 1973 to June 1987.

2. The results for India, Malaysia, Singapore and Thailand were taken from Table 2 of Chan, M. W. L., Khanthavit, A., and Thomas, H., (1996). The indices and time periods considered were Bombay Stock Exchange Sensitive Index (India) from April 1979 to December 1992, KLSE Composite Index (Malaysia) from January 1974 to December 1992, Straits Times Industrial Index (Singapore) from January 1969 to December 1992, and SET valued-weighted index (Thailand) from May 1975 to December 1991.

3. The results for South Korea and Taiwan were based on Tables 1 and 2 of Lee (1992). The indices and time periods considered were Korea Composite Stock Price Index (South Korea) from January 1975 to December 1989 and Taiwan Stock Exchange Index (Taiwan) from January 1970 to December 1989.

4. The results for Philippines were based on Table 6 of Ho (1990). The index and time period considered were Manila Mining Index from January 1976 to November 1987.

tax year-end. For South Korea where the tax year-end is December, the low
return in January seems to negate the tax loss hypothesis. In summary, the
empirical evidence for Asian emerging markets indicates that seasonality of
monthly returns occurred in only two of the seven countries examined and
there was no empirial support for the tax loss hypothesis in these markets.

4   Turn-of-the-Month Effect

The turn-of-the-month effect refers to the concentration of higher than average positive returns around the first few days of each month. Ariel (1987)
found that stock returns tended to be higher, on average, in the first half of
each month. Updated results on the turn-of-the-month effect for the US market appear in the paper of Hensel, Sick and Ziemba in this volume. Extending
Ariel's study to the stock markets of Singapore, Malaysia, Hong Kong, Taiwan and Thailand, Wong (1995) reported that the US-type of intra-month
effect in stock returns was almost non-existent in these Asian stock markets.
The stock returns in these markets seemed to be generated by a process which
was fairly independent of other major markets. A further investigation of the
intra-month effect in these Asian markets from 1975 to 1989 by Wong and
Ho (1997) indicated that unique intra-month effects did exist in some of the
Asian stock markets but they were mostly period specific. Four (Singapore,
Hong Kong, Malaysia and Taiwan) of the five markets exhibited significant
intra-month effects only for the period 1975 to 1979. There was strong evidence of the intra-month effect for the Thai stock market during 1980 to 1984
and the Hong Kong market from 1985 to 1989. These unique intra-month
effects detected in each market were found to be unstable over time. The
intra-month effect established for each market in one period tended to disappear in the following period. The January effect and the timing of dividend
announcements were not helpful in explaining the intra-month effect.

Further investigation of the intra-month effect by Lakonishok and Smidt
(1988) showed that the positive returns from the first half of each month
tended to be concentrated in the four-day period from the last trading day
of a month to the third trading day of the following month in the US market
from 1897 to 1996 for the Dow Jones Industrials. See the update for the
S&P 500 by Hensel, Sick and Ziemba in this volume. As shown by Jaffe and
Westerfield (1989) and other researchers, the turn-of-the-month effect was
not peculiar to the US market as it also occurred in Australia.

Little research has been done on the turn-of-the-month effect in Asian stock
markets except for Hong Kong and Singapore. The average daily returns for
the eight-day window surrounding the turn of the month are reported in
Table 5. None of the daily returns were significantly different from zero.
Furthermore, Agrawal and Tandon (1994) found that the aggregate return

Table 5. Tests for turn-of-the-month effect in Asian stock markets

	-4	-3	-2	-1	+1	+2	+3	+4	Ave.
Hong Kong	-0.138	0.271	0.06	-0.078	-0.105	0.080	0.176	0.174	0.041
Singapore	0.066	0.092	0.079	0.131	0.021	0.080	0.108	0.066	0.043**

Note: The table showed the mean daily percentage returns of Hong Kong and Singapore from the fourth last day of the month (Day −4) to the fourth day of the next month (Day +4). The results were drawn from Aggrawal and Tandon (1994). Figures marked with ** (*) were means which were significantly different from zero at the 1% (5%) level based on a 2-tail test. The last column of the table showed the average daily return. Those marked with ** (*) indicated a rejection of the null hypothesis of zero mean at the 1% (5%) level.

over the (−1, +3) window period for Hong Kong was lower than all other four-day window periods. Based on the limited evidence of Hong Kong and Singapore markets, there seemed to be little evidence of the turn-of-the-month effect in Asian markets.

5 Day-of-the-Week Effect

Researchers such as Cross (1973), French (1980), Gibbons and Hess (1981), and Agrawal and Tandon (1994) also found seasonality in daily stock returns. This means that the probability distributions of daily stock returns vary significantly from day to day. This phenomenon has been termed the day-of-the-week effect. The empirical evidence for the United States showed that average return on Monday was negative while that on Friday was abnormally large and positive, particularly in the earlier sample periods.

To detect the presence of seasonality of daily returns, researchers utilize the following regression model :

$$R_t = a_1 D_{1t} + a_2 D_{2t} + a_3 D_{3t} + a_4 D_{4t} + a_5 D_{5t} + \varepsilon_t \qquad (5.1)$$

where R_t is the return on day t, D_{kt}, $k = 1, 2, \ldots, 5$ is a dummy variable which equals 1 if the day is k and 0 otherwise. D_1 represents Monday, D_2 represents Tuesday and so on. The coefficient a_i, $i = 1, 2, \ldots, 5$ is the mean return for day i.

Agrawal and Tandon (1994) found that of the 18 countries they considered, 13 countries had negative returns on Monday (7 are statistically significant), 12 countries had negative returns on Tuesday (8 are statistically significant) and most countries had large positive returns from Wednesday to Friday. Although the Monday and Tuesday returns in Hong Kong, Japan and Singapore stock markets were negative, they were not significantly different from zero.

Table 6. Tests for the day-of-the-week effect in Asian stock markets

	Mon.	Tues.	Wed.	Thurs.	Fri.	Sat.	F-test for equal returns
Hong Kong[1]	−0.088	−0.157**	0.173**	0.092	0.176**		2.57**
India[2]	−0.038	0.155*	0.017	0.126	0.274**		2.50*
Malaysia[2]	−0.103*	−0.073	0.111*	0.097*	0.159**		7.24**
Philippines[3]	−0.165	−0.137	0.030	0.163*	0.171*		3.92**
Singapore[2]	−0.038	−0.072*	0.096**	0.121**	0.111**		7.11**
South Korea[3]	0.015	−0.042	0.111**	0.019	0.102**	0.215**	5.63**
Taiwan[3]	0.041	0.098	0.147**	0.045	0.073		0.58
Thailand[2]	0.029	0.018	0.085	0.107*	0.290**		5.09**

Table 6 showed the mean daily percentage returns of Hong Kong, India, Malaysia, Philippines, Singapore, South Korea, Taiwan and Thailand from Monday to Saturday. Figures marked with ** (*) were means which were significantly different from zero at the 1% (5%) level based on a 2–tail test. The last column of the table showed the F-statistic for a test of equal means across all the days of the week. Those marked with ** (*) indicated a rejection of the null hypothesis of equal means at the 1% (5%) level.

1 The results for Hong Kong were based on Table 2 of Aggrawal and Tandon (1994) for the Hang Seng Index from February 1973 to June 1987.

2. The results for India, Malaysia, Singapore and Thailand were taken from Table 1 of Chan, M. W. L., Khanthavit, A., and Thomas, H., (1996). The indices and time periods considered were Bombay Stock Exchange Sensitive Index (India) from April 1979 to December 1992, KLSE Composite Index (Malaysia) from January 1974 to December 1992, Straits Times Industrial Index (Singapore) from January 1969 to December 1992, and SET valued weighted index (Thailand) from May 1975 to December 1991.

3 The results for Philippines, South Korea and Taiwan were based on Ho (1990). The indices and time periods considered were Manila Mining Index (Philippines) from January 1976 to November 1987, Korea Composite Stock Price Index (South Korea) from January 1975 to November 1987, and Taiwan Weighted Index (Taiwan) from January 1975 to November 1987.

The negative Tuesday returns in the Hong Kong and Japan stock markets were significantly different from zero.

The empirical evidence for the day-of-the-week effect in Asian stock markets is reported in Table 6. Based on the F-test (last column) which tests the null hypothesis of equal returns across all five days of the week, there was empirical support for seasonality of daily returns in most of the Asian stock markets. For most countries there is rejection of the null hypothesis at the 1% significance level.

Stock markets in Hong Kong, Malaysia, Philippines and Singapore had negative returns for Monday and Tuesday and positive returns for Wednesday to Friday. These results are consistent with those reported by Aggrawal and Tandon (1994) for the 18 countries which they examined.

The results for India, Malaysia and Philippines were similar to those reported for the US market in that Monday registered the lowest negative return of the week while Friday registered the highest positive returns. The results for Hong Kong and Singapore were more akin to those reported by Jaffe and Westerfield (1985) and Ziemba (1993) for the Japanese stock market in that both Monday and Tuesday had negative returns with Tuesday the lowest return for the week. The return on Tuesday in South Korea and Thailand was also the lowest throughout the week although all the daily average returns were positive for Thailand while for South Korea only Tuesday return was negative.

What is also interesting about the results in Table 6 is that, with the exception of Singapore, South Korea and Taiwan, the other Asian stock markets reported the highest returns on Friday which were positive and significantly different from zero. For Singapore and South Korea, while Friday's return may not be the highest, it was relatively large, positive and significantly different from zero. Based on the trading hours of the various exchanges shown in Table 2, most of the Asian stock markets are trading approximately 12 hours ahead of the New York Stock Exchange. It can be argued that as economies of Asia are highly dependent on the US economy and that the currencies of these countries also tend to be directly or indirectly pegged to the US dollar, the US stock market and the Asian stock markets should be closely related. Hence, there may be a spillover effect from the US stock market to the Asian stock markets. This means that negative Monday returns in the US stock market may lead to negative Tuesday returns in the Asian stock markets. The large negative Tuesday returns for Hong Kong, Malaysia, Singapore and South Korea seemed consistent with such an argument.

To analyze the linkages between the US stock market and the Asian emerging stock markets, we tabulated the correlation coefficients in Table 7. Most of the correlation coefficients were relatively low, with Malaysia registering the highest correlation coefficient of 0.2165 at lag 1. The stock returns in Hong Kong, Malaysia, Singapore and Taiwan had the highest correlation with those of US at lag 1. The results for lag 1 were consistent with the time zone difference explanation for the negative Tuesday return observed for the various Asian stock markets.

However, the correlation between the Monday return in US and the returns in the Asian stock markets extended beyond lag 1 to longer lags. In particular, the correlation coefficients of the return of US market and the returns of Hong Kong, Malaysia, Singapore, Taiwan and Thailand in lag 2 were positive and significantly different from zero. These results were inconsistent with the negative Monday returns in US and the generally positive

Table 7. Correlation coefficients of returns between the US market (DJIA) and
the Asian emerging stock markets

Country	Lag 0	Lag 1	Lag 2	Lag 3	Lag 4
Hong Kong[1]	0.0901**	0.1449**	0.0907**	0.0538	0.0390*
Malaysia[1]	0.0056	0.2165**	0.0666**	0.0622**	0.0428*
Philippines[2]	−0.0163	−0.0062	0.0034	−0.0225	−0.0084
Singapore[1]	0.0246	0.1015**	0.0569**	0.0471**	0.0260
South Korea[2]	−0.0224	−0.0143	0.0173	−0.0085	0.0527**
Taiwan[1]	0.0535**	0.1120**	0.0685**	−0.0018	0.0285
Thailand[1]	−0.0085	0.0541**	0.0716**	0.0894**	0.0326

Table 7 showed the correlation coefficients of returns between US market
(DJIA) and the 7 Asian emerging stock markets of Hong Kong, Malaysia,
Philippines, Singapore, South Korea, Taiwan and Thailand. The figures for
Lag 0 (1, 2, 3, 4) were the correlation coefficients of the Monday return of
US market and the Monday (Tuesday, Wednesday, Thursday, Friday) return
of the Asian emerging stock markets. Figures marked with ** (*) were
correlation coefficients which were significantly different from zero at the 1%
(5%) level based on a 2-tail test.

1. The results for Hong Kong, Malaysia, Singapore, Taiwan and Thailand
were taken from Table 6 of Wong, K. A., T. K. Hui, and C. Y. Chan (1992).
The indices and time periods considered were the Hang Seng Index (Hong
Kong), KLSE Composite Index (Malaysia), the SES-All Share Index
(Singapore), the Taiwan Weighted Index (Taiwan) from January 1975 to
May 1988 and the SET valued weighted index (Thailand) from May 1975 to
May 1988.

2. The results for Philippines and South Korea were based on Ho (1990).
The indices and time periods considered were Manila Mining Index
(Philippines) from January 1976 to November 1987 and the Korea Composite
Stock Price Index (South Korea) from January 1975 to November 1987.

Wednesday returns in these 5 Asian stock markets. Furthermore, if the lag
1 correlation between US stock market and Asian stock markets was high,
we should also observe large Monday returns in Asian stock markets as the
Friday return in US was large and positive. However, the results in Table
6 show that Asian stock markets tended to have negative Monday returns.
These results showed that there was insufficient empirical support for the time
zone difference explanation for the day-of-the-week effect in Asian emerging
stock markets.

6   Holiday effect

The holiday effect refers to the phenomenon that average daily return on stocks is higher on the day before a holiday than those on the remaining days of the year. Ariel (1990) found that, for the period 1963 to 1986, approximately one-third of the total market return accrued on trading days preceding the eight holidays in the United States. Similar findings for the Christmas holiday and the New Year holiday in the US market were reported by Lakonishok and Smidt (1984), Keim (1983) and Roll (1983).

Studies by Cadsby and Ratner (1992) and Kim and Park (1994) showed that the holiday effect also occurred in Australia, Canada, Hong Kong, Japan and United Kingdom. The empirical evidence of the holiday effect in the Asian stock markets is reported in Table 8. The results showed that the pre-holiday mean daily returns were larger than the mean daily returns for the rest of the year in all the five Asian stock markets considered. The largest (smallest) difference of 0.55% (0.02%) in the mean daily returns between the pre-holidays and the other days of the year occurred in Taiwan (South Korea). The average difference in the mean daily returns between pre-holidays and the other days of the year of all 5 Asian countries was 0.32%.

To test for the holiday effect, the return on each country's stock index is regressed on a dummy variable Dt as follows:

$$R_t = a_0 + a_1 Dt + \varepsilon_t \qquad (6.1)$$

where D_t equals 1 if day t is the day before a holiday and 0 otherwise. The average return for all days excluding those preceding a holiday is given by the intercept a_0. The coefficient a_1 measures the difference in return between the pre-holiday days and all the other days. The null hypothesis is that there is no holiday effect or $a_1 = 0$. The t-statistic for the null hypothesis is given in column 4 of Table 8.

Except for Hong Kong and South Korea, the results generally supported the existence of a holiday effect. The holiday effect was particularly strong in Malaysia (pre-holiday average return of 0.39% versus the rest of the year return of 0.04%) and Taiwan (pre-holiday average return of 0.76% versus the rest of the year return of 0.11%).

Chan *et al.* (1996) found that the holiday effect occurred only on cultural holidays rather than on state holidays in three (India, Malaysia and Singapore) of the four countries they studied. Cultural holidays are those linked to religious celebrations or ethnic celebrations whereas state holidays are secular holidays. The cultural holidays include Chinese New Year, Islamic New Year, Deepavali, Aidilfitri, other Islamic holidays, Vesak Day, Christmas, Holi, Hindu holidays and Buddhist holidays. State holidays include Western New Year and holidays other than those mentioned earlier. In a study of the

Table 8. Means, standard deviations, and t-test statistic for holiday effect in
Asian stock markets

Country	Mean(Std Dev) of daily returns (%)		T-statistic of return difference
	Pre-holidays	Other days	
Hong Kong	0.42 (1.71)	0.07 (1.84)	1.83
Malaysia	0.39 (1.05)	0.04 (1.45)	2.29*
Singapore	0.29 (0.81)	0.05 (1.24)	1.96*
South Korea	0.09 (1.26)	0.07 (1.33)	0.19
Taiwan	0.76 (2.36)	0.11 (2.39)	2.33*

Source: Cervera and Keim (1999). Those marked with * indicated a
rejection of the null hypothesis of no difference in mean returns at the
1% significance level.

Lunar New Year holiday, Ho (1990) showed that prior to 1983, Hong Kong,
Taiwan and Malaysia exhibited a 'reverse' turn-of-the-lunar-year effect in that
the mean return for the first nine trading days of the lunar year was less than
that of the last nine trading days of the previous lunar year. However, for
the post-1983 period, the 'reverse' turn-of-the-lunar-year effect disappeared
and instead the turn-of-the-lunar-year effect was found in the Malaysia and
Singapore stock markets while there was no such effect in Hong Kong and
Taiwan stock markets. Lunar New Year is a major holiday for the Chinese
all over the world as they tend to celebrate the lunar new year rather than
the Gregorian new year due to cultural reason. These results showed that the
holiday effect may be period specific in Asian stock markets and more indepth
analysis should be carried out to determine if the anomaly is persistent over
time.

The empirical evidence indicates that the holiday effect is not an isolated
phenomenon in the United States. It seems to occur in stock markets of
developed as well as developing countries. What causes the holiday effect
to persist in various countries despite the widely differing dates of holidays
across the countries? Some researchers postulate that investors tend to close
out their short positions before a holiday so that they are not exposed to
adverse news when the market is closed. This explanation may make sense in
countries where short selling is legal. However, in countries such as Malaysia
and Singapore where short selling is forbidden by the authorities, the presence
of strong holiday effect negate such a hypothesis.

Another explanation for the holiday effect is the presence of a common risk
factor for all capital markets. If this hypothesis is true then the holiday effect

should be stronger for holidays which are common among countries compared to those non-common holidays. Common holidays in most countries include New Year's day, Christmas and Easter. Cervera and Keim (1999) tested this hypothesis and they found no evidence in support the hypothesis of common risk factor.

7 Firm-Size Effect

One of the earliest anomalies uncovered in the US stock market is the firm-size effect. The size effect refers to the phenomenon that small firms tend to earn higher average returns than large firms. Banz (1981) was among the first to document this anomalous deviation from the prescription of the CAPM. He postulated that the expected return on any security was determined by its beta risk (β_i) as well as its size (S_i) as measured by the market capitalization of its common equity. He estimated the following model for the period 1963 to 1975 and found that c_2 was statistically negative:

$$R_i = c_0 + c_1\beta_i + c_2S_i + e_i. \tag{7.1}$$

His finding was confirmed by Reinganum (1981, 1983) who found that portfolios of small firms had substantially higher average returns than portfolios of larger firms.

Other researchers found that the size effect was not confined to the US stock market. It also occurred in the Belgium, Irish, Japanese, Mexican, Spanish, Swiss and British stock markets as well. See Hawawini and Keim (1999) for details.

The empirical evidence for three Asian stock markets (Singapore, South Korea and Taiwan) are presented in Table 9. The size premium in column two is the difference between the average monthly return of the portfolio of stocks with the smallest market capitalization and that of the portfolio of stocks with the largest market capitalization. The size premium was positive in Singapore (0.42%) and Taiwan (0.57%) but negative in South Korea (–0.4%). Hence, the empirical evidence for size-effect in Asian stock markets was mixed with only Singapore and Taiwan stock markets exhibiting this anomaly.

To test for the presence of size effect in Singapore, Wong and Lye (1990) estimated the average risk-adjusted excess return (δ_p) of size-sorted portfolios using the regression model:

$$r_{p,t} - r_{f,t} = \delta_p + \beta_p(r_{m,t} - r_{f,t}) + e_{p,t} \tag{7.2}$$

where $r_{p,t}$ is the return on size-sorted portfolio p, $r_{f,t}$ is the return on the risk free asset, $r_{m,t}$ is the return on the market portfolio in month t and β_p is a measure of the systematic risk of portfolio p.

Table 9. The monthly size premium of Singapore, South Korea and Taiwan stock
markets

Country	Size premium (%)	Sample period	Number of portfolios
Singapore	0.42	1975–85	3
South Korea	-0.40	1984–88	10
Taiwan	0.57	1979–86	5

Source: Wong & Lye (1990) for the Singapore stock market, Kim *et al.*
(1992) for the South Korea stock market and Ma & Shaw (1990) for
the Taiwan stock market.

The average return on the smallest size portfolio MV1 (MV1*) was 1.569%
(1.487%) while the largest size portfolio MV3 (MV3*) was 1.149% (1.241%)
per month. The differential return between MV1* and MV3* of 0.246% was
smaller than the differential return between MV1 and MV3 of 0.42% per
month after controlling for the earnings-yield effect.

The null hypothesis of no size-effect was tested using the univariate t-test
(δ_p is equal to zero) and the multivariate Hotelling's T^2 test (vector of δ_p is
equal to zero). These tests were carried out for both size-sorted portfolios in
panel A and randomized size-sorted portfolios (marked in asterisks) in panel
B. The size-sorted portfolios were randomized to control for the confounding
effects of earnings yield. The results of the univariate t-test and the multivariate Hotelling T^2-test in Table 10 indicated that the null hypothesis of
no size effect was rejected for both the size-sorted portfolios and randomized
size-sorted portfolios at the 1% significance level.

8   Earnings-Yield Effect

The earnings-yield effect refers to the phenomenon that high earnings-yield
stocks tend to earn higher average returns compared to low earnings-yield
stocks. The earnings yield of a stock is defined as the ratio of the stock's
earnings per share to its market price per share. The high earnings-yield
strategy of picking stocks dates back to Graham and Dodd (1940) who recommended high earnings-yield as an important criterion for superior stock
selection. Although well-known by many analysts who subscribed to the fundamental analysis approach of stock investment, the earnings-yield effect was
given prominence by Basu (1977) recently who highlighted it as an anomaly
of CAPM. Basu (1977) documented a positive relationship between earnings-

Table 10. The size effect: Evidence from the Singapore stock market

Size-sorted portfolios	δ_p	$(t\delta_p)$	$(w\delta_p)$	$F(\delta)$
MV1	0.01569	3.24	1.5791	
MV2	0.01591	3.73	1.7948	
MV3	0.01149	2.81	1.8737	6.02
MV1*	0.01487	3.30	1.6962	
MV2*	0.01781	4.33	1.8571	
MV3*	0.01241	2.91	1.7955	10.22

The results were sourced from Wong and Lye (1990). MV1 to MV3 were size-sorted portfolios with MV1 being the portfolio of stocks with the smallest market capitalization. MV1* to MV3* were also size-sorted portfolios similar to MV1 to MV3 except that each portfolio was randomized to remove the confounding earnings-yield effect. δ_p was the estimated intercept of regression of excess portfolio return on excess market returns. It measured the average excess return after having adjusted for beta risk of the portfolio. $(t\delta_p)$ was the t-statistic for a test of $\delta_p = 0$, $(w\delta_p)$ was the vector of normalized weights associated with the Hotelling T^2 test for $\delta = 0$ and $F(\delta)$ was the F-value corresponding to the Hotelling T^2 statistic that the vector of $\delta s = 0$.

yield and average risk-adjusted returns of stocks. His findings were supported by Reinganum (1981) who considered a larger sample involving a different time period.

Empirical evidence concerning the earnings-yield effect for three Asian countries is presented in Table 11. The earnings-yield premium in column two is the difference between the average monthly return of the portfolio of stocks with the highest earnings yield and that of the portfolio of stocks with the lowest earnings yield. The earnings-yield premium was positive in Singapore (0.27%), Taiwan (0.85%) and South Korea (1.20%) for the earlier sub-period of 1980–83. However, the earnings-yield premium in South Korea for a later period of 1984–88 was zero. Hence, the empirical evidence for the earnings-yield effect in Asian stock markets was mixed with only Singapore and Taiwan stock markets exhibiting this anomaly.

In analyzing the earnings-yield effect in Singapore, Wong and Lye (1990) estimated the average risk-adjusted excess return (δ_p) of earnings-yield-sorted portfolios using the regression model in equation (7.2). The average return on the smallest earnings-yield portfolio EY1 (EY1*) was 1.164% (1.026%)

Table 11. The earnings-yield effect: Evidence from Asian stock markets

Country	Monthly earnings-yield premium (%)	Sample period	Number of portfolios
Singapore	0.27	1975–85	3
South Korea	1.20	1980–83	10
	0.00	1984–88	10
Taiwan	0.85	1979–86	5

The results were sourced from Wong and Lye (1990) for the Singapore market, Kim *et al.* (1992) for the South Korea stock market and Ma and Shaw (1990) for the Taiwan stock market. Chou and Johnson (1990), however, found the earnings-yield effect to be stronger than Ma and Shaw (1990). Using a sample of 5 portfolios from 1973 to 1988, they estimated the average risk-adjusted P/E premium to be 1.88%.

while the largest earnings-yield portfolio EY3 (EY3*) was 1.437% (1.951%) per month. The differential return between EY3* and EY1* of 0.925% was larger than the differential return between EY3 and EY1 of 0.273% per month after controlling for the size effect.

The null hypothesis of no earnings-yield effect was tested using the univariate t-test (δ_p is equal to zero) and the multivariate Hotelling's T^2 test (vector of δ_p is equal to zero). These tests were carried out for both the earnings-yield-sorted portfolios and randomized earnings-yield-sorted portfolios (marked in asterisks). The randomization of the earnings-yield-sorted portfolios was carried out to control for the confounding size-effect. The results of the univariate t-test and the multivariate Hotelling T^2-test in Table 12 showed that the null hypothesis of no earnings-yield effect was rejected at the 1% significance level.

9   Conclusion

Since 1980, researchers had uncovered several regular and persistent deviations of the static capital asset pricing model (CAPM). These anomalies were well documented for the US and European stock markets. This paper surveys the occurence of these anomalies in the Asian emerging stock markets. The month-of-the-year effect was not a common phenomenon in the Asian stock markets as only Malaysia and Singapore exhibited this anomaly. There was no evidence of the month-of-the-year effect in Hong Kong, India, Philippines, South Korea, Taiwan and Thailand. The unique institutional structures of the Asian stock markets demonstrated that the tax loss hypothesis cannot

Table 12. The earnings-yield effect: Evidence from the Singapore stock market

Earnings-yield-sorted	δ_p portfolios	$(t\delta_p)$	$(w\delta_p)$	$F(\delta)$
EY1	0.01164	2.40	1.5758	
EY2	0.01502	3.67	1.8699	
EY3	0.01437	2.46	1.3070	7.73
EY1*	0.01026	2.28	1.7025	
EY2*	0.01328	3.24	1.8681	
EY3*	0.01951	4.66	1.8256	16.48

The results were sourced from Wong and Lye (1990). EY1 to EY3 were earnings-yield-sorted portfolios with EY1 being the portfolio of stocks with the lowest earnings yield. EY1* to EY3* were also earnings-yield-sorted portfolios similar to EY1 to EY3 except that each portfolio was randomized to remove the confounding size effect. δ_p was the estimated intercept of regression of excess portfolio return on excess market returns. It measured the average excess return after having adjusted for beta risk of the portfolio. $(t\delta_p)$ was the t-statistic for a test of $\delta_p = 0$, $(w\delta_p)$ was the vector of normalized weights associated with the Hotelling T^2 test for $\delta = 0$ and $F(\delta)$ was the F-value corresponding to the Hotelling T^2 statistic that the vector of $\delta s = 0$.

explain the anomalies found in the Malaysia and Singapore stock markets as both countries did not impose taxes on capital gains.

Very little research had been done on the turn-of-the-year effect in Asian stock markets. Based on the limited research done on the Hong Kong and Singapore markets, there was no empirical support for this anomaly.

Most Asian bourses exhibited the day-of-the-week effect. However, the argument for a spillover effect from the US market to the Asian markets due to a difference in time-zone was not substantiated by the evidence provided in this paper.

The holiday effect was observed in the Malaysian, Singaporean and Taiwanese stock markets but not for the Hong Kong and South Korean stock markets. The argument that investors tended to close out their short positions before a holiday cannot explain the holiday effect observed in Malaysia and Singapore because both countries forbid short-selling of shares. Malaysia allowed the short-selling of shares on a small number of companies with effect from the middle of 1997 which was outside the sample period considered in the paper.

Based on the limited empirical results of three Asian countries, the firm-size and earnings-yield effects had been detected in the Singapore and Taiwan stock markets but not in the South Korean stock market.

References

Agrawal, A. and K. Tandon (1994) 'Anomalies or Illusions? Evidence from Stock Markets in Eighteen Countries', *Journal of International Money and Finance* **14** 83–106.

Ariel, R. (1987) 'A Monthly Effect in Stock Returns', *Journal of Financial Economics* **18** 161–174.

Ariel, R. (1990) 'High Stock Returns before Holidays: Existence and Evidence on Possible Causes', *Journal of Finance* **45** 1611–1626.

Banz, R.W. (1981) 'The Relationship between Return and Market Value of Common Stock', *Journal of Financial Economics* **9** 3–18.

Basu, S. (1977) 'Investment Performance of Common Stocks in Relation to their Price-earnings Ratios: A Test of the Efficient Market Hypothesis', *Journal of Finance* **32** 663–682.

Basu, S. (1983) 'The Relationship between Earnings' Yield, Market Value and Return for NYSE Common Stocks: Further Evidence', *Journal of Financial Economics* **12** 129–156.

Berges, A., J. McConnell, and G. Schlarbaum (1984) 'The-Turn-of-the-year in Canada', *Journal of Finance* **39** 185–92.

Black, F., M. Jensen, and M. Scholes (1972) 'The Capital Asset Pricing Model: Some Empirical Tests'. In *Studies in the Theory of Capital Markets*, M. Jensen (ed.), Praeger.

Blume, M., and I. Friend (1973) 'A New Look at the Capital Asset Pricing Model', *Journal of Finance* **28** 19–33.

Bowers, J. and E. Dimson (1988) 'Introduction'. In *Stock Market Anomalies*, E. Dimson (ed.), Cambridge University Press.

Cadsby, C.B. and M. Ratner (1992) 'Turn-of-month and Pre-holiday Effects on Stock returns: Some International Evidence', *Journal of Banking and Finance* **16** 497–509.

Cervera, A. and D.B. Keim (1999) 'High Stock Returns before Holidays: International evidence and additional tests', this volume, 512–531.

Chan, M.W.L., A. Khanthavit, and H. Thomas (1996) 'Seasonality and Cultural Influences on Four Asian Stock Markets', *Asia Pacific Journal of Management,* **13** 1–24.

Constantinides, G. (1984) 'Optimal Stock Trading with Personal Taxes: Implications for Prices and the Abnormal January Returns', *Journal of Financial Economics* **13** 65–89.

Cross, F. (1973) 'The Behaviour of Stock Prices on Fridays and Mondays', *Financial Analysts Journal* **29** 67–9.

Fama, E. (1991) 'Efficient Capital Market: II', *Journal of Finance* **46** 1575–1617.

Fama, E., and J. MacBeth (1973) 'Risk, Return and Equilibrium: Empirical Tests', *Journal of Political Economy* **71** 607–636.

French, K.R., 1980 'Stock Returns and the Weekend Effect', *Journal of Financial Economics* **8** 55–70.

Gibbons, M.R. and P. Hess (1981) 'Day of the Week Effects and Asset Returns', *Journal of Business* **54** 579–596.

Graham, B., and D. Dodd (1940) *Security Analysis: Principles and Technique*, McGraw-Hill.

Gultekin, M.N. and N.B. Gultekin (1983) 'Stock Market Seasonality: International Evidence', *Journal of Financial Economics* **12** 469–81.

Haugen, R.A. and J. Lakonishok (1988) *The Incredible January Effect*, Dow-Jones-Irwin, Homewood, IL.

Hawawini, G. and D.B. Keim (1999) 'The Cross Section of Common Stock Returns: A Review of The Evidence and Some New Findings', this volume, 3–43.

Hensel, C.R., G.A. Sick and W.T. Ziemba (1999) 'A Long Term Examination of the Turn-of-the-month Effect in the S&P 500', this volume, 218–246.

Ho, Y.K. (1990) 'Stock Return Seasonalities in Asia Pacific Markets', *Journal of International Financial Management and Accounting* **2** 47–77.

Jaffe, J. and R. Westerfield (1985) 'Patterns in Japanese Common Stock Returns: Day-of-the-Week and Turn-of-the-year Effects', *Journal of Financial and Quantitative Analysis* **20** 261–72.

Jaffe, J. and R. Westerfield (1985) 'The Week-end Effect in Common Stock Returns: The International Evidence', *Journal of Finance* **40** 433–454.

Jaffe, J., D.B. Keim and R. Westerfield (1989) 'Earnings Yields, Market Values, and Stock Returns', *Journal of Finance* **44** 135–148.

Jagadeesh, N. (1991) 'Seasonality in Stock Market Price Reversion: Evidence from the US and the UK', *The Journal of Finance* **46** 1427–44.

Kato, K. and J.S. Schallheim (1985) 'Seasonal and Size Anomalies in the Japanese Stock Market', *Journal of Financial and Quantitative Analysis* **20** 243–60.

Keim, D.B. (1983) 'Size-related Anomalies and Stock Return Seasonality: Further Empirical Evidence', *Journal of Financial Economics* **12** 13–32.

Keim, D.B. (1989) 'Trading Patterns, Bid-ask Spreads, and Estimated Security Returns: The Case of Common Stock Returns at the Turn of the Year', *Journal of Financial Economics* **25** 75–98.

Keim, D.B. and R.F. Stambaugh (1984) 'A Further Investigation of the Weekend Effect in Stock Returns', *Journal of Finance* **39** 819–835.

Kim, S.W. (1988) 'Capitalizing on the Weekend Effect', *Journal of Portfolio Management* **14** (3), 59–63.

Kim, C. and J. Park (1994) 'Holiday Effects and Stock Returns: Further Evidence', *Journal of Financial and Quantitative Analysis* **29** 145–157.

Lee, I. (1992) 'Stock Market Seasonality: Some Evidence from the Pacific-Basin Countries **19** 199–210.

Lintner, J. (1965) 'The Valuation of Risk Assets and the Selection of Risky Investment in Stock Portfolios and Capital Budgets', *Review of Economics and Statistics* **47** 13–37.

Mossin, J. (1966) 'Equilibrium in a Capital Asset Market', *Econometrica* **34** 768–783.

Reinganum, M.R. (1981) 'Misspecification of Capital Asset Pricing: Empirical Anomalies Based on Earnings Yields and Market Values', *Journal of Financial Economics* **9** 19–46.

Reinganum, M.R. (1983) 'The Anomalous Stock Market Behaviour of Small Firms in January: Empirical Tests for Tax-loss Selling Effects', *Journal of Financial Economics* **12** 89–104.

Rhee, S.G., R.P. Chang, and R. Ageloff (1990) 'An Overview of Equity Markets in Pacific-Basin Countries'. In *Pacific Basin Capital Markets Research Vol. II*, S. G. Rhee and R. P. Chang (eds.), North Holland.

Rhee, S.G. and R.P. Chang (1992) 'The Microstructure of Asian Equity Markets', *Journal of Financial Services Research*, 437–454.

Roll, R. (1983) 'Vas ist das? The Turn-of-the-Year Effect and the Return Premia of Small Firms', *Journal of Portfolio Management* **9** 18–28.

Rozeff, M., and W. Kinney (1976) 'Capital Market Seasonality: The Case of Stock Returns', *Journal of Financial Economics* **3** 379–402.

Schultz, P. (1983) 'Transaction Costs and the Small Firm Effect: A Comment', *Journal of Financial Economics* **12** 81–88.

Sharpe, W. (1964) 'Capital Asset Prices: A Theory of Market Equilibrium under Conditions of Risk', *Journal of Finance* **19** 425–442.

Tong, W.H.H. (1992) 'An Analysis of the January Effect of the United States, Taiwan and South Korean Stock Returns', *Asia Pacific Journal of Management* **9** 189–207.

Treynor, J. (1961) 'Toward a Theory of Market Value of Risky Assets', unpublished manuscript.

Wood, B.G. (1994) 'Seasonalities and the 1987 Crash: The International Evidence', *International Review of Financial Analysis* **3** 65–91.

Wong, K.A. (1995) 'Is There an Intra-month Effect on Stock Returns in Developing Stock Markets?', *Applied Financial Economics* **5** 285–289.

Wong, K.A. and P.C. Ho (1997) 'An Intra-month Effect in Stock Returns: Evidence from Developing Stock Markets', *Applied Financial Economics*, forthcoming.

Wong, K.A., T.K. Hui, and C.Y. Chan (1992) 'Day-of-the-week Effects: Evidence from Developing Stock Markets', *Applied Financial Economics* **2** 49–56.

Wong, K.A., and M.S. Lye (1990) 'Market Values, Earnings' Yields and Stock Returns: Evidence from Singapore', *Journal of Banking and Finance* **14** 311–326.

Wong, P.L., S.K. Neoh, K.H. Lee, and T.S. Thong (1990) 'Seasonality in the Malaysian Stock Market', *Asia Pacific Journal of Management* **7** 43–62.

Ziemba, W.T. (1989) 'Seasonality Effects in Japanese Futures Markets'. In *Research on Pacific Basin Security Markets Research*, S.G. Rhee and R.P. Chang (eds.), North Holland, 379–407.

Ziemba, W.T. (1991) 'Japanese Security Market Regularities: Monthly, Turn-of-the-month and Year, Holiday and Golden Week Effects', *Japan and World Economy* **3** 119–146.

Ziemba, W.T. (1993) 'Comment on "Why a Weekend Effect?" ', *Journal of Portfolio Management*, Winter, 93–99.

Ziemba, W.T. (1994) 'World Wide Security Market Regularities', *European Journal of Operational Research* **74** 198–229.

Ziemba, W.T., W. Bailey, and Y. Hamao (eds.) (1991) *Japanese Financial Market Research*, North Holland.

Japanese Security Market Regularities, 1990–1994*

Luis R. Comolli and William T. Ziemba

Abstract

This paper presents evidence on seasonal regularities on the Tokyo Stock Exchange during 1990–1994. Despite the bear market during this period, with a decline of over 50% in the Nikkei stock average, many of the effects found by Ziemba (1991) for 1949–1988 occurred with regularity. These include all gains are at night, the turn-of-the-month was positive, the January barometer was accurate and there were high returns during Golden Week. The day of the week effect changed from the earlier period largely due to the end of Saturday trading. Pre-holidays had positive returns but did not improve the day of the week returns. Volatility increased as prices declined.

Japanese security market regularities

Research on Japanese anomalies is recent. The thrust has been mainly to ascertain the similarities and differences with the analogous results in the US markets. Early work on seasonal and size anomalies appears in Jaffe and Westerfield (1985) and Kato and Schalheim (1985). Kato, Schwartz and Ziemba (1989) have surveyed the research and presented new results on day-of-the-week effects in Japanese security markets. Other aspects of Japanese security markets are discussed in Elton and Gruber (1989), Ziemba and Schwartz (1991) and Ziemba, Bailey and Hamao (1991). Ziemba (1991) provides a comprehensive analysis on Japanese security market regularities for the period 1949–1988. This paper updates Ziemba's study and investigates whether market anomalies held during the bear market of 1990–1994. We will also try to analyze possible effects of the 1990–1992 market correction, the suppression of Saturday trading, and some changes in Japanese holidays. The basic efficient markets working hypothesis is that all days are equivalent and have the same mean returns. The study by Ziemba (1991) provides the historical background for this work as well as a number of hypotheses that attempt to explain why such departures occur for the various effects. The

* This research was partially supported by the Centre for International Business Studies at the University of British Columbia, and the Social Sciences and Humanities Research Council of Canada. This paper is an update of Ziemba (1991) which was based on research originally conducted at the Yamaichi Research Institute, Tokyo. Thanks are due to D. Keim, S. Satchell and S. Schwartz for helpful comments on an earlier draft of this paper.

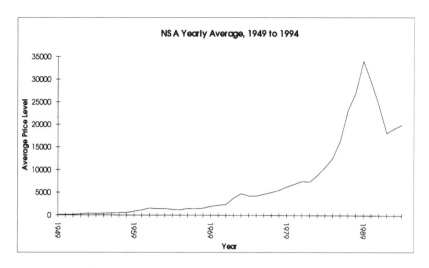

Figure 1. The NSA yearly averages, 1949–1994

answers seem to be a combination of cash flows, institutional and cultural factors and differences in risk. We attempt to check the validity of some of these hypothesis. This is possible for cases where specific institutional differences between the periods 1949–1988 and 1990–1994 have arisen.

1 The market correction of 1990–1992

One of the difficulties in comparing the recent market behavior with that observed during 1949–1988 is that the Japanese markets went through a serious correction during 1990–1992 when they declined almost 60% from their December 1989 historical peak. High interest rates in relation to earnings are at the root of the reason for the large decline; see Stone and Ziemba (1993) and Ziemba and Schwartz (1991) for analyses. The NSA is a price-weighted index of 225 stocks and the Topix is a market cap weighted index of all (1100+ stocks on the first section of the Tokyo stock exchange; see Ziemba and Schwartz (1991) for more details. Figures 1 and 2 show the NSA yearly average price level and the TOPIX end-year value, from 1949 to 1994.

Figure 3a shows the daily means and standard deviations for the NSA by decade during 1949–1988 and for the five-year period 1990–1994. Figure 3b shows that the mean return was strongly positive and the daily standard deviations below 1% for the 40-year rising market period. During the declining market period the standard deviation rose to 1.5%. This is consistent with US studies that indicate an inverse relationship between mean return and volatility, see e.g. Black (1976). Year by year results appear in Figure 4a,b.

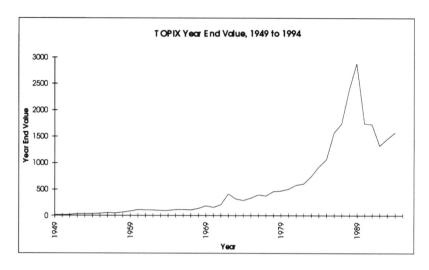

Figure 2. TOPIX, year end values, 1949–1994

2 The monthly effect on the NSA and TOPIX market indices, 1949–1994

Tables 1 and 2 give the monthly returns on the NSA from 1949 to 1988 and 1990 to 1994, respectively. Table 3 presents the TOPIX mean returns for the period 1990–1994. The mean return on the NSA was significantly positive during 1949–1988. All months had positive mean returns during this period. January had the highest returns, about four times the average. August and June had the second and third highest returns respectively, and May, July, September and October had very low returns. In contrast with these historical results, during the period 1990-1994 both the NSA and TOPIX had negative mean returns. January returns were indistinguishable from zero. May, October and December had the highest returns, in that order, but only May returns were significantly positive at a level of 10% for the NSA and 5% for the TOPIX. November, June and September had the most negative returns in that order, all at a significance level of 10% for the NSA and 5% or better for the TOPIX. The most striking differences between the historical period 1949–1988 and 1990–1994 were:

- Historically positive mean returns versus negative returns during 1990–1994.

- Strongly positive January returns versus January returns indistinguishable from zero.

(a) Daily mean returns

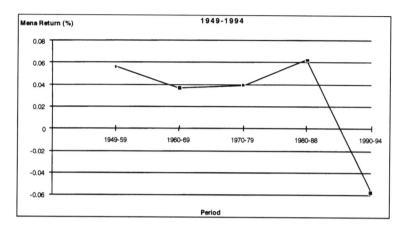

(b) Daily standard deviations of returns

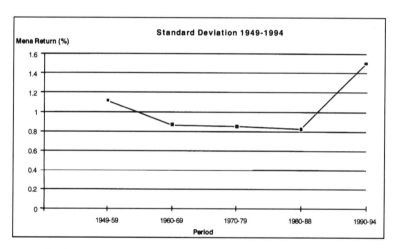

Figure 3. Daily means and standard deviations for the NSA for the period
1949–1988 by decade, and for the five-year period 1990–1994

- Historically strongly positive returns in June, August and March, versus negative returns during each of these months during 1990-1994.

- Historically, September and October returns have been indistinguishable from zero versus strongly negative September and strongly positive October returns during 1990-1994.

(a) Daily mean returns

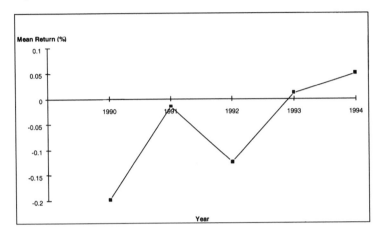

(b) Daily standard deviations of returns

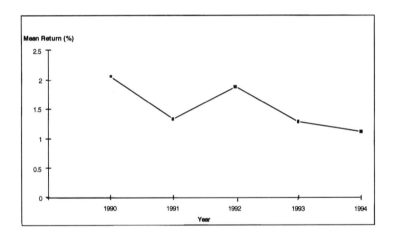

Figure 4. Daily means and standard deviations for the NSA for years
1990–1994

- Historically, positive November returns versus strongly negative November returns for 1990-1994.

The NSA and TOPIX had similar patterns of monthly mean returns during 1990–1994, but the TOPIX had less extreme variations and more reliability. Figure 5a–d shows the mean returns per month of the year for each year, from 1990 to 1994. The pattern drastically changes from 1990 to 1994. The period 1990–1992 had a strong market correction, while 1993 was a transition period,

Table 1. Mean daily returns, standard deviations, and percent positive
returns by month for the NSA during 1949–1988.

Month	Sample size	Mean return (%)	Standard deviation	t-statistic	P-value	% Positive returns
All Year	11529	0.048	0.093	5.56	0.0001	54.3
January	885	0.182	0.093	5.80	0.0001	60.7
February	905	0.055	0.089	1.87	0.0622	56.7
March	986	0.046	0.096	1.50	0.1335	55.0
April	947	0.062	0.091	2.10	0.0359	54.2
May	956	0.007	0.085	0.27	0.7843	53.1
June	1011	0.064	0.085	2.39	0.0169	56.8
July	1043	0.008	0.101	0.27	0.7904	52.3
August	1038	0.079	0.086	2.96	0.0032	55.2
September	931	0.006	0.082	−0.22	0.8252	51.6
October	998	0.009	0.115	0.24	0.8089	50.9
November	910	0.037	0.088	1.27	0.2031	52.4
December	919	0.047	0.104	1.37	0.1706	53.8

Source: Ziemba (1991)

and 1994 had a pattern similar to the historical one. Many differences with the historical patterns remain after the correction; the main ones being a very positive May, and negative June, March and November. Figure 6 compares the annualized daily standard deviations for the periods 1949–1988 and 1990–1988. The change of the market from a very negative 1990–1992 to a positive 1994 and the high volatility during the correction period result in volatility for 1990–1994 many times higher than during 1949–1988. Figure 7 compares nighttime (close to open) mean returns with daytime (open to close) mean returns for the period 1990–1994. As during previous decades, nighttime returns are highly positive. Only June and November had negative nighttime returns. It is the more negative daytime mean returns that determine the daily returns. Figure 8 compares January returns with the rest-of-the-year returns for the years 1990 to 1994. The January barometer, see Hirsch (1986) and Hensel and Ziemba (1995) has worked for each year, predicting the sign of the subsequent year mean returns.

There are differences between the small, medium and large components of the TOPIX index. For May and October, which are strongly positive, the small cap index has a higher percentage of positive returns than the large cap index. For the strongly negative November, the small index has a higher number of negative trading days. The small cap index responded more to extreme market conditions.

Table 2. Mean daily return, standard deviation, and percent positive
returns by month of the NSA, for the period 1990–1994.

Month	Sample size	Mean return (%)	Standard deviation	t–statistic	P–value	% Positive returns
All Year	1243	−0.058	1.574	−1.30	0.1900	47.1
January	95	0.052	1.788	0.28	0.7800	56.8
February	95	0.006	1.328	0.05	0.9600	53.7
March	107	−0.186	1.462	−1.31	0.1900	46.7
April	103	0.022	1.906	0.12	0.9100	46.6
May	98	0.202	1.127	1.78	0.0790	55.1
June	106	−0.323	1.250	−2.66	0.0090	40.6
July	111	0.030	1.381	0.23	0.8200	51.4
August	111	−0.079	2.047	−0.41	0.6800	46.9
September	98	−0.286	1.538	−1.84	0.0690	41.8
October	106	0.189	1.732	1.12	0.2600	51.9
November	99	−0.402	1.448	−2.76	0.0070	33.3
December	102	0.120	1.589	0.76	0.4500	53.9

Based on interviews with experienced investment professionals Ziemba
(1991) found out that the most important reasons behind the historical pattern of returns were:

- Large semi-annual bonuses paid by most Japanese companies in December and June. These bonuses were generally paid early in the month.

- Corporate officers make their earnings forecasts in May and financial
  analysts make theirs in March, June, September and December.

- There are typically large mutual fund investments in December, January
  and June.

- Individuals are net sellers, more so in December than in any other month
  of the year.

It would be worthwhile to study how and when these actions have been
taken during 1993 and 1994, and compare them with the actual market behavior presented here. June stayed negative in 1994, after the market correction
was over. What happened with the bonuses, earning forecasts, and Mutual
Fund investments in June 1994? May stayed strongly positive during 1994;
what were the earning forecasts at the time? Perhaps the market correction

Table 3. Mean daily return, standard deviation, and percent positive returns by month for the TOPIX, 1990–1994.

Month	Sample size	Mean return (%)	Standard deviation	t–statistic	P–value	% Positive returns
All Year	1232	−0.050	1.339	−1.31	0.1900	47.0
January	96	0.003	1.441	0.02	0.9900	45.8
February	95	0.014	1.137	0.12	0.9000	53.7
March	107	−0.151	1.321	−1.18	0.2400	43.9
April	103	0.060	1.788	0.34	0.7300	49.5
May	98	0.205	0.898	2.27	0.0260	58.2
June	106	−0.249	0.995	−2.57	0.0110	38.7
July	111	−0.004	1.083	−0.04	0.9700	48.6
August	111	−0.048	1.790	−0.28	0.7800	46.8
September	98	−0.314	1.235	−2.52	0.0140	39.8
October	106	0.169	1.303	1.34	0.1800	50.9
November	99	−0.385	1.306	−2.93	0.0042	35.4
December	102	0.096	1.334	0.73	0.4700	52.9

has induced some behavioral changes and the actions listed above are now performed at different times of the year. A closer look will provide a clearer understanding of the underlining causes of the monthly anomalies.

3 Turn-of-the-month effects

In the United States the returns on trading days −1 to +4 of each month have historically dominated the other days, particularly for small stocks. The day−1 refers to the last trading day of the previous month and +1 to the first trading day of the current month, etc. This is referred to as the turn-of-the-month effect. The return in this period, coupled with that in the second week of the month, trading days +5 to +8 or +10, essentially amounts to all the gains in the spot stock market in the period of the 1960s, 70s, and 80s. See Ariel (1987) for spot data for the period 1963–82 and Hensel, Sick and Ziemba's paper in this volume for an analysis of the S&P500 from 1928–1996 and Hensel, Sick and Ziemba (1994) for S&P500 and Value Line cash and futures data for 1982–92. In the latter paper the authors found similar results but there was some anticipation of the effect on days −4 to −2 that is, three days in advance. The returns in the rest of the month have been essentially noise and at best provided zero returns. The reasons for these effects are not fully known, but important ones are:

(a) First quarter

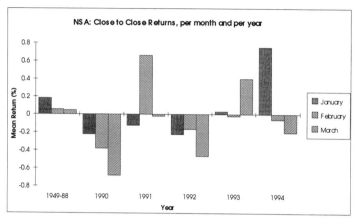

(b) Second quarter

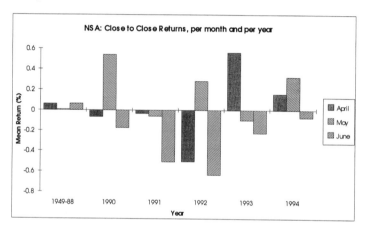

Figure 5. Mean Daily close-to-close returns by month on the NSA for
1949–1988 and the years 1990–1994.

- Salaries are received on or around the −1 day. People also receive their
  stock account statements at around this time, so they have funds to
  invest in stocks.

- There are portfolio re-balancing effects by large brokerage firms.

- Bad news is often delayed until the second half of the month.

- There is a large flow of funds into the stock market from cash flows and
  monetary action of agencies, such as: interest and principal payments on

(c) Third quarter

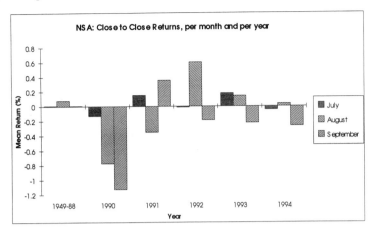

(d) Fourth quarter

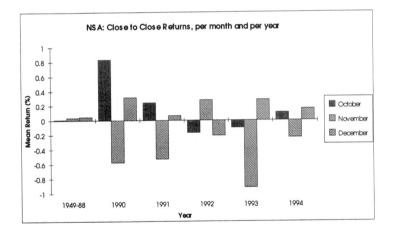

Figure 5. *cont.*

corporate debt; municipal debt; dividends on common stock, preferred
stock during the turn.

In Japan most firms pay salaries on the 25th of the month, and the turn-
of-the-month has historically, according to Ziemba (1991), been the seven-day
trading period from day −5 to day +2. Table 4 presents data for the turn-
of-the-month (TOM) on the NSA for the period 1949–1988. Tables 5 and 6
present data on the NSA and TOPIX respectively, for 1990–1994, and these

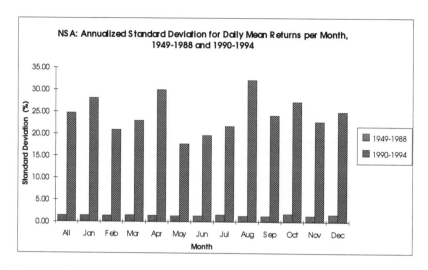

Figure 6. Annualized standard deviations of returns on the NSA by month
1949–1988 versus 1990–1944.

are compared visually in Figure 8.

During 1949–1988, returns around the turn-of-the-month were high, 0.10%
or more on average per day, and with a statistical significance at the 5% level
or better. The −1 day had very high returns, as in the US. The mean returns
were over 0.22%, making its effect about as strong as a pre-holiday, namely
providing returns about five times as large as on a typical trading day. The
first half-of-the-month effect, namely days −5 to +7, a twelve-day trading
period, has been present in previous decades. In contrast with this historical
pattern, during the period 1990–1994 only days −5, −1 and +2 had positive
returns, but none of them at a statistically significant level. Mean returns on
days −2 to +5 were still positive, in contrast with the negative mean returns
for the whole period. Most of the trading days from −5 to +17 had positive
nighttime returns. The more negative daytime mean returns determine the
daily pattern. The first-half-of-the-month effect was not present during 1990–
1994. The average of the mean daily returns over this period was negative.
Historically these days had been positive, providing most of the monthly
positive returns. Days +8 to +17 were strongly negative for the period 1990–
94, while they had been mainly noise or slightly negative during previous
decades. Figure 9 provides a comparison between the turn-of-the-month,
second week, and first-half-of-the-month for the periods 1949–88 and 1990–
94. Figure 10 shows the turn-of-the-month mean returns of the NSA and
TOPIX for each year from 1990 to 1994. Only 1990 had negative returns,
and 1992–94 returns were clearly positive. Figure 11 shows the mean return

(a) Close to open (night-time) daily returns on the NSA by month, 1990–1994

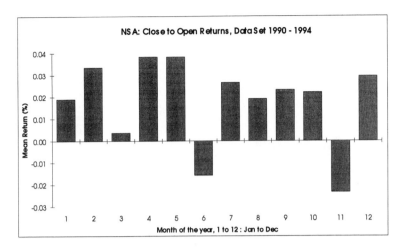

(b) Open to close (daytime) daily returns on the NSA by month, 1990–1994

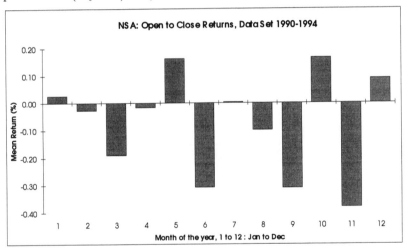

Figure 7

for trading days –5 to +17 per year from 1990 to 1994. The turn-of-the-month pattern of mean returns changes dramatically from 1990 to 1994. For 1990 both the one-week period –5 to +2 and the two-week period –5 to +8 had negative returns. For 1994 both the one-week and the two-week periods had positive returns analogous to their historical behavior. In contrast to their historical results, however, days –5 and –4 remain strongly negative in 1994, and no trading day in the first-half-of-the-month had statistically significant

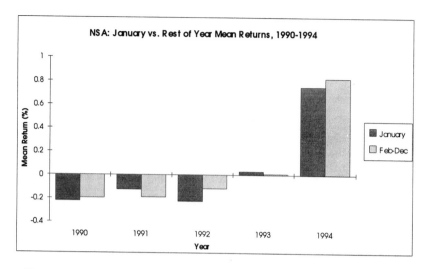

Figure 8a. January versus rest of the year mean returns on the NSA,
1990–1994.

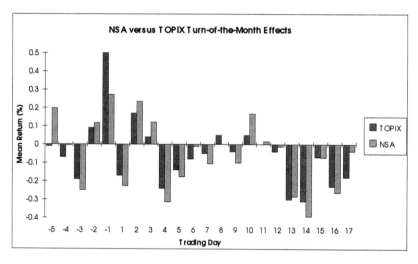

Figure 8b. Daily mean returns on the NSA and TOPIX at the
turn-of-the-month, 1990–1994

positive returns.

The turn-of-the-month effect is overlapped with the day-of-the-week effects, see Kato (1990) and Kato, Schwartz and Ziemba (1989). Ziemba and Schwartz (1991) present the results of a 52 variable model to separate out these and other effects to develop seasonality calendars to evaluate and rank all the days on a common basis. Since the sample sizes for the five-year period 1990–94 are relatively small, this overlap might strongly influence the results. Our results show, nonetheless, what the general pattern was during 1990–1994 and what the actual trend might be. Some of the plausible reasons for the turn-of-the-month effect in Japan, according to Ziemba (1991), are:

- Most salaries are paid on days 20–25 of the month with the 25th being the most common. Buying pressure starts, therefore, the 25th.

- There are portfolio adjustments or 'window dressing' on Day–1

- Security firms can invest for their own accounts based on their capitalization. Captialization usually rises each month, and is computed at the end of the month. There is buying pressure as early as Day–3.

- Large brokerage firms have a sales push that lasts 7 to 10 days and this starts on Day–3.

- Employment stock holding plans and mutual funds receive money in this period to invest, starting around Day–3.

- People buy mutual funds with their pay, received between the 15th and 25th of the month. Stock investments have a lag, and most of the buying occurs on Days –5 to +2.

- For low liquidity stocks buying occurs over several days by dealing in different accounts to minimize price pressure effects.

4 Holiday effects

There has been, historically, a holiday effect in Japanese securities similar to that in the US securities as documented by Ariel (1987), French and Roll (1986), Lakonishok and Smidt (1988), and Zweig (1986) for the US and Ziemba (1991) for Japan. The daily mean return of the NSA from 1949 to 1988 was 0.048%. On pre-holidays the return was about five times as high, 0.246%. Moreover, the risk measured by the daily standard deviation was also lower: 0.794% versus 0.979% per day, respectively. As in the US, there were no abnormal gains on the days around holidays except for the pre-holiday trading day. For example, the days following holidays had returns of 0.0067% per day on the NSA during 1949–88, which is less than a typical day.

We show below Ziemba's (1991) regression of the NSA daily mean returns on trading days –3 to +2 for the period 1949–88, and similar regressions of the NSA and TOPIX for the period 1990–94. R is the daily mean return, the effects of days –3 to +2 are separated out with $\{0,1\}$ variables, and the

Table 4. Mean daily return, maximum and minimum returns, standard deviations and t-values for the NSA around the turn-of-the-month, for the period 1949–1988.

Trading day	Sample size	Mean return (%)	Stand. dev.	Max. ret. (%)	Min. ret. (%)	Med. ret. (%)	t-stat.
−5	471	0.089	0.710	3.650	−2.578	0.083	2.75
−4	471	0.104	0.853	5.321	−2.853	0.120	2.65
−3	471	0.173	0.984	4.118	−4.216	0.237	3.82
−2	471	0.133	0.911	4.555	−6.734	0.139	3.18
−1	471	0.226	0.914	2.750	−8.686	0.301	5.36
1	471	0.098	0.875	4.501	−3.987	0.156	2.43
2	471	0.101	0.884	3.673	−4.849	0.131	2.47
3	471	0.307	0.924	5.633	−4.368	0.020	0.72
4	471	0.059	1.113	11.149	−9.997	0.044	1.15
5	471	0.036	0.957	6.307	−10.649	0.031	0.81
6	471	−0.001	0.811	4.410	−3.899	−0.009	−0.01
7	471	0.036	0.838	3.471	−4.600	0.057	0.09
8	471	−0.059	0.945	4.554	−8.218	0.011	−1.34
9	471	0.107	0.940	6.877	−4.716	0.109	2.46
10	471	0.062	0.852	4.795	−4.399	0.057	1.58
11	471	0.040	0.813	5.394	−4.025	0.057	1.05
12	471	−0.020	0.850	4.681	−6.970	0.053	−0.50
13	429	0.012	1.055	11.289	−7.680	0.020	0.24
14	350	−0.004	0.912	6.408	−3.542	0.022	−0.10
15	229	−0.031	1.237	6.138	−14.901	0.062	−0.54
16	118	0.072	1.148	9.888	−7.493	0.871	1.35
17	39	−0.498	0.888	4.716	−4.253	−0.038	−1.20

Source: Ziemba (1991)

constant term is the return on all other days. The notation −3 refers to the 3rd trading day before the holiday closure and +1 to the day after the holiday, etc. The t-statistics are shown in brackets below each estimated coefficient.

NSA 1949–1988:

$$R = \ 0.035 \ + 0.080\text{Day}_{-3} + 0.022\text{Day}_{-2} + 0.189\text{Day}_{-1} - 0.066\text{Day}_{+1} + 0.001\text{Day}_{+2}$$
$$ (3.745) \quad\ (1.491) \qquad\quad (0.424) \qquad\quad (3.709) \qquad\quad (-1.334) \qquad\quad (0.023)$$

Among the trading days around holidays, only pre-holidays with a mean return of 0.189% per day was statistically significant with a t-statistic of 3.709. The mean return on all other days was also positive at a statistically significant level, and day +1 had negative mean returns.

Table 5. Mean daily returns, maximum and minimum returns, and standard deviations and t- values for the NSA around the turn-of-the-month, for the period 1990–1994.

Trading day	Sample size	Mean return (%)	Stand. dev.	Max. ret. (%)	Min. ret. (%)	Med. ret. (%)	t-stat.	P-value
All	1319	−0.051	1.577	−6.827	12.430	−0.050	−1.17	0.240
−5	60	0.198	1.449	−2.167	4.721	0.176	1.06	0.290
−4	60	−0.004	1.188	−2.647	2.999	−0.018	−0.03	0.980
−3	60	−0.249	1.796	−4.863	5.946	−0.259	−1.07	0.290
−2	60	0.116	1.382	−3.949	3.611	0.193	0.65	0.520
−1	60	0.272	1.651	−3.688	7.551	0.255	1.28	0.210
1	60	−0.227	1.866	−6.827	4.288	−0.190	−0.94	0.350
2	60	0.234	2.073	−3.873	12.430	−0.025	0.88	0.380
3	60	0.121	1.552	−3.680	4.442	−0.004	0.60	0.550
4	60	−0.313	1.106	−3.610	1.820	−0.314	−2.19	0.032
5	60	−0.177	1.461	−3.560	4.208	−0.057	−0.94	0.350
6	60	−0.011	1.691	−3.524	5.037	−0.066	−0.05	0.960
7	60	−0.105	1.402	−3.419	3.301	−0.195	−0.58	0.570
8	60	0.000	1.602	−3.948	7.275	−0.026	0.00	1.000
9	60	−0.100	1.703	−4.327	4.377	−0.059	−0.46	0.650
10	60	0.167	1.426	−2.522	4.689	−0.052	0.91	0.370
11	60	0.016	1.556	−3.072	5.257	0.211	0.08	0.940
12	60	−0.012	1.327	−4.243	4.794	−0.159	−0.07	0.940
13	60	−0.285	1.553	−6.135	2.729	−0.163	−1.42	0.160
14	60	−0.393	1.652	−4.093	4.125	−0.439	−1.84	0.071
15	60	−0.072	1.592	−5.068	6.031	−0.255	−0.35	0.730
16	60	−0.263	1.435	−4.221	2.503	−0.246	−1.42	0.160
17	60	−0.038	1.905	−6.022	4.721	0.015	−0.15	0.880

NSA, 1990–1994:

$$R = -0.035 + 0.250\text{Day}_{-3} - 0.282\text{Day}_{-2} + 0.081\text{Day}_{-1} - 0.294\text{Day}_{+1} - 0.191\text{Day}_{+2}$$
$$\quad(-0.69)\qquad(1.16)\qquad\quad(-1.32)\qquad\quad(0.39)\qquad\quad(-1.40)\qquad\quad(-0.88)$$

TOPIX, 1990-1994:

$$R = -0.031 + 0.229\text{Day}_{-3} - 0.255\text{Day}_{-2} + 0.006\text{Day}_{-1} - 0.218\text{Day}_{+1} - 0.163\text{Day}_{+2}$$
$$\quad\textbf{(-0.71)}\qquad\textbf{(1.24)}\qquad\quad\textbf{(-1.38)}\qquad\quad\textbf{(0.03)}\qquad\quad\textbf{(-1.23)}\qquad\quad\textbf{(-0.88)}$$

During 1990–1994, only days −3 and −1 had positive mean returns. All the other days had negative mean returns, and none of the results has statistical significance.

Tables 7, 8 and 9 present the holiday effects on the NSA for the period 1949–88. When pre-holidays are lumped together with the last trading day of

Table 6. Mean daily returns, maximum and minimum returns, and standard deviations and t- values for the TOPIX around the turn-of-the-month, for the period 1990–1994.

Trading day	Sample size	Mean return (%)	Stand. dev.	Max. retd (%)	Min. retd (%)	Med. retd (%)	t-stat.	P-value
All	1320	−0.056	1.330	−7.365	9.116	−0.062	−1.53	0.130
−5	60	−0.012	1.049	−1.878	3.140	0.042	−0.09	0.930
−4	60	−0.073	1.077	−2.890	2.250	−0.060	−0.52	0.600
−3	60	−0.191	1.555	−4.238	4.819	−0.120	−0.95	0.350
−2	60	0.091	1.264	−5.339	2.906	0.141	0.56	0.580
−1	60	0.501	1.343	−3.091	6.426	0.401	2.89	0.005
1	60	−0.168	1.712	−7.365	4.956	−0.180	−0.76	0.450
2	60	0.168	1.653	−2.949	9.116	0.012	0.79	0.440
3	60	0.040	1.261	−3.098	3.478	−0.028	0.25	0.810
4	60	−0.241	0.966	−3.564	1.734	−0.231	−1.93	0.058
5	60	−0.139	1.310	−3.459	4.299	−0.191	−0.82	0.410
6	60	−0.075	1.495	−4.777	4.063	−0.012	−0.39	0.700
7	60	−0.052	1.039	−1.961	3.290	−0.160	−0.39	0.700
8	60	0.047	1.424	−3.353	6.972	−0.083	0.26	0.800
9	60	−0.038	1.295	−3.490	3.844	−0.037	−0.22	0.820
10	60	0.049	1.067	−2.414	2.531	−0.084	0.36	0.720
11	60	0.000	1.188	−2.318	3.843	0.058	0.00	1.000
12	60	−0.044	1.063	−2.686	4.003	−0.232	−0.32	0.750
13	60	−0.298	1.258	−5.382	1.834	−0.213	−1.84	0.071
14	60	−0.315	1.544	−3.594	4.091	−0.299	−1.58	0.120
15	60	−0.071	1.597	−4.618	7.284	−0.107	−0.35	0.730
16	60	−0.234	1.223	−3.521	3.408	−0.209	−1.48	0.140
17	60	−0.177	1.474	−5.869	2.464	0.026	−0.93	0.360

Table 7. Holiday effects on the NSA for the period 1949–88.

Day	Sample size	Min. ret. (%)	Max. ret. (%)	Mean ret. (%)	Stand. dev.	T-stat.
Pre–holiday	408	−5.4069	3.0631	0.2461	0.7943	6.26
Non–pre–holiday	7143	−14.9009	11.2893	0.0489	0.9786	4.22
Days after holiday	408	−8.6856	4.7122	0.0068	0.9893	0.14
Pre–holidays and last trading day of the week	2268	−6.6101	6.8765	0.1561	0.7242	10.27
Days after holidays and weekends	2268	−8.6856	5.3938	−0.0480	0.9473	−2.41

Source: Ziemba (1991)

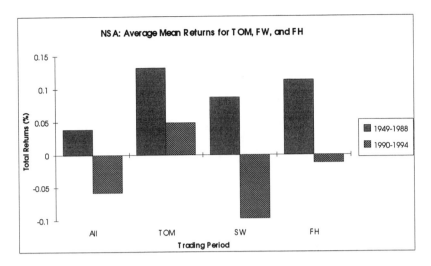

Figure 9. Daily mean returns during TOM, FW and FH on NSA, 1949–1988 versus 1990–1994.

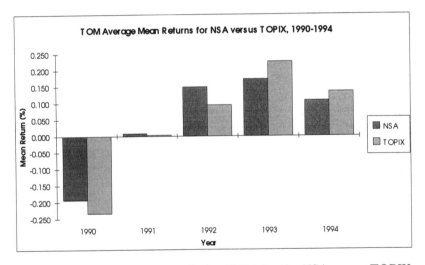

Figure 10. Daily mean returns during TOM for the NSA versus TOPIX, 1990–1994.

the week, Friday or Saturday, the average return is 0.156%, more than three times the average for all days. When after holidays are lumped together with the first trading day of the week, the mean return becomes more negative. This is mostly the effect of the negative Mondays. In Japan, Tuesdays were negative, on average, but not when there was Saturday trading the previous

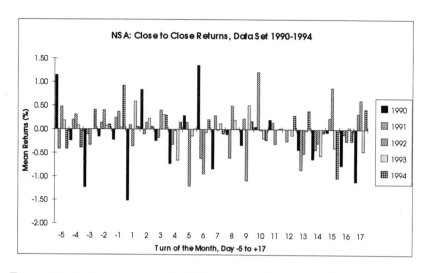

Figure 11: Daily returns on the NSA around the turn-of-the-month during
the years 1990 to 1994.

week. Then Monday is strongly negative, on average. Pre-holidays improve
every day of the week. Even Mondays and Tuesdays are positive if they are
pre-holidays. Wednesdays, Thursdays, Fridays and Saturdays have very high
returns if they are pre-holidays. Pre-holidays also increase the percentage of
positive returns for all the days of the week. A comparison between the day-
of-the-week effects and the pre-holidays effects shows that pre-holidays is the
largest. Wednesdays and Saturdays are also very positive and they represent
the bulk of the week's return.

Tables 10 to 13 present the holiday effects and pre-holidays mean returns
for the NSA during 1990–94. Figures 12–15a,b show the holiday effects on the
mean returns of the NSA during 1949–88 and on the mean returns of the NSA
and TOPIX during 1990–94. For the period 1990–94 pre-holidays plus the
last trading day of the week had negative mean returns, in contrast with the
historical results. After holidays plus first trading day of the week show highly
negative returns at a statistically significant level. The effect of pre-holidays
on the day of the week during 1990–94 is mixed. Pre-holidays make Mondays
and Fridays less negative, but the make Tuesdays and Wednesdays more negative. Only Thursdays show a significant improvement, with strongly positive
returns at a statistically significant level. Pre-holidays did not improve the
percentage of positive daily returns for any of the days of the week during
1990–94. Pre-holidays are not all the same, and they are also strongly influenced by the month of the year they belong to through the monthly effect. In
the period 1949–88 only one pre-holiday has negative mean returns, October

Table 8. Mean daily return, standard deviation and percent of positive returns on pre-holidays by days of the week of the NSA, for the period 1949–88.

Day	Sample size	Mean ret. (%)	Stand. dev.	T-stat.	P-value	Med. ret. (%)	Pos. ret. (%)
All Mondays	1937	−0.5890	0.9500	−2.73	0.0064	0.0189	50.9
All Tuesdays	1959	−0.0419	0.9586	−1.93	−0.0532	−0.0373	47.4
All Wednesdays	1960	0.1164	0.9722	5.30	0.0001	0.1212	56.9
All Thursdays	1959	0.0871	1.0234	3.77	0.0002	0.0676	55.1
All Fridays	1959	0.0544	0.9174	2.62	0.0087	0.0700	54.6
All Saturdays	1755	0.1409	0.7043	8.38	0.0001	0.1483	61.9
Mon. before holidays	59	0.3106	0.7698	3.10	0.0030	0.3788	76.3
Tues. before holidays	59	0.0103	0.7866	0.10	0.9198	0.1192	57.6
Wed. before holidays	58	0.1600	1.0388	1.17	0.2456	0.1955	69.0
Thur. before holidays	60	0.1868	0.7860	2.82	0.0064	0.2986	73.3
Fri. before holidays	71	0.3821	0.7944	4.05	0.0001	0.3411	74.6
Sat. before holidays	101	0.2758	0.6242	4.44	0.0001	0.2468	64.4

Source: Ziemba (1991)

Table 9. Day of the week and pre-holiday effects on the NSA, 1949–88.

Day effect	Mean return (%)	Stand. dev.	T-stat.	P-value
Monday	−0.0705	0.0214	−3.29	0.0010
Tuesday	−0.0435	0.0213	−2.04	0.0412
Wednesday	0.1151	0.0213	5.40	0.0001
Thursday	0.0807	0.0213	3.79	0.0002
Friday	0.0420	0.0214	1.97	0.0492
Saturday	0.1327	0.0229	5.81	0.0001
Pre–holiday	0.2461	0.0460	5.35	0.0001

Source: Ziemba (1991)

Table 10. Holiday effects on the NSA daily mean returns for the period 1990–94.

Day	Sample size	Min. ret. (%)	Max. ret. (%)	Mean ret.(%)	Stand. dev. (%)	t-test	P-value
PH –3	56	–3.654	4.794	0.215	1.313	1.23	0.220
PH –2	57	–4.327	3.455	–0.317	1.527	–1.57	0.120
PH	61	–3.150	2.647	0.046	1.250	0.29	0.770
AH	59	–3.948	3.495	–0.329	1.600	–1.58	0.120
AH +2	56	–4.863	4.377	–0.226	1.558	–1.08	0.280
PH + F	275	–4.327	7.275	–0.013	1.442	–0.15	0.880
AH + W1	278	–6.827	7.551	–0.338	1.842	–3.06	0.002
Non PH	1123	–6.827	12.430	–0.046	1.587	–0.97	0.330

Table 11. Mean daily return, standard deviation, t values and percent of positive returns on pre-holidays by days of the week of the NSA daily returns, for the period 1990–94.

Day	Sample size	Mean ret. (%)	Stand. dev. (%)	T-test	P-value	Med. ret. (%)	% Pos. Ret.
All Mon.	240	–0.306	1.872	–2.53	0.012	–0.306	40.4
All Tues.	254	–0.042	1.502	–0.44	0.660	0.071	46.9
All Wed.	252	–0.073	1.528	–0.76	0.450	–0.128	45.2
All Thurs.	250	0.156	1.465	1.68	0.094	0.147	56.0
All Fri.	248	–0.036	1.451	–0.39	0.700	0.089	46.8
PH Mon.	7	–0.066	1.926	–0.09	0.930	–0.120	42.9
PH Tues.	9	–0.208	0.836	–0.75	0.480	–0.294	44.4
PH Wed.	10	–0.176	1.426	–0.39	0.700	–0.119	50.0
PH Thurs.	9	0.821	1.021	2.41	0.042	0.750	55.6
PH Fri.	25	–0.023	1.165	–0.10	0.920	–0.385	40.0

10. During 1990–94, only three pre-holidays were highly positive, April 29, May 3–5, and September 15. Since September had very negative returns for 1990–94, this result for the September pre-holiday is remarkable. The period 1990–94 departs from the patterns observed in previous decades.

5   The Golden Week

In late April and early May in Japan there are three holidays, April 29, May 3 and May 5. These holidays fall in a one-week period referred to as the

Table 12. Day of the week and pre-holiday effects on the NSA, 1990–94.

Day	Sample size	Mean return (%)	Stand. dev. (%)	SE Mean	t-test	P-value
Mon.	218	−0.359	1.882	0.128	−2.82	0.010
Tues.	225	−0.010	1.553	0.104	−0.09	0.920
Wed.	232	−0.043	1.511	0.099	−0.44	0.660
Thurs.	221	0.188	1.462	0.098	1.91	0.060
Fri.	214	−0.030	1.495	0.102	−0.29	0.770
PH	61	0.046	1.250	0.160	0.29	0.770

Table 13. Mean daily returns and standard deviation on the NSA, 1990-1994.

Holiday	Sample size	Mean return (%)	Stand. dev.(%)	Min. ret. (%)	Max. ret. (%)	T-test	P-value
1 Jan	6	0.116	1.486	−2.109	2.406	0.19	0.860
15 Jan	5	0.122	1.370	−1.727	2.112	0.2	0.850
11 Feb	5	0.004	0.926	−1.310	0.792	0.01	0.990
21 Mar	5	−0.614	1.618	−2.085	2.107	−0.85	0.440
29 Apr	5	0.501	0.454	0.020	1.219	2.47	0.069
3/5 May	5	0.506	1.347	−0.789	2.246	0.84	0.450
15 Sep	5	0.468	1.709	−0.952	2.647	0.61	0.570
23 Sep	5	−0.074	1.052	−1.437	1.191	−0.16	0.880
10 Oct	5	0.038	1.145	−1.603	1.355	0.07	0.940
3 Nov	5	−0.348	0.528	−0.836	0.511	−1.47	0.210
23 Nov	5	−0.153	2.113	−3.150	2.524	−0.16	0.880
23 Dec	5	−0.027	1.286	−1.667	1.503	−0.05	0.960

Golden Week. During 1949–1988, NSA returns for April 28, May 2, and May 4 were strongly positive and, in the last two cases, statistically significant. After 1989 the May 3 and May 5 holidays were lumped together into a three-day holiday, so that there is no trading day in between. Consequently, there are now only two pre-holidays around these dates. During 1990–1994 both pre-holidays had strongly positive returns, but the small size of our sample precludes serious statistical analysis. Even April 30, a post-holiday, shows high positive mean returns, in contrast to was observed in previous decades. The reason for this might be the presence of the new 3-day holiday May 3–5 since 1989. Another difference with the historical results is that the post-holiday days May 6 and May 7 were strongly positive.

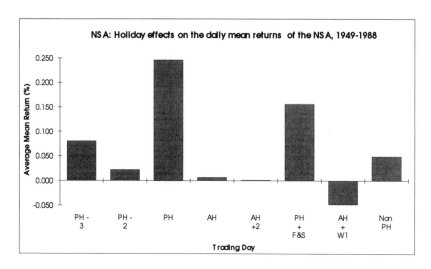

Figure 12. Holiday effects on the NSA, 1949–1988. Source: Ziemba (1991).

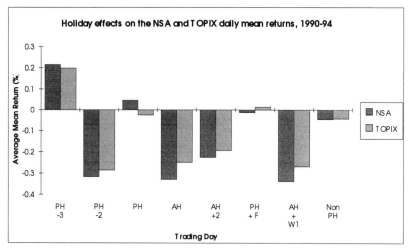

Figure 13. Holiday effects on the NSA and TOPIX, 1990–1994.

Mean daily returns for the three-week period from April 20 to May 21 are presented in Table 14. Despite the fact that these results have no statistical significance due to the small sample sizes, the mean return for this period was highly positive.

A comparison between golden week mean returns and all days mean returns for the NSA during 1990–1994 is presented in Table 15. Daily mean returns for the golden week period April 20–May 21 are strongly positive de-

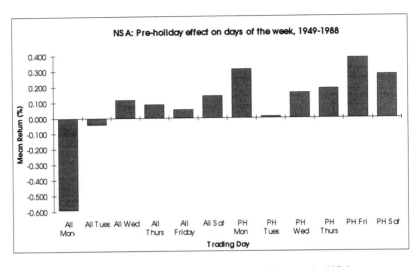

Figure 14a. Preholiday effects on individual days of the week, NSA returns, 1949–1988. Source: Ziemba (1991).

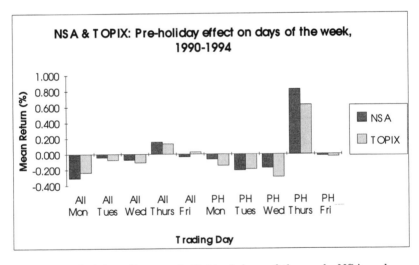

Figure 14b. Preholiday effects on individual days of the week, NSA and TOPIX returns, 1990–1994.

spite the negative mean returns for the whole period. Both nighttime and daytime returns are positive, contributing to the positive daily average. Risk, as measured by the standard deviation, was also lower during the time around the golden week than the rest of the year: 0.982% versus 1.576%, respectively.

(a) 1949–1988. Source: Ziemba (1991)

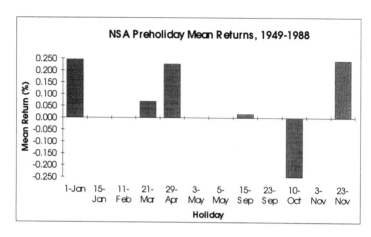

(b) 1990–1994

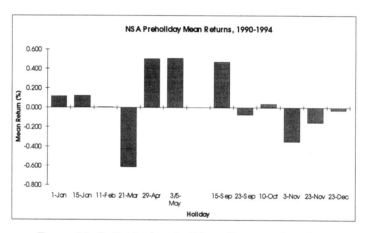

Figure 15. Individual preholiday effects on the NSA.

6   Turn-of-the-year effects

Around the world, equity returns in January have been historically high with
the highest returns during the January turn-of-the-month on trading days
−1 to +4; see e.g. Roll (1983) and Keim (1983) for US data. This is one
aspect of the phenomenon referred to as the turn-of-the-year effect. The
tendency of small capitalized stocks to outperform large capitalized stocks
in January is discussed in other papers in this volume by Booth and Keim,
Hensel and Ziemba, Dimson and Marsh and others. Ziemba (1991) found very

Table 14. Mean return per day around the time of the Golden Week for the NSA.

Date	Period 1949–1988				Period 1990–1994			
	Sample size	Mean return (%)	T-test	P-value	Sample size	Mean return (%)	T-test	P-value
20 April	32	−0.129	−0.73	0.469	3	−1.113	−2.97	0.097
21 April	32	−0.023	−0.18	0.862	3	−0.792	−1.78	0.220
22 April	33	0.058	0.34	0.696	4	−0.235	−0.49	0.660
23 April	34	−0.002	−0.01	0.989	4	1.076	1.34	0.270
24 April	33	0.113	1.23	0.229	3	−0.136	−0.29	0.800
25 April	34	0.134	0.92	0.366	3	−0.729	−1.54	0.260
26 April	33	0.292	2.10	0.043	4	−0.242	−1.27	0.290
27 April	33	0.128	0.74	0.464	4	0.864	1.18	0.320
28 April	34	0.215	1.41	0.169	3	0.546	1.51	0.270
30 April	31	−0.142	−0.47	0.645	3	0.472	0.52	0.660
1 May	33	0.431	2.69	0.011	3	0.429	0.76	0.520
2 May	34	0.536	4.47	0.000	3	0.262	0.37	0.750
4 May	30	0.326	2.96	0.006				
6 May	31	−0.098	−0.60	0.552	3	1.107	0.81	0.500
7 May	34	0.09	0.87	0.389	4	1.474	1.85	0.160
8 May	33	−0.133	−0.79	0.426	3	−0.089	−1.26	0.330
9 May	33	−0.117	−0.91	0.367	3	0.009	0.03	0.980
10 May	32	−0.035	−0.23	0.817	4	0.328	0.86	0.450
11 May	32	0.028	0.28	0.792	4	0.893	1.81	0.170
12 May	33	−0.07	−0.70	0.491	3	−0.578	−1.04	0.410
13 May	33	−0.134	−1.02	0.313	4	0.133	0.29	0.790
14 May	33	−0.242	−1.75	0.089	4	0.333	0.73	0.520
15 May	31	−0.211	−1.70	0.099	3	−1.635	−1.39	0.300
16 May	32	−0.064	−0.50	0.624	3	−0.559	−1.73	0.230
17 May	32	−0.221	−1.39	0.176	4	0.294	1.42	0.250
18 May	33	−0.321	−2.20	0.036	4	0.080	0.11	0.920
19 May	34	−0.289	−1.47	0.151	3	0.919	2.33	0.150
20 May	32	0.008	0.06	0.955	4	−0.193	−0.69	0.540
21 May	32	0.355	1.57	0.127	4	0.064	0.16	0.880

Source: Ziemba (1991)

high returns in late December into mid January during 1949–1988. Based on our discussion of the turn-of-the-month effect, one would expect the turn-of-the-year effect to begin around day −5 and run to the middle of January. The data shown in Table 16 indicates that for the period 1949–88 the effect seems to start on day −7, when the mean return is 0.142% per day. Then there were positive returns on every trading day until day +14. Many of the days had

Table 15. NSA mean returns around the time of the Golden Week versus returns, all trading days, period 1990–1994.

Night–time (Close to Open) Mean Returns

Day	Sample	Mean	Std. Dev.	SE Mean	T-Test	P-Value
All	1231	0.018	0.172	0.005	3.60	0.000
Golden Week	98	0.034	0.168	0.017	2.00	0.048

Daytime (Open to Close) Mean Returns

Day	Sample	Mean	Std. Dev.	SE Mean	T-Test	P-Value
All	1232	−0.073	1.516	0.043	−1.69	0.092
Golden Week	98	0.077	1.124	0.114	0.68	0.500

Daily (Close to Close) Mean Returns

Day	Sample	Mean	Std. Dev.	SE Mean	T-Test	P-Value
All	1232	−0.055	1.580	0.045	−1.23	0.220
Golden Week	98	0.111	1.194	0.121	0.92	0.360

extremely high mean returns: −4 had 0.468%, −3 had 0.441%, −1 had 0.246%, +3 had 0.381%, +5 had 0.407%, +7 had 0.255%, +9 had 0.401%, +10 had 0.274%, +11 had 0.224% and +14 had 0.218%. This period is shown inside a box in Table 16. During the 1990-94 bear market, this effect was not present. The daily mean return averaged over days −7 to +14 is negative, and only days −4 to +1 had consistently positive returns. The interval of trading days that show the turn-of-the-year effect for previous decades has in 1990–94 high volatility, with high positive and negative returns. Furthermore, there were only 10 positive versus 11 negative trading days.

A regression model that separates out the -1 day of the year effect from the −1 day of each month and pre-holidays for the NSA is shown below with t-statistics in parenthesis. All these effects were present for night-time, daytime and daily mean returns, with daily mean returns being the highest of all three.

R = Daily mean return

b_0 = constant, all other days

$b_1 = \{0, 1\}$: 1 on last trading day of the year, 0 otherwise

$b_2 = \{0, 1\}$: 1 on pre-holiday (except Jan 1), 0 otherwise

$b_3 = \{0, 1\}$: 1 on last trading day of month (non-ph), 0 otherwise

The regression equations on Daily Returns are:

Period 1949–1988:

$$R = 0.0341 + 0.2116\,b_1 + 0.2121\,b_2 + 0.1822\,b_3$$
$$(3.78) \quad (1.42) \quad\quad (4.30) \quad\quad (3.95)$$

Table 16. NSA daily mean returns for the turn-of-the-year effect.

Day	Period 1949–1988			Period 1990–1994		
	Mean return (%)	Stand. dev.	t-stat.	Mean return (%)	Stand. dev.	t-stat.
−15	0.0292	0.757	0.24	1.578	2.052	0.92
−14	−0.0603	0.983	−0.38	0.224	1.209	0.54
−13	−0.2631	1.266	−1.29	−0.122	1.171	0.52
−12	0.221	1.926	0.72	0.096	0.669	0.30
−11	−0.1036	0.721	−0.9	1.941	1.767	0.79
−10	0.0202	1.259	0.09	0.092	1.282	0.57
−9	−0.1197	1.382	−0.54	0.108	1.066	0.48
−8	0.0347	1.26	0.17	−0.233	2.072	0.93
−7	0.1413	1.168	0.76	−0.502	1.674	0.75
−6	0.1738	0.83	1.31	−0.371	0.885	0.40
−5	0.1303	0.753	1.08	−0.636	1.139	0.51
−4	0.4683***	0.653	4.48	0.079	2.206	0.99
−3	0.4413*	1.104	2.49	0.189	1.432	0.64
−2	0.0613	0.988	0.39	0.255	0.506	0.23
−1	0.2457	0.763	2.01	0.17	1.655	0.74
1	0.0369	1.059	0.22	0.805	1.607	0.72
2	0.1539	1.053	0.91	−0.413	1.558	0.70
3	0.3812	1.241	1.92	−1.405	2.064	0.92
4	0.246	1.052	1.46	0.497	1.062	0.48
5	0.4067	0.928	2.74	−0.536	1.824	0.82
6	0.0856	0.786	0.68	−0.214	1.719	0.77
7	0.2546	1.054	1.51	0.144	1.231	0.55
8	0.0985	0.834	0.74	−1.612	1.059	0.47
9	0.4007**	0.86	2.91	1.008	2.257	1.01
10	0.2736*	0.682	2.5	−0.271	1.403	0.63
11	0.2243	0.717	1.95	−0.391	1.182	0.53
12	0.0375	0.7265	0.32	0.993	2.093	0.94
13	0.1101	0.704	0.98	0.118	0.603	0.27
14	0.218*	0.667	2.04	−0.728	1.454	0.65
15	−0.0334	1.147	−0.14	−0.773	2.492	1.11

Source: Ziemba (1991) for 1949–88

Period 1990-1994:

$$R = -0.0765 + 0.193\,b_1 + 0.120\,b_2 + 0.390\,b_3$$
$$(-1.62) \quad (0.30) \quad (0.55) \quad (1.68)$$

Relative importance of each of the effects:

- 1949–1988: Pre-holiday > Turn of the Month > Turn of the Year

- 1990–1994: Turn of the Month > Turn of the Year > Pre-holiday

7   Day-of-the-week effects

Figure 16a,b shows the day-of-the-week effect on the US S&P500 index during 1970–83 and on the NSA mean daily returns during 1949–88. Both patterns are strikingly similar, except that in Japan there was Saturday trading during this period. If Saturdays and Fridays are lumped together for the NSA, the NSA and S&P day-of-the-week effects show very similar patterns. The day of the week effects on the NSA and TOPIX daily mean returns for the period 1990–1994 are presented in Tables 17 and 18 and Figures 16c and 17. Monday shows consistently negative returns for each year during this period while other days show a pattern that changes from 1990 to 1994. For 1990 all days of the week had negative returns, but in 1994 only Monday was negative. Thursday returns were strongly positive in 1992 and 1993, and only weakly positive in 1994. Years 1990 and 1992 had strongly negative returns for all days except Thursdays of 1992; 1993 shows significantly positive returns only for Thursdays, and 1994 shows returns that are not significantly different from zero. In consequence, when mean returns are averaged over the whole period 1990–1994, all days except Thursdays had negative returns. Monday returns were negative at a statistically significant level of 5% and Thursday returns were positive at a statistically significant level of 10%.

As observed for other effects, the day of the week effect for the NSA daily mean returns changes completely from 1990 to 1994 with 1993 representing a transition between 1990–1992 and 1994. The day of the week pattern observed for 1994 was close to the historically observed pattern, except that Thursdays in 1994 did not show very high returns, and Tuesdays were slightly negative instead of being slightly positive. A new pattern sets in, but more observations are needed in order to see how consistent it is. Part of the reason undoubtedly corresponds to the change from trading Saturdays to closed markets on Saturdays, and its effect on settlements and weekend anticipation.

We list below some of the most important reasons, according to Ziemba (1991), underlying the day-of-the-week effects in Japan:

(1) Negative Tuesday with Saturday Trading

    – Investors more likely to sell after thinking over the weekend

    – Lag from New York Monday fall via correlation effect

    – Sell on Tuesday if money is needed for Friday

    – It takes time to get the market again in the new week

(2) Negative Monday when Saturday is closed (now always)

Table 17. Day-of-the-week effects on the NSA mean daily returns, 1990–1994.

Day	Sample size	Mean return %	Standard deviation	SE Mean	T–statistic	P–value
All	1232	−0.055	1.580	0.045	−1.23	0.220
Mon.	240	−0.306	1.872	0.121	−2.53	0.012
Tues.	254	−0.042	1.502	0.094	−0.44	0.660
Wed.	252	−0.073	1.528	0.096	−0.76	0.450
Thurs.	250	0.156	1.465	0.093	1.68	0.094
Fri.	248	−0.036	1.451	0.092	−0.39	0.700

- Weekend to think over, time to get into new week, get receipts before end of the week
- Dealers have more difficulty in soliciting buy orders
- Individuals are net sellers, particularly on small cap stocks. Their sell orders accumulate over weekend, outweighing broker's buy orders and causing a net selling Monday.

(3) Positive Wednesdays

- Settlement hypothesis

(4) All gains overnight

- Orders are collected after the close of trading for execution at the open, with a buy bias.
- Cross selling late in the day to recover funds to match open purchases
- Speculative purchases

8 Conclusions

Our analysis of market regularities over the period 1990–1994 shows that during the market correction that occurred between 1990–1992 a period of apparent chaos set in. Most market regularities show, between 1990 and 1992, a behavior opposite to that observed during previous decades. The year 1994 behaved much in accordance to the historical patterns and 1993 seems to mark a transition period. During 1994, nonetheless, several differences with the historical patterns remain. More observations will be needed to determine

(a) S&P500, 1970–1983

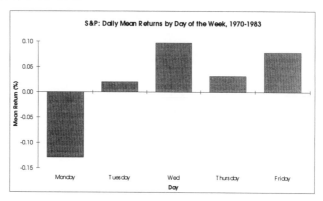

(b) NSA 1949–1988. Source: Ziemba (1991).

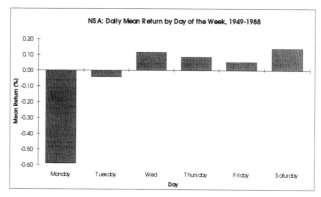

(c) NSA and TOPIX, 1990–1994

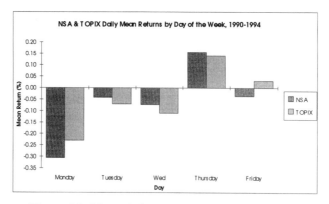

Figure 16. Mean daily return by day of the week

Table 18. Day–of–the–week effects on the TOPIX mean daily returns, 1990—-1994.

Day	Sample size	Mean return %	Standard deviation	SE Mean	T–statistic	P–value
All	1232	−0.050	1.339	0.038	−1.31	0.190
Mon.	242	−0.230	1.610	0.104	−2.26	0.025
Tues.	250	−0.070	1.204	0.076	−0.98	0.330
Wed.	248	−0.110	1.304	0.083	−1.28	0.200
Thurs.	248	0.140	1.230	0.078	1.73	0.085
Fri.	244	0.030	1.295	0.083	0.33	0.740

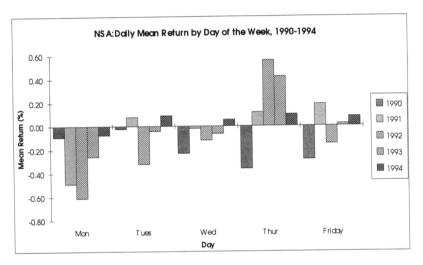

Figure 17. Daily mean returns by day of the week, 1990–1994

if the current behavior of the markets is essentially the same it has historically been, or if the correction produced 'institutional' changes big enough to cause a new and permanent pattern.

The monthly effects show that from 1990 to 1994 January was strongly positive and October and December weakly positive, in agreement with the historical results. However, May stays very high and June and November highly negative, in contrast with historical results.

The turn-of-the-month effect for the period 1990–1994 shows that the one-week period from day −5 to day +2 had positive returns in agreement with the historical results. However, none of these days had significantly positive daily returns, and there were days with negative mean returns, which contrast with the historical results. The first half of the month effect was not present at all. Mean returns during the period from day −5 to day +8 of the month were negative. The pattern changes drastically from 1990–1994, becoming closer to the historical results, but some differences remain.

The holiday effects were absent during 1990–1994. The small sample size does not allow reliable conclusions. However, the total pattern reversion for the period 1990–1994 respect to the previous decades.

The golden week effect was still present during the period of this study. Returns around the time of the golden week are strongly positive, and standard deviations lower than during the rest of the year.

The turn-of-the-year effects were not present. A regression showed that the effects of days −1 of the year, −1 of the month, and pre-holidays, were present. Again, the size of the sample was too small to provide solid conclusions respect to whether 1994 will mark the return to the previous pattern or to a new one. The day-of-week effects had considerable changes from 1990–1994. Mondays were consistently negative during the entire period, and Thursdays became strongly positive after 1992. During 1994 the day-of-the-week pattern was close to the historical pattern.

The following is a brief list of the main differences and similarities with respect to the period 1949–88 observed during the period 1990–1994:

- Most gains occurred at night, as before.

- The TOM was positive but the second week and ROM were negative.

- The January Barometer worked during each year from 1990–1994.

- Pre-holidays did not improve Day of the Week returns during 1990–1994, except for Thursdays returns.

- The Turn-of-the-Year effect was not working well, only days −4 to +1 were positive; volatility for the period of the historical turn-of-the-year was higher.

- Golden Week works very well, with higher returns than in the historical period.

- The days-of-the-week pattern of returns has changed considerably. The main reasons may lie in the consequences of the strong market correction, and in the new market closure on Saturdays.

- NSA and TOPIX had the same overall patterns, although some small differences were observed. The TOPIX had a smoother behaviour

References

Ariel, R.A. (1987) 'A monthly effect in stock returns', *Journal of Financial Economics* **18** 161–174.

Black, F. (1976) 'The pricing of commodity contracts', *Journal of Financial Economics* **3** 167–179.

Elton, E.J. and M.J. Gruber (eds.) (1989) *Japanese Capital Markets*, Harper and Row, New York.

French, K.R. and R. Roll (1986) 'Stock return variances: the arrival of information and the reaction of traders', *Journal of Financial Economics* **17** 5–26.

Hensel, C.R. and W.T. Ziemba (1995) 'The January barometer', *Journal of Investing* **4** (2) 67–70.

Hensel, C.R., G.A. Sick and W.T. Ziemba (1994) 'The turn of the month effect in the US index futures markets, 1982–92', *Review of Futures Markets* **13** (3) 827–856.

Hirsch, Y. (1986) *Don't Sell Stocks on Monday*, Facts on File Publications, New York.

Jaffe, J. and R. Westerfield (1985) 'Patterns in Japanese common stock returns: day of the week and turn of the year effects', *Journal of Financial and Quantitative Analysis* **20** 261–272.

Kato, K. (1990) 'Weekly patterns in Japanese stock returns', *Management Science* **36** 1031–043.

Kato, K. and J.S. Schalheim (1985) 'Seasonal and size anomalies in the Japanese stock market', *Journal of Financial and Quantitative Analysis* **20** 243–259.

Kato, K., S.L Schwartz and W.T. Ziemba (1989) 'Day of the week effects in Japanese stocks'. In Elton and Gruber (1989), 249–281.

Keim, D. (1983) 'Size related anomalies and stock market seasonality: further empirical evidence', *Journal of Financial Economics* 12–32.

Lakonishok, J. and S. Smidt (1988) 'Are seasonal anomalies real? A ninety-year perspective', *Review of Financial Studies* **1** 431–467.

Roll, R. (1983) 'Vas ist das? The turn of the year effect and the return premia of small firms', *Journal of Portfolio Management* **9** (Winter) 18–28.

Stone, D. and W.T. Ziemba (1993) 'Land and stock prices in Japan', *Journal of Economic Perspectives* **13** (Summer) 149–165.

Ziemba, W.T. (1991) 'Japanese security market regularities: monthly, turn of the month and year, holiday and Golden Week effects', *Japan and the World Economy* **3** 119–156.

Ziemba, W.T. and S.L. Schwartz (1991) *Invest Japan: The Structure, Performance and Opportunities of the Stock, Bond and Fund Markets*, Probus Publishing, Chicago, Il.

Ziemba, W.T., W. Bailey and Y. Hamao (eds.) (1991) *Japanese Financial Market Research*, North-Holland, Amsterdam.

Zweig, M. (1986) *Winning on Wall Street*, Warner Books, New York.

Predicting Returns on the Tokyo Stock Exchange*

Sandra L. Schwartz and William T. Ziemba

Abstract

This paper presents a multifactor model used to rank all 1229 (October 1990 count) stocks on the first section of the Tokyo Stock Exchange. The model uses cross section data on thirty fundamental factors most of which are updated monthly. The ranking system is useful for the construction of long and long-short portfolios. Out of sample simulation tests with re-estimated model coefficients indicate that the approach is able to predict differential future performance of different stocks. The model was estimated and tested in 1988–89 using 1979–89 data and used successfully in the 1990s in various forms in proprietary trading.

Introduction

A variety of approaches can be used to help to understand, evaluate, and predict stock prices. In the modern financial literature, the main approaches are those based on the capital asset and arbitrage pricing models. The CAPM was developed independently by Sharpe (1964), Lintner (1965) and Mossin (1966). Stock prices are subject to systematic, or market risk, and to unsystematic or diversification risk. The latter can be minimized with proper diversification.[1] The former, though, cannot, and this risk varies with how the stock price changes as the market moves. The mean return in a period of time from stock i equals the risk free rate of return plus the return the stock earns from bearing its market risk. Hence,

$$E(R_i) = R_f + \beta_i(R_m - R_f)$$

---

*This research was supported by the Centre for International Business Studies and the Program Financial Modeling, Faculty of Commerce, University of British Columbia and the Yamaichi Research Institute, Tokyo. The factor model results presented here were first reported in Ziemba (1990) and Ziemba and Schwartz (1991). Without implicating her, we would like to thank Asaji Komatsu for her help with the statistical calculations.

[1]For example, with 100 independent securities all having the same variance, the resulting portfolio of equally weighted securities has a systematic risk that is only 1% of that of each component stock.

where $E(R_i)$ is stock i's mean return in the time period, R_f is the risk free return in the same time period, $E(R_m)$ is the market's mean return in the same period, and β_i is the stock's beta or risk measure which is the correlation of the stock's return with the market, divided by the market variance.

How well does it work? We do not have space to fully answer this question. The papers by Roll(1988), Sharpe (1983), Ritter and Chopra (1989) and Cadsby (1992) provide insight as does the paper by Kothari and Shanken (1999) in this volume. Roll investigates how much of a stock's return variance the CAPM actually explains. He finds that over time it is about 20% for daily returns, and about 35% for monthly returns. Sharpe investigates how much of the monthly return variance the CAPM explained of US stock returns for 1931–1979. His average of 33.9% is similar to Roll's results. Ritter and Chopra, using data from 1935 to 1986, argue that the risk factor is priced, that is you actually receive extra return for bearing additional risk only during January and then only for small stocks. In months other than January one simply is not paid for the risk borne. Cadsby shows that the CAPM prices risk only during the anomalous periods of the year, namely at the turn of the month, before holidays, etc.

Figures 1 and 2h show the cumulative effect of investing in high beta stocks on the TSE from June 1979 to August 1989. The univariate effect of beta as a single predictor is slightly negative. Investing in stocks with betas one standard above the mean beta yields a total return over the period of about 16% less than the TOPIX. However, in a thirty factor model, once the other twenty-nine variables are separated out, the higher beta stocks outperformed the TOPIX by about 30% in the ten years. This reinforces the view that a high beta was not an especially useful attribute to look for in Japanese stocks, see Ziemba and Schwartz (1991).

From a statistical point of view, it is not surprising that one has trouble predicting stock prices well with only one variable, no matter how reasonable that variable might appear to be. A model that lends itself to many factors is the arbitrage pricing model of Ross (1976). This model is based on deeper mathematics and has many intricacies that require elaborate analysis and subtle interpretation. See Connor and Korajek (1995) for a recent comprehensive survey of research on the APM. The model suggests that mean stock returns are generated by a series of factors whose changes induce changes in the underlying stocks. A typical model is

$$E(R_i) = R_f + b_{1i}F_1 + b_{2i}F_2 + \cdots + b_{ni}F_n$$

where $E(R_i)$ is stock i's mean return in the time period, R_f is the risk free return in the same time period, $F_1, \ldots, F_n$ are the level of the factors $1, \ldots, n$ in this time period and they affect the mean return through the coefficients $b_{1i}, \ldots, b_{ni}$.

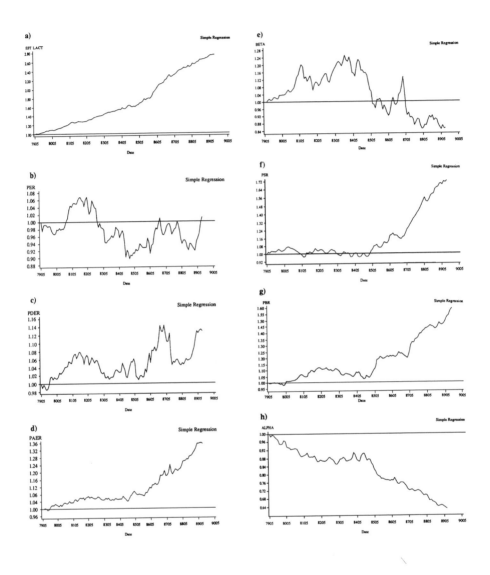

Figure 1. Cumulative Returns Relative to the TOPIX over the Period June 1979 to
August 1989. Eight Multivariate Variables: The *Naive* Effects. *Source*: Yamaichi
Research Institute reported in Ziemba (1990) and Ziemba and Schwartz (1991).

The theory does not specify the number of factors nor what their factors are. Hence, one can either devise factors that are independent using multivariate factor analysis procedures or specify factors that seem to be reasonable. The former approach was tested in Roll and Ross (1980), Dhrymes, Friend and Gultekin (1984) and other authors such as Chen, Roll and Ross (1986) using US data. Sharpe's (1983) study of US stock returns during 1931–79 indicated a set of five factors – yield, beta, size, bond beta and alpha – provides a model that explains 38.2% of the monthly variance of returns versus the 33.9% for beta. When one adds the sector factors – basic industries, capital goods, construction, consumer goods, energy, finance, transportation and utilities – the explanatory power increases slightly to 40.3%. Sharpe also investigated how much one could explain cross sectionally of the differences among stocks in a given month. This is a more difficult task than to explain a given stock's return over time. He found that beta explains only 3.7%, and common factors a further 7.7% for a total of only 10.4%.

A Japanese Factor Model

The factor[2] model developed by William Ziemba with the assistance of Asaji Komatsu at the Yamaichi Research Institute in 1989 followed the approach

---

[2]A detailed survey of other Japanese asset pricing factor models developed around 1988–89 appears in Ziemba and Schwartz (1991). These include: (a) the four factor model of Elton and Gruber (1988) which utilized a constructed factor model approach – linear combinations of fundamental factors – rather than specified individual fundamental factors; (b) Hamao's (1988) APT model that used the six prespecified factors: industrial production, inflation, investor confidence, interest rates, exchange rates and oil prices. This model is similar to the Chen, Roll and Ross (1986) study for the US; (c) Roll and Ross Asset Management used a confidential model that has some of its origins in the Chen, Roll and Ross (1986) and Hamao (1988) work, for asset management in the US, Japan and other localities. This model was used in connection with Daiwa International to jointly manage a multibillion dollar Japanese equity portfolio.. Their research suggested the following key factors: short term inflation measured by CPI changes, long term inflation: measured by changes in real short term risk free interest rates., business cycle: measured by changes in an industrial production index to reflect real output changes, interest rates: measured by long term government bond yields to capture investor's expectations of future interest rates, investor confidence: measured by the difference between high and low grade bonds; (d) CAPMD which was developed by Yamaichi Securities with the aid of consultant professors Steven Brown and Toshiyuki Otsuki. This model combines three separate models: the CAPM, the APT and the approaches of Rosenberg and his co-workers (Rosenberg and McKibben, 1973, and Rosenberg and Guy, 1976a,b, to relate betas and other factors to fundamental indicators that affect the financial environment of the firm. In the model the security return equals the risk-free rate, plus the risk premium and the excess return the model can provide. The risk premium is composed of macroeconomic, market and microeconomic factors. The excess return has theme and idiosyncratic (security specific component) factors. The macroeconomic risk factors are changes in interest rates and percentage changes in the money supply, production, price of oil, wholesale price index and the yen/dollar exchange rate are defined and related to industry returns. By disaggregating the level of the individual firm, the model assesses the sensitivity of the returns

of Jacobs and Levy (1988abc) Thirty factors were developed *ex-ante* to any data analysis after extensive discussions within Ziemba's study group on security market anomalies at YRI. Data was then collected by YRI staff for the 123 months from June 1979 to August 1989. Each of the thirty variables was standardized by substracting the mean and dividing by the standard deviation. Some variables are revised montly, others yearly. The variables used are defined in Table 1.

The returns are computed relative to the TOPIX. For each variable x_i, the standardized variable is $\frac{xi-\text{Mkt}x}{\text{std}x}$, where $\text{Mkt}x = \sum_{i=1}^{n} w_i x_i$, with the weights w_i = market value of i/total market value and

$$\text{std}x = \sqrt{\frac{\sum_{i=1}^{n} (x_i - \text{Mkt}x)^2}{n-1}},$$

and n = all TSE-I stocks, 1229 as of October 1990. Figure 1 shows these cumulative returns in a univariate one at a time sense for beta and several other factors.[3] Jacobs and Levy (1988abc) call these the *naive* effects. Figure 2 has the *pure* or multivariate effects which are the effects when the other 29 variables are separated from each factor for the seven factors with highest mean return (also highest absolute t values) labeled 1–7 in Table 3, for beta and several other factors. Tables 2 and 3 present the results from the univariate and multivariate regressions. The means of the monthly returns from each of the thirty naive and pure effects, plus the t and p statistics for a two-sided test of the hypothesis that there is an effect are shown.

The expected return of each stock i in the TSE-I universe was estimated using the 30 variable regression model

$$E(R_{i,t+1}) = \sum_{j=1}^{30} b_{ij} \overline{A}_{jt}$$

where $E(R_{i,t+1})$ is the expected return of the ith security during month $t+1$ estimated at the end of month t, and $\overline{A}_{jt}$ is the average over the past sixty months of the jth attributes's pure anomaly return, namely

$$\overline{A}_{tj} = \frac{1}{60} \sum_{s=0}^{5} 9 A_{jt-s}$$

where A_{jt} is the jth attribute's pure anomaly return in month t.

Some of the major conclusions are:

---

to these sources of macroeconomic risk. The model then determines the impact of market risk and risk factors for individual firms. This model was used as a valuation screen to identify potential buy and sell decisions, in portfolio management to allow the institution to offset the risk exposure generated by active management, and in portfolio management to attribute the performance to the investment decision process of an active manager.

[3]There is not enough space to show all the plots in Figures 1 and 2, and we focus on the key factors. For plots of missing factors, see Ziemba and Schwartz (1991).

Table 1. Definition of Fundamental and Technical Variables used in the Factor Model Study.

D=difference, A=acceleration, P=price, BV=book value, Div= dividend

	Variable	Definition
1.	PER	Most recent 1 year's earnings/price
2.	PDER	(Most recent 1 year's earnings – previous year's earnings)/price
3.	PAER	{(Most recent 1 year's earnings – previous year's earnings) – (previous year's earning – earnings of 2 year's ago)}/price
4.	PSR	Most recent 1 year's sales/price
5.	PDSR	(Most recent 1 year's sales – previous year's sales)/price
6.	PADR	{(Most recent 1 year's sales – previous year's sales) – (previous year's sales – sales of 2 year's ago)}/price
7.	PBR	Most recent BV/price
8.	PDBR	(Most recent BV – BV of 2 year's ago)/price
9.	PABR	{(Most recent BV – BV of 1 year ago) – BV of 1 year ago – BV of 2 years ago)}/price
10.	YIELD	Most recent 1 year's Div/price
11.	DYield	(Most recent 1 year's Div – previous year's Div)/price
12.	AYield	{(Most recent 1 year's Div– previous year's Div) – (previous year's Div – Div of 2 years ago)}/price
13.	Div-P	Most recent year's Div/most recent year's earnings
14.	DDiv-P	(Most recent year's Div/most recent year's earnings) – (previous year's Div/previous year's earnings)
15.	ADiv-P	{(Most recent year's Div/most recent year's earnings) – (previous year's Div/previous year's earnings)} – {(previous year's Div/most recent year's earnings) – (Div of 2 year's ago – earnings of 2 year's ago)}
16.	D-Zero	$\begin{cases} 1 & \text{if no dividends were paid during the most recent year,} \\ 0 & \text{otherwise} \end{cases}$
17.	TMVLOG	Log(capitalization)
18.	Shares	Outstanding shares
19.	Price50	Price converted to par value of Y50, measures low price effect
20.	IND-R	Shares held by individual investors/outstanding shares
21.	D-PAR	$\begin{cases} 1 & \text{if par value is at least Y500,} \\ 0 & \text{otherwise} \end{cases}$
22.	EST-LACT (growth)	(Earnings estimate of the current fiscal year – most previous year's earning)/price
Variables 23–27 are based on the previous 60 month's regression		
23.	Beta	Slope of time series regression
24.	Alpha	Intercept of the time series regression
25.	Sigma	Mean squared error of the time series regression
26.	EPS	Most recent month's residual
27.	EPS-1	Previous month's residual
28.	RELSTR	Relative strength measured by $0.4\text{Ret}_t + 0.2\text{Ret}_{t-1} + 0.2\text{Ret}_{t-2} + 0.2\text{Ret}_{t-3}$
29.	D-RELSTR	$\text{RELSTR}_t - \text{RELSTR}_{t-1}$
30.	R-MAX24	Price/highest price during the last 2 years

Source: Yamaichi Research Institute reported in Ziemba (1990) and Ziemba and Schwartz (1991).

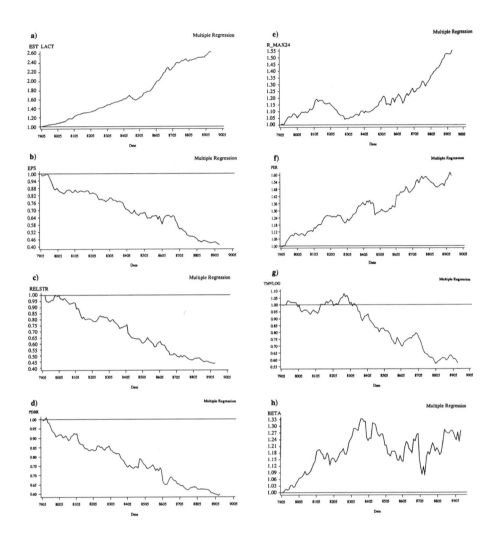

Figure 2. Cumulative Returns Relative to the TOPIX over the Period June 1979 to August 1989 for each of 30 Multivariate Variables: the Pure Effects.

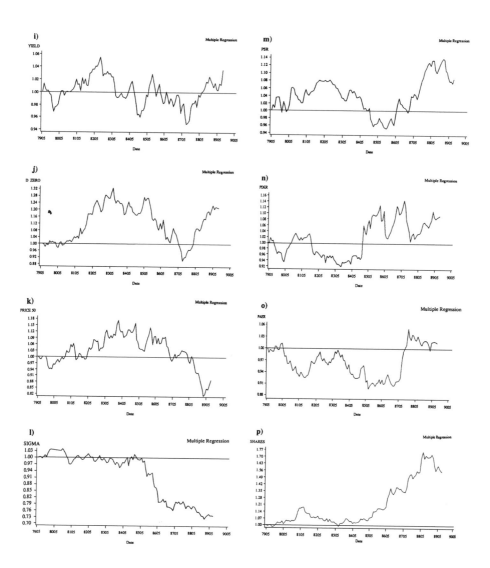

Figure 2 *continued*

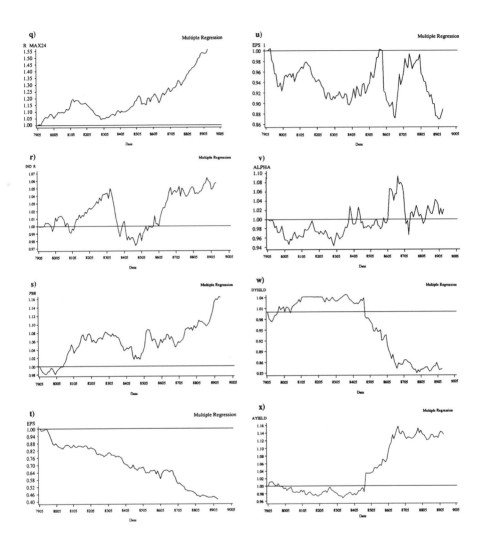

Figure 2 *continued*

Table 2. Univariate (Naive) Effects of 30 Factors on the TSE-I, June 1979 to August 1989.

	Factor	Monthly Mean Return from Factor, %	t-Statistic	p-Significance, 2-sided test
1	ADiv-P	0.00013	0.12	0.904
2	Alpha	−0.00384	−3.99	0.000
3	AYield	0.00065	1.31	0.191
4	Beta	−0.00104	−0.45	0.066
5	D-PAR	−0.00370	−0.02	0.980
6	D-RELSTR	−0.00233	−1.92	0.058
7	D-Zero	0.60613	1.88	0.063
8	DDiv-P	0.00110	0.94	0.347
9	Div-P	0.00033	0.22	0.826
10	DYield	−0.00122	−0.18	0.067
11	EPS	-0.00350	−3.04	0.003
12	EPS-1	−0.00195	−1.46	0.146
13	EST-LACT	0.00820	8.74	0.000
14	INP-R	0.00240	2.50	0.014
15	PABR	0.00017	0.19	0.852
16	PAER	0.00263	3.25	0.002
17	PASR	0.00084	0.95	0.344
18	PBR	0.00398	3.31	0.001
19	PDBR	−0.00165	−1.39	0.167
20	PDER	0.00094	0.96	0.338
21	PDSR	−0.00117	−1.29	0.200
22	PER	−0.00038	−0.30	0.768
23	Price50	−0.00719	−3.04	0.003
24	PSR	0.00416	2.98	0.003
25	R-MAX24	−0.00279	−1.47	0.144
26	RELSTR	−0.00649	−3.85	0.000
27	Shares	−0.00017	−0.07	0.944
28	Sigma	−0.00434	−3.02	0.003
29	TMVLOG	−0.00396	−2.24	0.027
30	YIELD	0.00233	2.50	0.014

Source: Yamaichi Research Institute reported in Ziemba (1990) and Ziemba and Schwartz (1991)

Table 3. Multivariate (Pure) Effects of 30 Factors on the TSE-I, Ranked by
t-Statistics, June 1979 to August 1989.

	Factor	Monthly Mean Return from Factor, %	t-Statistic	p-Significance, 2-sided test
1	EST-LACT	0.00788	7.85	0.000
2	EPS	-0.00693	−3.65	0.000
3	RELSTR	−0.00645	−3.42	0.001
4	PDBR	−0.00397	−2.87	0.005
5	R-MAX24	0.00363	2.86	0.005
6	PER	0.00385	2.69	0.008
7	TMVLOG	−0.00429	−2.21	0.029
8	Shares	0.00382	1.87	0.065
9	Sigma	−0.00234	−1.72	0.088
10	PBR	0.00124	1.68	0.095
11	AYield	0.00105	1.49	0.139
12	DYield	−0.00136	−1.45	0.149
13	DDiv-P	0.00281	1.35	0.179
14	D-RELSTR	0.00250	1.22	0.224
15	PDSR	−0.01645	−1.20	0.233
16	Beta	0.00223	1.11	0.268
17	ADiv-P	−0.00147	−1.00	0.318
18	PDSR	0.00109	0.93	0.356
19	IND-R	0.00046	0.84	0.404
20	D-Zero	0.00184	0.83	0.407
21	PSR	0.00072	0.67	0.507
22	EPS=1	−0.00846	−0.63	0.528
23	PABR	0.00097	0.60	0.550
24	PDER	0.00092	0.53	0.596
25	YIELD	0.00033	0.38	0.706
26	Price50	−0.00068	−0.34	0.737
27	Alpha	0.00025	0.19	0.846
28	PAER	0.00020	0.18	0.854
29	D-PAR	−0.00042	−0.09	0.930
30	Div-P	−0.00012	−0.07	0.941

Source: Yamaichi Research Institute reported in Ziemba (1990) and Ziemba and
Schwartz (1991).

- Low PER had no effect by itself (Figure 1c) but once the effects of other variables were eliminated the pure effect was quite strong (Figure 2f)

- Earnings changes and acceleration in earnings were very positive (Figures 2no)

- Low price to sales and price to book value ratios were very positive by themselves (Figures 1de) but the pure effect gave only a 10% edge over the TOPIX over the ten years (Figures 2ms).

- As in the US, stocks with higher yields had higher returns and the highest returns are those with no dividends at all (Figures 2ij)

- There was a strong small stock effect (Figure 2g). Low price stocks also had high returns as shown in Price 50 (Figure 2k)

- Stocks reacted positively to the number of outstanding shares (Figure 2p)

- Stocks with higher percentages owned by individuals outperformed those with lower (Figure 2r).

- By far the strongest positive variable with over a 250% edge on the TOPIX was earnings growth - EST-LACT[4] (Figure 2a).

- Beta had a negative univariate (Figure 1) but a positive multivariate effect (Figur 2h).

- There was mean reversion, negative alpha, in the univariate results but not in its pure effect (Figure 2r). But there was strong mean reversion of current and past residuals as well as relative return strength averaged over the past four months (Figures 2ctu).

- Sigma had a negative effect (Figure 2l).

---

[4]Darrough and Harris (1991) consider the effect of management forecasts of earnings on stock prices in Japan. Their empirical results demonstrate that: (1) management forecasts at both parent and consolidated earnings were generally more accurate than a random walk model using past earnings; (2) the analysts forecasts of parent-only earnings preceeding the announcement date was the most accurate measure of unexpected earnings and was most closely associated with unexpected returns; (3) it was difficult to distinguish between managment forecast, imputed (parent-earnings-based) and random walk measures of unexpected consolidated earnings relative to their associations with unexpected returns and (4) the sign of management forecasts – up or down – are associated with unexpected returns for both parent and consolidated earnings. This research suggests even more earnings variables to utilize in models. Whether or not they would improve on EST-LACT is not clear. What is clear is that these variables are strong predictors of future stock prices.

- Changes in price relative to its maximum price in the past two years was negative in univariate but strongly positive in its pure effect (Figures 2q).

A Simulation Test

Table 4 shows how one might have done with monthly revisions by choosing the best 50, 100, 200, 300, 400, 500 versus all the TSE-I equally weighted, the TOPIX (the TSE-I value weighted), and the NSA 225. This simulation re-estimated the model with new data from June 1979 to month t then the model selects the best stock for month $t + 1$ and a revision is made to choose the best stocks, equally weighted in the portfolio. One yen invested in the best 50 stocks would have grown to Y7.57 from January 1985 to December 1989. Meanwhile, the TOPIX and NSA were at Y2.96 and Y3.09, respectively.

The simulation test used all thirty factors to predict the expected return of each stock in each month for the years 1985 to September 1989. For each year's predictions, the model was re-estimated using data up to December of the previous year. The model was then used for each of the twelve months of the year in question and then re-estimated for the following year. The model portfolio was revised monthly with transactions costs estimated to be 1% round trip:

$$\sum_{j=1}^{30} b_{ij}\overline{A_j}$$

where Aj_t is the effect of factor j in period t, $\overline{A_j} = \sum_{t=1}^{T} A_{jt}$ is the average effect of factor j and b_{ij} the estimated factor weightings. Table 4 summarizes the results of the simulation of r the ranking strategy to determine the model's best 50, 100, 200, 300, 400, and 500 stocks. These are compared with three indices, the equally weighted TSE, the price weighted Nikkei 225 and the value weighted TOPIX. The first column has the final September 1989 values of each portfolio assuming an investment of Y1 at the beginning of January 1985. The TSE-I's Y4.44 versus the TOPIX's Y2.96 shows a positive small cap effect in this period. The ranking system worked well with monotone increases in returns for the higher ranked portfolios; there was an increase in standard deviation risk as well but CAPM risk adjusted the model's results did add value. Figure 3 visually displays the month by month results. The model was used after 1989 in proprietary trading by Yamaichi Securities and elements of this model have been used in proprietary trading by Buchanan Partners, a London based hedge fund.

Table 4. Performance of the Best 50, 100, 200, 300, 400, 500 and all Equally Weighted TSE-I Stocks versus the Topix and NSA225 with Monthly Revisions, January 1985 to September 1989.

Portfolio	Value of Y1 invested in Jan 85 at end of Sept 89	Annualized* Return in %	Monthly Mean Return, %	St Dev of Monthly Returns
Best 50	7.57	53.1	3.76	5.57
Best 100	7.12	51.2	3.64	5.36
Best 200	6.54	48.5	3.48	5.17
Best 300	6.16	46.6	3.37	5.13
Best 400	5.84	45.0	3.26	5.02
Best 500	5.54	43.4	3.16	4.88
Equally weighted TSE-I	4.44	36.8	2.80	4.17
Nikkei	3.09	26.8	2.09	4.42
TOPIX	2.96	25.7	2.04	4.96

Source: Yamaichi Research Institute reported in Ziemba (1990) and Ziemba and Schwartz (1991).

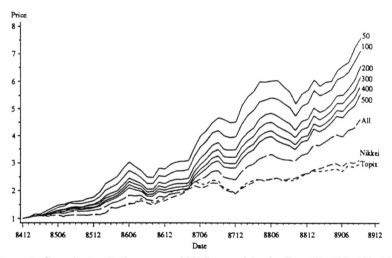

Figure 3. Cumulative Performance of Y1 Invested in the Best 50, 100, 200, 300, 400, 500 and all Equally Weighted TSE-I Stocks versus the Topix and NSA225, Monthly Revisions of Data and Model, January 1985 to September 1989. *Source*: Yamaichi Research Institute reported in Ziemba (1990) and Ziemba and Schwartz (1991).

Related work of Jacobs & Levy and Haugen

The factor model discussed in this paper is a Japanese version of the approach suggested by Jacobs and Levy (1988a). The study group on anomalies of the Yamaichi Research group led by William Ziemba used the JL work as a protoype for the Japanese application with factors develped specifically for the Japanese situation. JL used the 25 factors listed in Table 5. Their model utilized data from January 1978 to December 1986 of the 1500 lowest capitalized stocks. Their benchmark was the S&P500. Their estimates of the expected return contributions for the univariate and multivariate models are shown in Table 6 for the whole year and adjusted for January effects.

Haugen, as reported in Haugen and Baker (1996), in a paper presented at the Isaac Newton Institute Conference upon which this book is based, presents an approach independently developed that is similar to that of Jacobs and Levy for the US and that presented here for Japan.

Haugen separated the factors into the five groups:

I. Risk
  Market Beta
  APT Betas, Ind. Prod., Infl., T-Bill, Term, Default,
  Volatility of Total Return
  Residual Variance
  Debt/Equity
  Interest Coverage
  E.P.S. Std. Error

II. Liquidity
  Price-Per-Share
  Avg. Daily Volume/Market Cap.
  Market Cap.

III. Price Level
  Earnings to Price
  Cash Flow to Price
  Sales to Price
  Dividend to Price

IV. Growth Potential
  Profit Margin
  Capital Turnover
  Return on Assets
  Return on Equity
  Trailing Earnings Growth

V. Technical
  1, 3, 6, 12, 24, 60 Month Excess Returns

Table 5. Factors in Current and Lagged Time used in the Jacobs–Levy (1988a,b,c) Study.

Low P/E	Trailing year's fully diluted earnings per share divided by price
Book/Price	Common equity per share divided by price
Cash Flow/Price	Trailing year's earnings plus depreciation and deferred taxes per share divided by price
Sales/Price	Trailing year's earnings per share divided by price, relative to the capitalization-weighted acergae sales per share for that stock's industry
Yield	Indicated annual dividend divided by price, as well as a binary indicator of zero yield
Beta	Calculated quarterly froma rolling 60-month regression of stock excess (over Treasury bill) returns on S&P excess returns, with a Vasicek Bayesian adjustment
Coskewness	Calculated quarterly on a rolling 60-month as:

$$\frac{\sum (R_i - \overline{R}_i)(R_m - \overline{R}_m)^2}{\sum (R_m - \overline{R}_m)^2},$$

where R_i is stock excess (over Treasury bill) return, R_m is the S&P 500 excess return, and $\overline{R}_i$ and $\overline{R}_m$ are rolling 60-month arithmetic averages.

Sigma	Calculated as the standard error of estimate, or dispersion of error terms, from the beta regression
Small Size	$-\log(\text{market capitalization})$
Earnings Torpedo	The change from the latest earnings per share last reported to next year's consensus estimate, divided by stock price
Earnings	The standard deviation of next year's analyst's earnings
Controversy	estimates, divided by stock prices
Neglect	$-\log(1 +$ the number of security analysts following each stock)
Low price	$-\log(\text{stock price})$
Relative Strength	The intercept, or alpha, from the rolling 60-month beta regression
Residual reversal	Measured separately for each of the two most recently completed months as the residuals from the beta regression
Tax-Loss Measures	Proprietary models of potential short- and long-term tax-loss selling pressure for each stock
Trends in Analysts'	Measure separately for each of the three most recently
Earnings Estimates	completed months as the change in the next fiscal year's consensus estimate, divided by stock price
Earnings Surprise	Measured separately for each of the three most recently completed months as the difference between the announced earnings and the consensus estimate on that date, divided by the stock price

Table 6: Fundamental Variable Effects on US Equity Prices, Univariate and
Multivariate.

Return factor	Univariate effect Monthly mean	t-stat	Multivariate effect Monthly mean	t-stat	Multivariate effect Jan. mean	t-stat	Multivariate effect Non-Jan. mean	t-stat
Low P/E	0.59	3.4	0.46	4.7	0.09	0.5	0.49	4.7
Small Size	0.15	2.3	0.12	2.7	0.14	1.3	0.12	2.5
Yield	−0.01	− 0.1	0.03	0.5	0.67	3.4	−0.03	−0.4
Zero Yield	0.00	0.0	0.15	1.3	1.00	1.9	0.08	0.6
Neglect	0.14	1.9	0.10	1.7	0.36	1.8	0.08	1.3
Low Price	−0.01	−0.1	0.01	0.2	0.38	2.0	−0.02	−0.4
Book/Price	0.17	1.4	0.09	1.2	0.51	2.4	0.05	0.7
Sales/Price	0.17	3.1	0.17	3.7	0.05	0.2	0.18	4.1
Cash/Price	0.36	2.7	0.04	0.6	−0.15	−2.0	0.05	0.8
Volatility (S.D.)	0.16	0.6	0.07	0.6	0.62	2.1	0.02	0.2
Beta	−0.01	0.0	0.04	0.3	−0.05	−0.1	0.05	0.4
Coskewness	0.09	0.6	0.04	0.7	0.10	0.5	0.04	0.6
Controversy	−0.33	−2.1	−0.05	−0.8	−0.01	−0.1	−0.06	−0.8
Trend in Ear. Ests (−1)	0.48	4.8	0.51	8.1	0.60	3.8	0.50	7.5
Trend in Ear. Ests (−2)	0.40	4.4	0.28	4.9	0.25	1.6	0.29	4.7
Trend in Ear. Ests (−3)	0.29	3.0	0.19	3.8	0.13	0.6	0.19	3.8
Earn. Surprise (−1)	0.44	2.1	0.48	3.7	1.36	1.6	0.42	3.4
Earn. Surprise (−2)	0.47	1.8	0.18	0.8	0.14	2.0	0.18	0.7
Earn. Surprise (−3)	−0.03	−0.1	−0.21	−1.1	−0.01	0.0	−0.22	−1.1
Earn. Torpedo	0.00	−0.0	−0.10	−1.7	−0.08	0.3	−0.12	−1.9
Relative Strength	0.30	1.4	0.34	3.5	−0.13	−0.2	0.39	1.0
Res. Reversal (−1)	−0.54	−4.9	−1.08	−17.8	−1.38	−6.0	1.06	−16.8
Res. Reversal (−2)	−0.13	−1.4	−0.37	−8.1	−0.56	−2.5	−0.35	−7.7
Short Term Tax	−0.08	−0.4	−0.04	−0.4	0.38	1.8	−0.08	−0.7
Long Term Tax	−0.29	−1.6	0.00	−0.1	0.78	3.2	−0.07	−1.2

Source: Jacobs and Levy (1988a). −1 means one month lagged, etc. The effect is the
mean change in expected return from a one standard deviation tilt in the direction
of the given factor.

Table 7. Mean Returns and *t*-Statistics in the 12 most Important Factors, Japan and the US, 1985–93.[5]

	United States		Germany		France		United Kingdom		Japan	
	Ave.	*T*-stat	Ave.	*T*-stat	Ave.	*T*-stat	Ave.	*T*-stat	Ave.	*T*-stat
1. 1-Month Excess Return	-0.32%	-10.8	-0.26%	-8.8	-0.33%	-11.3	-0.22%	-7.6	-0.39%	-13.3
2. Book to Price	0.14%	4.7	0.16%	5.3	0.18%	6.1	0.12%	4.2	0.12%	4.2
3. 12-Month Excess Return	0.23%	7.8	0.08%	2.8	0.12%	4.2	0.21%	7.3	0.04%	1.5
4. Cash Flow to Price	0.18%	6.2	0.08%	2.7	0.15%	5.1	0.09%	3.1	0.05%	1.7
5. Earnings to Price	0.16%	5.5	0.04%	1.4	0. 13%	4.4	0.08%	2.7	0.05%	1.9
6. Sales to Price	0.08%	2.7	0.10%	3.3	0.05%	5.1	0.05%	1.7	0.13%	4.5
7. 3-Month Excess Return	-0.01%	-0.5	-0.14%	-4.7	-0.08%	-2.6	-0.08%	-2.6	-0.26%	-8.7
8. Debt to Equity	-0.06%	-2.1	-0.06%	-2.1	-0.09%	-3.1	-0.10%	-3.4	-0.01%	-0.4
9. Variance of Total Return	-0.06%	-1.9	-0.04%	-1.4	-0.12%	-4.1	-0.01%	-0.5	-0.11%	-3.8
10. Residual Variance	-0.08%	-2.6	-0.04%	-1.3	-0.09%	-3.1	-0.03%	-1.2	0.00%	-0.1
11. 5-Year Excess Return	-0.01%	-0.4	-0.02%	-0.7	-0.06%	-1.9	-0.06%	-2.1	-0.07%	-2.3
12. Return on Equity	0.11%	3.9	0.01%	0.4	0.10%	3.5	0.04%	1.3	0.0%	1.8

Source: Haugen and Baker (1996)

Table 7 shows the top twelve factors found by Haugen and Baker (1996) for the US and Japan and also Germany, France and the United Kingdom using data from 1985–93. Given the different data periods used by the three authors, there are some differences in the most important factors. However, all three models have been successfully used and have many common factors.

References

Cadsby, Charles B. (1992) 'The CAPM and the calendar: empirical anomalies and the risk-return relationship', *Management Science* **38** (11) 1543–1561.

Chen, Nai-fu, Richard Roll and Stephen Ross (1986) 'Economic forces and the stock market', *Journal of Business* **59** 383–403.

Connor, Greg and Robert Korajek (1995) 'The arbitrage theory and multifactor models of asset returns'. In *Finance*, R.A. Jarrow, V. Masimovic and W.T. Ziemba, eds., North-Holland, 87–144.

Darrough, Masako and Trevor Harris (1991) 'Do management forecasts of earnings affect stock prices in Japan?' In *Japanese Financial Market Research*, W.T. Ziemba, W. Bailey and Y. Hamao, eds., North-Holland, 197–229.

Dhrymes, Phoebus I., Irwin Friend, and N. Bulent Gultekin (1984) 'A critical reexamination of the empirical evidence on the arbitrage pricing theory', *Journal of Finance* **39** 323–346.

Elton, Edwin J. and Martin J. Gruber (1988) 'A multi-index risk model of the Japanese stock market', *Japan and the World Economy* **1** 21–44.

Hamao, Yasushi (1988) 'An empirical examination of the arbitrage pricing theory using Japanese data', *Japan and the World Economy* **1** 45–61.

Haugen, Robert and N. Baker (1996) 'Commonality in the determinants of expected stock returns', *Journal of Financial Economics* **41**(3) 401–439.

Jacobs, Bruce I. and Kenneth N. Levy (1988a) 'Disentangling equity return regularities: new insights and investment opportunities', *Financial Analysts Journal* **44** 18–43.

Jacobs, Bruce I. and Kenneth N. Levy (1988b) 'On the value of value', *Financial Analysts Journal* **44** 47–62.

Jacobs, Bruce I. and Kenneth N. Levy (1988c) 'Calendar anomalies: abnormal returns at calendar turning points', *Financial Analysts Journal* **44** 28–39.

Kothari, S.P. and J. Shanken (1999) 'Beta and Book-to-Market', this volume, 44–64.

Lintner, John (1965) 'The valuation of risky assets and the selection of risky investments in stock portfolios and capital budgets', *Review of Economics and Statistics* **47** 13–37.

Mossin, Jan (1966) 'Equilibrium in a capital asset market', *Econometrica* **34** 768–783.

Ritter, Jay and Nisan Chopra (1989) 'Portfolio rebalancing and the turn-of-the-year effect', *Journal of Finance* **44** 149–166.

Roll, Richard (1988a) 'The international crash of 1987'. In *Black Monday and the Future of Financial Markets*, R.W. Kamphuis, R.C. Kormendi and J.W.H. Watson, eds., Dow-Jones Irwin, Homewood, Ill.

Roll, Richard (1988b) R^2', *Journal of Finance* **63** 541–566.

Roll, Richard and Stephen Ross (1980) 'An empirical investigation of the arbitrage pricing theory', *Journal of Finance* **35** 1073–1103.

Rosenberg, Barr and James Guy (1976a) 'Prediction of Beta from Investment Fundamentals', *Financial Analysts Journal* **32** 62–70.

Rosenberg, Barr and James Guy (1976b) 'Prediction of beta from investment fundamentals; Part II', *Financial Analysts Journal* **32** 71–79.

Rosenberg, Barr and Walt McKibben (1973) 'The prediction of systematic and specific risk in common stocks', *Journal of Financial and Quantitative Analysis* **8** 317–333.

Ross, Stephen (1976) 'The arbitrage theory of capital asset pricing', *Journal of Economic Theory* **13** 341–360.

Sharpe, William F. (1964) 'Capital asset prices: a theory of market equilibrium under conditions of risk', *Journal of Finance* **19** 425–442.

Sharpe, William F. (1983) 'Factors in New York stock exchange returns: 1931–1979', *Journal of Portfolio Management* (Winter) 5–19.

Ziemba, William T. (1990) 'Cumulative effects of fundamental variables on the Tokyo Stock Exchange: 1979–89', working paper, University of British Columbia, Vancouver, Canada. Shortened version in *Investing*, presented at Berkeley Program in Finance, September 11, 1990.

Ziemba, William T. and Sandra L. Schwartz (1991) *Invest Japan*, Probus Publishing, Chicago.

High Stock Returns Before Holidays: International Evidence and Additional Tests*

Alonso Cervera and Donald B. Keim

Abstract

Stock markets around the world exhibit significantly higher than average returns on the trading days preceding holidays. Stocks advance with disproportionate frequency on pre-holiday trading days. The pre-holiday return is 5.5 times larger than the daily return on other days, on average, across the 17 (non-US) countries in our sample for the period 1980–1994. The largest difference is found in the Canadian stock market, and the Netherlands is the only country where pre-holiday returns are lower than the average return. These results are consistent with the results found for the US stock market.

1 Introduction

Previous research documents a 'holiday effect' in the US stock market – on the trading day before a holiday that closes the market, stock returns are economically and statistically greater than returns on the remaining days of the year. The earliest recording of this phenomenon was by Fields (1934) who noted a disproportionate number of positive returns for the Dow Jones Industrial Average (DJIA) on trading days preceding long holiday weekends during the period 1901–1932. Merrill (1966) finds similar evidence for the DJIA on days preceding holidays in the 1897–1965 period. Fosback (1976) reports high pre-holiday returns in S&P 500 index returns. Ariel (1990) recently updated the evidence for the value- and equal-weighted indexes constructed by the Center for Research in Security Prices (CRSP) for the 1963–1986 period. Ariel estimates that over one-third of the total market return was earned on the trading days that precede the eight holidays that result in market closings each year. Evidence in other US stock market research is consistent with these holiday findings: Lakonishok & Smidt (1984) report high stock returns on the last trading day before Christmas, and Keim (1983) and Roll (1983) find that US small stocks have disproportionately high returns on the last trading day before New Year's Day.

Recent studies by Ziemba (1991), Cadsby & Ratner (1992) and Kim & Park (1994) extend the analysis to a sample of non-US stock markets and

---

*We thank Mike Barclay and Bill Ziemba for helpful comments and Karla Cervera and Todd Solash for expert research assistance.

find evidence of a holiday effect in the UK, Japan, Canada, Hong Kong, and Australia. Cadsby & Ratner (1992) and Kim & Park (1994) conclude that because the dates of the market-closing holidays are different across markets, the separately-identified holiday effects in the various markets are unrelated (in contrast, say, to the Monday seasonal which shares a common timing across markets).

An explanation for the high stock returns before holidays is elusive. Popular explanations rely on price pressure stories in which prices are artificially bid up by excess buying demand. For example, speculators might cover their short positions prior to the holiday to avoid an 'excessively risky' exposure over the extended closed period during which adverse (relative to their position) news might be announced. Such an argument has been forwarded by Ritter (1988) who suggests that investors' decisions to either buy or sell might cluster at certain calendar turning points (e.g., before weekends, holidays, or other market closures), inducing irrational price movments. Consistent with such investor behavior, Keim (1989) reports a tendency for stocks to close at the ask on the last trading days before holidays in the US, indicating that the pre-holiday price movements might result from systematic buying and selling behavior of market participants on those days. Keim reports that a significant fraction of the pre-holiday returns results from intra-spread price movements that are not exploitable (for profit) by typical investors.

This paper expands on previous research by extending both the time period and the number of markets in which the pattern has been analyzed. We examine the existence of the holiday effect for the period 1980–1994 for a sample of seventeen countries. The holiday effect is particularly strong for non-European countries. We also find that a large component of the international holiday effect is attributable to large return preceding holidays that are common to many of the countries in our sample (e.g., Christmas, New Year's day, and Easter). However, our results show that a holiday effect remains in the data after accounting for the common holidays.

The paper is organized as follows. Section 2 describes the data. Section 3 describes the empirical tests and reports our findings. Section 4 describes some of the structural differences across the stock markets and suggests possible explanations for the large positive returns that accrue to stocks on the days preceding holidays. Section 5 concludes the paper.

2 Data

2.1 Stock Market Indexes

This section describes the stock market indexes used in this study.[1] The seventeen international stock market indexes are listed below by geographic

---

[1]The information presented in this section is from *The Guide to World Equity Markets 1992* edited by Stuart Allen and Selina O'Connor, published by Euromoney Publications

locale. The sample represents both developed and emerging capital markets, and is geographically diversified. In most cases, the daily data cover the period 1980–1994. Exceptions are indicated below. The sources of these indices were Datastream and the Bolsa Mexicana de Valores (Mexican Stock Exchange).

North America

1. **Canada** – the Toronto Stock Exchange (300) Composite Price Index is calculated from the capital-weighted average stock price of approximately 300 companies listed on the Toronto Stock Exchange representing 14 industry sectors.

2. **Mexico** – the Indice de Precios y Cotizaciones (IPC) (Index of Prices and Quotations) is a value-weighted average of approximately 30 stocks that are representative in the market.

South America

3. **Chile** – the Indice General de Precios de las Acciones (IGPA) (General Price Index) is a capitalization-weighted index of the majority of the companies traded on the Santiago Stock Exchange. The data cover the period 1987–1994.

Europe

4. **France** – Datastream's equal-weighted index of representative French stocks trading on the Paris Bourse.

5. **Germany** – the Frankfurt Commerzbank Price Index, a capitalization-weighted index of leading German blue-chip stocks.

6. **Italy** – Datastream's equal-weighted index of representative Italian stocks.

7. **The Netherlands** – the capitalization-weighted CBS All Share General Price Index includes all the ordinary shares of Dutch companies listed on the Amsterdam Stock Exchange.

8. **Spain** – the Madrid Stock Exchange General Price Index is computed for selected stocks trading on the Madrid Stock Exchange and represents over 80% of the total market capitalization. The data cover the period 1984–1994.

9. **Switzerland** – Datastream's equal-weighted index of representative Swiss stocks.

Plc. and *The Handbook of World Stock and Commodity Exchanges*, published by Blackwell Finance in 1994.

10. **United Kingdom** – the FTSE 100 Price Index is based on the 100 largest companies (by market capitalization) listed on the London Stock Exchange. The data cover the period 1984–1994.

Asia and Pacific Rim

11. **Australia** – the All Ordinaries Price Index is the primary market indicator for the Australian Stock Exchange. It contains 339 of the most active listed companies and represents 95.5% of the total capitalization of the market.

12. **Hong Kong** – the Hang Seng Price Index is a value-weighted index containing approximately 35 companies. It is the primary indicator of market performance in Hong Kong. It represents approximately 70% of the total market capitalization of the Hong Kong Stock Exchange.

13. **Japan** – the Nikkei 225 Price Index is a simple average of 225 top-rated stocks, the calculation of which is similar to that of the Dow Jones Industrial Average in the US

14. **Korea** – the Korea Composite Stock Price Index (KOSPI) is a capitalization-weighted index of the most representative companies in the market which make up 70% of the market value of the Korean Stock Exchange.

15. **Malaysia** – the Kuala Lumpur Composite Price Index is a value weighted average of the prices of approximately 100 stocks.

16. **Singapore** – the Straits Times Industrial Price Index is an equal-weighted average of the prices of 30 stocks.

17. **Taiwan** – the Weighted Price Index is a capitalization-weighted index of all listed common shares traded on the Taiwan Stock Exchange. The data cover the period 1986–1994.

2.2   Holidays

The legal holidays for each country were obtained from Puello (1993) and Thompson & Carlson (1994) and are reported in the Appendix. While these days are all classified as legal holidays, some do not result in a stock market closing. As a result, we treat such days in our analysis as ordinary trading days. The returns that accrue on the trading days prior to other events, like presidential elections, that cause the suspension of stock market activities were not treated as pre-holiday returns. Finally, in some cases there are local holidays that only apply to a certain region of the country in which the market operates. These cases, not shown in the Appendix, were treated as market holidays if they result in market closings.

3   Results

Table 1 presents estimates of the mean returns for pre-holiday days and for the other days of the year. In every country except the Netherlands, pre-holiday returns are higher than the mean returns for the rest of the days. The economic magnitude of this return differential is impressive. In 11 of the 17 markets in our sample, pre-holiday returns are 15 basis points (more than 37% annualized) greater than the average returns on the other trade days of the year. Stated differently, the mean pre-holiday return is three times greater than the mean return on the rest of the trade days in 15 of the 17 markets, and five times greater in 9 markets. The ratio of pre-holiday returns to returns on other days is greatest in Malaysia and Canada where pre-holiday average returns are about 10 times the mean return on all other days. On average, the pre-holiday returns are 5.5 times larger than the daily returns on other days for the 17 markets in our sample.

To test the statistical significance of this difference, we estimate the following regression for each stock market index:

$$R_t = a + bD_t + \epsilon_t \tag{1}$$

where D_t equals 1 if day t is the day before a holiday, and 0 otherwise. The intercept a measures the average return for all days except those preceding a holiday, and the coefficient b measures the difference in return between pre-holiday days and all other days. The T-statistic for the null hypothesis that this coefficient is zero tests the significance of the difference.

Estimated t-statistics are reported in the third column of Table 1 for the entire time period. In eight of the seventeen countries, the return differential is significantly different from zero at the .05 level. The exceptions are primarily European countries (France, Germany, Italy, the Netherlands, Spain, Switzerland and the United Kingdom) and Hong Kong and Korea.[2] For most of the emerging markets, the results indicate the existence of the holiday effect. The holiday effect is most significant in Mexico where the pre-holiday return of 0.67% exceeds the average return on all other trading days by 0.46% (T-statistic = 2.45).[3]

To assess whether the significantly higher mean returns before holidays are unduly influenced by a few outliers, we also conduct a nonparametric test.

---

[2]Cadsby & Ratner (1992) similarly find no holiday effect for the European markets in their sample.

[3]We also estimated a variant of regression model (1) that tested whether the average returns for pre-holiday trading sessions associated with holidays on Fridays or Mondays are statistically different from average pre-holiday returns for holidays on Tuesday or Wednesday, or Thursday. The idea is that holidays that occur on Mondays or Fridays result in a longer market closure because of the adjacent weekend. The results indicate that for most of the markets in our sample, the pre-holiday returns on Thursdays or Fridays are not significantly different from the pre-holiday returns on Mondays, Tuesdays and Wednesdays. The only exceptions are Korea and Mexico.

Table 1. Means, standard deviations, and frequency of positive returns for international stock indexes for pre-holidays and other days. (1980–1994)

Country	Mean (Std. Dev.) of Daily Returns (%)		T-Statistic of Return Difference[a]	Positive Returns on Pre-Holidays		χ^{2} [c]
	Pre-Holidays	Other Days		Observed Percentage	Expected Percentage[b]	
Australia	0.27 (0.71) 85[d]	0.04 (1.03) 3,688	2.08	65.88%	53.25%	5.10
Canada	0.21 (0.50) 118	0.02 (0.82) 3,651	2.57	67.80%	53.12%	9.57
Chile	0.55 (0.98) 68	0.14 (1.08) 1,909	3.06	75.00%	57.21%	7.53
France	0.07 (1.11) 112	0.05 (1.08) 3,661	0.24	61.61%	52.88%	3.23
Germany	0.07 (0.87) 121	0.04 (1.12) 3,590	0.31	52.07%	52.63%	0.01
Hong Kong	0.42 (1.71) 95	0.07 (1.84) 3,608	1.83	71.58%	53.23%	12.02
Italy	0.18 (0.90) 73	0.07 (1.38) 3,735	0.68	61.64%	52.91%	2.10
Japan	0.22 (1.16) 161	0.03 (1.17) 3,493	2.07	60.87%	52.98%	3.78
Korea	0.09 (1.26) 121	0.07 (1.33) 3,553	0.19	56.20%	49.62%	2.11
Malaysia	0.39 (1.05) 93	0.04 (1.45) 3,576	2.29	73.12%	53.45%	13.47
Mexico	0.67 (2.05) 132	0.21 (2.14) 3,540	2.45	66.67%	55.45%	5.99
Netherlands	0.00 (0.95) 86	0.05 (1.02) 3,531	−0.44	45.51%	53.36%	1.51
Singapore	0.29 (0.81) 99	0.05 (1.24) 3,630	1.96	64.65%	53.12%	4.95
Spain	0.24 (0.94) 107	0.06 (1.12) 2,576	1.62	60.75%	51.58%	3.48
Switzerland	0.19 (0.51) 81	0.03 (0.83) 3,661	1.74	69.14%	54.30%	6.56
Taiwan	0.76 (2.36) 75	0.11 (2.39) 2,053	2.33	70.67%	53.99%	7.72
United Kingdom	0.18 (0.81) 66	0.04 (0.97) 2,700	1.11	62.12%	53.11%	2.02

Notes to Table 1

[a]The t-statistic tests the significance (from zero) of the coefficient b in the following regression:

$$R_t = a + b\,D_t + e_t \;,$$

where D_t equals 1 if day t is the trade day before a holiday, zero otherwise. The coefficient b measures the difference in return between pre-holiday days and all other days.

[b]The expected percentage is based on the observed frequency of positive return days over the entire sample period for the particular country.

[c]The χ^2 statistic tests the hypothesis that the observed number of positive return days on pre-holiday days is equal to the expected number given the distribution across all trading days in the period. It is defined as:

$$\chi^2 = 2(O - E)^2/E \;,$$

where O represents the observed number and E represents the expected number.

[d]The values in this row are the number of observations in each category.

We test the hypothesis that the observed number of positive pre-holiday returns equals the expected number under the assumption that the pre-holiday returns are random draws from the entire sample of returns. The χ^2 statistic, which is a special case of Pearson's χ^2 test, is calculated as:

$$\chi^2 = 2(O - E)^2/E \tag{2}$$

where, O represents the *observed* number of positive returns on pre-holiday days; E represents the *expected* number of positive returns on pre-holiday days, calculated as the product of the fraction of all positive return days in the entire sample times the number of pre-holiday days. Rejection of the null hypothesis implies that the observed frequency of positive pre-holiday return days is higher than the observed frequency of positive return days for all trading days in the period.

The three rightmost columns in Table 1 contain information relating to the nonparametric tests: the observed percentage of positive pre-holiday returns; the expected percentage of positive pre-holiday returns which, under the null hypothesis, is simply the percentage of positive returns for the global sample for that individual market; and the χ^2 statistic. Note that, in column 5, the fraction of positive return days across all trading days in the sample is similar across countries, ranging between 50 and 58%. On average, 54% of the trading days exhibit positive returns. In contrast, all but three of the equity markets in our sample exhibit positive returns on more than 60% of the pre-holiday days, ranging as high as 75% for Chile, and as low as 47% for the Netherlands.

Based on the χ^2 statistics, ten of the seventeen countries reject (at the .05 level) the null hypothesis of equal positive return frequencies in the two subsamples in favor of the alternate hypothesis of more frequent pre-holiday advances. Those countries that do not exhibit statistically significant differences in the frequency of positive returns are in most cases the same ones for which we did not reject equality of mean returns for pre-holiday days and all other days. The two exceptions are Hong Kong and Switzerland. Thus, the results do not derive from a small number of extremely high pre-holiday returns.

The tests above implicitly assume independence of returns across markets, an assumption that we know is violated in the data.[4] To account for correlation across markets, we estimate the following variant of equation (1) using Zellner's seemingly unrelated regression technique (SUR)

$$R_{it} = a_i + b_i D_{it} + \epsilon_{it}, \quad i = 1, 17, \tag{3}$$

where the subscript i represents country i and the other variables are as defined previously. The SUR model estimates equation (3) for all 17 countries simultaneously and accounts for cross-market correlations in the residuals when estimating the parameters. As such, the SUR technique gives asymptotically more efficient estimators than OLS, and use of daily data here avoids concerns about potential small-sample problems. We estimate equation (3) for the time period January 5, 1987 to December 30, 1994 that is common to all our country samples. The inferences about the holiday effect from equation (3) are qualitatively the same as for the OLS estimation in equation (1), although the T-statistics for the estimated b_i are uniformly lower in equation (3).

To provide a more concise characterization of the holiday effect for our entire sample of 17 countries, we estimate the following SUR model

$$R_{it} = a + b D_{it} + \epsilon_{it}, \quad i = 1, 17. \tag{4}$$

Equation (4) differs from equation (3) in that we constrain the intercept and slope to be equal across all markets. This setup yields a single estimate of the holiday effect across all 17 markets. The estimates (T-values) of a and b from

---

[4]See, for example, the return correlations reported in Section 4.

equation (4) are 0.0466 (3.86) and 0.1313 (3.91) respectively, indicating the holiday effect found in the US data is also evident, on average, in international markets.

Finally, Table 1 shows that the standard deviation of pre-holiday returns tends to be lower than the standard deviation of returns for the rest of the days. The one exception is France. We estimate an F-statistic to test for equality of the pre-holiday and the non-pre-holiday variances. The F-statistics indicate rejection of equality for eight of the seventeen countries. Interestingly, in several markets (Switzerland, UK) where the returns across samples are not statistically different, the variance is statistically smaller for the pre-holiday returns.

3.1   Subperiod Results

We also estimate the difference-of-means test and the nonparametric test for each country for two subperiods to examine the consistency of the anomaly through the period. The results are in Table 2 and Table 3. In most cases, the full sample period (usually 1980–1994) was split in half (1980–1987 and 1988–1994). The four countries for which the subperiods differ are identified in the table.

The difference-of-means test in Table 2 indicates that only 3 of 17 countries exhibit a holiday effect for the first subperiod: Chile, Japan, and Mexico. The results for the non-parametric test for the first subperiod are somewhat different. Seven of the 17 countries showed evidence of a holiday effect. Interestingly, two of the markets that display a holiday effect according to the parametric test did not do so according to the nonparametric test. Most of the rejections for the nonparametric test in the first subperiod are for countries that didn't reject for the parametric test.

In the second subperiod, in Table 3, there is less evidence of a holiday effect. Three markets display a holiday effect according to the difference-in-means test. These same three markets also reject, along with Chile, with the nonparametric test. While a holiday effect is evident for the full sample period, the evidence is weaker in the subperiods.

The results in this section demonstrate the existence of a holiday effect across a broad cross section of world equity markets. The trading day prior to holidays, on average, exhibits high positive returns. This result is particularly interesting because the sample includes stock markets with important structural differences. A brief analysis on these differences is presented in the next section.

Table 2. Means, standard deviations, and frequency of positive returns for international stock indexes for pre-holidays and other days. (1980–1987)

Country	Mean (Std. Dev.) of Daily Returns (%)		T-Statistic of Return Difference[a]	Positive Returns on Pre-Holidays		$\chi^{2\,c}$
	Pre-Holidays	Other Days		Observed Percentage	Expected Percentage[b]	
Australia	0.35 (0.62) 45[d]	0.05 (1.16) 1,962	1.71	73.33%	54.51%	5.85
Canada	0.21 (0.57) 63	0.03 (0.98) 1,943	1.51	65.08%	52.74%	3.64
Chile[e]	0.64 (1.07) 36	0.14 (1.20) 947	2.44	72.22%	57.78%	2.60
France	−0.09 (1.34) 54	0.06 (1.16) 1,963	−0.91	61.11%	53.20%	1.27
Germany	0.08 (0.89) 68	0.03 (1.10) 1,901	0.35	51.47%	53.22%	0.08
Hong Kong	0.48 (2.12) 50	0.06 (2.08) 1,916	1.41	76.00%	53.20%	9.77
Italy	0.27 (0.88) 40	0.11 (1.54) 1,983	0.65	70.00%	53.53%	4.05
Japan	0.29 (1.19) 88	0.06 (0.92) 1,841	2.34	64.77%	55.21%	2.92
Korea	0.17 (1.02) 69	0.08 (1.20) 1,879	0.64	62.32%	51.49%	3.14
Malaysia	0.34 (1.15) 49	0.02 (1.60) 1,909	1.42	71.43%	51.69%	7.39
Mexico	1.03 (2.48) 70	0.23 (2.45) 1,869	2.70	72.86%	55.80%	7.30
Netherlands	0.09 (1.14) 48	0.05 (1.18) 1,872	0.26	43.75%	52.08%	1.28
Singapore	0.34 (0.88) 54	0.03 (1.38) 1,923	1.62	70.37%	53.06%	6.10
Spain[f]	0.24 (1.03) 61	0.12 (1.12) 1,385	0.82	63.93%	53.46%	2.50
Switzerland	0.24 (0.50) 41	0.02 (0.82) 1,943	1.68	68.29%	54.13%	3.04
Taiwan[g]	0.39 (2.40) 32	0.29 (2.17) 877	0.25	71.88%	59.85%	1.55
United Kingdom[f]	0.12 (0.75) 36	0.06 (1.06) 1,468	0.32	69.44%	56.18%	2.25

Notes to Table 2

[a]The t-statistic tests the significance (from zero) of the coefficient b in the following regression:

$$R_t = a + b\,D_t + e_t \;,$$

where D_t equals 1 if day t is the trade day before a holiday, zero otherwise. The coefficient b measures the difference in return between pre-holiday days and all other days.

[b]The expected percentage is based on the observed frequency of positive return days over the entire sample period for the particular country.

[c]The χ^2 statistic tests the hypothesis that the observed number of positive return days on pre-holiday days is equal to the expected number given the distribution across all trading days in the period. It is defined as:

$$\chi^2 = 2(O - E)^2/E \;,$$

where O represents the observed number and E represents the expected number.

[d]The values in this row are the number of observations in each category.

[e]The subperiod for Chile is 1987–1990.

[f]The subperiod for Spain and the United Kingdom is 1984–1989.

[g]The subperiod for Taiwan is 1986–1989.

---

4 Why a Holiday Effect?

The evidence on the holiday effect in our sample of international stock markets is particularly surprising since the sample includes markets with very different characteristics. Table 4 shows how these markets differ in four aspects: total market capitalization; the number of companies listed; the market capitalization to GDP ratios; and the average company size in each market.

As is evident in Table 4, the markets differ substantially in terms of capitalization and, therefore, liquidity. For example, the difference in market capitalization between Japan and Chile is large (more than $3.6 trillion), but most of the results found in this paper apply to both markets. The number of companies listed in the market also varies widely, although 9 out of 17 have less than 500. There are also significant differences across the countries in terms of the relative importance of the equity markets in their respective economies. In 1990, for example, the Gross Domestic Product (GDP) in Asian countries like Malaysia, Hong Kong, Japan and Singapore was very similar to the market capitalization of those respective equity markets. In that same year, however, the equity markets in countries like Germany, Italy, Mexico and Spain represented around 22%, on average, of GDP in those countires. As Table 4 shows, the differences in the ratios of market capitalization to GDP for the countries in our sample grew significantly between 1990 and 1995. However, the large differences across the stock markets in our sample make the task of explaining the existence of the holiday effect more difficult.

One explanation offered for the holiday effect is that investors cover short positions before the holiday on the presumption that the probability of the release of adverse news is greater over the holiday because of the extended market closure. The release of news may be asymmetric because companies prefer to reveal adverse information when the market is scheduled to be closed to give market participants time to digest the information and, thereby, avoid overreaction.[5] Interestingly, there are a number of markets in our sample in which participants are forbidden to sell short, casting doubt on this explanation. For instance, the Mexican and the Chilean markets both forbid short selling during the period of analysis, yet still exhibit a significant holiday effect. The British stock market, on the other hand, allows short selling of stocks but does not exhibit this calendar anomaly.

One influence on the existence of the holiday effect across international markets is the common global market factor in returns attributable to the increasing integration of capital markets. Table 5 contains correlation matrices based on the monthly returns (denominated in $US) of the indices used in this paper for two different sub-periods: 1976–1985 and 1986–1994.[6] The information contained in these correlation matrices is rich, but a detailed analysis of it is beyond the scope of this paper. There are, however, certain elements of the matrices that shed light on the existence, or lack thereof, of a holiday effect in certain markets. Recall that for the results reported in Section 3, four of the five markets that do not exhibit a holiday effect are European markets (France, the Netherlands, the United Kingdom and Germany). From the correlation matrices in Table 5 we observe that the correlations between France and Germany, and the United Kingdom and the Netherlands are higher than the average correlation.

To the extent that stock market returns are correlated across markets, the holiday effect associated with those holidays that are not common across countries should be lower than for the holidays that are common. To test this

---

[5]See Miller (1988) and Ziemba (1993) for similar explanations with regard to the weekend effect.

[6]The United States appears in these matrices for reference. The data for Malaysia and Taiwan start in 1985.

Table 3. Means, standard deviations, and frequency of positive returns for international stock indexes for pre-holidays and other days. (1988–1994)

Country	Mean (Std. Dev.) of Daily Returns (%)		T-Statistic of Return Difference[a]	Positive Returns on Pre-Holidays		$\chi^{2\,c}$
	Pre-Holidays	Other Days		Observed Percentage	Expected Percentage[b]	
Australia	0.18 (0.80) 40[d]	0.02 (0.85) 1,726	1.18	57.50%	51.81%	0.50
Canada	0.21 (0.42) 55	0.01 (0.57) 1,708	2.61	70.91%	53.55%	6.19
Chile[e]	0.46 (0.87) 32	0.15 (0.95) 962	1.81	78.13%	56.64%	5.22
France	0.22 (0.82) 58	0.03 (0.98) 1,698	1.42	62.07%	52.51%	2.02
Germany	0.05 (0.87) 53	0.04 (1.15) 1,689	0.08	52.83%	51.95%	0.02
Hong Kong	0.35 (1.11) 45	0.08 (1.54) 1,692	1.17	66.67%	53.25%	3.04
Italy	0.07 (0.93) 33	0.02 (1.16) 1,752	0.23	51.52%	52.21%	0.01
Japan	0.14 (1.13) 73	0.00 (1.40) 1,652	0.85	56.16%	50.49%	0.93
Korea	−0.02 (1.53) 52	0.05 (1.45) 1,674	−0.36	48.08%	47.51%	0.01
Malaysia	0.45 (0.92) 44	0.08 (1.26) 1,667	1.93	75.00%	55.46%	6.05
Mexico	0.27 (1.32) 62	0.19 (1.73) 1,671	0.36	59.68%	55.05%	0.48
Netherlands	−0.11 (0.65) 38	0.05 (0.78) 1,659	−1.30	50.00%	54.80%	0.32
Singapore	0.23 (0.73) 45	0.06 (1.05) 1,707	1.09	57.78%	53.20%	0.36
Spain[f]	0.24 (0.83) 46	−0.01 (1.12) 1,191	1.47	56.52%	49.39%	0.95
Switzerland	0.14 (0.52) 40	0.04 (0.85) 1,718	0.78	70.00%	54.49%	3.53
Taiwan[g]	1.04 (2.32) 43	−0.03 (2.54) 1,176	2.73	69.77%	49.63%	7.03
United Kingdom[f]	0.24 (0.88) 30	0.02 (0.85) 1,232	1.44	53.33%	49.45%	0.18

Notes to Table 3

[a]The t-statistic tests the significance (from zero) of the coefficient b in the following regression:

$$R_t = a + b\,D_t + e_t \,,$$

where D_t equals 1 if day t is the trade day before a holiday, zero otherwise. The coefficient b measures the difference in return between pre-holiday days and all other days.

[b]The expected percentage is based on the observed frequency of positive return days over the entire sample period for the particular country.

[c]The χ^2 statistic tests the hypothesis that the observed number of positive return days on pre-holiday days is equal to the expected number given the distribution across all trading days in the period. It is defined as:

$$\chi^2 = 2(O - E)^2/E \,,$$

where O represents the observed number and E represents the expected number.

[d]The values in this row are the number of observations in each category.

[e]The subperiod for Chile is 1991–1994.

[f]The subperiod for Spain and the United Kingdom is 1990–1994.

[g]The subperiod for Taiwan is 1990–1994.

---

hypothesis, we estimate the following variant of equation (4)

$$R_{it} = a + bD_{it} + cD_t^{\mathrm{comm}} + \epsilon_{it}, \quad i = 1, 9. \qquad (5)$$

where D_t^{comm} equals one if day t precedes one of three common holidays (Christmas, New Year's day and Easter), and the other variables are as defined previously. Equation (5) is estimated using SUR for the 9 markets in our sample that share these common holidays – Australia, Canada, France, Germany, Holland, Italy, Spain, Switzerland, and the UK. The estimate of a measures the difference between the average return on non-common holidays and common holidays. The estimates (T-values) for a, b and c from equation (5) are 0.0307 (2.81), 0.0718 (1.96), and 0.1296 (1.16). The coefficient a is positive but insignificant, indicating that the average returns preceding common holidays are statistically indistinguishible from the average returns preceding other holidays. However, after controlling for the common holidays

Table 4. Characteristics of the Equity Markets in Our Sample

Country	Market Cap. ($US, in bill)[1]	# of Companies listed[1]	Average Company Size ($US, in bill)[1]	Market Cap/ GDP (%)[2]	
				1990	1995
Australia	245	1,178	208	36	70
Canada	366	1,196	306	43	64
Chile	74	284	260	45	121
France	522	450	1160	26	34
Germany	577	678	852	23	24
Hong Kong	304	518	586	112	211
Italy	210	250	838	14	19
Japan	3,667	2,263	1,621	98	72
Korea	182	721	252	44	40
Malaysia	223	529	421	114	261
Mexico	91	185	490	13	36
Netherlands	356	387	921	42	90
Singapore	148	212	698	94	177
Spain	198	362	546	23	35
Switzerland	434	233	1,861	71	144
Taiwan	187	347	539	n.a.	n.a.
UK	1,408	2,078	677	87	127

[1]At the end of 1995. Source: Emerging Stock Markets Factbook 1996, International Finance Corporation.

[2]GDP is Gross Domestic Product. Source: 1997 World Development Indicators, The World Bank.

n.a. - Not available.

in the sample we find that the holiday effect, as measured by coefficient a, is smaller than when estimated over all holidays, although it is still marginally significant. When equation (4) is estimated for these 9 markets, the estimate of b is 0.0821 with a T-value of 2.31.

5   Conclusions

The results in this paper demonstrate the existence of a holiday effect across a broad cross section of world equity markets. The trading day prior to holidays, on average, exhibits high positive returns. This result is particularly interesting because the sample includes stock markets with important structural differences. A brief analysis of these differences is presented in Section 4. While a holiday effect is evident for the full sample period, the evidence is weaker in the subperiods.

The results by no means imply that the return patterns documented imply profit opportunities. Transaction costs and market liquidity (which reflects the bid-ask spread) have to be taken into consideration. Such costs are high, especially in some of the less-developed markets where we observe this pattern, and likely offset the implied average price movements. Nevertheless, these findings suggest an opportunity for investors who plan to trade anyway

Table 5. Correlation matrix of monthly returns (in $US) for international equity market indexes.

A. 1976–1985

	AUS	CAN	CHI	SPA	GER	HK	NETH	ITA	JAP	KOR	MEX	SIN	SWI	FRA	UK	USA
AUS	1.00	0.480	0.126	0.130	0.064	0.213	0.264	-0.210	0.223	-0.044	0.110	0.372	0.286	0.117	0.304	0.337
CAN		1.000	-0.034	0.029	0.068	0.128	0.329	-0.249	0.216	0.091	0.174	0.246	0.317	0.043	0.369	0.718
CHI			1.000	-0.016	-0.033	0.009	-0.017	-0.054	0.056	0.012	0.036	-0.042	-0.002	0.077	-0.004	-0.131
SPA				1.000	0.093	0.067	0.153	-0.207	0.277	-0.009	0.144	-0.036	0.051	0.176	0.164	-0.007
GER					1.000	-0.074	0.327	-0.299	0.258	-0.072	0.069	-0.009	0.489	0.397	0.058	0.118
HK						1.000	0.277	-0.235	0.251	-0.001	0.045	0.401	0.154	0.118	0.182	0.152
NETH							1.000	-0.373	0.341	0.046	0.183	0.229	0.621	0.311	0.479	0.442
ITA								1.000	-0.097	0.014	-0.184	-0.143	-0.326	-0.154	-0.312	-0.205
JAP									1.000	-0.023	0.186	0.158	0.269	0.169	0.330	0.164
KOR										1.000	0.025	0.070	0.178	0.145	0.036	0.065
MEX											1.000	0.120	0.082	0.069	0.216	0.121
SIN												1.000	0.121	0.153	0.156	0.306
SWI													1.000	0.344	0.380	0.352
FRA														1.000	0.167	0.086
UK															1.000	0.382
USA																1.000

B. 1986–1994

	AUS	CAN	CHI	SPA	GER	HK	NETH	ITA	JAP	KOR	MEX	SIN	SWI	FRA	UK	MAL	TAI	USA
AUS	1.000	0.593	0.183	0.437	-0.050	0.581	0.585	-0.268	0.253	0.038	0.399	0.498	0.500	-0.077	0.553	0.315	0.290	0.458
CAN		1.000	0.287	0.445	-0.101	0.552	0.553	-0.293	0.312	0.247	0.359	0.521	0.526	-0.097	0.594	0.507	0.200	0.769
CHI			1.000	0.332	-0.054	0.377	0.137	0.000	0.025	0.118	0.313	0.245	0.516	0.044	0.163	0.149	0.256	0.291
SPA				1.000	0.122	0.404	0.505	-0.491	0.495	0.184	0.365	0.400	0.090	0.190	0.554	0.173	0.205	0.470
GER					1.000	-0.019	0.153	-0.163	0.154	0.077	0.067	0.099	0.399	0.737	0.040	-0.044	-0.120	-0.007
HK						1.000	0.440	-0.203	0.211	0.124	0.413	0.576	0.771	-0.087		0.531	0.317	0.433
NETH							1.000	-0.455	0.402	0.148	0.317	0.539	-0.458	0.123	0.667	0.314	0.163	0.544
ITA								1.000	-0.098	-0.169	-0.034	-0.280	0.412	-0.156	-0.387	-0.103	-0.148	-0.227
JAP									1.000	0.209	0.046	0.264	0.055	0.148	0.370	0.098	0.141	0.211
KOR										1.000	0.172	0.161	0.352	0.156	0.158	0.136	0.055	0.437
MEX											1.000	0.525	0.405	0.002	0.310	0.408	0.344	0.607
SIN												1.000		0.017	0.577	0.772	0.185	0.526
SWI													1.000	0.049	0.597	0.232	-0.108	-0.004
FRA														1.000	0.040	-0.095	0.190	0.649
UK															1.000	0.433	0.264	0.499
MAL																1.000		0.170
TAI																	1.000	
USA																		1.000

and who have some discretion as to the timing of the trade. To wit, instead of selling early on the pre-holiday, one should wait until the end of the trading day to sell. We caution with the usual caveat, though: That these patterns have persisted in the past does not guarantee that they will hold in the future.

We do not provide an explanation of the holiday effect. However, we do provide evidence of its existence in a number of developed and emerging markets, and an invitation for further research that may lead to a consistent economic explanation.

Appendix

Holidays that Result in Closing of Local Equity Market

USA

New Year	1 Jan
Martin Luther King's B Day	January, 3rd Monday
Washington B Day	February, 3rd Monday
Memorial Day	May, last Monday
Independence Day	4 Jul
Labor Day	September, 1st Monday
Columbus Day	October, 2nd Monday
Veterans Day	11 Nov
Thanksgiving Day	November, 4th Thursday
Christmas	25 Dec

Canada

New Year	1 Jan
Good Friday	varies
Easter Monday	varies
Victoria Day	20 May
Canada Day	1 Jul
Labor Day	2 Sep
Thanksgiving	Oct, 2nd Monday
Remembrance Day	11 Nov
Christmas	25 Dec
Boxing Day	26 Dec

Mexico

New Year	1 Jan
Constitution Day	5Feb
Benito Juarez Day	21 Mar
Jueves Santo	(Thursday before Easter)
Viernes Santo	(Good Friday)
Labor Day	1 May
Batalla de Puebla	5 May
Independence Day	16 Sep
Descubrimiento de America	12 Oct
Saints Day	2 Nov
Revolution Day	20 Nov
Virgen de Guadalupe	12 Dec
Christmas	25 Dec

Chile

New Year	1 Jan
Good Friday	
Labor Day	1 May
Navy Day	21 May
Assumption Day	15 Aug
Independence Day	18 Sep
Army Day	19 Sep
Columbus Day	12 Oct
All Saints Day	1 Nov
Immaculate Conception	8 Dec
Christmas	25 Dec
St. Patrick and St. Paul	29 Jun

Holidays that Result in Closing of Local Equity Market

United Kingdom		**Germany**	
New Year	1 Jan	New Year	1 Jan
St. Patrick's (N. Ireland)	17 Mar	Epiphany	6Jan
Good Friday		Good Friday	
Easter Monday		Easter Monday	
May Day Holiday	First Mon. in May	Labor Day	1 May
		Ascension Day	April/June
Spring Bank Holiday	Last Mon. in May	Whit Monday	May/June
		Day of Unity	17 Jun
Holiday (N.Ireland)	12 Jul	Assumption Day	15 Aug
Bank Holiday (Scotland)	5 Aug	All Saints DAy	1 Nov
Late Summer Holiday	Last Mon. in Aug.	Day of Prayer and Repentance	20 Nov
Christmas	25 Dec	Christmas	25 Dec
Boxing Day	26 Dec		26 Dec

France		**Switzerland**	
New Year	1 Jan	New Year	1 Jan
Easter Day		Easter Day	
Easter Monday		Easter Monday	
Labor Day	1 May	Ascension Day	
VE Day	8 May	Whit Monday	
Ascension Day		Swiss National Day	1 Aug
Whit Monday		Christmas	25 Dec
Bastille Day	14 Jul		26 Dec
Assumption Day	15 Aug		
All Saints Day	1 Nov		
Armistice Day	11 Nov		
Christmas	25 Dec		

Netherlands		**Italy**	
New Year	1 Jan	New Year	1 Jan
Good Friday		Epiphany	6 Jan
Easter Monday		Easter Monday	
Queen's B Day	30 Apr	Liberation Day	25 Apr
Liberation Day	5 May	Labor Day	1 May
Ascension Day		Assumption Day	15 Aug
Whit Monday		All Saints Day	1 Nov
Christmas	25 Dec	Immaculate Conception	8 Dec
	26 Dec	Christmas	25 Dec
			26 Dec

Holidays that Result in Closing of Local Equity Market

Japan

New Year	1–3 Jan
Adult's Day	15 Jan
Foundation Day	11 Feb
Vernal Equinox Day	21 Mar
Greenery Day	29 Apr
Constitution Day	3 May
National Holiday	4 May
Children's Day	5 May
Respect for aged day	15 Sep
Autumnal Equinox Day	24 Sep
Health & Sports Day	10 Oct
Culture Day	3 Nov
Labor Thanksgiving Day	23Nov
Emperor's B Day	23 Dec

Korea

New Year	1–3 Jan
Folklore Day	15 Feb
Independence Movement	1 Mar
Arbor Day	5 Apr
Childrens Day	5 May
Buddha's Day	21 May
Memorial Day	6 Jun
Constitution Day	17 Jul
Liberation Day	15 Aug
Thanksgiving Day	22 Sep
Armed Forces Day	1 Oct
National Foundation Day	3 Oct
Korean Alphabet Day	9 Oct
Christmas	25 Dec

Spain

New Year	1 Jan
Epiphany	6 Jan
St. Joseph's Day	19 Mar
Good Friday	
Easter Monday	
Labor Day	1 May
St. James' Day	25 Jul
Assumption Day	15 Aug
All Saints Day	1 Nov
Immaculate Conception	8 Dec
Christmas	25 Dec

Taiwan

Adult's Day	15 Jan
Foundation Day Rep. of China	1 Jan
Youth Day	29 Mar
Tomb Sweeping Day	5 Apr
Dragon Boat Festival	28 May
Confucius' Birthday	28 Sep
National Day	10 Oct
Taiwan Retrocession Day	25 Oct
President's B Day	31 Oct
Dr. Sun Yat Sen's B Day	12 Nov
Constitution Day	25 Dec

Hong Kong

New Year	1 Jan
Lunar New Year	Feb 14–16
Good Friday	
Easter Monday	
Queen's B Day	June 15–17
Liberation Day	26 Aug
Christmas	25 Dec
	26 Dec

Singapore

New Year	1 Jan
Chinese New Year	Feb 15–16
Good Friday	
Han Raya Puasa	16 Apr
Labor Day	1 May
Vesak Day	April/May
Hari Raya Haji	23 Jun
National Day	9 Aug
Christmas	25 Dec

Australia

New Year	1 Jan
Australia Day	Jan 26 & 29
Good Friday	
Easter Monday	
Anzac Day	25 Apr
Christmas	25 Dec
	26 Dec

Malaysia

Chinese New Year	Feb 15–16
Labor Day	1 May
National Day	31 Aug
Christmas	25 Dec

References

Ariel, R. (1990) 'High stock returns before holidays: existence and evidence on possible causes', *Journal of Finance* **45** 1611–1626.

Cadsby, C.B. and M. Ratner (1992) 'Turn-of-month and pre-holiday effects on stock returns: some international evidence', *Journal of Banking and Finance* **16** 497–509.

Fields, M.J. (1934) 'Security prices and stock exchange holidays in relation to short selling', *Journal of Business* **7** 328–338.

Fosback, N.G. (1976) *Stock Market Logic*, The Institute for Economic Research, Fort Lauderdale, Florida.

Keim, D.B. (1983) 'Size-related anomalies and stock return seasonality: further empirical evidence,' *Journal of Financial Economics* **12** 13–32.

Keim, D.B. (1989) 'Trading patterns, bid-ask spreads, and estimated security returns: the case of common stocks at the turn of the year', *Journal of Financial Economics* **25** 75–98.

Kim, C. and J. Park (1994) 'Holiday effects and stock returns: further evidence', *Journal of Financial and Quantitative Analysis* **29** 145–157.

Lakonishok, J. and S. Smidt (1984) 'Volume and turn-of-the-year behavior', *Journal of Financial Economics* **13** 435–456.

Merrill, Arthur A. (1966) *Behavior of Prices on Wall Street*, The Analysis Press, Chappaqua, NY.

Miller, E.M. (1988) 'Why a weekend effect?', *Journal of Portfolio Management* **15** 42–48.

Puello, A.D. (1993) *Public Holidays in Latin America*, International Business Report.

Ritter, J.R. (1988) 'The buying and selling behavior of individual investors at the turn of the year', *Journal of Finance* **43** 701–717.

Roll, R. (1983) 'Vas ist das? The turn-of-the-year effect and the return premia of small firms', *Journal of Portfolio Management* **9** (Winter) 18–28.

Thompson, S.E. and B.W. Carlson (1994) *Festivals and Celebrations of the World Dictionary*, Omnigraphics Inc.

Ziemba, W.T. (1991). 'Japanese security market regularities, monthly, turn-of-the-month and year, holiday and Golden Week effects', *Japan and the World Economy* **3** 119–146.

Ziemba, W.T. (1993) 'Comment on *Why a weekend effect?*', *Journal of Portfolio Management* **19** 93–99.